Neuroendocrinology

H.-W. Korf K.-H. Usadel (Eds.)

Springer

Berlin
Heidelberg
New York
Barcelona
Budapest
Hong Kong
London
Milan
Paris
Santa Clara
Singapore
Tokyo

H.-W. Korf • K.-H. Usadel (Eds.)

Neuroendocrinology

Retrospect and Perspectives

With 138 Figures and 17 Tables

Springer

PROF. DR. HORST-WERNER KORF
Dr. Senckenbergische Anatomie
Klinikum der J.W. Goethe-Universität
Theodor-Stern-Kai 7
D-60590 Frankfurt/Main, Germany

PROF. DR. KLAUS-HENNING USADEL
Medizinische Klinik I
Klinikum der J.W. Goethe-Universität
Theodor-Stern-Kai 7
D-60590 Frankfurt/Main, Germany

ISBN-13: 978-3-642-64605-8 e-ISBN-13: 978-3-642-60915-2
DOI: 10.1007/978-3-642-60915-2

CIP Data applied for
Die Deutsche Bibliothek — CIP-Einheitsaufnahme
Neuroendocrinology : retrospect and perspectives ; with 17 tables / H.-W. Korf ; K. H. Usadel. - Berlin ; Heidelberg ; New York ; Barcelona ; Budapest ; Hong Kong; London ; Milan ; Paris ; Santa Clara ; Singapore ; Tokyo, Springer 1997

Softcover reprint of the hardcover 1st edition 1997

Cover design: design & production GmbH, Heidelberg
Typesetting: Camera ready by the editors
SPIN 10629571 31/3137 5 4 3 2 1 0 - Printed on acid free paper

Preface

Contemporary neuroendocrinology rests upon the concept of neurosecretion created by Ernst and Berta Scharrer. Crucial milestones for this concept were laid in the period between 1933-1937 when Ernst and Berta Scharrer were associated with the Neurologisches Institut of Ludwig Edinger in Frankfurt/ Main. In 1937, Ernst and Berta Scharrer published a first synopsis of their results on gland-nerve cells and neurosecretory organs of both vertebrates and invertebrates. This paper can be considered as the foundation of modern neuroendocrinology; it defines the functional role of secretory neurons as mediators that convey information between the nervous system, on the one hand, and the endocrine system, on the other hand. Ernst and Berta Scharrer's concept on neurosecretion was of fundamental importance for the definition of the peptidergic neuron presented by Wolfgang Bargmann, Kiel, in 1967. Peptidergic neurons synthesize and secrete neuropeptides and employ them as neurohormones, neuromodulators, or neurotransmitters. The high phylogenetic age of secretory neurons and neuropeptides and the widespread distribution of these elements in the brain, peripheral nervous system and other organs such as the intestine underline their fundamental role for the integration of body functions. There is now compelling evidence that peptidergic neurons and neuropeptides not only form an important interface between the nervous and the endocrine system but also connect these two with the immune system.

In recognition of the outstanding contributions of Ernst and Berta Scharrer as pioneers in neuroendocrinology, the Ludwig Edinger-Stiftung, Frankfurt/Main, and the Fachbereich Humanmedizin of the Johann Wolfgang Goethe-Universität, Frankfurt/Main, have founded the Ernst and Berta Scharrer Medal. The first Ernst and Berta Scharrer Medal was awarded to Professor Tomas Hökfelt, Stockholm, during an international Ernst and Berta Scharrer Symposium on Neuroendocrinology held in Frankfurt/Main from 14-15 February, 1997. This volume comprises the contributions to the symposium which cover the multiple facets of contemporary neuroendocrinology and reflect the rapid progress in this ever-growing biomedical discipline at the systemic, cellular, and molecular levels.

We are most grateful to Drs. Barbara and Helmut Wicht who efficiently assisted us in editing this book. With sincere gratitude we acknowledge financial support from the Ludwig Edinger-Stiftung, Dr. Senckenbergische Stiftung, Walter Artelt- und Edith Heischkel-Artelt-Stiftung, Deutsche Gesellschaft für Klinische Pharmakologie und Therapie, Rhein-Main Arbeitsgemeinschaft für Endokrinologie, Diabetes und Stoffwechsel, Fachbereich Humanmedizin of the Johann Wolfgang Goethe-Universität (all Frankfurt/Main), and Dr. W. Vogel, Hofgeismar.

Frankfurt am Main, July 25, 1997

Horst-Werner Korf
Klaus-Henning Usadel

Contents

1 Ernst and Berta Scharrer - Pioneers in Neuroendocrinology

A. Oksche

Department of Anatomy and Cell Biology, Justus Liebig University of Giessen, 123 Aulweg, D-35392 Giessen, Germany; Phone: +49 641 9947030, Fax: +49 641 9947039

This book contains the contributions to a symposium dedicated to the memory of Ernst and Berta Scharrer, two eminent scientists whose discoveries marked the advent of a new discipline in neurosciences - namely, neuroendocrinology. I remember them with deep gratitude as my highly esteemed scientific mentors. The ingenious scientific work of Ernst and Berta Scharrer reflects an unusual perception into the basis of natural science and is evidence for two exceptional personalities. Curiosity, imagination, and persistence were prerequisites for their highly revolutionary investigations. The concept of neurosecretion developed by Ernst and Berta Scharrer between 1928 and 1937, and later extended in close cooperation with Wolfgang Bargmann, forms the foundation for contemporary neuroendocrinology, particularly the concept of peptidergic neurons in vertebrates and invertebrates.

In 1928, Ernst Scharrer discovered colloid-like inclusions in the magnocellular preoptic nucleus of a teleost, the European minnow. He interpreted this phenomenon as a manifestation of endocrine activity, which was further supported by the rich vascularization of the area and an obvious correlation with the hypophysis. Furthermore, he showed that the pineal organ of the minnow is an extraocular photoreceptive structure.

In Munich, both Ernst Scharrer and Berta Vogel had been disciples of Karl von Frisch, the later Nobel Laureate (1973). They married in 1934 and moved from Munich to Frankfurt, where they became associated with the Neurology Institute (Edinger Institute). For the development of the concept of neurosecretion, the years in Frankfurt (1934-1937) were of extraordinary importance. During this period, Ernst and Berta Scharrer became a congenial team, dividing the Animal Kingdom such that Ernst studied vertebrates and Berta invertebrates. In 1937, Berta Scharrer published a representative paper on neurosecretion in the central nervous system of different invertebrates. This paper was followed by a comprehensive review by Ernst and Berta Scharrer on gland-like nerve cells and neurosecretory organs both in vertebrates and invertebrates. These two papers can be regarded as the conceptual basis of neuroendocrinology.

In 1937, Ernst and Berta Scharrer left Germany. Chicago, New York, Cleveland and Denver became their further academic stations. The idea of a secretory activity of nerve cells was so revolutionary that it originally met with considerable opposition. The breakthrough for Ernst and Berta Scharrer occurred

Ernst Scharrer
(August 1, 1905 - April 29, 1965)

Berta Scharrer
(December 1, 1906 - July 23, 1995)

during their Denver period, in the early 1950s. In 1953, they organized the first Symposium on Neurosecretion at Naples. In 1954, they published several important surveys on neurosecretion and hormones produced by neurosecretory cells; these reviews belong to the classics in neuroendocrinology.

In 1955, Ernst and Berta Scharrer founded the Department of Anatomy at the Albert Einstein College of Medicine in New York. In 1963, they published a comprehensive monograph on *Neuroendocrinology*, a classic document of the state-of-the-art in the field. In 1964, Ernst Scharrer presented an updated concept of photoneuroendocrine systems. This concept deals with the influence of environmental factors on neuroendocrine functions. Karl von Frisch and Ernst Scharrer can be regarded as the founders of photoneuroendocrinology.

After the sudden, tragic death of Ernst Scharrer in 1965, Berta Scharrer continued to interpret the role of neurosecretory cells in the central nervous system. She extended the concept of neurosecretion to an overall concept of peptidergic neurons. Furthermore, Berta Scharrer was one of the early defenders of the unifying concept of a diffuse neuroendocrine system. It is amazing to see to what extent new molecular evidence can be integrated into the original framework of the concept of neurosecretion.

During the last decade of her life, Berta Scharrer was fascinated by the interrelationships between the neuroendocrine system and the immune apparatus.

Again, the comparative approach proved to be very fruitful; remarkable parallelisms between invertebrates and vertebrates could be established. A comprehensive monograph, published in 1994, is an impressive document of this effort. Berta Scharrer continued her research work long past retirement age; her last research paper dealing with neuroimmunological questions was submitted for publication a few days before her death.

Ernst and Berta Scharrer were eminent scientists who were admired for their integrity, warmth and willingness to help. It is a privilege to have known them. They will long be remembered by the international scientific community. The Ernst and Berta Scharrer medal of the Ludwig Edinger Foundation and the University of Frankfurt will represent a lasting sign of deep respect, recognition and admiration.

Bibliography

Bargmann W, Oksche A, Polenov A, Scharrer B (eds) (1978) Neurosecretion and neuroendocrine activity. Evolution, structure, and function. Springer, Berlin, Heidelberg, New York, pp 411

Oksche A (1987) Das Gehirn als hormonbildendes Organ - Durchbruch und Irrwege der Konzepte. In: Czeschlik D (ed) Irrtümer in der Wissenschaft. Springer, Berlin, Heidelberg, New York, pp 29-47

Oksche A (1995) In memoriam Berta Scharrer 1906-1995. Cell Tissue Res 282: 1-2

Oksche A (1996) In memoriam Berta Scharrer (1906-1995) Ann Anat 178: 293-298

Scharrer B (1937) Über sekretorisch tätige Nervenzellen bei wirbellosen Tieren. Naturwissenschaften 25: 131-138

Scharrer B (1941) Endocrines in invertebrates. Physiol Rev 21: 383-409

Scharrer B (1975) Neurosecretion and its role in neuroendocrine regulation. In: Meites J, Donovan BT (eds) Pioneers in neuroendocrinology. Plenum Press, New York, pp 257-265

Scharrer B (1978) Peptidergic neurons: facts and trends. Gen Comp Endocrinol 34: 50-62

Scharrer B (1982) Peptidergic neurons. Acta Morphol Neerl-Scand 20: 219-223

Scharrer B (1985) Neurosecretion: the development of a concept. In: Lofts B, Holmes WN (eds) Proc Ninth Int Symp Comp Endocrinol, vol I. Hong Kong University Press, Hong Kong, pp 23-27

Scharrer B (1990) The neuropeptide saga. Am Zool 30: 887-895

Scharrer B, Korf HW, Hartwig HG (eds) (1987) Functional morphology of neuroendocrine systems. Springer, Berlin, Heidelberg, New York, pp 1-229

Scharrer B, Smith EM, Stefano GB (1994) Neuropeptides and immunoregulation. Springer, Berlin, Heidelberg, New York, pp 1-174

Scharrer B, Paemen L, Smith EM, Hughes TK, Liu Y, Pope M, Stefano GB (1996) The presence and effects of mammalian signal molecules in immunocytes of the insect *Leucophaea maderae*. Cell Tissue Res 283: 93-97

Scharrer E (1928) Die Lichtempfindlichkeit blinder Elritzen. Untersuchungen über das Zwischenhirn der Fische I. Z Vgl Physiol 7: 1-38

Scharrer E (1954) Das Hypophysen-Zwischenhirnsystem der Wirbeltiere. Verh Anat Ges 51 (Anat Anz Suppl 100): 5-29

Scharrer E (1962/1995) Geschichte des Neurologischen Instituts 1933-1937. In: Doerr HW, Korf H-W (eds) Berühmte Ärzte und Forscher in Frankfurt am Main. Alpha, Lampertheim, pp 84-90

Scharrer E (1964) Photo-neuro-endocrine systems: General concepts. Ann NY Acad Sci 117: 13-22

Scharrer E, Scharrer B (1937) Über Drüsen-Nervenzellen und neurosekretorische Organe bei Wirbellosen und Wirbeltieren. Biol Rev 12: 185-216

Scharrer E, Scharrer B (1953) Symposium on Neurosecretion at Naples, Italy, May 11-16. Science 118: 579-580

Scharrer E, Scharrer B (1954) Neurosekretion. In: Bargmann W (ed) Handbuch der mikroskopischen Anatomie des Menschen, Bd VI/5. Springer, Berlin, Heidelberg, New York, pp 953-1066

Scharrer E, Scharrer B (1963) Neuroendocrinology. Columbia Univ Press, New York, pp 289

2 The Ups and Downs of Neuropeptides

T. Hökfelt, X. Zhang, Z.Q. Xu, R.R. Ji, T. Shi, J. Corness, N. Kerekes, M. Landry, M. Rydh-Rinder, J. Kopp, K. Holmberg, and C. Broberger

Department of Neuroscience, Karolinska Institutet, S-171 77 Stockholm, Sweden; Phone: +46 8 7287070, Fax. +46 8 33 16 92, E-mail: Tomas.Hokfelt@neuro.ki.se

Abstract

Peptides are not only hormones in endocrine and in hypothalamic neurosecretory cells; they are also present in many neurons in the central and peripheral nervous system. Whereas their role in the two former systems seems to be well established, their function in the nervous system proper is more uncertain. Over the last 25 years or so, extensive studies using various methodologies have been carried out to improve our understanding of neuropeptides. Their distribution has been mapped, their intracellular localization has been defined, and results from functional experiments demonstrating pronounced effects in various models have been reported. More recently, at least one receptor of the seven transmembrane-spanning, G-protein-coupled type has been identified for almost every peptide, strongly supporting a role of neuropeptides in nervous system function. We have studied dorsal root ganglion neurons as a model system to probe peptide functions, with particular focus on their possible role in pain mechanisms. These neurons express robust levels of several peptides and peptide receptors. Moreover, peripheral nerve injury (axotomy) causes dramatic changes in peptide and peptide receptor expression, resulting in a virtually new phenotype with regard to these molecules. Some mechanisms which possibly underlie these phenotypic changes, as well as their functional significance are discussed in this article. Various growth factors as well as the neuroimmune molecule leukemia inhibitory factor may be important for the regulation of peptide synthesis in dorsal root ganglion neurons. Moreover, we have hypothesized that galanin can be an endogenous analgesic compound, which is recruited especially after nerve injury (neuropathic pain). In fact, our and others' studies suggest the existence of two defense systems against pain, one localized at the level of the dorsal horn, where local neurons release inhibitory peptides including opioid peptides. This system responds to inflammatory pain. The second defense system is within the primary sensory neurons themselves and reacts towards neuropathic pain, for example by upregulating galanin. Another important aspect of peptide function is their involvement in trophic events, and there is now evidence that, for example, galanin and neuropeptide Y can influence regeneration and neurite out-growth. Thus, neuropeptides may exert manifold actions, but their roles in most cases still have to be defined and established as being physiologically and/or pathophysiologically significant.

Introduction

Only a few scientists are granted the privilege of initiating a new discipline in biomedical sciences. Ernst and Berta Scharrer belong to this small group of pioneering scientists, since they indeed started the field of neuroendocrinology, which today is a thriving discipline with strong basic and clinical research and with national and international societies holding meetings at regular intervals. In 1928 Ernst Scharrer suggested that hypothalamic neurosecretory neurons are important for peripheral hormonal systems (Scharrer 1928), a concept that he subsequently developed with his wife Berta and others (see Bargmann 1949; Scharrer 1952; Harris 1955; Scharrer and Scharrer 1963). The seminal finding that the two posterior pituitary hormones are peptides (Du Vigneaud 1956), and in particular the subsequent discovery by Guillemin, Schally, Vale and their collaborators that hypothalamic factors regulating anterior pituitary hormone secretion are also mostly peptides (Guillemin 1978; Schally 1978), firmly connected neuroendocrinology with peptide research. Some aspects of this topic have been summarized by Berta Scharrer in her article *The Neuropeptide Saga* (1990). Subsequent work has expanded the role of peptides from initially being peripheral hormones and neuroendocrine messengers to also being present in neurons (and glial cells) in the entire nervous system. It can, however, be noted that the initial excitement **(ups!)** in the peptide field in the 1970s has been followed by a more balanced attitude **(downs!)** with regard to the functional significance of neuropeptides in extrahypothalamic systems.

Neuropeptides

The early discoveries of peptides were based on collection of huge amounts of tissues, for example millions of fragments from the basal hypothalamus, for the purification of the hypothalamic releasing and inhibitory factors (see Guillemin 1978; Schally 1978), or of intestinal tissue for the isolation of cholecystokinin (CCK) and many more peptides (see Mutt 1987). A novel approach was to use C-terminal amidation as a biochemical marker for isolation of neuropeptides, as originally described by Tatemoto and Mutt (1981), leading to the discovery of numerous neuropeptides including peptide histidine isolucine (PHI) (Tatemoto and Mutt 1981) and neuropeptide tyrosine (NPY) (Tatemoto et al. 1982). Another rich source was frog skin which was exploited by Erspamer and his colleagues, leading to the discovery of numerous peptides (see Erspamer 1991). More recently, molecular biological approaches have been used, first to discover the calcitonin gene-related peptide (CGRP) (Rosenfeld et al. 1983). The 17 amino

Table 1. Some mammalian neuropeptides and neuropeptide families

Neuropeptide family Neuropeptide	Abbreviation	No. of amino acid residues
Hypothalamic hormones		
Oxytocin		9
Vasopressin		9
Hypothalamic releasing and inhibiting hormones		
Thyrotropin releasing hormone	TRH	3
Luteinizing hormone releasing hormone	LHRH	10
Somatostatin growth hormone release inhibiting hormone		14, and other forms
Corticotropin releasing hormone	CRH	41
Growth hormone releasing hormone	GHRH	44
Tachykinins		
Substance P	SP	11
Neurokinin A, substance K		10
Neurokinin B		10
Neuropeptide K		36
Opioid peptides		
Met- and leu-enkephalin		5
Dynorphin		17, and other forms
β-endorphin		30
NPY and related peptides		
Neuropeptide tyrosine	NPY	36
Pancreatic polypeptide	PPP	36
Peptide tyrosine-tyrosine	PYY	36
VIP-glucagon family		
Vasoactive intestinal polypeptide	VIP	28
Peptide histidine-isoleucine	PHI	27
Pituitary adenylate cyclase activating peptide	PACAP	27 or 38
Glucagon-like peptide-1	GLP-1	29
Other peptides		
Cholecystokinin	CCK	8
Neurotensin		13
Calcitonin gene-related peptide, α and β-form	CGRP	37
Galanin	GAL	29 or 30
Parathyroid hormone related protein		34 or 37
Brain natriuretic peptide		32
Islet amyloid polypeptide, amylin	IAPP	37
Novel neuropeptides		
Secretoneurin		33
Urocortin		40; 45% sequence identity with CRH
Nociceptin, orphanin FQ		17

acid peptide nociceptin or orphanin FQ was found as the endogenous agonist for a previously described orphan receptor, opioid receptor-like 1 (Meunier et al. 1995; Reinscheid et al. 1995). Thanks to these and other techniques, numerous peptides are now known, and often they can be grouped into families. In Table 1 many, but not all of the mammalian neuropeptides have been listed (see Hökfelt and Mutt 1997). We apologize for not giving original references for all peptides mentioned in Table 1. Also, we have not taken into consideration here the numerous peptides present in non-mammalian species, which in many cases are homologous to mammalian peptides. They include both the above-mentioned frog peptides, and insect peptides which are also of great interest.

It may be mentioned that peptides are expressed in neurons in at least three types of mode (see also Tohyama 1992; Table 2). Some peptides are expressed at high levels under normal circumstances, which indicates that they are functionally available at any time. A second type of peptide is expressed at low or undetectable levels under normal circumstances. They are then upregulated under certain conditions, for example after nerve injury. Thus a specific stimulus is required for such a peptide to become functionally relevant. The third type of peptide is expressed during development, often only prenatally, and is then downregulated postnatally. There seems to be a connection between the second and third type in that many of the peptides expressed transiently during ontogenesis can be "reactivated" in adulthood. It is important to note that one and the same peptide can belong to all three groups, since it is the type of neuron in which the peptide is expressed that determines the classification (Table 2).

Table 2. Peptides are expressed in 3 modes

Mode	Occurrence
Mode 1: Substantial levels synthesized and stored under normal conditions	SP and CGRP in primary sensory neurons GAL in hypothalamic neurons VIP and NPY in cortical neurons
Mode 2: Very low levels under normal conditions. Stimulus for upregulation of synthesis required	VIP, GAL and NPY in sensory neurons
*Mode 3 :*Transient expression during development	SOM in many central systems SP in spinal guiding neurons GAL in sensory neurons

Coexistence of Neuropeptides with Classic Transmitters and other Messenger Molecules

Some twenty years ago it was observed, using immunohistochemistry and double-staining techniques for cellular analysis, that a peptide could be present in the same neuron as a classic transmitter, for example somatostatin and noradrenaline in many sympathetic neurons in the rat (Hökfelt et al. 1977). Subsequent analyses in many laboratories have shown that this seems to be a general phenomenon, and numerous combinations of various messenger molecules have been described (see Hökfelt 1991). This indicates that peptides are either complementary to classic transmitters, for example modulating their actions, or play a completely different role in nervous system function, such as exerting trophic actions, as has been shown for example for vasoactive intestinal polypeptide (VIP) (Gozes and Brenneman 1990; see also Strand et al. 1991).

The coexistence of neuropeptides and classic transmitters seems in fact to be so frequent that the question has been posed as to whether or not there are any peptide neurons lacking a classic transmitter. The best candidates for this would be the hypothalamic magnocellular neurons, which contain a host of peptides with oxytocin and vasopressin as principal messengers (Brownstein and Mezey 1986). So far, no classic transmitter has been identified with certainty in these neurons, and it seems fair to say that vasopressin and oxytocin at least must be the main messenger molecules. Also, in the parvocellular hypothalamic neurons, peptides are presumably most important, except for some tubero-infundibular neurons which may have dopamine as the main messenger (see Fuxe et al. 1974). In the corticotropin releasing factor (CRF) neurons in the paraventricular nucleus, some neurons express glutamic acid decarboxylase (GAD) and GABA (Meister et al. 1988). However, it seems highly likely that in this case CRF still is the main hormone and that GABA may modulate, for example, the release of CRF and/or is involved in other mechanisms related to the CRF neuron. Thus, hypothalamic neurosecretory cells, brain children of Ernst and Berta Scharrer, represent excellent systems to elucidate the role of peptides and peptidergic mechanisms. However, they may not be typical for the majority of peptide expressing neurons in the central and peripheral nervous system, since, as discussed above, in the latter the peptides are not the main messengers. Also, the peptides are released into the vascular system, making them a special type of neuron. For these reasons, we have studied many other peptide systems in parallel, for example dorsal root ganglion (DRG) neurons.

Neuropeptide Receptors

It was not until the end of the 1980s that the first neuropeptide receptor could be cloned, and Masu et al. (1987) first showed that an intestinal substance K receptor was of the seven transmembrane (7-TM), G-protein-coupled receptor type. Subsequently, numerous receptors have been cloned, and it is now clear that there exists at least one receptor for every neuropeptide, and that in some cases there may be half a dozen subtypes for a single neuropeptide (see Meyerhof et al. 1993; Schwartz et al. 1995).

Expression and Regulation of Peptides

The focus in this chapter is on plasticity in expression of peptides and peptide receptors in DRG neurons. It should, however, be pointed out that this phenomenon is by no means confined to DRG neurons. Thus, to mention a few examples related to the research of Berta and Ernst Scharrer, dramatic changes have been observed in the hypothalamic magnocellular neurons with marked increases in galanin and CCK mRNA after hypophysectomy (Villar et al. 1990) and in CCK_B receptor mRNA after hypophysectomy and salt load (Meister et al. 1994).

DRG neurons express robust levels of several peptides including substance P and calcitonin gene-related peptide (CGRP) under normal circumstances (see Dalsgaard 1988). There is strong evidence that glutamate and perhaps ATP are principal transmitters in subpopulations of these neurons (see Salt and Hill 1983; Rustioni and Weinberg 1989). Interestingly, some other peptides have been shown to be expressed at very low or undetectable levels in such neurons under normal circumstances, but at high levels after different types of manipulations (**ups!**), such as nerve injury. There are also dramatic changes in neuropeptide receptor expression after such manipulation, whereas substance P and CGRP are down-regulated after nerve injury (**downs!**) (see Hökfelt et al. 1994). Neurons in spinal ganglia lend themselves to this type of experiment in an advantageous way, since their pseudounipolar nature allows manipulation of one branch and analysis of events in the cell soma in the DRG as well as of changes in the central branches in the dorsal horn. Possible secondary changes in local dorsal horn neurons and other components of the dorsal horn such as glial cells can also be studied.

The first peptide to be shown to be upregulated in DRG neurons was VIP (Shehab and Atkinson 1986). Subsequently, both galanin (Hökfelt et al. 1987; Villar et al. 1989) and neuropeptide tyrosine (NPY) (Wakisaka et al. 1991) as well as pituitary adenylate cyclase activating peptide (PACAP) (Zhang et al. 1995a, 1996) were also shown to be strongly and selectively upregulated. PACAP was increased in the highest proportion of DRG neurons, amounting to around 70% of all DRG neuron profiles one week after axotomy as compared to 40-50% for galanin and lower percentages for NPY and VIP. In fact, the upregulation of NPY, VIP and galanin often occurred in those neurons that expressed PACAP. However, upregulation of PACAP was transient in contrast to galanin, which remained upregulated for up to 7 months, the longest period studied, provided that regeneration was prevented. A further difference was the size of the neuron showing upregulation, NPY being mostly found in large neurons, galanin mainly in small but also in some large neurons, VIP mainly in small ones and PACAP in both small and large neurons. These findings suggest that the peptides may have different roles, although there is also a possibility that they serve similar roles, but in different neuron populations.

Although axotomy seems to be the most studied experimental model, it has been shown that galanin is influenced by a number of other manipulations, for example herpes zoster and other neuropathic models (see Table 3). In the case of peptides, many studies have been carried out both with in situ hybridization and immunohistochemistry, which compare changes in mRNA and peptide levels. Mostly there is good agreement, suggesting that increased mRNA levels are translated into increased peptide levels.

Table 3. Experimental procedures that cause upregulation of galanin in DRG neurons

Procedure	Source
Axotomy (nerve injury)	Hökfelt et al. 1987; Villar et al. 1989, 1991; Xu et al. 1990; Kashiba et al. 1992b
Nerve crush	Villar et al. 1989
Local application of vinblastine	Kashiba et al. 1992a
Systemic administration of resiniferatoxin	Farkas-Szallasi et al. 1995
Herpes simplex infection	Henken and Martin 1992a, b
Chronic constriction injury (Bennett model)	Nahin et al. 1994
Nerve ligation (Seltzer model)	Shi et al. 1997 (in prep.)

Expression and Regulation of Peptide Receptors

So far, our knowledge of regulation of peptide receptors in sensory neurons is limited. However, CCK_B receptor mRNA (Kopin et al. 1992) levels are normally low in rat DRG neurons, but after peripheral nerve injury more than half of all neurons express this receptor mRNA (Zhang et al. 1993a). Neurotensin receptor mRNA (Tanaka et al. 1990) was detected in about 30% of all small DRG neurons, and injury caused a distinct downregulation (Zhang et al. 1995c). The galanin-R1 receptor (Burgevin et al. 1995; Parker et al. 1995) is expressed in a population of DRG neurons which also are CGRP-positive, and nerve injury causes downregulation of this receptor (Xu et al. 1996b).

We have specially focused on NPY receptors. Thus, NPY Y1 receptor mRNA was found in a large population of mainly small DRG neurons which, in virtually all cases, were CGRP-positive (Zhang et al. 1994b). The Y1-R mRNA levels decreased after sciatic nerve transection, although the total number of profiles expressing Y1 mRNA was increased. The Y1 R mRNA positive neurons represented a different population to the neurotensin-receptor mRNA positive ones. Also, the recently cloned NPY Y2 receptor (Gerald et al. 1995; Rose et al. 1995; Gehlert et al. 1996) has been shown to be present in rat DRG neurons (Gustafson et al. 1997; Zhang et al. 1997). Around 15% of all DRG neuron profiles were Y2 R mRNA positive (Fig. 1a), and these neurons were also CGRP mRNA positive, but of a medium or a large size (Fig. 1c). Thus, CGRP neurons can be subdivided on the basis of presence of Y1-R and Y2-R, the former being small and the latter large CGRP neurons. After injury, there was a marked increase in Y2-R mRNA positive profiles (up to almost 35%) as well as elevated mRNA levels (Fig. 1b, d). After axotomy Y2-R mRNA positive neurons often coexpressed NPY mRNA suggesting the possibility of autoreceptor mechanisms. Monkey (*M. mulatta*) neurons were also analyzed, showing that some small neuron profiles expressed Y2-R mRNA and that there was a marked increase after nerve injury (from 10 to almost 60% of all neuron profiles).

Sensory ganglia have been analyzed with ligand binding studies and autoradiography, and evidence for the presence of both Y1 and Y2 receptors has been obtained (Walker et al. 1988; Ghilardi et al. 1994). In our own studies (Zhang et al. 1995b), using ^{125}I-PYY as ligand, a marked increase mainly in Y2 binding was observed over axons in DRGs, over dorsal roots and in the dorsal horn after peripheral axotomy. The Y1 receptor has also been studied with immunohistochemistry using antiserum against a peptide fragment of the C-terminal part of the receptor protein (Zhang et al. 1994a). These results strongly suggest that the Y1 receptor is a somatic receptor incorporated into the cell membrane of the DRG neurons, but not transported centrifugally to the dorsal horn or to peripheral tissues. Thus, taken together with the ligand binding studies referred to above, it seems as if it is the Y2 receptor that is transported into the axons and acts as a presynaptic receptor and that the Y1 receptor in this system represents a somatic ("postsynaptic") receptor. These findings are in good agreement with the original concept by Wahlestedt et al. (1986) of Y1 being a postsynaptic and Y2 a presynaptic receptor.

Fig. 1. Dark- (**a, b**) and bright- (**c, d**) field micrographs of control (**a, c**) and ipsilateral (**b, d**) rat L5 DRGs hybridized with probes for Y2-R mRNA 14 days after unilateral sciatic nerve cut. Several large (*arrows*) and some medium-sized (*arrowheads*) DRG neurons are Y2-R mRNA-positive in control ganglia (**a, c**). After axotomy there is an increase in the number of positive neurons and of Y2-R mRNA levels, including some labeled small neurons (*curved arrows*) (**b, d**). *Bars:* 250 µm in **a** and **b** and 50 µm in **c** and **d**. Reprinted from Zhang et al. (1997) with permission of the National Academy of Sciences, U.S.A.

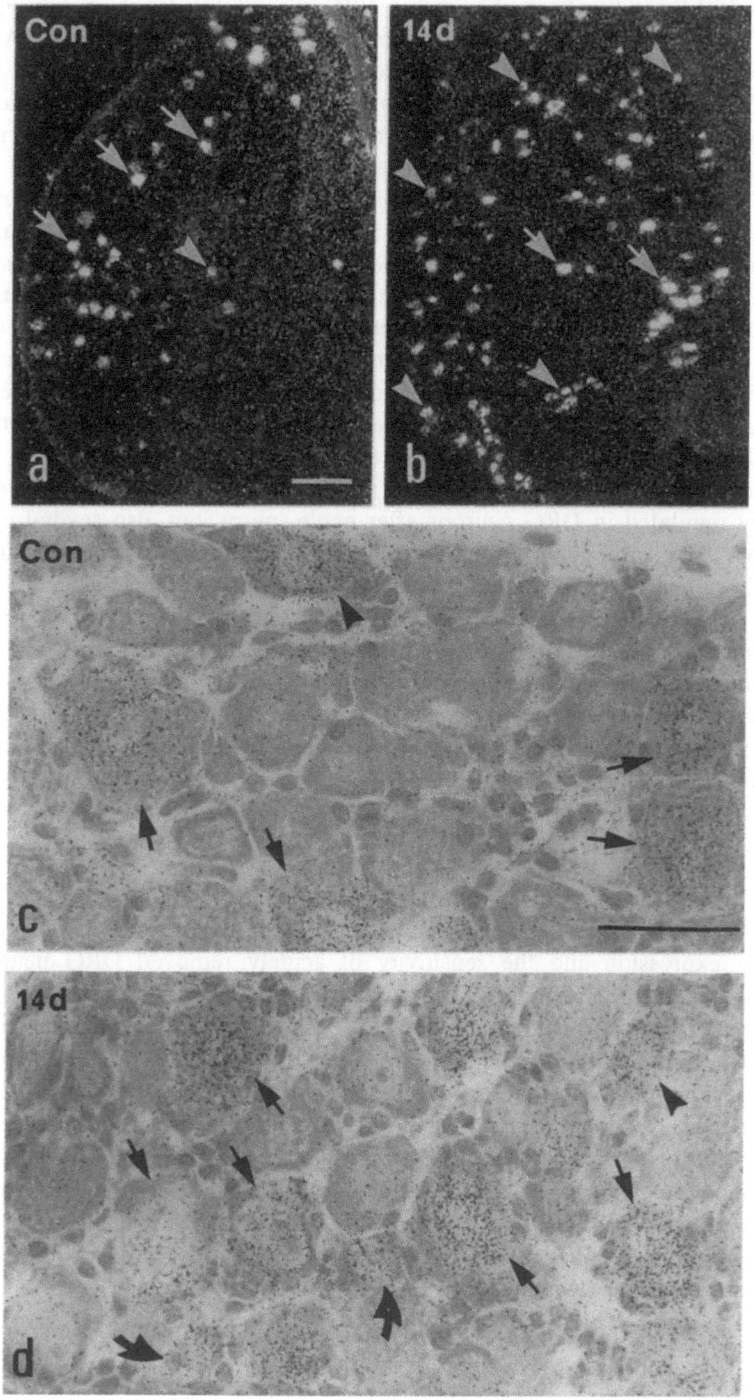
Con
14d
a
b
Con
c
14d
d

Functionally it has been shown that NPY can inhibit depolarization-induced release of substance P from DRG neurons in vitro (Walker et al. 1988). Also, in vivo, NPY can inhibit substance P release in the dorsal horn (Duggan et al. 1991), and NPY has, via Y2 receptors, been shown to inhibit Ca^{2+} influx into cultured rat DRG neurons (Ewald et al. 1988; Bleakman et al. 1991). Our histochemical data support the view that these effects are exerted via a Y2 receptor.

An interesting question is the origin of NPY which, under normal circumstances, activates the Y1 receptor on the DRG neuronal somata. Normally, DRG neurons do not express NPY (Gibson et al. 1984), although there is a strong upregulation after axotomy (Wakisaka et al. 1991). Also, normally there are only low numbers of NPY-positive nerve endings around blood vessels, and it is unlikely that NPY released from such nerves would influence the DRG neurons to a major degree. We have proposed that circulating NPY from other sources such as distant sympathetic nerves, adrenal medulla and perhaps thrombocytes may act upon DRG neurons.

Taken together, results obtained so far suggest that not only are peptides strongly regulated in DRG neurons after nerve injury, but also several of their receptors. The same peptides and receptors are in addition often present in local dorsal horn neurons. This makes it difficult to interpret the mechanism and site of action of a peptide, for example NPY, injected intrathecally, and to make deductions on the physiological role of this peptide. Principally, this NPY could act at presynaptic receptors on primary afferents, local dorsal horn neurons and/or on descending systems, as well as at postsynaptic receptors on dorsal horn neurons. Moreover, there may be additional NPY receptor subtypes in these regions.

Electrophysiological Studies

Recent studies in our laboratory have attempted to compare the changes in receptor expression and the sensitivity of acutely removed DRG neurons to exogenously applied peptides (Xu and Hökfelt, unpubl.). For example, we have shown that DRG neurons normally are not sensitive to CCK octapeptide, but after axotomy, a large number of neurons are depolarized by CCK, in good agreement with the upregulation of CCK_B receptors after axotomy (Zhang et al. 1993a). Furthermore, no effect of galanin could be recorded in normal DRG neurons, but after axotomy this peptide caused depolarization, in agreement with an in vitro study showing that galanin has an excitatory effect on cultured DRG neurons (Puttick et al. 1994). The two studies seem to agree, since it has been proposed that DRG neurons in culture with regard to peptide expression represent axotomized neurons (Kerekes et al. 1997).

Species Differences

Most studies on plasticity of peptide and peptide receptor expression in DRG neurons have been carried out on rats. It is, however, important to perform comparative analyses, in particular when considering pain mechanisms and pain treatment. Focusing on the three peptides galanin, NPY and substance P (Arvidsson et al. 1991; Zhang et al. 1993b; Corness et al. 1996; Rydh-Rinder et al. 1996), it has been shown that the same changes as described above in rats also occur in mice, that is upregulation of galanin and NPY and downregulation of substance P. In the guinea pig, there is downregulation of substance P and a strong upregulation of NPY but only a weak upregulation of galanin. The galanin upregulation in cats is also very weak. In the monkey (*M. mulatta*), substance P and galanin follow the patterns in the rat, whereas NPY is not upregulated. Moreover, we have recently been able to analyze two human DRGs extirpated for therapeutic purposes and frozen rapidly after surgery (Landry, Åman, Dostrowski, Lozano, Zhang, Wiesenfeld-Hallin and Hökfelt, unpublished). Galanin was expressed in both ganglia, one of which was from a patient with herpes zoster. In both ganglia, galanin was strongly expressed, but in the zoster ganglion, galanin and CGRP were present in essentially different cell populations. This could be compatible with the idea that herpes zoster causes an axotomized-like state (see Table 3), resulting in downregulation of CGRP and upregulation of galanin which thus appear in different neuron populations. Under all circumstances, the results show that human DRGs can express galanin at substantial levels, and they may indicate that there is a regulation in a similar manner as described above in rats. Galaninergic mechanisms may therefore operate at the spinal level of humans, and could be involved in the processing of pain stimuli.

Mechanisms Underlying Axotomy-Induced Peptide Regulation

Nerve growth factor (NGF) has been shown to be involved in the regulation of substance P and CGRP in DRG neurons (Kessler and Black 1980; Otten et al. 1980; Fitzgerald et al. 1985; Lindsay and Harmar 1989; Lindsay et al. 1989; Mulderry 1994; Verge et al. 1995). Briefly, NGF deficiency causes a marked downregulation of substance P and CGRP synthesis, and substitution with NGF completely restores peptide levels. A possible mechanism for regulation of galanin is therefore that NGF normally suppresses galanin synthesis, and that axotomy removes this suppression, leading to increased galanin levels. Some support for this hypothesis was obtained in experiments with intrathecal application of NGF and in vitro DRG neuron cultures (Verge et al. 1995; Kerekes et al. 1997), whereas NGF applied locally via a capsule onto the transected sciatic nerve did not influence galanin upregulation, but fully counteracted substance P downregulation (Ji et al. 1996). Thus it is unlikely that NGF is **the** critical factor involved in galanin regulation. More recently, leukemia inhibitory factor (LIF; also called cholinergic differentiation factor, CDF) (Fukada 1985; Yamamori et al. 1989) has emerged as a key molecule for control of synthesis of some peptides in autonomic and sensory neurons, in particular after nerve injury (Rao et al. 1993; Banner and Patterson 1994; Patterson 1994). Thus, in LIF deficient mice the axotomy-induced upregulation of galanin is markedly attenuated both in sympathetic (Rao et al. 1993), and in DRG neurons as shown both with radioimmunoassay (Sun and Zigmond 1996) and immunohistochemistry (Corness et al. 1996).

It is noteworthy that other growth factors may participate. For example, it has been shown that both acidic fibroblast growth factor (FGF) and basic FGF strongly counteract the lesion-induced upregulation of NPY in DRG neurons (Ji et al. 1996). However, these two growth factors did not influence the nerve injury induced downregulation of substance P (Ji et al. 1996).

Functional Aspects

A general hypothesis for the significance of the dramatic changes in the peptidergic phenotype of DRG neurons after peripheral nerve injury would be attenuation of transmission in the dorsal horn, in this way preventing excessive pain sensations to reach higher centers. A second purpose could be to enhance survival and regeneration. Downregulation of the two excitatory peptides, substance P and CGRP, would fit into this scheme, and there is some evidence that NPY and galanin can exert inhibitory actions in the dorsal horn. In fact, it has been postulated that galanin may be an endogenous analgesic compound, of particular importance after nerve injury (Wiesenfeld-Hallin et al. 1992). However, trophic actions should also be seriously considered. Thus, White and Mansfield (1996) have shown that NPY enhances neurite elongation, and in galanin deficient mice, regeneration is impaired (Wynick, pers. comm.). Such a role for galanin is also supported by the fact that galanin expression occurs in DRG neurons very early on in embryogenesis (Xu et al. 1996a). Thus, preprogalanin mRNA and galanin/GMAP-LI can be demonstrated in DRG (Fig. 2a) and trigeminal (Fig. 2b) neurons and in the dorsal horn from E14/E15 and onwards. Moreover, during embryogenesis galanin message has been observed in the retina, the inner ear and in the olfactory mucosa as well as in bone anlage. All these findings naturally suggest that galanin may play a role during development. It is therefore possible that the marked increase after nerve injury represents a replication of an ontogenetic process and that galanin does not necessarily have a distinct function in adulthood. However, galanin receptors are strongly expressed in the dorsal horn in adult rats, both as revealed with ligand binding autoradiography (Melander et al. 1988) and in situ hybridization (Burgevin et al. 1995; Parker et al. 1995; Gustafson et al. 1996). These receptors should therefore be available for galanin

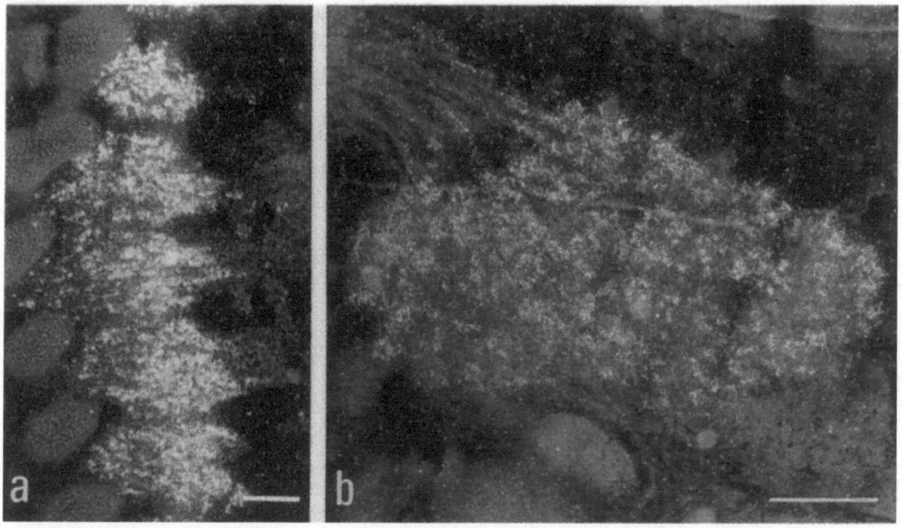

Fig. 2. Dark-field micrographs of dorsal root ganglia (**a**) and a trigeminal ganglion (**b**) of E17 embryos after hybridization with probe complementary to preprogalanin mRNA. Note strong signal over all dorsal root ganglia and the trigeminal ganglion at this stage. *Bar:* 100 μm

presumably released from primary afferents in increased amounts after nerve injury.

In support of an analgesic role of galanin in the dorsal horn are findings with a putative galanin antagonist (M35) which, upon chronic administration, causes a marked increase in autotomy (Verge et al. 1993), a behavior which may be related to painful sensations (Wall et al. 1979; Wiesenfeld-Hallin 1984; Coderre et al. 1986; Devor 1991). Moreover, administration of galanin antisense probes to the transected sciatic nerve via a capsule markedly reduces galanin levels in DRG neuronal cell bodies and causes a strong increase in autotomy (Ji et al. 1994). However, these experiments are still open to criticism, and the true role of galanin in dorsal horn function remains to be proven.

Concluding Remarks

Using DRG neurons as a model system we have attempted to understand the functional roles of some neuropeptides. These neurons express certain peptides at high levels under normal circumstances, and another group of peptides are markedly upregulated after peripheral nerve injury. Peptide receptors in these neurons are regulated by nerve injury. Studies carried out in others' and our own laboratory indicate that the first group of peptides have excitatory actions in the dorsal horn, whereas those upregulated after nerve injury may attenuate dorsal horn transmission as well as exert trophic actions. We have proposed that peptides upregulated in DRG neurons after nerve injury belong to a pain defense system associated with neuropathic pain. A second pain defense system is represented by local dorsal horn neurons which produce, for example, opioid peptides such as enkephalin and dynorphin, and which react towards inflammatory pain (Dubner and Ruda 1992; Weihe et al. 1994). Thus, neuropeptides may have multifacetted functions, and much work remains to firmly establish their role during normal and pathological conditions.

Acknowledgments

These studies were supported by Marianne and Marcus Wallenbergs Stiftelse, the Swedish MRC (04X2887), Gustaf V:e och Drottning Victorias Stiftelse and Astra Pain Control AB.

References

Arvidsson U, Ulfhake B, Cullheim S, Bergstrand A, Theodorsson E, Hökfelt T (1991) Distribution of ^{125}I-galanin and its coexistence with 5-hydroxytryptamine in the cat spinal cord: biochemical, histochemical and experimental studies at the light and electron microscopic level. J Comp Neurol 308: 115-138

Banner LR, Patterson PH (1994) Major changes in the expression of the mRNAs for cholinergic differentiation factor/leukemia inhibitory factor and its receptor after injury to adult peripheral nerves and ganglia. Proc Natl Acad Sci USA 91: 7109-7113

Bargmann W (1949) Über die neurosekretorische Verknüpfung von Hypothalamus und Neurohypophyse. Z Zellforsch 34: 610-634

Bleakman D, Colmers WF, Fournier A, Miller RJ (1991) Neuropeptide Y inhibits Ca^{2+} influx into cultured dorsal root ganglion neurons of the rat via a Y2 receptor. Br J Pharmacol 103: 1781-1789

Brownstein MJ, Mezey E (1986) Multiple chemical messengers in hypothalamic magnocellular neurons. In: Hökfelt T, Fuxe K, Pernow B (eds) Progress in brain research, vol 68. Elsevier, Amsterdam, pp 161-168

Burgevin M-C, Loquet I, Quarteronet D, Habert-Ortoli E (1995) Cloning, pharmacological characterization, and anatomical distribution of a rat cDNA encoding for a galanin receptor. J Mol Neurosci 6: 33-41

Coderre TJ, Grimes RW, Melzack R (1986) Deafferentation and chronic pain in animals. An evaluation of evidence suggesting autotomy is related to pain. Pain 26: 61-84

Corness J, Shi T-J, Xu Z-Q, Brulet P, Hökfelt T (1996) Influence of leukemia inhibitory factor on galanin/GMAP and neuropeptide Y expression in mouse primary sensory neurons after axotomy. Exp Brain Res 112: 79-88

Dalsgaard C-J (1988) The sensory system. In: Björklund A, Hökfelt T, Owman C (eds) Handbook of chemical neuroanatomy, vol 6. The peripheral nervous system. Elsevier Science, Amsterdam, pp 599-667

Devor M (1991) Sensory basis of autotomy in rats. Pain 45: 109-110

Du Vigneaud V (1956) Hormones of the posterior pituitary gland: oxytocin and vasopressin. In: The Harvey lectures 1954-55. Academic Press, New York, pp 1-26

Dubner R, Ruda MA (1992) Activity-dependent neuronal plasticity following tissue injury and inflammation. TINS 15: 96-103

Duggan AW, Hope PJ, Lang CW (1991) Microinjection of neuropeptide Y into the superficial dorsal horn reduces stimulus-evoked release of immunoreactive substance P in the anesthetized cat. Neuroscience 44: 733-740

Erspamer V (1991) The tachykinin peptide family. TINS 4: 267-269

Ewald DA, Matthies HJG, Perney TM, Walker MW, Miller RJ (1988) The effect of down regulation of protein kinase C on the inhibitory modulation of dorsal root ganglion neuron Ca^{2+} currents by neuropeptide Y. J Neurosci 8: 2447-2451

Farkas-Szallasi T, Lundberg JM, Wiesenfeld-Hallin Z, Hökfelt T, Szallasi A (1995) Increased levels of GMAP, VIP, and nitric oxide synthase, and their mRNAs, in lumbar dorsal root ganglia of the rat following systemic resiniferatoxin treatment. NeuroReport 6: 2230-2234

Fitzgerald M, Wall PD, Goedert M, Emson PC (1985) Nerve growth factor counteracts the neurophysiological and neurochemical effects of chronic sciatic nerve section. Brain Res 332: 131-141

Fukada K (1985) Purification and partial characterization of a cholinergic neuronal differentiation factor. Proc Natl Acad Sci USA 82: 8795-8799

Fuxe K, Goldstein M, Hökfelt T, Jonsson G, Lidbrink P (1974) Dopaminergic involvement in hypothalamic function: extrahypothalamic and hypothalamic control. A neuroanatomical analysis. Adv Neurol, vol 5. Raven Press, New York, pp 405-419

Gehlert DR, Beavers LS, Johnson D, Gackenheimer SL, Schober DA, Gadski RA (1996) Expression cloning of human brain neuropeptide Y Y2 receptor. Mol Pharmacol 49: 224-228

Gerald C, Walker MW, Vaysse PJ-J, He C, Branchek TA, Weinshank RL (1995) Expression cloning and pharmacological characterization of a human hippocampal neuropeptide Y/peptide YY Y2 receptor subtype. J Biol Chem 270: 26758-26761

Ghilardi JR, Allen CJ, Vigna SR, McVey DC, Mantyh PW (1994) Cholecystokinin and neuropeptide Y receptors on single rabbit vagal afferent ganglion neurons: site of prejunctional modulation of visceral sensory neurons. Brain Res 633: 33-40

Gibson SJ, Polak JM, Allen JM, Adrian TE, Kelly JS, Bloom SR (1984) The distribution and origin of a novel brain peptide, neuropeptide Y, in the spinal cord of several mammals. J Comp Neurol 227: 78-91

Gozes I, Brenneman DE (1990) VIP: molecular biology and neurobiological function. Mol Neurobiol 3: 201-236

Guillemin R (1978) Peptides in the brain: the new endocrinology of the neuron. Science 202: 390-402

Gustafson EL, Smith KE, Durkin MM, Gerald C, Branchek TA (1996) Distribution of a rat galanin receptor mRNA in rat brain. NeuroReport 7: 953-957

Gustafson EL, Smith KE, Durkin MM, Walker MW, Gerald C, Weinshank R, Branchek TA (1997) Distribution of the neuropeptide Y Y2 receptor mRNA in rat central nervous system. Mol Brain Res 46: 223-235

Harris GW (1955) Neuronal control of the pituitary gland. Edward Arnold Ltd, London

Henken DB, Martin JR (1992a) Herpes simplex virus infection induces a selective increase in the proportion of galanin-positive neurons in mouse sensory ganglia. Exp Neurol 118: 195-203

Henken DB, Martin JR (1992b) The proportion of galanin-immunoreactive neurons in mouse trigeminal ganglia is transiently increased following corneal inoculation of herpes simplex virus type-1. Neurosci Lett 140: 177-180

Hökfelt T (1991) Neuropeptides in perspective: the last ten years. Neuron 7: 867-879

Hökfelt T, Mutt V (1997) Neuropeptides. In: Adelman G, Smith B (eds) Encyclopedia of neuroscience, 2nd edn. CD-ROM version. Elsevier, Amsterdam

Hökfelt T, Elfvin LG, Elde R, Schultzberg M, Goldstein M, Luft R (1977) Occurrence of somatostatin-like immunoreactivity in some peripheral sympathetic noradrenergic neurons. Proc Natl Acad Sci USA 74: 3587-3591

Hökfelt T, Wiesenfeld-Hallin Z, Villar MJ, Melander T (1987) Increase of galanin-like immunoreactivity in rat dorsal root ganglion cells after peripheral axotomy. Neurosci Lett 83: 217-220

Hökfelt T, Zhang X, Wiesenfeld-Hallin Z (1994) Messenger plasticity in primary sensory neurons following axotomy and its functional implications. TINS 17: 22-30

Ji R-R, Zhang Q, Pettersson RF, Hökfelt T (1996) aFGF, bFGF and NGF differentially regulate neuropeptide expression in dorsal root ganglia after axotomy and induce autotomy. Reg Pep 66: 179-189

Ji RR, Zhang Q, Bedecs K, Arvidsson J, Zhang X, Xu XJ, Wiesenfeld-Hallin Z, Bartfai T, Hökfelt T (1994) Galanin antisense oligonucleotides reduce galanin levels in dorsal root ganglia and induce autotomy in rats after axotomy. Proc Natl Acad Sci USA 91: 12540-12543

Kashiba H, Senba E, Kawai Y, Ueda Y, Tohyama M (1992a) Axonal blockade induces the expression of vasoactive intestinal polypeptide and galanin in rat dorsal root ganglion neurons. Brain Res 577: 19-28

Kashiba H, Senba E, Ueda Y, Tohyama M (1992b) Co-localized but target-unrelated expression of vasoactive intestinal polypeptide and galanin in rat dorsal root ganglion neurons after peripheral nerve crush injury. Brain Research 582: 47-57

Kerekes N, Landry M, Rydh-Rinder M, Hökfelt T (1997) The effect of NGF, BDNF and bFGF on expression of galanin in cultured rat dorsal root ganglia. Brain Res 754: 131-141

Kessler JA, Black IB (1980) Nerve growth factor stimulates the development of substance P in sensory ganglia. Proc Natl Acad Sci USA 77: 649-652

Kopin AS, Lee YM, McBride EW, Miller LJ, Lu M, Lin HY, L.F. K, Beiborn M (1992) Expression cloning and characterization of the canine parietal cell gastrin receptor. Proc Natl Acad Sci USA 89: 3605-3609

Lindsay RM, Harmar AJ (1989) Nerve growth factor regulates expression of neuropeptide genes in adult sensory neurons. Nature 337: 362-364

Lindsay RM, Lockett C, Sternberg J, Winter J (1989) Neuropeptide expression in cultures of adult sensory neurons: modulation of substance P and calcitonin gene-related peptide levels by nerve growth factor. Neuroscience 33: 53-65

Masu Y, Nakayama K, Tamaki H, Harada Y, Kuno M, Nakanishi S (1987) cDNA cloning of bovine substance-K receptor through oocyte expression system. Nature 329: 836-838

Meister B, Broberger C, Villar MJ, Hökfelt T (1994) Cholecystokinin B receptor gene expression in hypothalamic neurosecretory neurons after experimental manipulations. Neuroendocrinology 60: 458-469

Meister B, Hökfelt T, Geffard M, Oertel W (1988) Glutamic acid decarboxylase- and γ-aminobutyric acid-like immunoreacitivites in corticotropin-releasing factor-containing parvocellular neurons of the hypothalamic paraventricular nucleus. Neuroendocrinology 48: 516-526

Melander T, Köhler C, Nilsson S, Hökfelt T, Brodin E, Theodorsson E, Bartfai T (1988) Autoradiographic quantification and anatomical mapping of ^{125}I-galanin binding in the rat central nervous system. J Chem Neuroanat 1: 213-233

Meunier J-C, Mollereau C, Toll L, Suaudeau C, Moisand C, Alvinerie P, Butour J-L, Guillemot J-C, Ferrare P, Monsarrat B, Mazaguil H, Vassart G, Parmentier M, Costentin J (1995) Isolation and structure of the endogenous agonist of opioid receptor-like ORL_1 receptor. Nature 377: 532-535

Meyerhof W, Darlison MG, Richter D (1993) The elucidation of neuropeptide receptors and their subtypes through the application of molecular biology. In: Hucho F (ed) New comprehensive biochemistry, Volume 25. Elsevier, Amsterdam, pp 335-353

Mulderry PK (1994) Neuropeptide expression by newborn and adult rat sensory neurons in culture: effects of nerve growth factor and other neurotrophic factors. Neuroscience 59: 673-688

Mutt V (1987) Chemical properties of some polypeptides affecting vascular function. In: Nobin A, Owman C, Arneklo-Nobin B (eds) Neuronal messengers in vascular function. Elsevier, Amsterdam, pp 125-198

Nahin RL, Marino De León KR, Ruda M (1994) Primary sensory neurons exhibit altered gene expression in a rat model of neuropathic pain. Pain 58: 95-108

Otten U, Goedert M, Mayer N, Lembeck F (1980) Requirement of nerve growth factor for development of substance P-containing sensory neurons. Nature 287: 158-159

Parker EM, Izzarelli DG, Nowak HP, Mahle CD, Iben LG, Wang J, Goldstein ME (1995) Cloning and characterization of rat GALR1 galanin receptor from RIN14B insulinoma cells. Mol Brain Res 34: 179-189

Patterson PH (1994) Leukemia inhibitory factor, a cytokine at the interface between neurobiology and immunology. Proc Natl Acad Sci USA 91: 7833-7835

Puttick RM, Pinnock RD, Woodruff GN (1994) Galanin-induced membrane depolarization of neonatal rat cultured dorsal root ganglion cells. Eur J Pharmacol 254: 303-306

Rao MS, Sun Y, Escary JL, Perreau J, Tresser S, Patterson PH, Zigmond RE, Brulet P, Landis SC (1993) Leukemia inhibitory factor mediates an injury response but not a target-directed developmental transmitter switch in sympathetic neurons. Neuron 11: 1175-1185

Reinscheid RK, Nothacker H-P, Bourson A, Ardati A, Henningsen RA, Bunzow JR, Grandy DK, Langen H, Monsma JFJ, Civelli O (1995) A neuropeptide that activates an opioid like G protein coupled receptor. Science 270: 792-794

Rose PM, Fernandes P, Lynch JS, Fraxier ST, Fisher SM, Kodukula K, Kienzle B, Seethala R (1995) Cloning and functional expression of a cDNA encoding a human type 2 neuropeptide Y receptor. J Biol Chem 270: 22661-22664

Rosenfeld MG, Mermod JJ, Amara SG, Swanson LW, Sawchenko PE, Rivier J, Vale WW, Evans RM (1983) Production of a novel neuropeptide encoded by the calcitonin gene via tissue-specific RNA processing. Proc Natl Acad Sci USA 304: 129-135

Rustioni A, Weinberg RJ (1989) The somatosensory system. In: Björklund A, Hökfelt T, Swanson LW (eds) Handbook of chemical neuroanatomy, vol 7. Intergrated systems of the CNS, Part II. Central visual, auditory, somatosensory, gustatory. Elsevier, Amsterdam, pp 219-321

Rydh-Rinder M, Holmberg K, Elfvin L-G, Wiesenfeld-Hallin Z, Hökfelt T (1996) Effects of peripheral axotomy on neuropeptides and nitric oxide synthase in dorsal root ganglia and spinal cord of the guinea pig: an immunohistochemical study. Brain Res 707: 180-188

Salt TE, Hill RG (1983) Neurotransmitter candidates of somatosensory primary afferent fibers. Neuroscience 10: 1083-1103

Schally AV (1978) Aspects of hypothalamic regulation of the pituitary gland. Its implications for the control of reproductive processes. Science 202: 18-28

Scharrer B (1990) The neuropeptide saga. Am Zool 30: 887-895

Scharrer E (1928) Die Lichtempfindlichkeit blinder Elritzen. Untersuchungen über das Zwischenhirn der Fische. Z Vergl Physiol 7: 1-38

Scharrer E (1952) The general significance of the neurosecretory cell. Scientia 46: 177-183

Scharrer E, Scharrer B (1963) Neuroendocrinology. Columbia University Press, New York

Schwartz TW, U. G, Schambye HT, Hjorth SA (1995) Molecular mechanism of action of non-peptide ligands for peptide receptors. Current Pharm Design 1: 325-342

Shehab SA, Atkinson ME (1986) Vasoactive intestinal polypeptide (VIP) increases in the spinal cord after peripheral axotomy of the sciatic nerve originate from primary afferent neurons. Brain Res 372: 37-44

Strand FL, Rose KJ, Zuccarelli LA, Kume J (1991) Neuropeptide hormones as neurotrophic factors. Physiol Rev 71: 1017-1037

Sun Y, Zigmond RE (1996) Leukaemia inhibitory factor induced in the sciatic nerve after axotomy is involved in the induction of galanin in sensory neurons. Eur J Neurosci 8: 2213-2220

Tanaka K, Masu M, Nakanishi S (1990) Structure and functional expression of the cloned rat neurotensin receptor. Neuron 4: 847-854

Tatemoto K, Mutt V (1981) Isolation and characterization of the intestinal peptide porcine PHI (PHI127), a new member of the glucagon-secretion family. Proc Natl Acad Sci USA 78: 6603-6607

Tatemoto K, Carlquist M, Mutt V (1982) Neuropeptide Y, a novel brain peptide with structural similarities to peptide YY and pancreatic polypeptide. Nature 296: 659-660

Tohyama M (1992) An overview of the ontogeny of neurotransmitters and neuromodulators in the central nervous system. In: Björklund A, Hökfelt T, Tohyama M (eds) Handbook of chemical neuroanatomy. Ontogeny of transmitters and peptides in the CNS, vol 10. Elsevier, Amsterdam, pp 647-650

Verge VMK, Richardson PM, Wiesenfeld-Hallin Z, Hökfelt T (1995) Differential influence of nerve growth factor on neuropeptide expression in vivo: a novel role in peptide suppression in adult sensory neurons. J Neurosci 15: 2081-2096

Verge VMK, Xu X-J, Langel Ü, Hökfelt T, Wiesenfeld-Hallin Z, Bartfai T (1993) Evidence for an endogenous inhibitory control on autotomy, a behavioral sign of neuropathic pain, by galanin in the rat after sciatic nerve section: demonstrated by chronic intrathecal infusion of M-35, a newly developed high affinity galanin receptor antagonist. Neurosci Lett 149: 193-197

Villar MJ, Cortés R, Theodorsson E, Wiesenfeld-Hallin Z, Schalling M, Fahrenkrug J, Emson PC, Hökfelt T (1989) Neuropeptide expression in rat dorsal root ganglion cells and spinal cord after peripheral nerve injury with special reference to galanin. Neuroscience 33: 587-604

Villar MJ, Meister B, Cortés R, Schalling M, Morris M, Hökfelt T (1990) Neuropeptide gene expression in hypothalamic magnocellular neurons of normal and hypophysectomized rats: a combined immunohistochemical and in situ hybridization study. Neuroscience 36: 181-99

Villar MJ, Wiesenfeld-Hallin Z, Xu XJ, Theodorsson E, Emson PC, Hökfelt T (1991) Further studies on galanin-, substance P-, and CGRP-like immunoreactivities in primary sensory neurons and spinal cord: effects of dorsal rhizotomies and sciatic nerve lesions. Exp Neurol 112: 29-39

Wahlestedt C, Yanaihara N, Håkanson R (1986) Evidence for different pre- and post-junctional receptors for neuropeptide Y and related peptides. Reg Pep 13: 307-318

Wakisaka S, Kajander KC, Bennett GJ (1991) Increased neuropeptide (NPY)-like immunoreactivity in rat sensory neurons following peripheral axotomy. Neurosci Lett 124: 200-203

Walker ME, Ewald DA, Perney TM, Miller RG (1988) Neuropeptide Y modulates neurotransmitter release and Ca^{2+} currents in rat sensory neurons. J Neurosci 8: 2438-2446

Wall PD, Devor M, Inbal R, Scadding JW, Schonfield D, Seltzer Z, Tomkiewiocz MM (1979) Autotomy following peripheral nerve lesions: experimental anaesthesia dolorosa. Pain 7: 103-113

Weihe E, Schäfer MK-H, Nohr D, Persson S (1994) Expression of neuropeptides, neuropeptide receptors and neuropeptide processing enzymes in spinal neurons and peripheral non-neural cells and plasticity in models of inflammatory pain. In: Hökfelt T, Schaible H-G, Schmidt RF (eds) Neuropeptides, nociception and pain. Chapman & Hall, London, pp 43-69

White DM, Mansfield K (1996) Vasoactive intestinal polypeptide and neuropeptide Y act indirectly to increase neurite outgrowth of dissociated dorsal root ganglion cells. Neuroscience 73: 881-887

Wiesenfeld-Hallin Z (1984) The effect of intrathecal morphine and naltrexone on autotomy in sciatic nerve sectioned rats. Pain 18: 267-278

Wiesenfeld-Hallin Z, Bartfai T, Hökfelt T (1992) Galanin in sensory neurons in the spinal cord. Front Neuroendocrinol 13: 319-343

Xu X-J, Wiesenfeld-Hallin Z, Villar MJ, Fahrenkrug J, Hökfelt T (1990) On the role of galanin, substance P and other neuropeptides in primary sensory neurons of the rat: studies on spinal reflex excitability and peripheral axotomy. Eur J Neurosci 2: 733-743

Xu Z-Q, Shi T-J, Hökfelt T (1996a) Expression of galanin and a galanin receptor in several sensory systems and bone anlage of rat embryos. Proc Natl Acad Sci USA 93: 14901-14905

Xu Z-Q, Shi T-J, Landry M, Hökfelt T (1996b) Evidence for galanin receptors in primary sensory neurons and effect of axotomy and inflammation. NeuroReport 8: 237-242

Yamamori T, Fukada K, Aebersold R, Korsching S, Fann MJ, Patterson PH (1989) The cholinergic neuronal differentiation factor from heart cells is identical to leukemia inhibitory factor. Science 246: 1412-1416

Zhang X, Dagerlind Å, Elde RP, Castel M-N, Broberger C, Wiesenfeld-Halin Z, Hökfelt T (1993a) Marked increase in cholecystokinin B receptor messenger RNA levels in rat dorsal root ganglia after peripheral axotomy. Neuroscience 57: 227-233

Zhang X, Ju G, Elde R, Hökfelt T (1993b) Effect of peripheral nerve cut on neuropeptides in dorsal root ganglia and the spinal cord of monkey with special reference to galanin. J Neurocytol 22: 342-381

Zhang X, Bao L, Xu Z-Q, Kopp J, Arvidsson U, Elde R, Hökfelt T (1994a) Localization of neuropeptide Y Y1 receptors in the rat nervous system with special reference to somatic receptors on small dorsal root ganglion neurons. Proc Natl Acad Sci USA 91: 11738-11742

Zhang X, Wiesenfeld-Hallin Z, Hökfelt T (1994b) Effect of peripheral axotomy on expression of neuropeptide Y receptor mRNA in rat lumbar dorsal root ganglia. Eur J Neurosci 6: 43-57

Zhang Q, Shi T-J, Ji R-R, Zhang Y-T, Sundler F, Hannibal J, Fahrenkrug J, Hökfelt T (1995a) Expression of pituitary adenylate cyclase-activating polypeptide in dorsal root ganglia following axotomy: time course and coexistence. Brain Res 705: 149-155

Zhang X, Ji R-R, Nilsson S, Villar M, Ubink R, Ju G, Wiesenfeld-Hallin Z, Hökfelt T (1995b) Neuropeptide Y and galanin binding sites in rat and monkey lumbar dorsal root ganglia and spinal cord and effect of peripheral axotomy. Eur J Neurosci 7: 367-380

Zhang X, Xu Z-Q, Bao L, Dagerlind Å, Hökfelt T (1995c) Complementary distribution of receptors for neurotensin and NPY in small neurons in rat lumbar DRGs and regulation of the receptors and peptides after peripheral axotomy. J Neurosci 15: 2733-2747

Zhang Y-Z, Hannibal J, Zhao Q, Moller K, Danielsen N, Fahrenkrug J, Sundler F (1996) Pituitary adenylate cyclase activating peptide (PACAP) expression in the rat dorsal root ganglia: up-regulation after peripheral nerve injury. Neuroscience 74: 1099-1110

Zhang X, Shi T-J, Holmberg K, Landry M, Huan W, Xiao H, Ju G, Hökfelt T (1997) Expression and regulation of the neuropeptide Y Y2 receptor in sensory and autonomic ganglia. Proc Natl Acad Sci USA 94: 729-734

3 Neurohormones and Neurohormone Receptors in Invertebrates

C.J.P. Grimmelikhuijzen and F. Hauser

Department of Cell Biology and Anatomy, Zoological Institute, University of Copenhagen, Universitetsparken 15, DK-2100 Copenhagen Ø, Denmark;
Phone: +45 35 32 12 27, Fax: +45 35 32 12 00

Abstract

Neuropeptides and their receptors play an extremely important role in invertebrates. From the lowest invertebrate group having a nervous system, the cnidarians, we isolated about 30 different neuropeptides, and this can be regarded as being the tip of an iceberg. The cnidarian neuropeptides are located in dense-cored neurosecretory vesicles and are biologically active as neurotransmitters or neurohormones. We have also cloned the preprohormones of the cnidarian neuropeptides. These preprohormones have a very high copy number (up to 36) of the immature neuropeptide sequence. In addition to well-known "classical" processing enzymes, novel processing enzymes must be present in cnidarian neurones, making neuropeptide biosynthesis in cnidarians more complex than in higher organisms. In 1993, in an attempt to clone the receptors for the cnidarian neurohormones, we found that sea anemones produce a receptor that shows a striking structural homology with members of the mammalian glycoprotein hormone (TSH, FSH, LH/CG) receptor family. This was an exciting finding, since glycoprotein hormone receptors had, at that time, only been cloned from mammals. Our results suggested that glycoprotein hormones and possibly all the processes that are mediated by these types of receptors are conserved throughout the animal kingdom, from cnidarians to mammals. We then focused on insects and cloned a *Drosophila* receptor that, again, showed a strong homology, both at the protein and gene levels, with the mammalian glycoprotein hormone receptors. The *Drosophila* receptor gene is strongly expressed starting 8-16 h after oviposition and the expression stays high until after pupation. This suggests that the receptor is involved in insect development. From these examples in cnidarians and insects, it is clear that several hormonal systems, such as the neuropeptide/receptor and glycoprotein hormone/receptor couples have a long evolutionary history.

Introduction

When the structures of the first mammalian neuropeptides were elucidated in the 1950s and 1960s and especially during the "boom" of neuropeptide discoveries in the seventies and eighties (Acher and Chauvet 1953; Du Vigneaud et al. 1953a,b; Guillemin and Burgus 1972; Guillemin 1978; Schally 1978; Hökfelt et al., this Vol., Chap. 2), it was sometimes suggested that neuropeptides were needed for the fine-tuning of the highly complicated mammalian brain, or in other words, that peptides made the mammalian brain what it is: the crown on nervous system evolution. In the last 10-15 years, however, it has become increasingly clear that neuropeptides have a long phylogenetic history and that these substances play important roles in all invertebrates. Nowadays, large numbers of biologically active neuropeptides have been isolated from molluscs, insects, crustaceans, annelids, nematodes, plathyhelminths and cnidarians. In the first part of this review we would like to focus on cnidarians, because they are the lowest animal group in the Animal Kingdom having a nervous system, and it probably was within cnidarians, or a related ancestor phylum that nervous systems first evolved (Mackie 1990). In the second part, we would like to give an example of a neurohormone receptor in insects, since they comprise by far the largest group of invertebrate animals.

Cnidarians

Cnidarians are primitive animals such as sea anemones, corals, jellyfishes and the freshwater polyp *Hydra*. Taxonomists have subdivided the phylum Cnidaria into four classes: Hydrozoa (such as the freshwater polyp *Hydra),* Cubozoa (box jellyfishes), Scyphozoa (true jellyfishes) and Anthozoa (animals such as corals, sea anemones and sea pansies). Cnidarians are not only beautiful animals, but they are also attractive models for biologists. This is for the following reasons: (1) cnidarians have a simple nervous system and a simple behaviour and, therefore, can be used as a model system for electrophysiologists to determine the cellular basis of behaviour (Anderson and Schwab 1982; Spencer and Arkett 1984; Anderson and Spencer 1989); (2) cnidarians have an impressive regeneration capacity. When, for example, *Hydra* is cut into 10-20 slices, each slice will develop into a new animal within a few days. *Hydra* can even be dissociated into its separate cells and, after reaggregation of the cells by centrifugation, develop into new animals from the cell clumps (Gierer et al. 1972; Gierer 1977). *Hydra* has less than ten different cell types. After elimination of the stem cells for most of these cell types, it is possible to obtain *Hydra* that only consist of one type of cells, the epitheliomuscular cells (Campbell 1976; Sugiyama and Fujisawa 1978). By introducing stem cells into these "epithelial" *Hydra,* it is possible to follow the differentiation of a completely new nervous system in an originally "nerve-free" animal (Minobe et al. 1995). Thus, it is easy to manipulate cnidarians and for this reason, *Hydra* and other hydrozoan polyps are widely used as model systems by developmental biologists to study processes such as growth, cell differentiation and pattern formation (Gierer 1977; Bode 1992).

Neurohormones in Cnidarians

The basic organization of the cnidarian nervous system is that of a nerve net. This net can be condensed and form a nervous plexus in some areas, for example in the head and foot regions of *Hydra*. Sometimes, the neuronal processes fuse with each other to make fast-conducting, large diameter nerve tracts ("giant axons") or nerve rings (Mackie 1973, 1984, 1989; Spencer and Arkett 1984; Grimmelikhuijzen et al. 1986). However, structures such as ganglia or brains do not exist in cnidarians. At the ultrastructural level, many of the cnidarian neurones appear to be multifunctional: they have sensory cilia (are "sensory neurones"), make synaptic contacts with two or more other neurones (are "interneurones"), make synaptic contacts with muscle cells (are "motor neurones"), and have dense-cored secretory vesicles at nonsynaptic release sites (are "neurosecretory neurones") (Westfall 1973; Westfall and Kinnamon 1978). Westfall proposes that these primitive, multifunctional neurones are the ancestors of the more specialized neurones that we find in higher animals today.

The existence of synapses and nonsynaptic release sites in the cnidarian nervous systems have been known for more than 35 years (Horridge and Mackay 1962), but for a long time uncertainty has remained on the nature of the neurotransmitters or neurohormones. We, for example, have been unable to demonstrate acetylcholine, catecholamines or serotonin in the nervous system of *Hydra* and the same negative results have been obtained by other workers in the field (O. Koizumi, pers. comm.). However, if one looks at the morphology of the presynaptic vesicles (most of them are dense-cored, with a diameter of 100-200 nm), one realizes that neuropeptides could be good candidates for being the long-searched for transmitters, and this is exactly what we have found. Using different radioimmunoassays, we have isolated 16 different neuropeptides from a single sea anemone species, *Anthopleura elegantissima* (13 are given in Table 1), and about 15 neuropeptides have been purified from species belonging to the other cnidarian classes (10 are given in Table 2) (Grimmelikhuijzen and Graff 1986; Grimmelikhuijzen and Groeger 1987; Grimmelikhuijzen et al. 1988, 1990, 1992; Graff and Grimmelikhuijzen 1988a, b; Nothacker et al. 1991a, b; Carstensen et al. 1992, 1993; Moosler et al. 1996). Because of the specificities of the radioimmunoassays used, all isolated neuropeptides have an Arg-X-NH_2 or Lys-X-NH_2 C terminus. Many peptides are protected at their N terminus by the well-known <Glu group, but some peptides are blocked by an unusual L-3-phenyllactyl group (Table 1).

Using immunocytochemistry, it was found that all peptides mentioned in Tables 1 and 2 are neuropeptides and that peptides belonging to one group (for example the Antho-RWamides, see Table 1) are produced by one specific set of sea anemone neurones, whereas peptides belonging to another group (for example the Antho-RPamides) are produced by another specific set of neurones (Grimmelikhuijzen 1985; Grimmelikhuijzen et al. 1988, 1996; Carstensen et al. 1992; Nothacker et al. 1991b). Therefore, sea anemones have at least six populations of neurones that are neurochemically and probably also functionally different. Thus, multifunctionality (as explained above) does not mean that all cnidarian neurones are the same. The cnidarian neuropeptides (at least the Hydra-RFamides, Antho-RFamide and the Antho-RWamides) are located in neuronal dense-cored vesicles (Koizumi et al. 1989; Westfall and Grimmelikhuijzen 1993; Westfall et al. 1995). Low concentrations (10^{-9}-10^{-8} M) of the sea anemone neuropeptides have an excitatory or inhibitory action on intact sea anemones, sea anemone muscle preparations or isolated muscle cells (McFarlane et al. 1987, 1991, 1992, 1993; Anctil and Grimmelikhuijzen 1989; McFarlane and Grimmelikhuijzen 1991; Nothacker et al. 1991b; Carstensen et al. 1992, 1993). All of these data suggest that the cnidarian neuropeptides are neurotransmitters or neuromodulators.

We have also cloned the preprohormones for most of the neuropeptides given in Tables 1 and 2. From the preprohormones in higher animals it is known that an

Table 1. Neuropeptide families in anthozoans

Species	Structure	Name
A. elegantissima	L-3-phenyllactyl-Phe-Lys-Ala-NH_2	Antho-KAamide
A. elegantissima	L-3-phenyllactyl-Tyr-Arg-Ile-NH_2	Antho-RIamide I
A. elegantissima	Tyr-Arg-Ile-NH_2	Antho-RIamide II
A. elegantissima	L-3-phenyllactyl-Leu-Arg-Asn-NH_2	Antho-RNamide I
A. elegantissima	Leu-Arg-Asn-NH_2	Antho-RNamide II
A. elegantissima	<Glu-Ser-Leu-Arg-Trp-NH_2	Antho-RWamide I
A. elegantissima	<Glu-Gly-Leu-Arg-Trp-NH_2	Antho-RWamide II
A. elegantissima	<Glu-Asn-Phe-His-Leu-Arg-Pro-NH_2	Antho-RPamide II
A. elegantissima	Leu-Pro-Pro-Gly-Pro-Leu-Pro-Arg-Pro-NH_2	Antho-RPamide I
A. elegantissima	Gly-Pro-Hyp-Ser-Leu-Phe-Arg-Pro-NH_2	Antho-RPamide IV
A. elegantissima	<Glu-Val-Lys-Leu-Tyr-Arg-Pro-NH_2	Antho-RPamide III
A. elegantissima	Tyr-Arg-Pro-NH_2	Antho-RPamide V
A. elegantissima	<Glu-Gly-Arg-Phe-NH_2	Antho-RFamide
Renilla köllikeri	<Glu-Gly-Arg-Phe-NH_2	Antho-RFamide

A. elegantissima: *Anthopleura elegantissima*

Table 2. The RFamide neuropeptide family in cnidarians

Species	Structure	Name
A. elegantissima	<Glu-Gly-Arg-Phe-NH_2	Antho-RFamide
Renilla köllikeri	<Glu-Gly-Arg-Phe-NH_2	Antho-RFamide
Cyanea lamarckii	Gly-Arg-Phe-NH_2	Cyanea-RFamide III
Cyanea lamarckii	<Glu-Pro-Leu-Trp-Ser-Gly-Arg-Phe-NH_2	Cyanea-RFamide II
Polyorchis penicillatus	<Glu-Leu-Leu-Gly-Gly-Arg-Phe-NH_2	Pol-RFamide I
Hydra magnipapillata	<Glu-Trp-Leu-Gly-Gly-Arg-Phe-NH_2	Hydra-RFamide I
Hydra magnipapillata	<Glu-Trp-Phe-Asn-Gly-Arg-Phe-NH_2	Hydra-RFamide II
Polyorchis penicillatus	<Glu-Trp-Leu-Lys-Gly-Arg-Phe-NH_2	Pol-RFamide II
Cyanea lamarckii	<Glu-Trp-Leu-Arg-Gly-Arg-Phe-NH_2	Cyanea-RFamide I
Hydra magnipapillata	Lys-Pro-His-Leu-Arg-Gly-Arg-Phe-NH_2	Hydra-RFamide III
Hydra magnipapillata	His-Leu-Arg-Gly-Arg-Phe-NH_2	Hydra-RFamide IV

A. elegantissima: *Anthopleura elegantissima*

immature neuropeptide sequence is flanked, both at its N and C termini, by basic amino acid residues, which function as the primary cleavage signals. After endoproteolytic cleavage at these residues by a prohormone convertase (PC1/PC3 or PC2) -related enzyme (Smeekens and Steiner 1990; Seidah et al. 1990, 1991; Smeekens et al. 1991), other processing enzymes convert the immature peptide sequence into a biologically active compound (Fig. 1). If the biologically active peptide is protected by a C-terminal amide or an N-terminal <Glu group, Gly and Gln residues are found at the corresponding positions in the prohormone (Fig. 1).

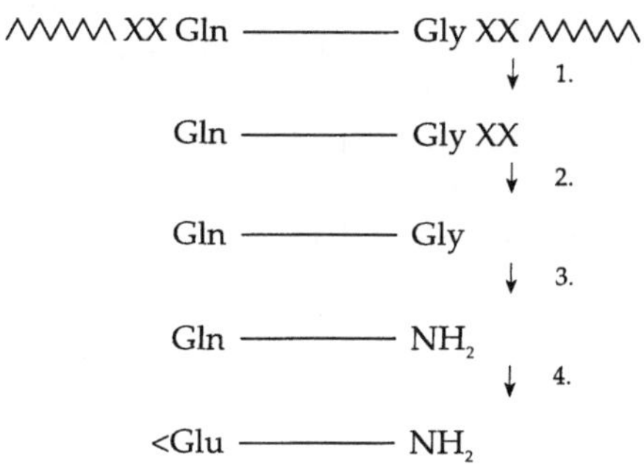

Fig. 1. Neuropeptide precursor processing in higher invertebrates and vertebrates. *Top* The precursor protein contains one or more copies of an immature neuropeptide sequence, which are flanked by basic residues (marked by *XX*). The cleavage or processing steps are catalyzed by the following enzymes: **1** An endoproteinase cleaving at the C-terminal side of pairs or single basic residues. **2** A carboxypeptidase B-like enzyme. **3** Peptidyl-glycine hydroxylase (also called peptidylglycin α-hydroxylating monooxygenase or PHM) and peptidyl-hydroxyglycine N-C lyase (also called peptidyl α-hydroxyglycine α-amidating lyase or PAL). **4** Glutaminyl cyclase. (From Grimmelikhuijzen et al. 1996)

In this review, we will discuss the preprohormone for the *Renilla* (sea pansy) neuropeptide <Glu-Gly-Arg-Phe-NH_2 (Antho-RFamide; Table 1) as an example of a cnidarian preprohormone. We would expect that this sea pansy preprohormone would have one or more copies of the immature sequence Gln-Gly-Arg-Phe-Gly, flanked by basic residues. Figure 2 shows the Antho-RFamide precursor in sea pansies. It contains 36 Gln-Gly-Arg-Phe-Gly copies, which are distributed in a very regular fashion over the whole precursor molecule. The unusually high neuropeptide copy number in the precursor suggests that the biosynthesis of Antho-RFamide is carried out in a very efficient way illustrating, once again, that neuropeptides must play an important role in primitive nervous systems.

The immature Antho-RFamide sequences have exactly the same structure as we predicted, indicating that the enzymes catalyzing steps 3 and 4 in Fig. 1 must have their counterparts in cnidarians. Furthermore, each immature Antho-RF-amide sequence is followed by basic amino acid residues, suggesting also that processing enzymes similar to the ones catalyzing steps 1 and 2 in Fig. 1 must be present in cnidarians. We also expected that basic amino acid residues would be present at the N-terminal side of each immature Antho-RFamide sequence (Fig. 1). It was, therefore, a surprise for us to find that nearly all Antho-RFamide sequences were preceded by acidic (Glu) residues (Fig. 2). This finding implies that a novel processing enzyme must exist in cnidarian neurones cleaving at the C-terminal sides of acidic residues. Two immature Antho-RFamide sequences are preceded by a Lys residue (at amino acid positions 128-132 and 298-302 of Fig. 2) and, of course, one could argue that only these sequences are cleaved from the precursor by a conventional processing enzyme, whereas the rest of the precursor is not used. However, cloning work on the precursors of the other cnidarian neuropeptides (Tables 1, 2) gives examples of preprohormones having only one copy of the immature neuropeptide in question, as is the case with Antho-RPamide I (Table 1), where the immature peptide sequence together with its flanking regions have the following structure: ...Gln-Tyr-Asp-Gly-Pro-Glu-Gly-Asp-Tyr-Glu-Asp-**Leu-Pro-Pro-Gly-Pro-Leu-Pro-Arg-Pro-Gly**-Arg-Lys-Arg (Carstensen and Grimmelikhuijzen, unpubl.). Here, it is clear that there is no other way to produce mature Antho-RPamide I (which we have isolated and sequenced) from its precursor than by cleaving at the C-terminal side of acidic residues. The novel processing enzyme that is responsible for this cleavage is, based on the structure of the N-terminal flanking sequence, most likely to be an endoproteinase.

Fig. 2. cDNA and deduced amino acid sequence of the *Renilla* Antho-RFamide precursor. Nucleotide residues are numbered from 5' to 3' end, and amino acid residues are numbered starting with the first ATG codon in the open reading frame. Neuropeptide sequences are *underlined*. Polyadenylylation signals are marked by *broken lines*. (Modified from Reinscheid and Grimmelikhuijzen 1994)

```
                                                                 TTTAGTATTGCGTCGTGTGTGTGTGTGTGTGTG   35
TTTGGCTCATTTTCTACGATCAACGTTGCTCTACACATTTGTATGGAGAAAACCTGTCAGTCTATAACGTATATACTTC  114
AAAATAATGGACTTAGTAATATAAGTATCAACTGCATTATTGTTGTCAATGTCCA ATG GTT AGT CTG GGT TTT  187
                                                        Met Val Ser Leu Gly Phe    6

TTC GTT CGT GAT GTT ACT CCA ACA TTC ATT GTA GAT CAT ATG TTC TAC ATG CTA TTC CCG  247
Phe Val Arg Asp Val Thr Pro Thr Phe Ile Val Asp His Met Phe Tyr Met Leu Phe Pro   26

ATG GAT TTT ACG TGT TAC GTT GCC GGG CTG TTG TTG ATA TTA AAT ACT TAC AGT TTG GCC  307
Met Asp Phe Thr Cys Tyr Val Ala Gly Leu Leu Leu Ile Leu Asn Thr Tyr Ser Leu Ala   46

GGG CCC TCA ACT AGC GAA GGA CTA AAC GAA CGG AAC TTG CTG GAT AAA ACA GAG TTG TCG  367
Gly Pro Ser Thr Ser Glu Gly Leu Asn Glu Arg Asn Leu Leu Asp Lys Thr Asp Leu Ser   66

ATA AAT GAC GAG ATA TTT AGC GAA GAT GAT GAT ATG CTG GCA AGG GAC GCT GAA GAC AAA  427
Ile Asn Asp Glu Ile Phe Ser Glu Asp Asp Asp Met Leu Ala Arg Asp Ala Glu Asp Lys   86

CAA GGA CGA TTT AAT CGT AAA TTA AAT AAC AAG TTG AAT GAA GCG GTA CAA GGA CGC TTC  487
Gln Gly Arg Phe Asn Arg Lys Leu Asn Asn Lys Leu Asn Glu Ala Val Gln Gly Arg Phe  106

GGG AGA AAT GAG AGA AAA GAA TCG GAA GAA GAA CAA GGA AGG TTC GGG CGA GAA AAT GAA  547
Gly Arg Asn Glu Arg Lys Glu Ser Glu Glu Glu Gln Gly Arg Phe Gly Arg Glu Asn Glu  126

AAA CAA GGA AGA TTT GGA AGG GAA AGC GAG GAG CAA GGG AGA TTT GGA CGA GAA AAC AAA  607
Lys Gln Gly Arg Phe Gly Arg Glu Ser Glu Glu Gln Gly Arg Phe Gly Arg Glu Asn Lys  146

GAA CAA GGA AGA TTT GGA AGG GAA AAC AAA GAA CAA GGA AGG TTT GGA CGA GAA AAC GAG  667
Glu Gln Gly Arg Phe Gly Arg Glu Asn Lys Glu Gln Gly Arg Phe Gly Arg Glu Asn Glu  166

GAA CAA GGA AGA TTT GGA AGG GAA AGC GAG GAG CAA GGG AGA TTT GGA AGA GAA AAC GAA  727
Glu Gln Gly Arg Phe Gly Arg Glu Ser Glu Glu Gln Gly Arg Phe Gly Arg Glu Asn Glu  186

GAA CAA GGA AGA TTT GGA CGA GAA AAC GAA GAA CAA GGA AGA TTT GGA CGA GAA AAC GAA  787
Glu Gln Gly Arg Phe Gly Arg Glu Asn Glu Glu Gln Gly Arg Phe Gly Arg Glu Asn Glu  206

GAA CAA GGA AGG TTT GGA CGA GAA AAC GAG GAG CAA GGG AGA TTT GGA AGG GAG AAC GAA  847
Glu Gln Gly Arg Phe Gly Arg Glu Asn Glu Glu Gln Gly Arg Phe Gly Arg Glu Asn Glu  226

GTT CAA GGA AGA TTT GGA CGA GAG AAC GAG GAG CAA GGA AGA TTT GGA CGA GAA AAC GAA  907
Val Gln Gly Arg Phe Gly Arg Glu Asn Glu Glu Gln Gly Arg Phe Gly Arg Glu Asn Glu  246

GAA CAA GGA AGG TTT GGA AGA GAA AAC GAA GAA CAA GGA CGA TTT GGA CGA GAA AAC GAG  967
Glu Gln Gly Arg Phe Gly Arg Glu Asn Glu Glu Gln Gly Arg Phe Gly Arg Glu Asn Glu  266

GAG CAA GGA AGA TTT GGA CGA GAA AAT GAA GAA CAA GGA AGA TTT GGA CGA GAA AAC GAG 1027
Glu Gln Gly Arg Phe Gly Arg Glu Asn Glu Glu Gln Gly Arg Phe Gly Arg Glu Asn Glu  286

GAG CAA GGA AGA TTT GGA CGA GAA AAC GAA AAA CAA GGA CGA TTT GGA CGA GAA AAC GAA 1087
Glu Gln Gly Arg Phe Gly Arg Glu Asn Glu Lys Gln Gly Arg Phe Gly Arg Glu Asn Glu  306

GAA CAA GGA AGA TTT GGA CGA GGA AAC GAG GAG CAA GGG AGA TTT GGA AGG GAA AAC GAA 1147
Glu Gln Gly Arg Phe Gly Arg Gly Asn Glu Glu Gln Gly Arg Phe Gly Arg Glu Asn Glu  326

GAA CAA GGA AGA TTT GGA CGA GAA AAC GAA GAA CAA GGA AGA TTT GGA CGA GAA AAC GAA 1207
Glu Gln Gly Arg Phe Gly Arg Glu Asn Glu Glu Gln Gly Arg Phe Gly Arg Glu Asn Glu  346

GAA CAA GGA CGA TTT GGG CGA GAA AAC GAA GAA CAA GGA AGG TTT GGG CGA GAA AAC GAA 1267
Glu Gln Gly Arg Phe Gly Arg Glu Asn Glu Glu Gln Gly Arg Phe Gly Arg Glu Asn Glu  366

GAA CAA GGA AGG TTT GGA AGG GAG AAC GAG AAG CAA GGG AGA TTT GGA AGA GGG GAC GAA 1327
Glu Gln Gly Arg Phe Gly Arg Glu Asn Glu Lys Gln Gly Arg Phe Gly Arg Gly Asp Glu  386

GAA CAA GGA AGG TTT GGA AGG GAG AAC GAG GAG CAA GGG AGA TTT GGA AGA GGG GAC GAA 1387
Glu Gln Gly Arg Phe Gly Arg Glu Asn Glu Glu Gln Gly Arg Phe Gly Arg Gly Asp Glu  406

GAA CAA GGA AGA TTT GGA CGA GAA AAC GAA GAG CAA GGG AGA TTT GGA AGA GAA AAC AAA 1447
Glu Gln Gly Arg Phe Gly Arg Glu Asn Glu Glu Gln Gly Arg Phe Gly Arg Glu Asn Lys  426

GAA CAA GGA AGA TTT GGA AGA GAA AAC GAA GAA CAA GGA AGA TTT GGA AGG GGG AAC AAA 1507
Glu Gln Gly Arg Phe Gly Arg Glu Asn Glu Glu Gln Gly Arg Phe Gly Arg Gly Asn Lys  446

GAA CAA GGA AGA TTT GGA CGA GAA AAC GAA CAA GGA AGA TTT GGA AGA GAA AAC GAA GTT 1567
Glu Gln Gly Arg Phe Gly Arg Glu Asn Glu Gln Gly Arg Phe Gly Arg Glu Asn Glu Val  466

CAA GGA AGG TTT GGA AGA TTC AGT CGG GAG TTG GCG AAA GGT TTA AAG ATT GAC GAT GTT 1627
Gln Gly Arg Phe Gly Arg Phe Ser Arg Glu Leu Ala Lys Gly Leu Lys Ile Asp Asp Val  486

CTC TGA  CAATGAACTAATTACGTGAATTACTAGAAACAAACTTAGAGAAAGTTGTTTATGAATCTATCAATAGTAT 1703
Leu  *                                                                             487

TTAAAAAGCGTTTCCAAATATTTAGTGTGAAATGATATTTTAAAAAAATATACACCGAAAAAAAAAAAAAGGCCGATCG 1782
GGCCG(A)32                                                                      1819
```

Cloning of the preprohormones for the other cnidarian neuropeptides showed that, in addition to acidic amino acid residues, many more sequences or residues must function as novel processing signals (Darmer et al. 1991; Schmutzler et al. 1992, 1994; Reinscheid and Grimmelikhuijzen 1994; Leviev and Grimmelikhuijzen 1995; Grimmelikhuijzen et al. 1996; Leviev et al. 1997). These additional processing sites are Thr, Ser, Asn and Tyr residues and X-Pro or X-Ala sequences. All these novel processing signals are located at the N-terminal sides of the immature neuropeptide sequences, and the enzymes responsible for cleavage could, in the case of Thr, Ser, Asn and Tyr residues, either be endoproteinases or aminopeptidases. The enzyme that is responsible for the removal of N-terminally located X-Pro or X-Ala sequences is most probably dipeptidyl aminopeptidase (Leviev and Grimmelikhuijzen 1995).

The picture that emerges from this work on the biosynthesis of neuropeptides in cnidarians, is the following: (1) many of the cnidarian preprohormones contain a very high copy number of the immature neuropeptide sequences, suggesting a very efficient neuropeptide biosynthesis; (2) the processing enzymes that we know from higher animals (Fig. 1) are probably also present in cnidarians. However, in addition to these four to five known processing enzymes, a variety of novel processing enzymes must be involved in the maturation of the cnidarian neuropeptides. It is amusing that this last point appears to be a paradox: cnidarians are supposed to have a very simple nervous system, but in reality, the biosynthesis of their neuropeptides appears to be more complex than in higher organisms.

The paradox is supported by our recent cloning of the cnidarian enzyme catalyzing the first step of peptide amidation (step 3 of Fig. 1). In mammals, two enzymes are responsible for C-terminal peptide amidation. The first enzyme converts peptidylglycine into peptidylhydroxyglycine and is either called peptidylglycine α-hydroxylating monooxygenase (PHM) or peptidylglycine hydroxylase. The second enzyme cleaves peptidylhydroxyglycine into peptidylamide and glyoxylic acid and is either called peptidyl α-hydroxyglycine α-amidating lyase (PAL) or peptidylhydroxyglycine N-C lyase (Bradbury et al. 1982; Kato et al. 1990; Katopodis et al. 1990, 1991; Perkins et al. 1990; Bradbury and Smyth 1991; Eipper et al. 1992). In vertebrates, both enzymatic activities are contained within one bifunctional enzyme, peptide α-amidating monooxygenase (PAM). In mammals only one gene for PAM exists. In the sea anemone *Calliactis parasitica,* however, we found two, clearly different PHM genes (Hauser and Grimmelikhuijzen, unpubl.). No cDNA coding for a bifunctional PAM could be identified, and we expect that PAL is encoded by a separate (yet unidentified) gene. The two types of sea anemone PHM show 40% amino acid sequence identity among each other, which is the same homology (about 40%) that each type shows with mammalian PHM. The existence of two rather different types of PHM suggests that these enzymes may have different catalytic activities, and this could point to the use of different substrates (peptidyl-Xaa' s) for α-amidation in cnidarians. Again, more processing enzymes appear to be involved in the peptide amidation step in cnidarians than in mammals, making neuropeptide biosynthesis in this primitive animal group more complex than in higher animals.

Neurohormone Receptors in Cnidarians

If one is interested in neuropeptides, one should also be interested in neuropeptide receptors. In order to find the receptors for the neurohormones listed in Table 1, we applied PCR using cDNA from *A. elegantissima* as a template and primers coding for conserved sequences of G protein-coupled receptors. This approach yielded a receptor that clearly was not a receptor for one of the small sea anemone neuropeptides, but instead, showed a striking structural homology with members of the mammalian glycoprotein hormone (TSH, FSH, LH/CG) receptor family (Nothacker and Grimmelikhuijzen 1993). This was an exciting finding, since glycoprotein hormone receptors had, in 1993, only been cloned from mammals and not from other vertebrates or invertebrates. Our findings in cnidarians suggested that glycoprotein hormone receptors and possibly all the processes that are mediated by these types of receptors (such as certain steps in reproduction) are conserved in the Animal Kingdom, from cnidarians to mammals. Our results also meant that the endocrinology of sea anemones must already be quite complex, involving at least 16 neuropeptides and, in addition, a new class of presumed glycoprotein hormones. As there are no endocrine cells in cnidarians, but only neurones, these presumed glycoprotein hormones are likely to be neurohormones.

Neurohormone Receptors in Insects

Insects are very successful animals and are of extreme ecological and economical importance, since about 70% of all flowering plants depend on insects for their pollination and insects can be severe pest animals. Despite the importance of insects, however, the molecular basis of their reproduction is not well understood. This is in contrast to the situation in mammals, where reproduction is controlled by (1) steroid hormones, (2) glycoprotein hormones (LH, FSH and CG), and (3) hypothalamic releasing hormones that control the release of LH and FSH from the anterior pituitary.

When we found that cnidarians produce putative glycoprotein hormone receptors, we realized that this might give us a clue to elucidate the molecular basis of reproduction in insects. Therefore, we focused on insects and, using oligonucleotide probes derived from consensus sequences for glycoprotein hormone receptors, cloned a putative glycoprotein hormone receptor from *Drosophila melanogaster* (Hauser et al. 1997). This *Drosophila* receptor is structurally related to the mammalian glycoprotein hormone receptors, because of the following observations: (1) the *Drosophila* receptor is a seven-transmembrane receptor having a very large, extracellular N terminus that comprises about half of the receptor molecule (Figs. 3A, 4). This overall structure is similar to that of the mammalian glycoprotein hormone receptors; (2) the *Drosophila* receptor has 50-53% sequence identity in the transmembrane region with the rat TSH, FSH, LH/CG receptors (Fig. 4). This is very high if one takes into account that the sequence identity between the transmembrane regions of the rat LH/CG and TSH receptors is 70% and that between the rat TSH and FSH receptor is 67% (Salesse et al. 1991); (3) there is only 20% sequence identity in the N terminus between the *Drosophila* and the mammalian receptors (Fig. 4). However, the N termini of the mammalian glycoprotein hormone receptors have a special feature that is also present in the *Drosophila* receptor, namely the presence of Leu(Ile/Val)-rich repeats (Fig. 5). The function of the Leu-rich repeats in N termini of the mammalian receptors has not been fully established yet. Recently, however, several research groups have proposed three-dimensional models for the N termini of the mammalian glycoprotein hormone receptors based on the crystal structure of another protein, the porcine ribonuclease/nuclease inhibitor complex, where the inhibitor also has Leu-rich repeats that function as binding motifs for the ribonuclease (Kobe and Deisenhofer 1993, 1995; Jiang et al. 1995; Kajava et al. 1995). In these models for the receptor N termini, the Leu-rich repeats are arranged in a horseshoe-like structure, where each repeat has an inwardly directed β-sheet (at the concave surface of the horseshoe) and an outwardly directed α-helix (at the convex surface of the horseshoe). The inwardly directed β-sheets are supposed to bind the ligand. The *Drosophila* receptor has exactly the same distribution of the aliphatic residues within each Leu-rich repeat as the mammalian receptors (see alignments in Fig. 5). This means that the N terminus of the *Drosophila* receptor probably also has a horseshoe-like structure and that the inside surface of the horseshoe binds the presumed glycoprotein hormone ligand.

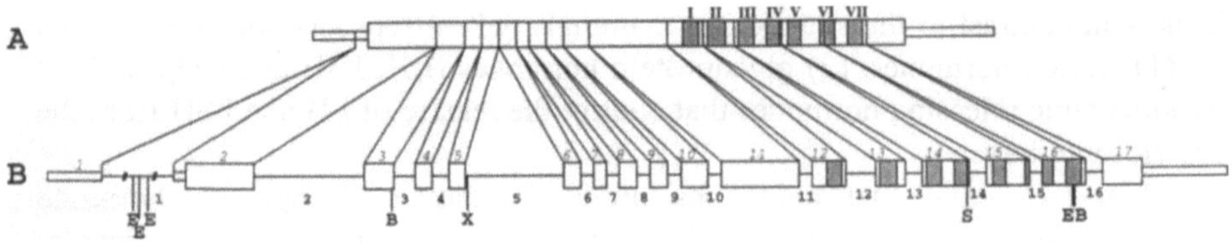

Fig. 3. Schematic representation of the cDNA and the gene coding for the glycoprotein hormone receptor from *Drosophila:* **A** The receptor cDNA. The *broad bar* represents the coding region, whereas the two *narrow bars* represent the 5'- and 3'- noncoding domains. The boxes I-VII highlighted in *gray* indicate the seven membrane-spanning domains. **B** The receptor gene. The introns are given as lines and numbered 1-16, whereas the exons are represented by boxes and numbered *1-17* (exon numbers are in *italic type)*. *E Eco*RI*; B Bam*HI; *X Xba*I; *S Sal*I restriction sites. (Modified from Hauser et al. 1997)

We have also investigated the gene for the *Drosophila* receptor. This gene consists of 17 exons and 16 introns (Fig. 3B). Seven of these introns (introns 3-9) occur at exactly the same positions in the *Drosophila* gene as the introns in the genes for the mammalian receptors. The triplet coding for the aliphatic residues at these seven intron-exon transitions (marked by filled circles in Fig. 4) are always interrupted after the second nucleotide (intron phase = 2). The same intron phasing is found in the corresponding introns of the mammalian receptor genes (Gross et al. 1991; Koo et al. 1991; Tsai-Morris et al. 1991; Heckert et al. 1992). All of these data strongly suggest that the *Drosophila* receptor is evolutionarily related to the TSH, FSH and LH/CG receptors from mammals.

When we started our cloning work on the *Drosophila* glycoprotein hormone receptor, it was our expectation that this receptor would be functionally related to the mammalian FSH and LH receptors and that we, therefore, would be able to obtain new insights into the molecular mechanisms of insect reproduction. However, when we looked at the expression of the receptor at different developmental stages of *Drosophila,* we found to our surprise that the receptor was already strongly expressed 8-16 h after oviposition and that the expression stayed high until after pupation. Adult male flies express high levels of receptor mRNA, but female flies about 6 times less (Hauser et al. 1997). Thus, the expression pattern in embryos and larvae indicate that the receptor must be involved in insect development, although a role of the receptor in reproduction, especially in males cannot be excluded. To investigate the role of the receptor, we are presently raising transgenic flies (that produce large amounts of free circulating receptor N terminus without its seven-transmembrane domain - in order to deplete the animals of their receptor ligands) and mutant flies where the receptor gene has been "knocked out" (interrupted by P-element insertion). In another approach, we are isolating the ligand and looking for its effects.

Mammals have three glycoprotein hormone (TSH, FSH, LH/CG) receptors. We were, therefore, impressed to find that *Drosophila* also has a second glycoprotein hormone receptor (Hauser, Eriksen, and Grimmelikhuijzen, unpubl.). This second receptor is rather different from the first *Drosophila* receptor, but has 47-51%

Fig. 4. Amino acid sequence comparison of the *Drosophila* receptor *(GHRD),* the rat TSH *(TSHR),* FSH *(FSHR),* and LH *(LHR)* receptors and the putative glycoprotein hormone receptor from the sea anemone *Anthopleura elegantissima (GHRA).* Amino acid residues that are identical between the *Drosophila* receptor and at least one of the other receptors are *boxed.* Known intron-exon transitions in the genes coding for four of the receptors are highlighted in *gray* at the corresponding amino acid residues. The positions of the aliphatic residues characteristic for the Leu-rich repeats in the *Drosophila* receptor (see Fig. 5), are marked by *asterisks* or *filled circles.* The *filled circles* also mark intron-exon transitions in the *Drosophila* receptor gene that occur at the same positions and have the same intron phasing as in the mammalian receptor genes. The seven membrane-spanning domains are indicated by I-VII. *Dashed lines* represent spaces introduced to optimize alignment. The amino acid positions are given at the right. (From Hauser et al. 1997)

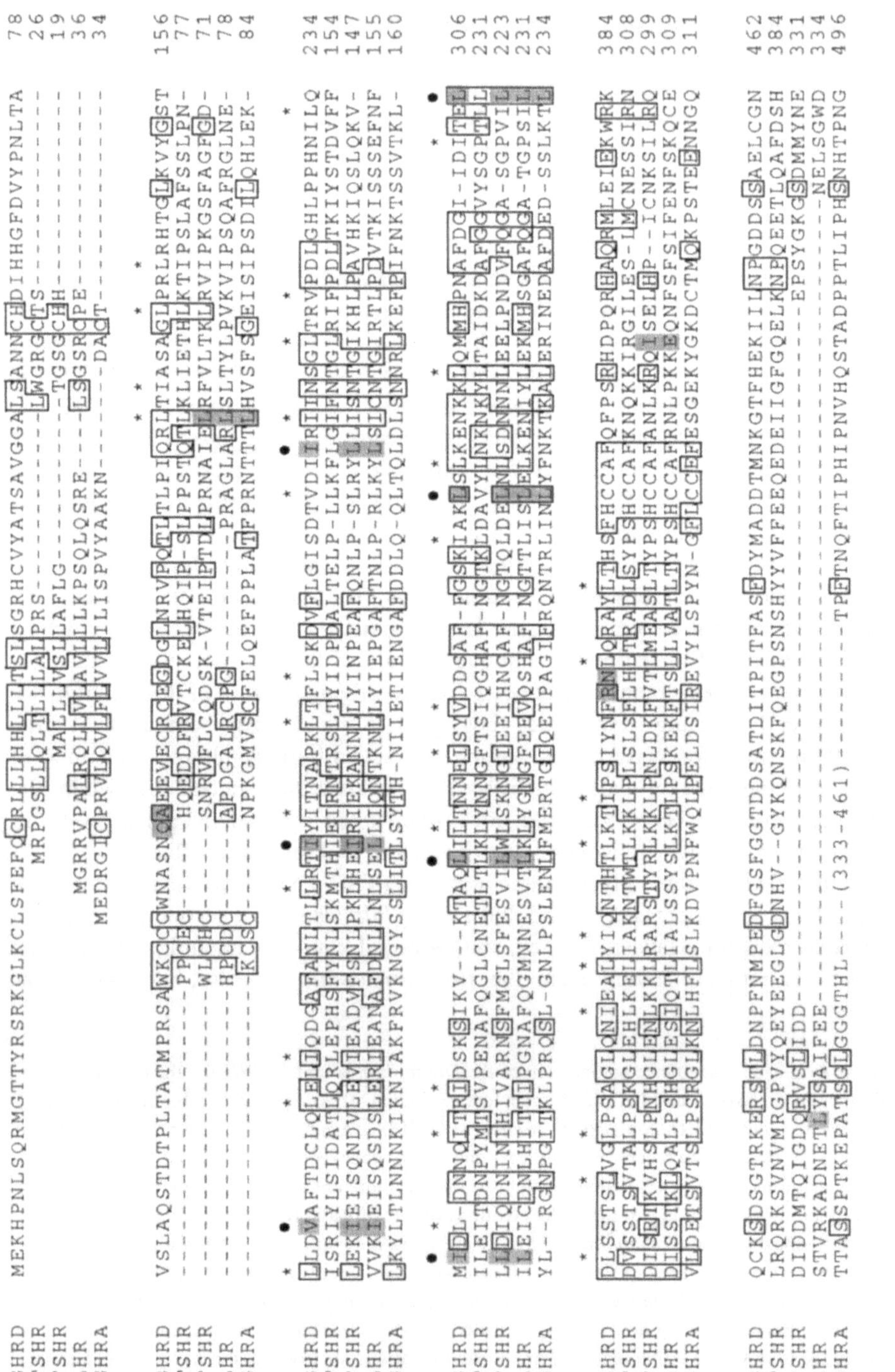

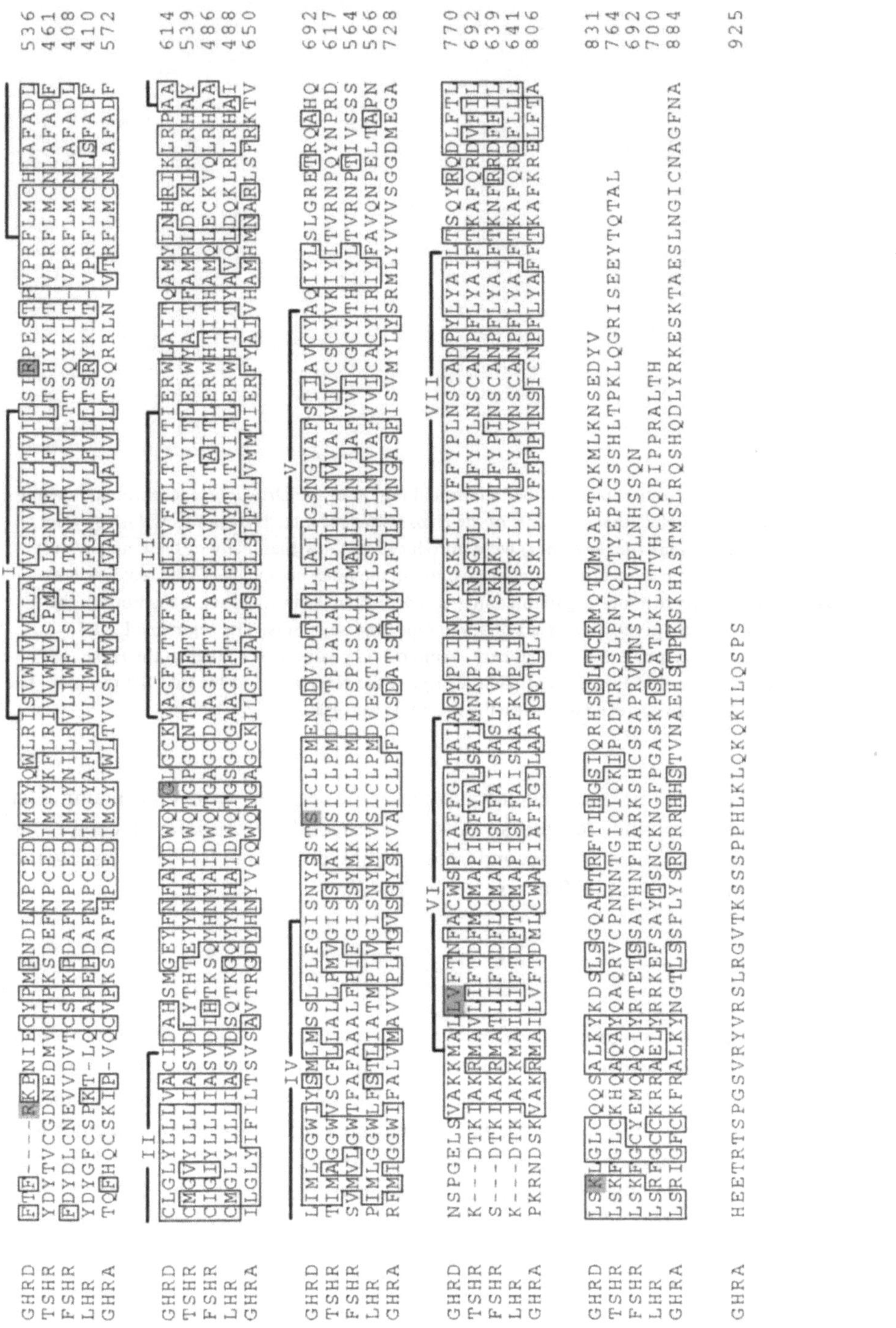

A

```
L1   L T I A S A G - L P R L R H T G L K V Y G S T L L D   159
L2   V A F T D C L Q L E L I Q D G A F A N L T - L L R T   184
L3   I Y I T N A P K L T F L S K D V F L G I S D T V D I   210
L4   I R I I N S G - L T R V P D L G H L P P H N I L Q M   235
L5   I D L D N N Q - I T R I D S K S I K V - - - K T A Q   257
L6   L I L T N N E - I S Y V D D S A F F G S - - K I A K   280
L7   L S L K E N K K L Q M M H P N A F D G I I - D I T E   305
L8   L D L S S T S - L V G L P S A G L Q - - - - N I E A   326
L9   L Y I Q N T H T L K T I P S I Y N F - - - R N L Q R   349
L10  A Y L T H S F H C C A F Q - - - - F P S R H D P Q R   371
```

B

```
L1   L R F V L T K - L R V I P K G S F A G F G - D L E K    74
L2   I E I S Q N D V L E V I E A D V F S N L P - K L H E    99
L3   I R I E K A N N L L Y I N P E A F Q N L P - S L R Y   124
L4   L L I S N T G - I K H L P A V H K I Q S L Q K V - L   148
L5   L D I Q D N I N I H I V A R N S F M G L S F E S V I   174
L6   L W L S K N G - I E E I H N C A F N G T - - Q L D E   197
L7   L N L S D N N N L E E L P N D V F Q G A S - G P V I   222
L8   L D I S R T K - V H S L P N H G L E - - - - N L K K   243
L9   L R A R S T Y R L K K L P N L D K F V - - - T L M E   266
L10  A S L T Y P S H C C A F A N L K R Q I - - S E L H P   290
```

Fig. 5. Leu (Ile/Val)-rich repeats in the N-terminal region of the *Drosophila* receptor and the rat FSH receptor. **A** Consecutive segments of the *Drosophila* receptor N terminus (L1-L10) were aligned, and small gaps (*dashed lines*) were introduced to show that many of the Leu (Ile/Val) residues in one segment occur at a similar position in the other segments. These residues are *boxed*. The aliphatic residues highlighted in *gray* correspond to intron-exon transitions in the receptor gene and are given at the start of seven repeating segments. The segments L2-L7, therefore, each correspond to a distinct exon. Amino acid positions are given at the *right*.
B A similar alignment of Leu(Ile/Val)-rich repeats of the N terminus of the rat FSH receptor. The segments L1-L7 each correspond to a distinct exon. (From Hauser et al. 1997)

sequence identity in the seven-transmembrane region with the mammalian glycoprotein hormone receptors. Also the second *Drosophila* receptor has Leu-rich repeats. Since this second receptor has only recently been discovered, nothing is known about its function in *Drosophila*.

Conclusion

Primitive nervous systems, such as those of cnidarians, use a large variety of neuropeptides as transmitters, suggesting that neuropeptides were among the earliest neurotransmitters used in evolution. The biosynthesis of neuropeptides in cnidarians appears to be very efficient and requires more processing enzymes than we know from higher animals. In addition to neuropeptides, other hormonal systems, such as the glycoprotein hormone/receptor couple have a long evolutionary history. All of this shows that the endocrinological systems in invertebrates are more complex and sophisticated than we might have previously assumed.

Acknowledgments

We want to thank Astrid Juel Jensen for typing the manuscript, and the Danish Natural Science Research Council, the Danish Biotechnological Research and Developmental Program of the Danish Research Councils, Læge- og Naturvidenskabelige Komité of the Novo Nordisk Foundation, and Carlsberg Foundation for financial support. F.H. was recipient of a postdoctoral fellowship from the Max Planck Society (Germany).

References

Acher R, Chauvet J (1953) La structure de la vasopressine de boeuf. Biochim Biophys Acta 12: 487-488

Anctil M, Grimmelikhuijzen CJP (1989) Excitatory action of the native neuropeptide Antho-RFamide on muscles in the pennatulid *Renilla kölleri.* Gen Pharm 20: 381-384

Anderson PAV, Schwab WE (1982) Recent advances and model systems in coelenterate neurobiology. Progr Neurobiol 19: 213-236

Anderson PAV, Spencer AN (1989) The importance of cnidarian synapses for neurobiology. J Neurobiol 20: 435-457

Bode HR (1992) Continuous conversion of neuron phenotype in *Hydra.* Trends Genet 8: 279-284

Bradbury AF, Smyth DG (1991) Peptide amidation. Trends Biochem Sci 16:112-115

Bradbury AF, Finnie MDA, Smyth DG (1982) Mechanism of C-terminal amide formation by pituitary enzymes. Nature 298: 686-688

Campbell RD (1976) Elimination of *Hydra* interstitial and nerve cells by means of colchicine. J Cell Sci 21: 1-14

Carstensen K, Rinehart KL, McFarlane ID, Grimmelikhuijzen CJP (1992) Isolation of Leu-Pro-Pro-Gly-Pro-Leu-Pro-Arg-Pro-NH_2 (Antho-RPamide), an N-terminally protected, biologically active neuropeptide from sea anemones. Peptides 13: 851-857

Carstensen K, McFarlane ID, Rinehart KL, Hudman D, Sun F, Grimmelikhuijzen CJP (1993) Isolation of <Glu-Asn-Phe-His-Leu-Arg-Pro-NH_2 (Antho-RPamide II), a novel, biologically active neuropeptide from sea anemones. Peptides 14: 131-135

Darmer D, Schmutzler C, Diekhoff D, Grimmelikhuijzen CJP (1991) Primary structure of the precursor for the sea anemone neuropeptide Antho-RFamide (<Glu-Gly-Arg-Phe-NH_2). Proc Natl Acad Sci USA 88: 2555-2559

Du Vigneaud V, Ressler C, Trippett S (1953a) The sequence of amino acids in oxytocin, with a proposal for the structure of oxytocin. J Biol Chem 205: 949-957

Du Vigneaud V, Lawler HC, Popenoe EA (1953b) Enzymatic cleavage of glycinamide from vasopressin and a proposed structure for this pressor-antidiuretic hormone of the posterior pituitary. J Am Chem Soc 75: 4880-4881

Eipper BA, Stoffers DA, Mains RE (1992) The biosynthesis of neuropeptides: peptide α-amidation. Ann Rev Neurosci 15: 57-85

Gierer A (1977) Biological features and physical concepts of paffern formation exemplified by *Hydra.* Curr Top Dev Biol 11:17-59

Gierer A, Berking S, Bode H, David CN, Flick K, Hansmann G, Schaller H, Tenckner E (1972) Regeneration of *Hydra* from reaggregated cells. Nature New Biol 239: 98-101

Graff D, Grimmelikhuijzen CJP (1988a) Isolation of <Glu-Ser-Leu-Arg-Trp-NH_2, a novel neuropeptide from sea anemones. Brain Res 442: 354-358

Graff D, Grimmelikhuijzen CJP (1988b) Isolation of <Glu-Gly-Leu-Arg-Trp-NH_2 (Antho-RWamide II), a novel neuropeptide from sea anemones. FEBS Lett 239:137-140

Grimmelikhuijzen CJP (1985) Antisera to the sequence Arg-Phe-amide visualize neuronal centralization in hydroid polyps. Cell Tissue Res 241:171-182

Grimmelikhuijzen CJP, Graff D (1986) Isolation of <Glu-Gly-Arg-Phe-NH_2 (Antho-RFamide), a neuropeptide from sea anemones. Proc Natl Acad Sci USA 83: 9817-9821

Grimmelikhuijzen CJP, Groeger A (1987) Isolation of the neuropeptide <Glu-Gly-Arg-Phe-amide from the pennatulid *Renilla köllikeri.* FEBS Lett 211:105-108

Grimmelikhuijzen CJP, Spencer AN, Carré D (1986) Organization of the nervous system of physonectid siphonophores. Cell Tissue Res 246: 463-479

Grimmelikhuijzen CJP, Hahn M, Rinehart KL, Spencer AN (1988) Isolation of <Glu-Leu-Leu-Gly-Gly-Arg-Phe-NH_2 (Pol-RFamide), a novel neuropeptide from hydromedusae. Brain Res 475: 198-203

Grimmelikhuijzen CJP, Rinehart KL, Jacob E, Graff D, Reinscheid RK, Nothacker H-P, Staley AL (1990) Isolation of L-3-phenyllactyl-Leu-Arg-Asn-NH_2 (Antho-RNamide), a sea anemone neuropeptide containing an unusual amino-terminal blocking group. Proc Natl Acad Sci USA 87: 5410-5414

Grimmelikhuijzen CJP, Rinehart KL, Spencer AN (1992) Isolation of the neuropeptide <Glu-Trp-Leu-Lys-Gly-Arg-Phe-NH_2 (Pol-RFamide II) from the hydromedusa *Polyorchis penicillatus.* Biochem Biophys Res Commun 183: 375-382

Grimmelikhuijzen CJP, Leviev I, Carstensen K (1996) Peptides in the nervous systems of cnidarians: structure, function and biosynthesis. Int Rev Cytol 167: 37-89

Gross B, Misrahi M, Sar S, Milgrom E (1991) Composite structure of the human thyrotropin receptor gene. Biochem Biophys Res Commun 177: 679-687

Guillemin R (1978) Peptides in the brain: the new endocrinology of the neuron. Science 202: 390-402

Guillemin R, Burgus R (1972) The hormones of the hypothalamus. Sci Am 227: 24-33

Hauser F, Nothacker H-P, Grimmelikhuijzen CJP (1997) Molecular cloning, genomic organization and developmental regulation of a novel receptor from *Drosophila melanogaster* structurally related to members of the thyroid-stimulating hormone, follicle stimulating hormone, luteinizing hormone/choriogonadotropin receptor family from mammals. J Biol Chem 272:1002-1010

Heckert LL, Daley IJ, Griswold MD (1992) Structural organization of the follicle-stimulating hormone receptor gene. Mol Endiocrinol 6: 70-80

Horridge GA, Mackay B (1962) Naked axons and symmetrical synapses in coelenterates. Q J Micr Sci 103: 531-541

Jiang X, Dreano M, Buckler DR, Cheng S, Ythier A, Wu H, Hendrickson WA, El Tayar N (1995) Structural predictions for the ligand-binding region of glycoprotein hormone receptors and the nature of hormone-receptor interactions. Structure 3: 1341-1353

Kajava AV, Vassart G, Wodak SJ (1995) Modeling of the three-dimensional structure of proteins with the typical leucine-rich repeats. Structure 3: 867-877

Kato I, Yonekura H, Tajima M, Yanagi M, Yamamoto H, Okamoto H (1990) Two enzymes concerned in peptide hormone α-amidation are synthesized from a single mRNA. Biochem Biopys Res Commun 172: 197-203

Katopodis AG, Ping D, May SW (1990) A novel enzyme from bovine neurointermediate pituitary catalyzes dealkylation of α-hydroxyglycine derivates, thereby functioning sequentially with peptidylglycine α-amidating monooxygenase in peptide amidation. Biochemistry 29: 6115-6120

Katopodis AG, Ping D, Smith CE, May SW (1991) Functional and structural characterization of peptidylamidoglycolate lyase, the enzyme catalysing the second step in peptide amidation. Biochemistry 30: 6189-6194

Kobe B, Deisenhofer J (1993) Crystal structure of porcine ribonuclease inhibitor, a protein with leucine-rich repeats. Nature 366: 751-756

Kobe B, Deisenhofer J (1995) A structural basis of the interactions between leucine-rich repeats and protein ligands. Nature 374: 183-186

Koizumi O, Wilson JD, Grimmelikhuijzen CJP, Westfall JA (1989) Ultrastructural localization of RFamide-like peptides in neuronal dense-cored vesicles in the peduncle of *Hydra*. J Exp Zool 249:17-22

Koo YB, Ji I, Slaughter RG, Ji TH (1991) Structure of the luteinizing hormone receptor gene and multiple exons of the coding sequence. Endocrinology 128: 2297-2308

Leviev I, Grimmelikhuijzen CJP (1995) Molecular cloning of a preprohormone from sea anemones containing numerous copies of a metamorphosis inducing neuropeptide: a likely role for dipeptidyl aminopeptidase in neuropeptide precursor processing. Proc Natl Acad Sci USA 92: 11647-11651

Leviev I, Williamson M, Grimmelikhuijzen CJP (1997) Molecular cloning of a preprohormone from *Hydra magnipapillata* containing multiple copies of Hydra-LWamide (Leu-Trp-NH_2) neuropeptides: evidence for processing at Ser and Asn residues. J Neurochem 68: 1319-1325

Mackie GO (1973) Report on giant nerve fibres in *Nanomia*. Publ Seto Marine Lab 20: 745-756

Mackie GO (1984) Fast pathways and escape behavior in Cnidaria: In: Eaton RC (ed) Neural mechanisms of startle behavior. Plenum Press, New York, pp 15-42

Mackie GO (1989) Evolution of cnidarian giant axons: In: Anderson PAV (ed) Evolution of the first nervous systems. Plenum Press, New York, pp 395-407

Mackie GO (1990) The elementary nervous system revisited. Amer Zool 30: 907-920

McFarlane ID, Grimmelikhuijzen CJP (1991) Three anthozoan neuropeptides, Antho-RFamide and Antho-RWamides I and II, modulate spontaneous tentacle contractions in sea anemones. J Exp Biol 155: 669-673

McFarlane ID, Graff D, Grimmelikhuijzen CJP (1987) Exitatory actions of Antho-RFamide, an anthozoan neuropeptide, on muscles and conducting systems in the sea anemone *Calliactis parasitica*. J Exp Biol 133: 157-168

McFarlane ID, Anderson PAV, Grimmelikhuijzen CJP (1991) Effects of three anthozoan neuropeptides, Antho-RWamide I, Antho-RWamide II and Antho-RFamide, on slow muscles from sea anemones. J Exp Biol 156: 419-431

McFarlane ID, Reinscheid RK, Grimmelikhuijzen CJP (1992) Opposite actions of the anthozoan neuropeptide Antho-RNamide on antagonistic muscle groups in sea anemones. J Exp Biol 164: 295-299

McFarlane ID, Hudman D, Nothacker H-P, Grimmelikhuijzen CJP (1993) The expansion behaviour of sea anemones may be coordinated by two inhibitory neuropeptides, Antho-KAamide and Antho-RIamide. Proc R Soc B (Lond) 253:183-188

Minobe S, Koizumi O, Sugiyama T (1995) Nerve cell differentiation in epithelial *Hydra* from precursor cells introduced by grafting. I. Tentacles and hypostome. Dev Biol 172: 170-181

Moosler A, Rinehart KL, Grimmelikhuijzen CJP (1996) Isolation of four novel neuropeptides, the Hydra-RFamides I-IV, from *Hydra magnipapillata*. Biochem Biophys Res Commun 229: 596-602

Nothacker H-P, Grimmelikhuijzen CJP (1993) Molecular cloning of a novel, putative G protein-coupled receptor from sea anemones structurally related to members of the FSH, TSH, LH/CG receptor family from mammals. Biochem Biophys Res Commun 197: 1062-1069

Nothacker H-P, Rinehart KL, Grimmelikhuijzen CJP (1991a) Isolation of L-3-phenyllactyl-Phe-Lys-Ala-NH_2 (Antho-KAamide), a novel neuropeptide from sea anemones. Biochem Biophys Res Commun 179: 1205-1211

Nothacker H-P, Rinehart KL, McFarlane ID, Grimmelikhuijzen CJP (1991b) Isolation of two novel neuropeptides from sea anemones: the unusual, biologically active L-3-phenyllactyl-Tyr-Arg-Ile-NH_2 and its des-phenyllactyl fragment Tyr-Arg-Ile-NH_2. Peptides 12: 1165-1173

Perkins SN, Husten EJ, Eipper BA (1990) The 108 kDA peptidylglycine α-amidating monooxygenase precursor contains two separable enzymatic activities involved in peptide amidation. Biochem Biophys Res Commun 171: 926-932

Reinscheid RK, Grimmelikhuijzen CJP (1994) Primary structure of the precursor for the anthozoan neuropeptide Antho-RFamide from *Renilla köllikeri:* evidence for unusual processing enzymes. J Neurochem 62: 1214-1222

Salesse R, Remy JJ, Levin JM, Jallal B, Garnier J (1991) Towards understanding the glycoprotein hormone receptors. Biochimie 73:109-120

Schally AV (1978) Aspects of hypothalamic regulation of the pituitary gland. Science 202: 18-28

Schmutzler C, Darmer D, Diekhoff D, Grimmelikhuijzen CJP (1992) Identification of a novel type of processing site in the precursor for the sea anemone neuropeptide Antho-RFamide (<Glu-Gly-Arg-Phe-NH_2) from *Anthopleura elegantissima.* J Biol Chem 267: 22534-22541

Schmutzler C, Diekhoff D, Grimmelikhuijzen CJP (1994) The primary structure of the Pol-RFamide neuropeptide precursor protein from the hydromedusa *Polyorchis penicillatus* indicates a novel processing proteinase activity. Biochem J 299: 431-436

Seidah NG, Gaspar L, Mion P, Marcinkiewicz M, Mbikay M, Chrétien M (1990) cDNA sequence of two distinct pituitary proteins homologous to Kex2 and furin gene products: tissue-specific mRNAs encoding candidates for pro-hormone processing proteinases. DNA Cell Biol 9: 415-424

Seidah NG, Marcinkiewicz M, Benjannet S, Gaspar L, Beaubien G, Mattei MG, Lazure C, Mbikay M, Chrétien M (1991) Cloning and primary sequence of a mouse candidate prohormone convertase PC1 homologous to PC2, furin and Kex2: distinct chromosomal localization and messenger RNA distribution in brain and pituitary compared to PC2. Mol Endocrinol 5: 111-122

Smeekens SP, Steiner DF (1990) Identification of a human insulinoma cDNA encoding a novel mammalian protein structurally related to the yeast dibasic processing protease Kex2. J Biol Chem 265: 2997-3000

Smeekens SP, Avruch AS, LaMendola J, Chan SJ, Steiner DF (1991) Identification of a cDNA encoding a second putative prohormone convertase related to PC2 in AtT20 cells and islets of Langerhans. Proc Nati Acad Sci USA 88: 340-344

Spencer AN, Arkett SA (1984) Radial symmetry and the organization of central neurones in a hydrozoan jellyfish. J Exp Biol 110: 69-90

Sugiyama T, Fujisawa T (1978) Genetic analysis of developmental mechanisms in *Hydra.* II. Isolation and characterization of an interstitial cell deficient strain. J Cell Sci 29: 35-52

Tsai-Morris CH, Buczko E, Wang W, Xie X-Z, Dufau ML (1991) Structural organization of the rat luteinizing hormone (LH) receptor gene. J Biol Chem 266:11355-11359

Westfall JA (1973) Ultrastructural evidence for a granule-containing sensory-motor interneuron in *Hydra littoralis.* J Ultrastruct Res 42: 268-282

Westfall JA, Grimmelikhuijzen CJP (1993) Antho-RFamide immunoreactivity in neuronal synaptic and nonsynaptic vesicles of sea anemones. Biol Bull 185:109-114

Westfall JA, Kinnamon JC (1978) A second sensory-motor-interneuron with neurosecretory granules in *Hydra.* J Neurocytol 7: 365-379

Westfall JA, Sayyar KL, Elliott CF, Grimmelikhuijzen CJP (1995) Ultrastructural localization of Antho-RWamide I and II at neuromuscular synapses in the gastrodermis and oral sphincter muscle of the sea anemone *Calliactis parasitica.* Biol Bull 189: 280-287

4 Components of the Neuroendocrine Exocytosis Apparatus

M. Gratzl and B. Höhne-Zell

Anatomisches Institut der Technischen Universität München,
Biedersteiner Straße 29, D-80802 München, Germany;
Phone: +49 89 4140 3150, Fax: +49 89 397035,
E-mail: gratzl@lrz.tu-muenchen.de

Abstract

Exocytosis by neurons and permeabilized endocrine cells is inhibited by clostridial neurotoxins. This observation suggested that the mechanism of exocytosis is similar in both cell types although the process by which the neurotoxins gain access to the cell interior may be different. The present contribution summarizes the basic observations which led to the identification of the targets of the neurotoxins shared by neurons and endocrine cells. Furthermore, the probable function of the interaction of the identified membrane proteins synaptobrevin, synaptosome associated protein of 25 kDa (SNAP-25) and syntaxin in endocrine cells and the possible role of the calcium binding synaptotagmin and of the N-ethylmaleimide sensitive factor (NSF) and soluble NSF attachment proteins (SNAPs) is discussed.

Introduction

Exocytosis of synaptic vesicles (release of neurotransmitters) and of secretory vesicles (hormone secretion) is triggered by a rise in cytoplasmic free calcium ions caused by an influx of calcium ions from the extracellular space and/or release from intracellular stores. Following exocytosis, the containers of the secretory products, the vesicle membranes, are retrieved by endocytosis.

Although the principal involvement of calcium ions during exocytosis has long been recognized, the specific function of calcium ions within the cells and subsequent processes leading to exocytotic membrane fusion are still unknown. However, improvements in techniques suitable for following exocytosis, the identification of the target proteins of clostridial neurotoxins which block neurotransmitter release and the analysis of the interactions of proteins identified with further plasmalemmal secretory/synaptic vesicle membrane proteins and soluble proteins have yielded, as yet, an incomplete view of the mechanism of exocytosis. Interestingly enough, observations in a variety of systems have shown that some components of the exocytosis apparatus are shared by neurons and endocrine cells. Moreover, some of the proteins participating in regulated neurotransmitter and hormone release have also been detected to participate in constitutive membrane fusion processes. These findings may indicate that calcium ion-regulated exocytosis of neurotransmitters and hormones belongs to a group of basic cellular membrane fusion mechanisms adapted to the needs of neural or endocrine communication.

Analysis of the mechanism of exocytosis and its control by intracellular regulatory substances has long been hampered by the lack of suitable experimental approaches to study the interaction of the cytoplasmic surfaces of the plasmalemma with those of the membranes of intracellular storage organelles. Besides electrophysiological techniques, cell permeabilization techniques became instrumental in studies on the mechanism and the control of exocytosis. High voltage discharges (Knight and Baker 1982; Baker and Knight 1983), detergents (Wilson and Kirshner 1983; Dunn and Holz 1983) and bacterial pore forming substances (Ahnert-Hilger et al. 1985; Bader et al. 1986) have successfully been used to gain access to the cytoplasm of secretory cells. Whereas the holes generated by high voltage discharges and detergents like digitonin are not exactly defined in size, pores of known diameters can be produced by staphylococcal α-toxin and streptococcal streptolysin O (SLO), which allow either small molecules or small and large molecular weight substances to pass through the plasmalemma (Ahnert-Hilger et al. 1985; Bader et al. 1986; Ahnert-Hilger et al. 1987; Ahnert-Hilger et al. 1993; Bhakdi et al. 1993).

Permeabilization by SLO and digitonin has become a widely used technique to study the effects of neurotoxins on exocytosis of secretory granules by chromaffin and other endocrine cells, neurohypophysial and cortical terminals (Ahnert-Hilger et al. 1989a, b; Bittner et al. 1989a, b; Stecher et al. 1989a, b; Lazarovici et al. 1989; Dayanithi et al. 1990; McInnes and Dolly 1990; Dayanithi et al. 1992; Stecher et al. 1992b; Höhne-Zell et al. 1993, 1994; Ahnert-Hilger and Weller 1993; Lawrence et al. 1994; Sadoul et al. 1995; Regazzi et al. 1995; Foran et al. 1996; Höhne-Zell and Gratzl 1996). Intracellular surrogate media used in such studies often contain potassium glutamate, buffer substances for protons and divalent cations, ATP, GTP and cyclic nucleotides. Clearly, the presence of chelating substances in the media used and in the permeabilized cells, requires the precise adjustment of trigger calcium during the experiments (Föhr et al. 1993).

Identification of the Core Proteins of the Exocytosis Apparatus Using Clostridial Neurotoxins

Tetanus toxin in humans causes generalized spasms because it preferentially blocks transmitter release from glycinergic inhibitory interneurons in the spinal cord. Application of tetanus toxin to neurons in culture, synaptosomes and neurohypophysial nerve endings prevented exocytosis of both synaptic vesicles and secretory granules (Halpern et al. 1990; Williamson et al. 1996; McMahon et al. 1992). Injection of tetanus toxin and studies with permeabilized cells revealed that tetanus toxin acts within chromaffin cells, pancreatic B-cells, and nerve terminals. Moreover, it was found that tetanus toxin inhibits exocytosis of secretory granules at a step downstream from the elevation of intracellular calcium (Penner et al. 1986; Ahnert-Hilger et al. 1989a, b; Lazarovici et al. 1989; Bittner et al. 1989b; Dayanithi et al. 1992; Stecher et al. 1992b; Regazzi et al. 1995). In addition, permeabilization studies revealed that separation of the constituent light and heavy chains of tetanus toxin is a prerequisite for the inhibition of exocytosis and that the light chain of tetanus toxin alone is able to block exocytosis of secretory granules (Ahnert-Hilger et al. 1989a, b; Bittner et al. 1989b; Dayanithi et al. 1992; Stecher et al. 1992b). Thus a prerequisite for the inhibition of exocytosis by tetanus toxin is reductive separation of the disulfide-linked light and heavy chains and translocation of the light chains into the cytoplasm.

The light chain of tetanus toxin binds zinc (Wright et al. 1992) and shares the metal binding motif HELIH with a number of metalloendoproteases (Jongeneel et al. 1989; Niemann 1991). Moreover, histidine-specific reagents and chelators for divalent cations abrogate the effectiveness of tetanus toxin (Schiavo et al. 1992a, b; Höhne-Zell et al. 1993). In addition, mutational modification of the histidine residues within the zinc-binding motif caused inactivation of the neurotoxin when tested in permeabilized chromaffin cells or neurohypophysial nerve terminals (Höhne-Zell et al. 1993; Dayanithi et al. 1994). Finally, removal of zinc from the tetanus toxin light chain by dipicolinic acid reversibly abolished its activity (Höhne-Zell et al. 1993). Taken together, these findings suggest that inhibition of exocytosis of secretory granules by the light chain of tetanus toxin is caused by zinc-dependent proteolysis.

The protein cleaved by the tetanus toxin light chain within neurons and endocrine cells turned out to be **synaptobrevin** (Schiavo et al. 1992a, b; Link et al. 1992; Höhne-Zell et al. 1994; Regazzi et al. 1995). Synaptobrevin (also referred to as vesicle-associated membrane protein, VAMP) is a protein present in membranes of synaptic vesicles and secretory granules (Trimble et al. 1988; Baumert et al. 1989; Höhne-Zell et al. 1994; Regazzi et al. 1995). Cleavage of synaptobrevin of isolated secretory granules by tetanus toxin is inhibited by peptides spanning the cleavage site of synaptobrevin (Höhne-Zell et al. 1994). In the posterior pituitary, the effect of tetanus toxin on secretory granule exocytosis is abolished by these peptides (Dayanithi et al. 1994). These observations imply that synaptobrevin is centrally involved in exocytosis of secretory granules. Moreover, they are consistent with the idea that tetanus toxin acts primarily as a zinc-dependent protease.

Sensitivity to tetanus toxin can be used as a tool to investigate whether synaptobrevin is involved in exocytosis in particular types of cells. However, as recently reported, the two known forms of synaptobrevin, synaptobrevin 1 and synaptobrevin 2, are differentially distributed in neurons and endocrine cells, and are found also in a variety of other cell types (Rossetto et al. 1996; Volchuk et al.

1994; Cain et al. 1992). In addition, synaptobrevin 1 appears to be less susceptible to cleavage by tetanus toxin than synaptobrevin 2 (Schiavo et al. 1992a). Thus it is not surprising that exocytosis by pancreatic acinar cells, which contain mainly synaptobrevin 1 and only small amounts of synaptobrevin 2 (Rossetto et al. 1996) is only partly, if at all, inhibited by tetanus toxin (Stecher et al. 1992a; Gaisano et al. 1994).

Also, botulinum neurotoxin A has successfully been used for the identification of components of the exocytosis machinery. In vivo, botulinum neurotoxin A blocks acetylcholine release at motor endplates, which results in a progressive flaccid neuromuscular paralysis. In permeabilized adrenal chromaffin cells, inhibition of exocytosis by botulinum neurotoxin A, similar to the situation seen with tetanus toxin, requires separation of the constituent chains before the light chain becomes active (Stecher et al. 1989a, b; Bittner et al. 1989a; Dayanithi et al. 1990). However, exocytosis of secretory granules by adrenal chromaffin cells (Stecher et al. 1989a, b; Bittner et al. 1989a; Lawrence et al. 1994), by pheochromocytoma cells (McInnes and Dolly 1990; Lomneth et al. 1991; Banerjee et al. 1993; Ahnert-Hilger and Weller 1993) and neurohypophysial terminals (Dayanithi et al. 1990) is not completely blocked by the light chain of botulinum neurotoxin A. Consistent with this observation, **SNAP-25** (synaptosome associated protein of 25 kDa), the target of botulinum neurotoxin A in neurons (Schiavo et al. 1993; Blasi et al. 1993) and in endocrine cells (Sadoul et al. 1995; Höhne-Zell and Gratzl 1996), is only incompletely cleaved in permeabilized adrenal chromaffin cells and in pancreatic B-cells (Sadoul et al. 1995; Höhne-Zell and Gratzl 1996). One possibile explanation of these observations is that binding of other proteins of the exocytosis apparatus prevents complete cleavage of SNAP-25 in endocrine cells (see below).

Syntaxin (also called HPC-1; Inoue et al. 1992) as well as SNAP-25 is cleaved by botulinum neurotoxin C1 in neurons and endocrine cells (Blasi et al. 1993; Williamson et al. 1996; Foran et al. 1996). Botulinum neurotoxin C1 treatment results in blockade of neurotransmitter and hormone release. Because intracellularly applied antibodies directed against syntaxin or different portions of the cytoplasmic domain of syntaxin also inhibit exocytosis, syntaxin obviously also plays an important role in exocytosis of secretory granules (Bennett et al. 1993; Gutiérrez et al. 1995).

Interactions of the Core Proteins of the Exocytosis Apparatus

The proteins identified as components of the exocytosis apparatus, using tetanus and botulinum toxins as tools, form, when extracted from brain with detergents, a heterotrimeric complex. This complex in addition binds two cytosolic proteins (Söllner et al. 1993a, b). Specifically, the complex between synaptobrevin, SNAP-25 and syntaxin binds NSF (N-ethylmaleimide sensitive factor) and SNAPs (soluble NSF attachment proteins), cytosolic proteins involved in a variety of constitutive cellular membrane fusion processes (Rothman and Wieland 1996). Three species termed α-, β- and γ-**SNAP** have been purified from the brain, while NSF, an ubiquitous ATPase, exists only in one known form. The heterotrimeric complex associated with NSF and α-SNAP dissociates upon addition of ATP. Heterotrimeric complexes have been found in solubilized adrenal chromaffin and pheochromocytoma (PC12) cells (Roth and Burgoyne 1994; Oho et al. 1995; Chilcote et al. 1995), and upon incubation of the recombinant proteins in media containing nonionic detergents (Pellegrini et al. 1994, 1995; Hayashi et al. 1994; Söllner et al. 1993a; Pevsner et al. 1994). The complex between the vesicular synaptobrevin and the plasmalemmal proteins syntaxin and SNAP-25, the association of soluble SNAPs and the ATPase NSF, and the disassembly of the complex by ATP, may reflect stages of docking and fusion during exocytosis. Functional data on the involvement of α-SNAP in exocytosis of synaptic vesicles and secretory granules have been presented (DeBello et al. 1995; Morgan and Burgoyne 1995). In adrenal chromaffin cells, α-SNAP is required for the ATP-dependent priming step and not for the subsequent release process triggered by calcium (Chamberlain et al. 1995; Barnard et al. 1996).

SNAP-25 and syntaxin are localized mainly at the plasmalemma of neurons and chromaffin cells (Oyler et al. 1989; Bennett et al. 1992b; Gutiérrez et al. 1995; Sanna et al. 1991; Roth and Burgoyne 1994). However, a fraction of syntaxin and SNAP-25 also exists in synaptic vesicle and in secretory granule membranes (Walch-Solimena et al. 1995; Tagaya et al. 1995; Höhne-Zell and Gratzl 1996). Subfractionation studies revealed syntaxin and stable SNAP-25/syntaxin heterodimers in chromaffin vesicle membranes suggesting that the heterodimers and syntaxin may serve as receptors of soluble NSF attachment proteins (SNAP receptors) during chromaffin vesicle exocytosis (Höhne-Zell and Gratzl 1996).

Role of Calcium Ions and ATP During Exocytosis

Besides the core of the exocytosis machinery consisting of synaptobrevin, SNAP-25 and syntaxin, several further proteins have been suggested as participating in exocytosis. However, evidence for regulatory or functional involvement of proteins is often difficult to obtain. From studies in permeabilized cells, it is evident that calcium and ATP are of critical importance for exocytosis. Thus the possible role of these substances is discussed separately.

To date, synaptotagmin is the most likely candidate to serve during exocytosis as an intracellular calcium sensor (Brose et al. 1992). Synaptotagmin was the first membrane protein detected in both synaptic vesicles and secretory granules (Matthew et al. 1981). Its presence in chromaffin granules and in neurohypophysial granules (Schmidle et al. 1991; Trifaró et al. 1989) has been confirmed (Walch-Solimena et al. 1993). Synaptotagmin is characterized by several interesting properties. Its ability to bind calcium ions in the presence of phospholipids suggests a role as a calcium sensor during exocytosis (Brose et al. 1992; Li et al. 1995). The cytoplasmic part of synaptotagmin contains two C2 regulatory domains with different properties. While the domain C2A binds primarily calcium, C2B interacts with higher phosphorylated inositol polyphosphates such as IP4, IP5 and IP6 (Mikoshiba et al. 1995; Fukuda et al. 1995). These intracellular messengers are metabolites of IP3, which triggers release of calcium from intracellular stores. While calcium is the major trigger substance for exocytosis inositolpolyphosphates have been shown to inhibit exocytosis (Llinás et al. 1994, Ohara-Imaizumi et al. 1997). Thus, binding of two opposing effectors would allow efficient control of exocytosis via synaptotagmin.

Peptides, corresponding to cytoplasmic domains of synaptotagmin, inhibit exocytosis of both synaptic vesicles and secretory granules (Bommert et al. 1993; Elferink et al. 1993). The peptides either influence the regulatory domains or interfere with interactions between synaptotagmin and other proteins of the exocytosis apparatus. Synaptotagmin interacts not only in a calcium-dependent way with itself, but also with syntaxin (Yoshida et al. 1992; Bennett et al. 1992a; Sugita et al. 1996; Chapman et al. 1996). Consequently, the putative calcium sensor is spatially and probably also functionally linked to the core proteins of the exocytosis apparatus identified by their susceptibility to proteolysis by clostridial neurotoxins. Moreover, synaptotagmin/syntaxin has been found to interact with calcium channels at the plasma membrane, an observation that would dock the secretory organelles at the site of calcium influx that triggers exocytosis (Yoshida et al. 1992; O'Connor et al. 1993; Leveque et al. 1994).

Synaptotagmin, bound to the complex between synaptobrevin, syntaxin and SNAP-25 in the absence of α-SNAP and NSF (Söllner et al. 1993a) is displaced from the complex by α-SNAP, suggesting that these two proteins share binding sites at the complex. Only β-SNAP binds synaptotagmin, and the complex between synaptobrevin, syntaxin and SNAP-25 (Schiavo et al. 1995). This observation may indicate that the proposed calcium sensor, synaptotagmin, may be linked to the complex via the soluble protein β-SNAP, in addition to the direct binding with syntaxin described above.

The exact role of ATP during exocytosis is unknown. However, in adrenal chromaffin cells, exocytosis has been resolved into sequential ATP- and calcium-dependent steps (Holz et al. 1989; Hay and Martin 1992; Heinemann et al. 1993). ATP could be required to transport chromaffin secretory vesicles scattered throughout the cell body to the plasmalemma where they become docked and fuse as a consequence of calcium influx. However, ATP may also be involved in modifications of the proteins of the core complex by phosphorylation. Interestingly, the ATP-dependent "priming" step was found to be inhibited by botulinum neurotoxin A while subsequent catecholamine release triggered by calcium ionsis blocked by tetanus toxin (Bittner and Holz 1993). Thus SNAP-25, the target of botulinum neurotoxin A, may participate in ATP-dependent "priming" whereas synaptobrevin attacked by tetanus toxin takes part in the calcium-dependent step of the release process. The recent observation, that SNAP-25 can be phosphorylated at its C-terminal and between the cleavage sites of botulinum neurotoxin A and E by protein kinase C suggests that this process may be involved in "priming" (Shimazaki et al. 1996). In addition, ATP is required during the same stage as a substrate of NSF, a soluble ATPase, activated by α-SNAP (Morgan and Burgoyne 1995; Chamberlain et al. 1995) and also linked to the core proteins of the exocytosis machinery (see Sect. on interactions of the core proteins as described above).

Acknowledgments

Work in the authors' laboratory was supported by the Deutsche Forschungsgemeinschaft (Gr 681 and SFB 391) and Fonds der Chemischen Industrie.

References

Ahnert-Hilger G, Bhakdi S, Gratzl M (1985) Minimal requirements for exocytosis. A study using PC 12 cells permeabilized with staphylococcal α-toxin. J Biol Chem 260: 12730-12734

Ahnert-Hilger G, Bräutigam M, Gratzl M (1987) Ca^{2+}-stimulated catecholamine release from alpha-toxin permeabilized PC12 cells: biochemical evidence for exocytosis and its modulation by protein kinase C and G-proteins. Biochemistry 26: 7842-7848

Ahnert-Hilger G, Bader M-F, Bhakdi S, Gratzl M (1989a) Introduction of macromolecules into bovine adrenal medullary chromaffin cells and rat pheochromocytoma cells (PC12) by permeabilization with streptolysin O: inhibitory effect of tetanus toxin on catecholamine secretion. J Neurochem 52: 1751-1758

Ahnert-Hilger G, Weller U, Dauzenroth M-E, Habermann E, Gratzl M (1989b) The tetanus toxin light chain inhibits exocytosis. FEBS Lett 242: 245-248

Ahnert-Hilger G, Stecher B, Beyer C, Gratzl M (1993) Exocytotic membrane fusion as studied in toxin-permeabilized cells. Methods Enzymol 221: 139-149

Ahnert-Hilger G, Weller U (1993) Comparison of the intracellular effects of clostridial neurotoxins on exocytosis from streptolysin-permeabilized rat pheochromocytoma (PC12) and bovine adrenal chromaffin cells. Neuroscience 53: 547-552

Bader M-F, Thiersé D, Aunis D, Ahnert-Hilger G, Gratzl M (1986) Characterization of hormone and protein release from α-toxin-permeabilized chromaffin cells in primary culture. J Biol Chem 261: 5777-5783

Baker PF, Knight DE (1983) High-voltage techniques for gaining access to the interior of cells: application to the study of exocytosis and membrane turnover. In: Fleischer S, Fleischer B (eds) Methods in enzymology. Academic Press, New York, pp 28-37

Banerjee A, Martin TFJ, DasGupta BR (1993) Nerve growth factor induces sensitivity to botulinum neurotoxin type A in norepinephrine-secreting PC12 cells. Neurosci Lett 164: 93-96

Barnard RJO, Morgan A, Burgoyne RD (1996) Domains of α-SNAP required for the stimulation of exocytosis and for N-ethylmalemide-sensitive fusion protein (NSF) binding and activation. Mol Biol Cell 7: 693-701

Baumert M, Maycox PR, Navone F, De Camilli P, Jahn R (1989) Synaptobrevin: an integral membrane protein of 18 000 daltons present in small synaptic vesicles of rat brain. EMBO J 8: 379-384

Bennett MK, Calakos N, Kreiner T, Scheller RH (1992a) Synaptic vesicle membrane proteins interact to form a multimeric complex. J Cell Biol 116: 761-775

Bennett MK, Calakos N, Scheller RH (1992b) Syntaxin: a synaptic protein implicated in docking of synaptic vesicles at presynaptic active zones. Science 257: 255-259

Bennett MK, Garcia-Arrarás JE, Elferink LA, Peterson K, Fleming AM, Hazuka CD, Scheller RH (1993) The syntaxin family of vesicular transport receptors. Cell 74: 863-873

Bhakdi S, Weller U, Walev I, Martin E, Jonas D, Palmer M (1993) A guide to the use of pore-forming toxins for controlled permeabilization of cell membranes. Med Microbiol Immunol 182: 167-175

Bittner MA, DasGupta BR, Holz RW (1989a) Isolated light chains of botulinum neurotoxins inhibit exocytosis. J Biol Chem 264: 10354-10360

Bittner MA, Habig WH, Holz RW (1989b) Isolated light chain of tetanus toxin inhibits exocytosis: studies in digitonin-permeabilized cells. J Neurochem 53: 966-968

Bittner MA, Holz RW (1993) Protein kinase C and clostridial neurotoxins affect discrete and related steps in the secretory pathway. Cell Mol Neurobiol 13: 649-664

Blasi J, Chapman ER, Yamasaki S, Binz T, Niemann H, Jahn R (1993) Botulinum neurotoxin C1 blocks neurotransmitter release by means of cleaving HPC-1/syntaxin. EMBO J 12: 4821-4828

Bommert K, Charlton MP, DeBello WM, Chin GJ, Betz H, Augustine GJ (1993) Inhibition of neurotransmitter release by C2 domain peptides implicates synaptotagmin in exocytosis. Nature 363: 163-165

Brose N, Petrenko AG, Südhof TC, Jahn R (1992) Synaptotagmin: a calcium sensor on the synaptic vesicle surface. Science 256: 1021-1025

Cain CC, Trimble WS, Lienhard GE (1992) Members of the VAMP family of synaptic vesicle proteins are components of glucose transporter-containing vesicles from rat adipocytes. J Biol Chem 267: 11681-11684

Chamberlain LH, Roth D, Morgan A, Burgoyne D (1995) Distinct effects of α-SNAP, 14-3-3 proteins and calmodulin on priming and triggering of regulated exocytosis. J Cell Biol 130: 1063-1070

Chapman ER, An S, Edwardson JM, Jahn R (1996) A novel junction for the second C2 domain of synaptotagmin. J Biol Chem 271: 5844-5849

Chilcote TJ, Galli T, Mundigl O, Edelmann L, McPherson PS, Takei K, De Camilli P (1995) Cellubrevin and synaptobrevins: similar subcellular localization and biochemical properties in PC12 cells. J Cell Biol 129: 219-231

Dayanithi G, Ahnert-Hilger G, Weller U, Nordmann JJ, Gratzl M (1990) Release of vasopressin from isolated permeabilized neurosecretory nerve terminals is blocked by the light chain of botulinum A toxin. Neuroscience 39: 711-715

Dayanithi G, Weller U, Ahnert-Hilger G, Link H, Nordmann JJ, Gratzl M (1992) The light chain of tetanus toxin inhibits calcium dependent vasopressin release from permeabilized nerve endings. Neuroscience 46: 489-493

Dayanithi G, Stecher B, Höhne-Zell B, Yamasaki S, Binz T, Weller U, Niemann H, Gratzl M (1994) Exploring the functional domain and the target of the tetanus toxin light chain in neurohypophysial terminals. Neuroscience 58: 423-431

DeBello WM, O'Connor V, Dresbach T, Whiteheart SW, Wang SS-H, Schweizer FE, Betz H, Rothman JE, Augustine GJ (1995) SNAP-mediated protein-protein interactions are essential for neurotransmitter release. Nature 373: 626-630

Dunn LA, Holz RW (1983) Catecholamine secretion from digitonin-treated adrenal medullary chromaffin cells. J Biol Chem 258: 4989-4993

Elferink LA, Peterson MR, Scheller RH (1993) A role for synaptotagmin (p65) in regulated exocytosis. Cell 72: 153-159

Foran P, Lawrence GW, Shone CC, Foster KA, Dolly JO (1996) Botulinum neurotoxin C1 cleaves both syntaxin and SNAP-25 in intact and permeabilized chromaffin cells: correlation with its blockade of catecholamine release. Biochemistry 35: 2630-2636

Föhr KJ, Warchol W, Gratzl M (1993) Calculation and control of free divalent cations in solutions used for membrane fusion studies. Methods Enzymol 221: 149-157

Fukuda M, Kojima T, Aruga J, Niinobe M, Mikoshiba K (1995) Functional diversity of C2 domains of synaptotagmin family. Mutational analysis of inositol high polyphosphate binding domain. J Biol Chem 270: 26523-26527

Gaisano HY, Sheu L, Foskett JK, Trimble WS (1994) Tetanus toxin light chain cleaves a vesicle-associated membrane protein (VAMP) isoform 2 in rat pancreatic zymogen granules and inhibits enzyme secretion. J Biol Chem 269: 17062-17066

Gutiérrez LM, Cànaves JM, Ferrer-Montiel AV, Reig JA, Montal M, Viniegra S (1995) A peptide that mimics the carboxy-terminal domain of SNAP-25 blocks Ca^{2+}-dependent exocytosis in chromaffin cells. FEBS Lett 372: 39-43

Halpern JL, Habig WH, Trenchard H, Russell JT (1990) Effect of tetanus toxin on oxytocin and vasopressin release from nerve endings of the neurohypophysis. J Neurochem 55: 2072-2078

Hay JC, Martin TFJ (1992) Resolution of regulated secretion into sequential MgATP-dependent and calcium-dependent stages mediated by distinct cytosolic proteins. J Cell Biol 119: 139-151

Hayashi T, McMahon H, Yamasaki S, Binz T, Hata Y, Südhof TC, Niemann H (1994) Synaptic vesicle membrane fusion complex: action of clostridial neurotoxins on assembly. EMBO J 13: 5051-5061

Heinemann C, von Rüden L, Chow RH, Neher E (1993) A two-step model of secretion control in neuroendocrine cells. Pflügers Arch 424: 105-112

Holz RW, Bittner MA, Peppers SC, Senter RA, Eberhard DA (1989) MgATP-independent and MgATP-dependent exocytosis. J Biol Chem 264: 5412-5419

Höhne-Zell B, Stecher B, Gratzl M (1993) Functional characterization of the catalytic site of the tetanus toxin light chain using permeabilized adrenal chromaffin cells. FEBS Lett 336: 175-180

Höhne-Zell B, Ecker A, Weller U, Gratzl M (1994) Synaptobrevin cleavage by the tetanus toxin light chain is linked to the inhibition of exocytosis in chromaffin cells. FEBS Lett 355: 131-134

Höhne-Zell B, Gratzl M (1996) Adrenal chromaffin cells contain functionally different SNAP-25 monomers and SNAP-25/syntaxin heterodimers. FEBS Lett 394: 109-116

Inoue A, Obata K, Akagawa K (1992) Cloning and sequence analysis of cDNA for a neuronal cell membrane antigen, HPC-1. J Biol Chem 267: 10613-10619

Jongeneel CV, Bouvier J, Bairoch A (1989) A unique signature identifies a family of zinc-dependent metallopeptidases. FEBS Lett 242: 211-214

Knight DE, Baker PF (1982) Calcium-dependence of catecholamine release from bovine adrenal medullary cells after exposure to intense electric fields. J Membr Biol 68: 107-140

Lawrence GW, Weller U, Dolly JO (1994) Botulinum A and the light chain of tetanus toxins inhibit distinct stages of Mg/ATP-dependent catecholamine exocytosis from permeabilised chromaffin cells. Eur J Biochem 222: 325-333

Lazarovici P, Fujita K, Contreras ML, DiOrio JP, Lelkes PI (1989) Affinity purified tetanus toxin binds to isolated chromaffin granules and inhibits catecholamine release in digitonin-permeabilized chromaffin cells. FEBS Lett 253: 212-128

Leveque C, El Far O, Martin-Moutot N, Sato K, Kato R, Takahashi M, Seagar MJ (1994) Purification of the N-type calcium channel associated with syntaxin and synaptotagmin. A complex implicated in synaptic vesicle exocytosis. J Biol Chem 269: 6306-6312

Li C, Ullrich B, Zhang JZ, Anderson RGW, Brose N, Südhof TC (1995) Ca^{2+}-dependent and -independent activities of neural and non-neural synaptotagmins. Nature 375: 594-599

Link E, Edelmann L, Chou JH, Binz T, Yamasaki S, Eisel U, Baumert M, Südhof TC, Niemann H, Jahn R (1992) Tetanus toxin action: inhibition of neurotransmitter release linked to synaptobrevin proteolysis. Biochem Biophys Res Commun 189: 1017-1023

Llinás R, Sugimori M, Lang EJ, Morita M, Fukuda M, Niinobe M, Mikoshiba K (1994) The inositol high-polyphosphate series blocks synaptic transmission by preventing vesicular fusion: a squid giant synapse study. Proc Natl Acad Sci USA 91: 12990-12993

Lomneth R, Martin TFJ, DasGupta BR (1991) Botulinum neurotoxin light chain inhibits norepinephrine secretion in PC12 cells at an intracellular membranous or cytoskeletal site. J Neurochem 57: 1413-1421

Matthew WD, Tsavaler L, Reichardt LF (1981) Identification of a synaptic vesicle-specific membrane protein with a wide distribution in neuronal and neurosecretory tissue. J Cell Biol 91: 257-269

McInnes C, Dolly JO (1990) Calcium-dependent noradrenaline release from permeabilised PC12 cells is blocked by botulinum neurotoxin A or its light chain. FEBS Lett 261: 323-326

McMahon HT, Foran P, Dolly JO, Verhage M, Wiegant VM, Nicholls DG (1992) Tetanus toxin and botulinum toxins type A and B inhibit glutamate, γ-aminobutyric acid, aspartate, and met-enkephalin release from synaptosomes. J Biol Chem 267: 21338-21343

Mikoshiba K, Fukuda M, Moreira JE, Lewis FMT, Sugimori M, Niinobe M, Llinás R (1995) Role of the C2A domain of synaptotagmin in transmitter release as determined by specific antibody injection into the squid giant synapse preterminal. Proc Natl Acad Sci USA 92: 10703-10707

Morgan A, Burgoyne RD (1995) A role for soluble NSF attachment proteins (SNAPs) in regulated exocytosis in adrenal chromaffin cells. EMBO J 14: 232-239

Niemann H (1991) Molecular biology of clostridial neurotoxins. In: Alouf JE, Freer JH (eds) Sourcebook of bacterial protein toxins. Academic Press, New York, pp 299-344

O'Connor VM, Shamotienko O, Grishin E, Betz H (1993) On the structure of the 'synaptosecretosome'. Evidence for a neurexin/synaptotagmin/syntaxin/Ca^{2+} channel complex. FEBS Lett 326: 255-260

Ohara-Imaizumi M, Fukuda M, Niinobe M, Misonou H, Ikeda K, Murakami T, Kawasaki M, Mikoshiba K, Kumakura K (1997) Distinct roles of C2A and C2B domains of synatotagmin in the regulation of exocytosis in adrenal chromaffin cells. Proc Natl Acad Sci USA 94: 287-291

Oho C, Seino S, Takahashi M (1995) Expression and complex formation of soluble N-ethyl-maleimide-sensitive factor attachment protein (SNAP) receptors in clonal rat endocrine cells. Neurosci Lett 186: 208-210

Oyler GA, Higgins GA, Hart RA, Battenberg E, Billingsley M, Bloom FE, Wilson MC (1989) The identification of a novel synaptosomal-associated protein, SNAP-25, differentially expressed by neuronal subpopulations. J Cell Biol 109: 3039-3052

Pellegrini LL, O'Connor VM, Betz H (1994) Fusion complex formation protects synaptobrevin against proteolysis by tetanus toxin light chain. FEBS Lett 353: 319-323

Pellegrini LL, O'Connor V, Lottspeich F, Betz H (1995) Clostridial neurotoxins compromise the stability of a low energy SNARE complex mediating NSF activation of synaptic vesicle fusion. EMBO J 14: 4705-4713

Penner R, Neher E, Dreyer F (1986) Intracellularly injected tetanus toxin inhibits exocytosis in bovine adrenal chromaffin cells. Nature 324: 76-78

Pevsner J, Hsu S-C, Braun JEA, Calakos N, Ting AE, Bennett MK, Scheller RH (1994) Specificity and regulation of a synaptic vesicle docking complex. Neuron 13: 353-361

Regazzi R, Wollheim CB, Lang J, Theler J-M, Rossetto O, Montecucco C, Sadoul K, Weller U, Palmer M, Thorens B (1995) VAMP-2 and cellubrevin are expressed in pancreatic b-cells and are essential for Ca^{2+}- but not for GTPγS-induced insulin secretion. EMBO J 14: 2723-2730

Rossetto O, Gorza L, Schiavo G, Schiavo N, Scheller RH, Montecucco C (1996) VAMP/synaptobrevin isoforms 1 and 2 are widely and differentially expressed in nonneuronal tissues. J Cell Biol 132: 167-179

Roth D, Burgoyne RD (1994) SNAP-25 is present in a SNARE complex in adrenal chromaffin cells. FEBS Lett 351: 207-210

Rothman JE, Wieland FT (1996) Protein sorting by transport vesicles. Science 272: 227-234

Sadoul K, Lang J, Montecucco C, Weller U, Catsicas S, Wollheim CB, Halban PA (1995) SNAP-25 is expressed in islets of Langerhans and is involved in insulin release. J Cell Biol 128: 1019-1028

Sanna PP, Bloom FE, Wilson MC (1991) Dibutyryl-cAMP induces SNAP-25 translocation into the neurites in PC12. Dev Brain Res 59: 104-108

Schiavo G, Benfenati F, Poulain B, Rossetto O, de Laureto PP, DasGupta BR, Montecucco C (1992a) Tetanus and botulinum-B neurotoxins block neurotransmitter release by proteolytic cleavage of synaptobrevin. Nature 359: 832-835

Schiavo G, Poulain B, Rossetto O, Benfenati F, Tauc L, Montecucco C (1992b) Tetanus toxin is a zinc protein and its inhibition of neurotransmitter release and protease activity depends on zinc. EMBO J 11: 3577-3583

Schiavo G, Rossetto O, Catsicas S, Polverino de Laureto P, DasGupta BR, Benfenati F, Montecucco C (1993) Identification of the nerve terminal targets of botulinum neurotoxin serotypes A, D, and E. J Biol Chem 268: 23784-23787

Schiavo G, Gmachl MJS, Stenbeck G, Söllner TH, Rothman JE (1995) A possible docking and fusion particle for synaptic transmission. Nature 378: 733-736

Schmidle T, Weiler R, Desnos C, Scherman D, Fischer-Colbrie R, Floor E, Winkler H (1991) Synaptin/synaptophysin p65 and SV2: their presence in adrenal chromaffin granules and sympathetic large dense core vesicles. Biochim Biophys Acta 1060: 251-256

Shimazaki Y, Nishiki T, Omori A, Sekiguchi M, Kamata Y, Kozaki S, Takahashi M (1996) Phosphorylation of 25-kDa synaptosome-associated protein. J Biol Chem 271: 14548-14553

Söllner T, Bennett MK, Whiteheart SW, Scheller RH, Rothman JE (1993a) A protein assembly-disassembly pathway in vitro that may correspond to sequential steps of synaptic vesicle docking, activation, and fusion. Cell 75: 409-418

Söllner T, Whiteheart SW, Brunner M, Erdjument-Bromage H, Geromanos S, Tempst P, Rothman JE (1993b) SNAP receptors implicated in vesicle targeting and fusion. Nature 362: 318-324

Stecher B, Gratzl M, Ahnert-Hilger G (1989a) Reductive chain separation of botulinum A toxin - a prerequisite to its inhibitory action on exocytosis in chromaffin cells. FEBS Lett 248: 23-27

Stecher B, Weller U, Habermann E, Gratzl M, Ahnert-Hilger G (1989b) The light chain but not the heavy chain of botulinum A toxin inhibits exocytosis from permeabilized adrenal chromaffin cells. FEBS Lett 255: 391-394

Stecher B, Ahnert-Hilger G, Weller U, Kemmer TP, Gratzl M (1992a) Amylase release from streptolysin O-permeabilized pancreatic acinar cells. Effects of Ca^{2+}, guanosine 5'-[γ-thio]triphosphate, cyclic AMP, tetanus toxin and botulinum A toxin. Biochem J 283: 899-904

Stecher B, Hens JJH, Weller U, Gratzl M, Gispen WH, De Graan PNE (1992b) Noradrenaline release from permeabilized synaptosomes is inhibited by the light chain of tetanus toxin. FEBS Lett 312: 192-194

Sugita S, Hata Y, Südhof TC (1996) Distinct Ca^{2+}-dependent properties of the first and second C2-domains of synaptotagmin I. J Biol Chem 271: 1262-1265

Tagaya M, Toyonaga S, Takahashi M, Yamamoto G, Fujiwara T, Akagawa K, Moriyama Y, Mizushima S (1995) Syntaxin 1 (HPC-1) is associated with chromaffin granules. J Biol Chem 270: 15930-15933

Trifaró J-M, Fournier S, Novas ML (1989) The p65 protein is a calmodulin-binding protein present in several types of secretory vesicles. Neuroscience 29: 1-8

Trimble WS, Cowan DM, Scheller RH (1988) VAMP-1: a synaptic vesicle-associated integral membrane protein. Proc Natl Acad Sci USA 85: 4538-4542

Volchuk A, Mitsumoto Y, He L, Liu Z, Habermann E, Trimble WS, Klip A (1994) Expression of vesicle-associated membrane protein 2 (VAMP-2)/synaptobrevin II and cellubrevin in rat skeletal muscle and in a muscle cell line. Biochem J 304: 139-145

Walch-Solimena C, Takei K, Marek KL, Midyett K, Südhof TC, De Camilli P, Jahn R (1993) Synaptotagmin: a membrane constituent of neuropeptide-containing large dense-core vesicles. J Neurosci 13: 3895-3903

Walch-Solimena C, Blasi J, Edelmann L, Chapman ER, Fischer von Mollard G, Jahn R (1995) The t-SNAREs syntaxin 1 and SNAP-25 are present on organelles that participate in synaptic vesicle recycling. J Cell Biol 128: 637-645

Williamson LC, Halpern JL, Montecucco C, Brown JE, Neale EA (1996) Clostridial neurotoxins and substrate proteolysis in intact neurons. Botulinum neurotoxin C acts on synaptosomal-associated protein of 25 kDa. J Biol Chem 271: 7694-7699

Wilson SP, Kirshner N (1983) Calcium-evoked secretion from digitonin permeabilized adrenal medullary chromaffin cells. J Biol Chem 258: 4994-5000

Wright JF, Pernollet M, Reboul A, Aude C, Colomb MG (1992) Identification and partial characterization of a low affinity metal-binding site in the light chain of tetanus toxin. J Biol Chem 267: 9053-9058

Yoshida A, Oho C, Omori A, Kuwahara R, Ito T, Takahashi M (1992) HPC-1 is associated with synaptotagmin and omega-conotoxin receptor. J Biol Chem 267: 24925-24928

5 Neuroendocrine Cells Revisited: A System for Studying Subcellular mRNA Compartmentalization

E. Mohr and D. Richter

Universitätskrankenhaus Eppendorf, Institut für Zellbiochemie und klinische Neurobiologie, Martinistraße 52, D-20246 Hamburg, Germany;
Phone: +49 40 4717 4553, Fax: +49 40 4717 4541, E-mail: emohr@uke.uni-hamburg.de

Abstract

A number of mRNAs are sorted to dendrites and, in some cases, to axons of nerve cells. The protein synthesizing capacity of dendrites is well documented. mRNAs targeted to this domain might, therefore, be locally translated. The physiological role of axonal transcripts is more enigmatic, since mammalian axons lack a protein synthesizing machinery. The peptide hormone genes encoding the vasopressin and oxytocin precursors are expressed in hypothalamic magnocellular neurons, and substantial amounts of their mRNAs are detectable in both types of neurites. In order to characterize the molecular determinants involved in subcellular vasopressin and oxytocin mRNA sorting, eukaryotic expression vector constructs have been injected into the cell nuclei of cultured neurons which do not express the peptide hormone genes endogenously. Subsequent in situ hybridization analyses revealed that vasopressin and oxytocin sense transcripts are targeted to dendrites while the corresponding anti-sense RNAs remained confined to the cell somata. The results indicate that the RNA nucleotide sequence or a secondary structure formed by that sequence is implicated in subcellular transport. Mapping of the targeting elements revealed two segments within vasopressin mRNA that are able to confer dendritic compartmentalization to α-tubulin transcripts which are normally restricted to the cell bodies. These segments span part of the coding region as well as the 3'-untranslated region. In contrast, localization of vasopressin and oxytocin transcripts to axons of cultured nerve cells has not been observed. Thus, factors directing mRNAs to this domain might be more cell-specific than those mediating dendritic targeting.

Introduction

In order to allow nerve cells to respond appropriately within the neuronal network to external stimuli, their macromolecular constituents have to be sorted to defined subcellular microdomains that may be located at considerable distance from the cell somata. Transport of organelles and proteins to their ultimate destinations at pre- and postsynaptic sites has long been thought to constitute the sole determinants of nerve cell polarity and complexity. This view has changed, however, since the discovery of certain mRNA species in the neurites of a variety of nerve cells.

Table 1. RNAs located in axons and dendrites of various neuronal tissues

	Species/Tissue	Reference
Dendritic RNAs		
Arc/arg3.1	Rat/Dentate gyrus	Lyford et al. 1995; Link et al. 1995
BC1	Rat/Hippocampus	Tiedge et al. 1991
BDNF	Rat/Hippocampus	Dugich-Djordjevic et al. 1992
CaMKIIα	Rat/Hippocampus	Burgin et al. 1990
GAP 43	Rat/Brain	Landry et al. 1994
Glycine receptor α-subunit	Rat/Spinal Cord	Racca et al. 1997
Glutamate receptors	Rat /Hippocampus	Miyashiro et al. 1994
InsP3 receptor	Mouse/Cerebellum	Furuichi et al. 1993
L 7	Mouse/Cerebellum	Bian et al. 1996
MAP 2	Rat/Hippocampus	Garner et al. 1988
Neurogranin	Rat/Brain	Landry et al. 1994
Oxytocin	Rat/Hypothalamus	Mohr et al. 1995
SSTR 1	Rat/Cortex	Busch et al. 1997
Vasopressin	Rat/Hypothalamus	Mohr et al. 1995
Axonal RNAs		
BC 1	Rat/Hypothalamus	Tiedge et al. 1993
CaBP	Mouse/Cerebellum	Bian et al. 1996
CDCH	Snail/CNS	van Minnen 1994
Dynorphin	Rat/Hypothalamus	Mohr und Richter 1992
ELH	Snail/CNS	Dirks et al. 1989
MAP tau	Rat/Cortex	Litman et al. 1993
MIP	Snail/CNS	van Minnen 1994
NF	Squid/Stellate ganglion	Giuditta et al. 1991
NF-L	Rat/Hypothalamus	Mohr and Richter 1992
Odorant receptors	Mouse/Olfactory bulb	Ressler et al. 1994
OMP	Mouse/Olfactory bulb	Vassar et al. 1994
Oxytocin	Rat/Hypothalamus	Mohr et al. 1991
Vasopressin	Rat/Hypothalamus	Mohr et al. 1991

Arc, activity-regulated cytoskeleton-associated protein; *arg*, activity-regulated gene; *BC*, brain cytosolic; *BDNF*, brain-derived neurotrophic factor; *CaBP*, calbindin; *CaMKIIα*, Ca^{2+}/calmodulin-dependent protein kinase;*CDCH*, caudo-dorsal cell hormone;*ELH*, egg-laying hormone; *GAP*, growth-associated protein;*InsP3*, inositol trisphosphate;*MAP*, microtubule-associated protein;*MIP*, molluscan insulin-related peptide;*NF*, neurofilament; *NF-L*, low molecular weight neurofilament;*OMP*, olfactory marker protein;*SSTR*, somatostatin receptor

Initially, protein-encoding transcripts such as microtubule-associated protein (MAP) 2 mRNA (Garner et al. 1988) have been detected in dendrites of hippocampal neurons, providing the missing link to earlier observations, namely the presence of ribosomes and polysomes beneath postsynaptic sites (reviewed in Steward et al. 1995). These and other findings have lead to the hypothesis of local on-site synthesis of a subset of dendritic proteins, thus adding a new dimension to the function of nerve cells. The list of mRNAs residing in dendrites is still growing (Table 1) and the phenomenon is observed in different neuronal cell populations both in brain sections and in cultured neurons of the central and peripheral nervous system (reviewed in Steward et al. 1995).

In subsequent years it became apparent that in rare cases, the axon of nerve cells also harbor certain types of mRNAs (Table 1). In this case, the physiological meaning was less clear since axons, with the exception of the initial unmyelinated segment, are believed to lack protein synthesizing capacity, at least in mammals (reviewed in Mohr and Richter 1995). Studies aimed at investigating the subcellular localization of mRNAs encoding the vasopressin (VP) and oxytocin (OT) precursors in magnocellular neurons of the rat hypothalamo-neuro-hypophysial tract have led to the so far unprecedented observation of mRNA compartmentalization in both types of neurites. The present review will summarize the biochemical and morphological characteristics of subcellular VP and OT mRNA transport in these neurons. Furthermore, expression of the peptide hormone mRNAs in cultured neurons has given initial insights into the molecular determinants of mRNA targeting which will be discussed in a subsequent section.

Characteristics of Axonal and Dendritic VP and OT mRNAs

Axonal mRNAs

Since the discovery of VP- (Lehmann 1988) and, later on, also OT-encoding transcripts (reviewed in Mohr and Richter 1995) in rat posterior pituitary tissue, initial work has focused on elucidating the cellular origin of these mRNAs. The peptide hormone transcripts are located in the axonal compartment of hypothalamic magnocellular neurons rather than being expressed in the closely associated glial cells, the pituicytes, as has been shown by biochemical (Mohr et al. 1991) and morphological studies (Trembleau et al. 1994). Subsequent characterization revealed several interesting features. By employing RNA filter hybridization analyses it was evident that the axonal RNAs exhibited shorter poly(A) tracts than their counterparts in the cell bodies. Even though the poly(A) tail length of both VP and OT mRNAs in the cell bodies may vary considerably in response to external physiological stimulation, for instance following osmotic challenge, no such variation has been observed for these RNAs in the distal parts of axons in the posterior pituitary. However, when experimental animals were subjected to osmotic stress there was a substantial increase in the levels of axonal VP mRNA (approximately 17-fold) that exceeds the 2- to 3-fold increase observed in the cell bodies. In contrast, a three-fold accumulation of axonal OT-encoding transcripts resembled that measurable in the perikarya (reviewed in Mohr and Richter 1995). Morphological investigations have revealed that the peptide hormone VP and its mRNA are not co-localized. Axonal varicosities immunoreactive for VP are devoid of VP mRNA and vice versa (Trembleau et al. 1996). Similar data for the oxytocinergic system are not available. The relatively lower amounts of this RNA species are probably not sufficient to allow for detection in situ. Nonetheless, the data suggest that axonal mRNAs are not passively directed towards the axonal compartment as a consequence of the secretory activity of magnocellular neurons. In searching for a function, the question of whether the transcripts may be locally translated has been addressed. This is clearly not the case since an association with ribosomes or polysomes could not be verified (Mohr et al. 1995).

Storage of a subset of mRNAs outside the cell body is also conceivable. In principle, transcripts could be directed towards the axon prior to or subsequent to translation. Evidence for the latter stems from investigating the subcellular distribution of VP-encoding mRNA in homozygous and heterozygous Brattleboro (BB) rats. Deletion of a single nucleotide in the VP gene leads to a frame-shift mutation and ultimately to a very inefficient translation of the mutant VP precursor protein. No VP transcripts are detectable in the axonal compartment of affected homozygous animals. In heterozygous animals in which both the wild-type and the mutant alleles are transcribed, wild-type mRNA is transported to axons whereas mutant transcripts, as in homozygous BB rats, remain confined to the cell bodies (Mohr et al. 1995). Presumably, translational termination of the mutant transcripts is impaired due to the lack of an in-frame stop-codon. This in turn might lead to local degradation at the rough endoplasmic reticulum (RER) and thus to a failure of axonal targeting.

Despite extensive biochemical characterization, the physiological role of mRNAs localized to the axonal domain remains unknown. It is noteworthy, however, that this phenomenon is not only observed in neurosecretory cells such as hypothalamic magnocellular neurons. Neurons of the olfactory bulb harbor various mRNA species in distal axonal segments - among them transcripts encoding the olfactory marker protein and several odorant receptors while equally abundant RNAs are restricted to the perikarya (Ressler et al. 1994; Vassar et al. 1994; Wensley et al. 1995). The question of whether or not axonal mRNAs in nerve cells of the olfactory bulb are translated has not been addressed. In view of the assumption of the absence of translation in mammalian axons of the central (and peripheral) nervous system (Lasek and Brady 1981) this appears to be unlikely. An exception to this rule is the initial axonal segment which is equipped with components of the translational machinery (reviewed in Steward et al. 1995). Thus, delivery of MAP tau mRNA to the initial segment in cultured rat cortical neurons (Litman et al. 1993) may indeed allow local on-site synthesis of the corresponding protein. Together, the available data suggest that axonal mRNA compartmentalization is not an unspecific process. Yet, the physiological meaning of the majority of mRNAs in this cellular compartment is not related to local protein synthesis and has still to be defined.

Dendritic mRNAs

More precise intracellular RNA localization was possible when non-radioactive in situ hybridization techniques became applicable. Various mRNAs, among them VP- and OT-encoding transcripts, have been detected in neurites of hypothalamic magnocellular neurons with the morphological appearance of dendrites (Bloch et al. 1990). RNA filter hybridization of biochemically purified dendrites has confirmed these earlier results and enabled characterization of dendritic mRNAs in greater detail. It turned out that compared to mRNAs residing in the axonal compartment, dendritic transcripts exhibited different characteristics. The length of VP and OT mRNA poly(A) tails was always of the same size as that of transcripts extracted from the cell bodies. Notably, VP mRNA is located in dendrites in homozygous BB rats further underscoring that mRNA targeting to axons and dendrites, respectively, are likely to be brought about by distinct molecular mechanisms (Mohr et al. 1995). The protein synthesizing capacity of dendrites has been noted for some time (Tiedge and Brosius 1996). Importantly in this respect, VP and OT are dendritically released into the central nervous system where they presumably participate in processes related to neuronal plasticity as a consequence of external stimuli (reviewed in Morris et al. 1993). Immuno-histochemical studies at the ultrastructural level have confirmed synthesis of the VP and OT precursors in small cisternae of RER in dendrites (Morris et al., Chap. 6, this Vol.). This is in line with data obtained in our laboratory, namely detection of VP mRNA by electron microscopic in situ hybridization in dendritic segments containing RER (Prakash et al. 1997). A currently unresolved problem in terms of local dendritic synthesis of secretory and membrane-associated proteins is the apparent lack of Golgi-like structures. Even though Golgi marker proteins are detectable immunocytochemically, these molecules are located only in some dendrites of a given neuron and most often in parts proximal to the cell body (Krijnse-Locker et al. 1995).

Molecular Determinants of Subcellular mRNA Targeting

Subcellular mRNA Transport is Mediated by Cis- and Trans-Acting Signals

The non-random distribution of particular mRNA species has been described for many non-neuronal cell types such as fibroblasts, myoblasts, oligodendrocytes, astrocytes as well as for developing systems, e.g. oocytes and early embryos of

Drosophila and *Xenopus*. Extensive characterization of mRNA targeting in non-neuronal systems, in particular in *Drosophila*, has shed some light on the molecular mechanisms involved. Since these are believed to act in a very similar way in nerve cells, they deserve to be mentioned briefly.

In the *Drosophila* oocyte, various maternally synthesized mRNAs, among them *bicoid* and *oskar* transcripts, are specifically targeted to the anterior and posterior pole, respectively, in a strictly controlled spatial and temporal manner. Impaired sorting of these mRNAs severely interferes with correct body pattern formation. Genetic and molecular biological investigations have identified sequences residing in the 3'-untranslated region (3'-UTR), designated as *cis*-acting elements, of targeted mRNAs as well as several proteins (*trans*-acting factors) that are indispensable for subcellular RNA sorting to take place. In some instances these proteins specifically interact with the *cis*-acting elements (reviewed in St. Johnston 1995). An example is the RNA-binding protein Staufen which is involved in distinct steps of *bicoid* mRNA transport and which binds to a well characterized segment within the *bicoid* 3'-UTR containing the localizer element (Ferrandon et al. 1994).

The integrity of the cytoskeletal network is likewise required for mRNA sorting both in non-neuronal cells and in neurons (reviewed in Bassell and Singer 1997). Hence, RNA transport is thought to be mediated by binding of a combination of *trans*-acting factors to the *cis*-acting elements of mRNAs to be transported in order to guide these molecules either directly or via interaction with additional motor proteins along cytoskeletal elements to their ultimate destination within the cell (reviewed in St. Johnston 1995; Bassell and Singer 1997). RNA transport includes a component of translational control. The *bicoid* mRNA is not

Fig. 1. A Schematic representation of eukaryotic expression vector constructs designed for nuclear injections of primary cultured superior cervical ganglion (SCG) neurons. Expression of inserted cDNAs is driven by the cytomegalovirus (CMV) promoter. The 55 bp double-stranded oligonucleotide (β-gal sequence) corresponds to part of the coding region of the bacterial β-galactosidase gene. This sequence forms part of the 5'-UTR of any mRNA resulting from transcription of cDNAs inserted downstream. Detection of vector-expressed RNAs in situ may be accomplished by using a β-galactosidase-specific anti-sense oligonucleotide. The bovine growth hormone (BGH) polyadenylation signal ensures processing and polyadenylation of vector-expressed RNAs. **B-E** In situ hybridization of SCG neurons with digoxigenin-labeled VP- and OT-specific riboprobes following nuclear injections of constructs expressing VP sense RNA (**B**), VP anti-sense RNA (**C**), OT sense RNA (**D**), and OT anti-sense RNA (**E**). Hybridization was detected with an alkaline phosphatase-conjugated sheep anti-digoxigenin antibody and addition of appropriate enzyme substrates that give rise to a colour reaction (for experimental details see Prakash et al. 1997). The cell bodies are intensely labeled. Note that VP and OT sense but not anti-sense transcripts are targeted to dendrites. The RNAs are not evenly distributed within dendrites but appear to form granules at distinct sites. The same results were obtained when the cells were hybridized with the β-galactosidase-specific oligonucleotide (data not shown). *OT* Oxytocin; *VP* vasopressin. *Bar:* 30 μm

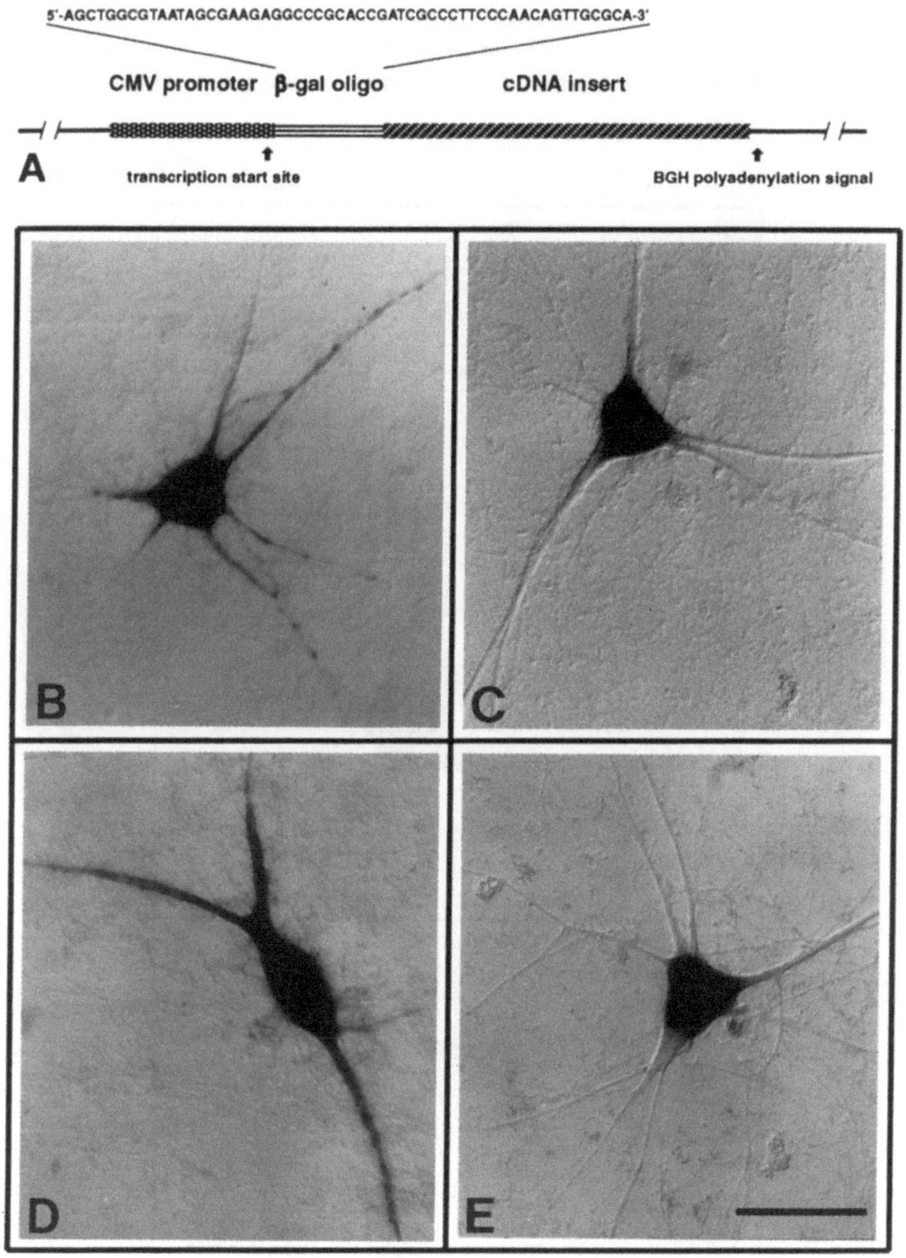
5'-AGCTGGCGTAATAGCGAAGAGGCCCGCACCGATCGCCCTTCCCAACAGTTGCGCA-3'
CMV promoter
β-gal oligo
cDNA insert
A
transcription start site
BGH polyadenylation signal
B
C
D
E

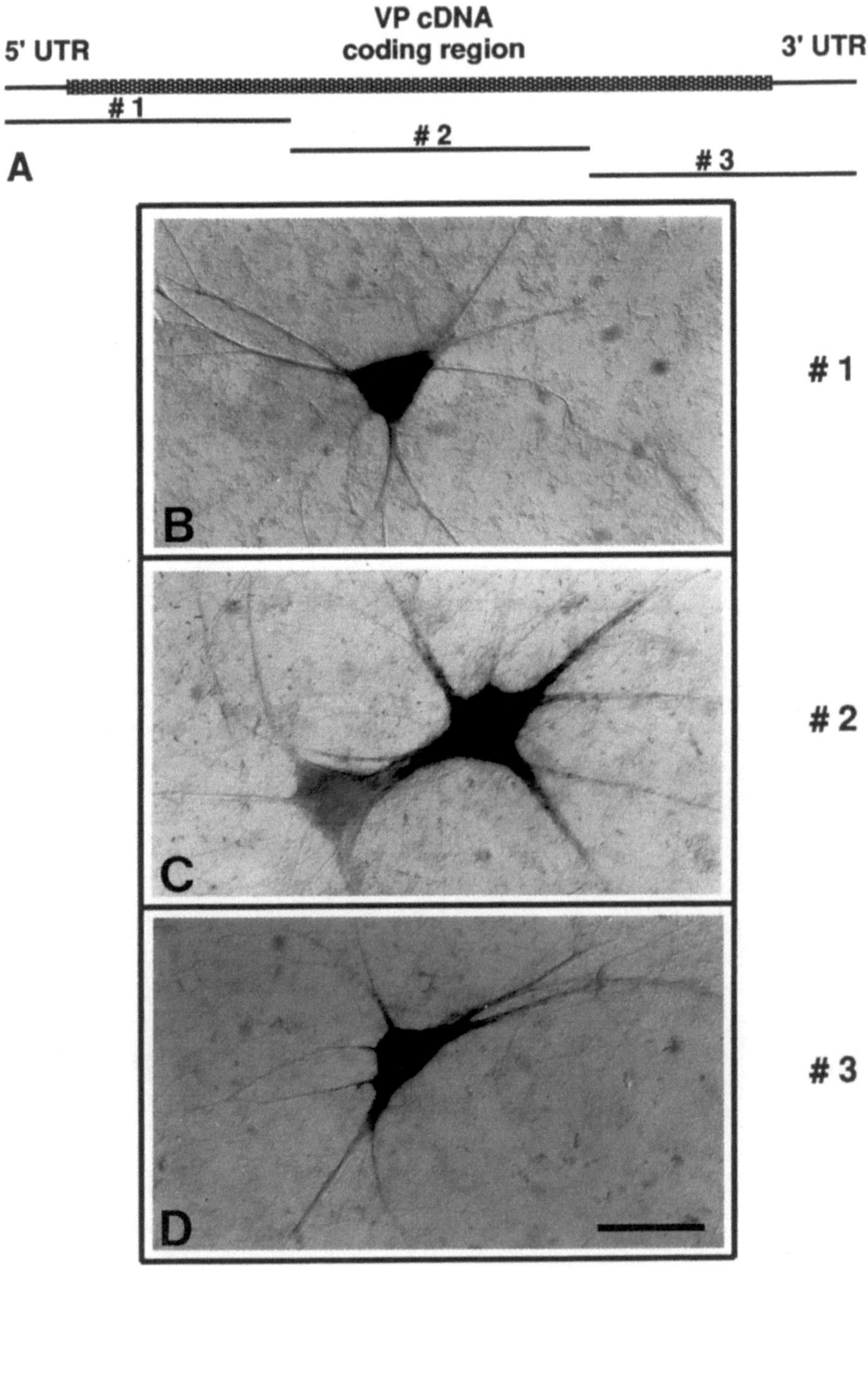
VP cDNA
coding region
5' UTR
3' UTR
1
2
3
A
B
C
D
1
2
3

Fig. 2. A Schematic representation of various parts of the rat VP cDNA that have been individually ligated to the 3'-end of a eukaryotic expression vector (for details see Fig. 1A), containing the complete rat α-tubulin cDNA. The VP sequence is expressed as part of the 3'-UTR of the recombinant RNA. **B-D** In situ hybridization of SCG neurons with a digoxigenin-labeled VP-specific riboprobe following nuclear injection of expression vectors containing VP cDNA fragment #1 (**B**), VP cDNA fragment #2 (**C**), and VP cDNA fragment #3 (**D**). Hybridization detection was done as described in the legend to Fig. 1. The VP RNA segments differ in their capacity to direct the vector-expressed RNAs to dendrites. Dendritic compartmentalization is seen with segments #2 and #3, but not with segment #1. *SCG* Superior cervical ganglion; *UTR* untranslated region; *VP* vasopressin. *Bar:* 30 μm

translated until it is delivered appropriately. Local synthesis of Bicoid protein, which is a transcriptional activator, at the anterior pole thus generates a protein gradient with diminishing concentrations towards the posterior pole. Only then is balanced gene expression along the axis of the embryo attained in a way which allows correct body pattern formation (reviewed in Whittaker and Lipshitz 1995). It might be interesting to note that Bicoid has a dual role in that it is not only a transcription factor but, in addition, an RNA-binding protein involved in translational repression. By specific interaction of Bicoid with sequences in the 3'-UTR of *caudal* mRNA which is evenly distributed in the *Drosophila* oocyte, its translation is inhibited in the anterior part and a posterior-to-anterior gradient of the Caudal morphogen is generated (Rivera-Pomar et al. 1996; Dubnau and Struhl 1996). Thus, the asymmetric distribution of proteins within polarized cells is likely to be brought about by three distinct mechanisms: (1) protein targeting, (2) mRNA targeting and local translation, and (3) regional translational inhibition of mRNAs.

With a few exceptions, the molecular determinants of mRNA sorting in neurons are unknown. Data obtained by expressing the genes encoding MAP2 and the α-subunit of the Ca^{2+}/calmodulin dependent protein kinase II (CaMKIIα) in transgenic mice indicate the importance of the 3'-UTR for dendritic mRNA targeting. In the absence of these sequences the corresponding mRNAs remain located in the perikarya (Marsden et al. 1996; Mayford et al. 1996). Likewise, in cultured neurons the 3'-UTR is implicated in sorting of transcripts encoding the rat somatostatin receptor subtype 1 (SSTR1; Busch et al. 1997) and MAP tau (Behar et al. 1995) to dendrites and the initial axonal segment, respectively. Yet, the *trans*-acting factors required for mRNA sorting to neurites have still to be identified.

Identification of Cis-Acting Elements in VP mRNA

In order to define the *cis*-acting sequences that may mediate VP and OT mRNA targeting to the neuronal processes, eukaryotic expression vectors (schematically depicted in Fig. 1A) have been injected into the nuclei of primary cultured neurons derived from embryonic rat superior cervical ganglia (SCG; Prakash et al. 1997). After injection of constructs containing the cDNAs in sense orientation, VP and OT transcripts were detectable throughout the cell somata as well as in dendrites (Fig. 1B, D). The mRNA residing in the dendritic compartment of SCG neurons appeared to be organized in a granular form. A similar observation has been reported in cultured oligodendrocytes microinjected with in vitro synthesized RNA encoding myelin basic protein (MBP) which is subsequently transported to the cell processes in the periphery (Ainger et al. 1993). Interestingly, these granules are composed of components of the translational machinery such as ribosomes, elongation factors and aminoacyl-tRNA-synthetases in addition to MBP mRNA (Barbarese et al. 1995). Hence, subcellular mRNA targeting may be mediated by the formation of distinct transport units.

In contrast, nuclear injection of constructs expressing the RNAs in anti-sense orientation failed to give rise to dendritic targeting, in spite of an abundant expression in the cell somata (Fig. 1C, E). These data support the current view that the nucleotide sequence or a secondary structure formed by a particular sequence of the mRNA contains elements necessary for subcellular mRNA targeting. The dendritic targeting capacity of various segments spanning the VP RNA has been analyzed by ligating the corresponding cDNA sequences individually to the 3'-end of α-tubulin cDNA such that they form part of the 3'-UTR. α-Tubulin transcripts have been shown to be expressed in the perikarya only, regardless of whether they were derived by transcription of the endogenous genome or by that of a microinjected expression vector construct (Prakash et al. 1997). As demonstrated in Fig. 2, two segments of VP mRNA spanning approximately 200 nucleotides each were able to direct the recombinant mRNAs to dendrites while one segment was ineffective in this respect. Subsequent mapping of those signals at a finer scale, as summarized in Table 2, revealed that several parts, for instance the 3'-UTR alone, possess dendritic targeting information. Thus, the dendritic localizer elements within VP mRNA are redundant and they are also located in the coding region. Moreover, the data underscore that primary cultured SCG neurons are equipped with the machinery necessary for dendritic compartmentalization of mRNAs, even if these, like VP (and OT) transcripts, are not endogenously expressed. Consequently, molecules acting on mRNA localization do not appear to be cell-specific. These data are consistent with observations made in non-neuronal systems. For instance, mRNAs encoding various actin isoforms are targeted to different subcellular locations in several cell types. Transfection studies have demonstrated correct sorting of the α-, β- and γ-actin transcripts regardless of the cell type employed for transfection and whether or not a particular isoform was endogenously expressed in that cell type (Kislauskis et al. 1995).

As outlined above, in hypothalamic magnocellular neurons, VP and OT mRNAs are localized in dendrites as well as in axons. However, in SCG neurons, axonal compartmentalization of VP and OT transcripts has not been observed (Prakash et al. 1997). Only in rare cases did cells contain VP transcripts in the very proximal axonal segment. It may be argued that, if the mRNA levels in this compartment are low, they might escape detection more easily, since the axons of SCG neurons are very long and form a dense network. On the other hand, the molecules necessary for sorting of distinct mRNA species to the axonal domain may be present in only a restricted number of cell types and may be lacking in SCG neurons.

Table 2. Dendritic targeting capacity of various rat VP mRNA segments. Individual segments of the rat VP cDNA were ligated to the 3'-end of α-tubulin cDNA and cloned into a eukaryotic expression vector (for details see Fig. 2). The resulting constructs have been injected into the cell nuclei of cultured neurons derived from embryonic rat superior cervical ganglia. The intracellular distribution of the vector-expressed RNAs has subsequently been assessed by non-radioactive in situ hybridization analyses as described (Prakash et al. 1997).

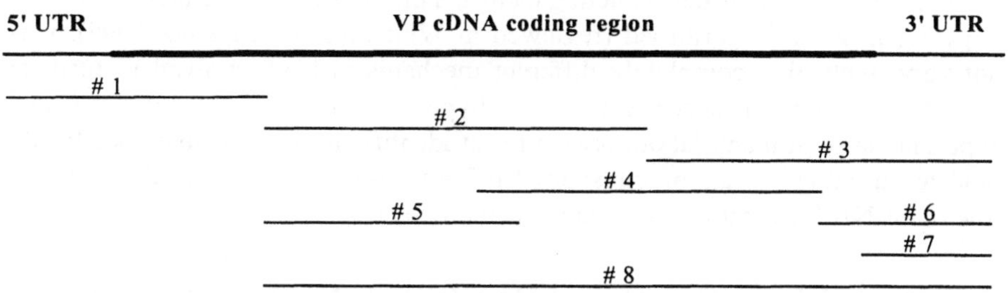

Fragment	Nucleotide residues	Amount of RNA in dendrites	Dendritic localization
# 1	1-200	-	---
# 2	201-410	++	proximal
# 3	411-595	++	proximal
# 4	312-511	(+)	very proximal
# 5	201-320	(+)	very proximal
# 6	512-595	+	proximal
# 7	539-595	+	proximal
# 8	201-595	+++	distal

UTR Untranslated region; *VP* vasopressin

Are There Alternative Mechanisms of Intracellular mRNA Distribution?

Is it also possible that mechanisms exist which prevent mRNAs from being localized in neurites, rather than specifically delivering them to destinations outside the cell body? Certainly, this view cannot be fully discounted, and there are in fact arguments in favor of it. For instance, application of protein synthesis inhibitors such as cycloheximide or puromycin to cultured neurons did not alter targeting of those mRNAs that are transported to dendrites in non-treated cultures

(Kleiman et al. 1993). However, transcripts encoding GAP 43, tubulin and actin that are confined to the somata in untreated cells were detectable in dendrites (Kleiman et al. 1993). Data obtained in our laboratory suggest that, if mechanisms should exist that prevent RNAs from being sorted outside the perikaryon, they are apparently overridden by dendritic targeting signals. As has been mentioned above, α-tubulin transcripts are not transported to dendrites of cultured SCG neurons whereas coupling of a targeting element of VP mRNA to the α-tubulin transcript clearly leads to dendritic sorting of the recombinant RNA (Prakash et al. 1997). mRNA targeting in non-neuronal cells has been investigated in more detail. The view emerging from these systems is most consistent with the assumption that a variety of protein factors bind to distinct structural motifs within targeted transcripts to direct them in a correct spatial and temporal manner to their final subcellular destinations via interaction with the cytoskeletal network. In the absence of either the *cis*-acting elements or the *trans*-acting factors, mRNA targeting is severely perturbed (reviewed in Whittaker and Lipshitz 1995). It is not very likely that completely different mechanisms have evolved to mediate mRNA transport in nerve cells. The hypothesis that is currently persued experimentally in many laboratories aims at identifying proteins that specifically bind to targeting signals of transported mRNAs in order to define their precise role in mRNA localization to the nerve cell processes.

Functional Significance of Dendritic mRNA Targeting

mRNA targeting to dendrites has been observed in numerous neuronal cell types. Several lines of evidence have suggested local translation of dendritic transcripts. For instance, proteins are synthesized in dendrites that have been physically detached from their cell bodies (Torre and Steward 1992) as well as in biochemically purified synapto-dendrosome preparations (Rao and Steward 1991). However, it is now becoming increasingly clear that the dendritic localization of a protein encoded by an mRNA which is subjected to dendritic compartmentalization is not exclusively dependent on prior mRNA targeting. That is, even if the targeting elements within mRNAs such as MAP2 (Marsden et al. 1996), CaMKIIα (Mayford et al. 1996) and SSTR1 (Busch et al. 1997) have been removed, which prevent localization of these transcripts in dendrites (see above), the corresponding proteins are still detectable in dendrites. Why should nerve cells have developed elaborate and energy-dependent transport mechanisms for RNA molecules (and components of the translational machinery) if the information for protein sorting to destinations outside the cell body is obviously intrinsic to these molecules? A final answer to this question is currently lacking. Apparently, dendritic transcripts fulfill more subtle functions in cell signaling. Protein synthesis in dendrites is, for instance, stimulated in guinea pig hippocampal brain slices by application of the cholinergic agonist, carbachol, combined with patterned Schaffer collateral stimulation (Feig and Lipton 1993).

Likewise, extracellular application of brain-derived neurotrophic factor or neurotrophin-3 to rat hippocampal slices enhanced synaptic transmission in CA1 pyramidal neurons which was markedly reduced in the presence of protein synthesis inhibitors. This sensitivity persisted even in slices isolated from pre- and postsynaptic cell bodies.

The data suggest the neurotrophin-induced synthesis of proteins in the neuropil layer containing the dendrites of pyramidal neurons (Kang and Schuman 1996). Recently, activation-dependent dendritic translation of a defined mRNA has been demonstrated. Within minutes after metabotropic glutamate receptor 1 stimulation of synapto-dendrosome preparations, elevated levels of a rat homologue of fragile X mental retardation protein are synthesized (Weiler et al. 1997). It may be speculated that dendritic mRNAs are translationally dormant or translated at very low levels until synaptic activation requires synthesis of particular proteins at defined sites. This may allow spatial restriction of distinct phases of synaptic plasticity to the site of the initial synaptic stimulation.

Acknowledgments

We thank Dr. Hartwig Schmale for critical comments on the manuscript and the Deutsche Forschungsgemeinschaft for financial support.

References

Ainger K, Avossa D, Morgan F, Hill SJ, Barry C, Barbarese E, Carson JH (1993) Transport and localization of exogenous myelin basic protein mRNA microinjected into oligodendrocytes. J Cell Biol 123: 431-441

Barbarese E, Koppel DE, Deutscher MP, Smith CL, Ainger K, Morgan F, Carson JH (1995) Protein translation components are colocalized in granules in oligodendrocytes. J Cell Sci 108: 2781-2790

Bassell G, Singer RH (1997) mRNA and cytoskeletal filaments. Curr Opin Cell Biol 9: 109-115

Behar L, Marx R, Sadat E, Barg J, Ginzburg I (1995) *Cis*-acting signals and *trans*-acting proteins are involved in tau mRNA targeting into neurites of differentiating neuronal cells. Int J Dev Neurosci 13: 113-127

Bian F, Chu T, Schilling K, Oberdick J (1996) Differential mRNA transport and the regulation of protein synthesis: selective sensitivity of Purkinje cell dendritic mRNAs to translational inhibition. Mol Cell Neurosci 7: 116-133

Bloch B, Guitteny AF, Normand E, Chouham S (1990) Presence of neuropeptide messenger RNAs in neuronal processes. Neurosci Lett 109: 259-264

Burgin KE, Waxham MN, Rickling S, Westgate SA, Mobley WC, Kelly PT (1990) In situ hybridization histochemistry of Ca^{2+}/calmodulin-dependent protein kinase in developing rat brain. J Neurosci 10: 1788-1798

Busch A, Kreienkamp H-J, Mohr E, Richter D (1997) Dendritic targeting of somatostatin receptor subtype 1 (SSTR1) mRNA and protein in rat cortical and cultured neurons. Eur J Neurosci (in press)

Dirks RW, Raap AK, van Minnen J, Vreugdenhil E, Smit AB, van der Ploeg M (1989) Detection of mRNA molecules coding for neuropeptide hormones of the pond snail *Lymnaea stagnalis* by radioactive and non-radioactive in situ hybridization: a model study for mRNA detection. J Histochem Cytochem 37: 7-14

Dubnau J, Struhl G (1996) RNA recognition and translational regulation by a homeodomain protein. Nature 379: 694-699

Dugich-Djordjevic MM, Tocco G, Willoughby DA, Najm I, Pasinetti G, Thompson RF, Baudry M, Lapchak PA, Hefti F (1992) BDNF mRNA expression in the developing rat brain following kainic acid-induced seizure activity. Neuron 8: 1127-1138

Feig S, Lipton P (1993) Pairing the cholinergic agonist carbachol with patterned Schaffer collateral stimulation initiates protein synthesis in hippocampal CA1 pyramidal cell dendrites via a muscarinic, NMDA-dependent mechanism. J Neurosci 13: 1010-1021

Ferrandon D, Elphick L, Nüsslein-Volhard C, St Johnston D (1994) Staufen protein associates with the 3' UTR of bicoid mRNA to form particles that move in a microtubule-dependent manner. Cell 79: 1221-1232

Furuichi T, Simon-Chazottes D, Fujino I, Yamada N, Hasegawa M, Miyawaki A, Yoshikawa S, Guenet J-L, Mikoshiba K (1993) Widespread expression of inositol 1,4,5-trisphosphate receptor type 1 gene (Insp3rl) in the mouse central nervous system. Receptors and Channels 1: 11-24

Garner CC, Tucker PR, Matus A (1988) Selective localization of messenger RNA for cytoskeletal protein MAP2 in dendrites. Nature 336: 674-677

Giuditta A, Menichini E, Perrone-Capano C, Langella M, Martin R, Castigli E, Kaplan BB (1991) Active polysomes in the axoplasm of the squid giant axon. J Neurosci Res 28: 18-28

Kang H, Schuman EM (1996) A requirement for local protein synthesis in neurotrophin-induced hippocampal synaptic plasticity. Science 273: 1402-1406

Kislauskis EH, Ross A, Latham VM, Zhu X, Bassell G, Taneja KL, Singer RH (1995) The mechanisms of RNA localization: its effect on cell polarity. In: Lipshitz HD (ed) Localized RNAs. Springer, Berlin, Heidelberg, New York, pp 185-195

Kleiman R, Banker G, Steward O (1993) Inhibition of protein synthesis alters the subcellular distribution of mRNA in neurons but does not prevent dendritic transport of RNA. Proc Natl Acad Sci USA 90: 11192-11196

Krijnse-Locker J, Parton RG, Fuller SD, Griffiths G, Dotti CG (1995) The organization of the endoplasmic reticulum and the intermediate compartment in cultured rat hippocampal neurons. Mol Biol Cell 6: 1315-1332

Landry CF, Watson JB, Kashima T, Campagnoni AT (1994) Cellular influences on RNA sorting in neurons and glia: an in situ hybridization histochemical study. Mol Brain Res 27: 1-11

Lasek RJ, Brady ST (1981) The axon: a prototype for studying expressional cytoplasm. Cold Spring Harb Symp Quant Biol 46: 113-124

Lehmann E (1988) Untersuchungen zur Expression des Vasopressingens in der Ratte. Doctoral Thesis, University of Heidelberg, Germany

Link W, Konietzko U, Kauselmann G, Krug M, Schwanke B, Frey U, Kuhl D (1995) Somatodendritic expression of an immediate early gene is regulated by synaptic activity. Proc Natl Acad Sci USA 92: 5734-5738

Litman P, Barg J, Rindzoonski L, Ginzburg I (1993) Subcellular localization of tau mRNA in differentiating neuronal cell culture: implications for neuronal polarity. Neuron 10: 627-638

Lyford GL, Yamagata K, Kaufmann WE, Barnes CA, Sanders LK, Copeland NG, Gilbert DJ, Jenkins NA, Lanahan AA, Worly PF (1995) Arc, a growth factor and activity-regulated gene, encodes a novel cytoskeleton-associated protein that is enriched in neuronal dendrites. Neuron 14: 433-445

Marsden KM, Doll T, Ferralli J, Botteri F, Matus A (1996) Transgenic expression of embryonic MAP2 in adult mouse brain: implications for neuronal polarization. J Neurosci 16: 3265-3273

Mayford M, Baranes D, Podsypanina K, Kandel E (1996) The 3'-untranslated region of CaMKIIα is a *cis*-acting signal for the localization and translation of mRNA in dendrites. Proc Natl Acad Sci USA 93: 13250-13255

Miyashiro K, Dichter M, Eberwine J (1994) On the nature and differential distribution of mRNAs in hippocampal neurites: implications for neuronal functioning. Proc Natl Acad Sci USA 91: 10800-10804

Mohr E, Richter D (1992) Diversity of mRNAs in the axonal compartment of peptidergic neurons in the rat. Eur J Neurosci 4: 870-876

Mohr E, Richter D (1995) mRNAs in extrasomal domains of rat hypothalamic peptidergic neurons. In: Lipshitz HD (ed) Localized RNAs. Springer, Berlin, Heidelberg, New York, pp 275-287

Mohr E, Fehr S, Richter D (1991) Axonal transport of neuropeptide encoding mRNAs within the hypothalamo-neurohypophyseal tract of rats. EMBO J 10: 2419-2424

Mohr E, Morris JF, Richter D (1995) Differential subcellular mRNA targeting: deletion of a single nucleotide prevents the transport to axons but not to dendrites of rat hypothalamic magnocellular neurons. Proc Natl Acad Sci USA 92: 4377-4381

Morris JF, Pow DV, Sokol HW, Ward A (1993) Dendritic release of peptides from magnocellular neurons in normal rats, Brattleboro rats and mice with hereditary nephrogenic diabetes insipidus. In: Gross P, Richter D, Robertson GL (eds) Vasopressin. John Libbey Eurotext, Paris, pp 171-182

Prakash N, Fehr S, Mohr E, Richter D (1997) Dendritic localization of rat vasopressin mRNA: ultrastructural analysis and mapping of targeting elements. Eur J Neurosci 9: 523-532

Racca C, Gardiol A, Triller A (1997) Dendritic and postsynaptic localizations of glycine receptor α subunit mRNAs. J Neurosci 17: 1691-1700

Rao A, Steward O (1991) Evidence that protein constituents of postsynaptic membrane specializations are locally synthesized: analysis of proteins synthesized within synaptosomes. J Neurosci 11: 2881-2895

Ressler KJ, Sullivan SL, Buck LB (1994) Information coding in the olfactory system: evidence for a stereotyped and highly organized epitope map in the olfactory bulb. Cell 79: 1245-1255

Rivera-Pomar R, Niessing D, Schmidt-Ott U, Gehring WJ, Jäckle H (1996) RNA binding and translational repression by bicoid. Nature 379: 746-749

Steward O, Kleiman R, Banker G (1995) Subcellular localization of mRNA in neurons. In: Lipshitz HD (ed) Localized RNAs. Springer, Berlin, Heidelberg, New York, pp 235-255

St Johnston D (1995) The intracellular localization of messenger RNA. Cell 81: 161-170

Tiedge H, Brosius J (1996) Translational machinery in dendrites of hippocampal neurons in culture. J Neurosci 15: 7171-7181

Tiedge H, Fremeau RT, Weinstock PH, Arancio O, Brosius J (1991) Dendritic location of neural BC1 RNA. Proc Natl Acad Sci USA 88: 2093-2097

Tiedge H, Zhou A, Thorn NA, Brosius J (1993) Transport of BC1 RNA in hypothalamo-neurohypophyseal axons. J Neurosci 13: 4214-4219

Torre ER, Steward O (1992) Demonstration of local protein synthesis within dendrites using a new cell culture system that permits the isolation of living axons and dendrites from their cell bodies. J Neurosci 12: 762-772

Trembleau A, Morales M, Bloom FE (1994) Aggregation of vasopressin mRNA in a subset of axonal swellings in the median eminence and posterior pituitary: light and electron microscopic evidence. J Neurosci 14: 39-53

Trembleau A, Morales M, Bloom FE (1996) Differential compartmentalization of vasopressin messenger RNA and neuropeptide within the rat hypothalamo-neurohypophysial axonal tracts: light and electron microscopic evidence. Neuroscience 70: 113-125

Van Minnen J (1994) RNA in the axonal domain: a new dimension in neuronal functioning? Histochem J 26: 377-391

Vassar R, Chao SK, Sitcheran R, Nuñez JM, Vosshall LB, Axel R (1994) Topographic organization of sensory projections to the olfactory bulb. Cell 79: 981-991

Weiler IJ, Irwin SA, Klintsova AY, Spencer CM, Brazelton AD, Miyashiro K, Comery TA, Patel B, Eberwine J, Greenough WT (1997) Fragile X mental retardation protein is translated near synapses in response to neurotransmitter activation. Proc Natl Acad Sci USA 94: 5395-5400

Wensley CH, Stone DM, Baker H, Kauer JS, Margolis FL, Chikaraishi DM (1995) Olfactory marker protein mRNA is found in axons of olfactory receptor neurons. J Neurosci 15: 4827-4837

Whittaker KL, Lipshitz HD (1995) Mechanisms and functions of RNA localization to the anterior pole of the *Drosophila* oocyte and early embryo. In: Lipshitz HD (ed) Localized RNAs. Springer, Berlin, Heidelberg, New York, pp 41-61

6 Peptide-Secreting Dendrites: New Controls for Neuroendocrine Neurons

J.F. Morris, D. Ma, D.V. Pow, H. Wang, and A. Ward

Department of Human Anatomy, University of Oxford,
South Parks Road, Oxford OX1 3QX, UK;
Phone: +44 1865 272164, Fax: +44 1865 272164,
E-mail: john.morris@anat.ox.ac.uk

Abstract

Peptidergic neurons are well known to synthesize and package peptides in their perikarya, and transport the peptide-containing dense-cored vesicles along their axons for release at the axon terminals. Immunocytochemistry and electron microscopy reveal that the dendrites of peptidergic neurons also contain substantial amounts of peptide-containing dense-cored vesicles in addition to receiving the synaptic boutons afferent to the neurons. The release of peptides from dendrites and the ability of dendrites to synthesize proteins has been investigated, using the magnocellular neurosecretory system of the hypothalamus, which has already proved itself a robust model for investigation of peptidergic neurons. Wild-type and homozygous Brattleboro rats which have hyperactive magnocellular neurons were studied. The dendrites contain all the cellular machinery needed to synthesize proteins, but there is little evidence of classical Golgi-cisterns to package them. It seems unlikely that the vasopressin or oxytocin packaged in dense-cored vesicles are synthesized in the dendrites, but mRNAs encoding receptor and microtubule proteins are present in the dendrites and could be translated there to provide finer control of afferent inputs. Neurotransmitters which stimulate magnocellular neurons elicit the exocytosis of vasopressin and oxytocin from both the dendrites and cell bodies, though that from dendrites is quantitatively the greater. Oestradiol and oxytocin are also capable of stimulating dendritic peptide release. There is increasing evidence that the release of dendritic peptide has important local effects on the magnocellular neurons themselves and on surrounding glia. Dendritic peptide thus provides local autocrine and paracrine controls on the activity of neurosecretory neurons.

Introduction

The seminal studies of Ernst and Berta Scharrer (e.g. Scharrer and Scharrer 1940, 1945) initially revealed neurosecretory material in the perikarya of hypothalamic neurons, but the material was also found in coarse and fine processes of the

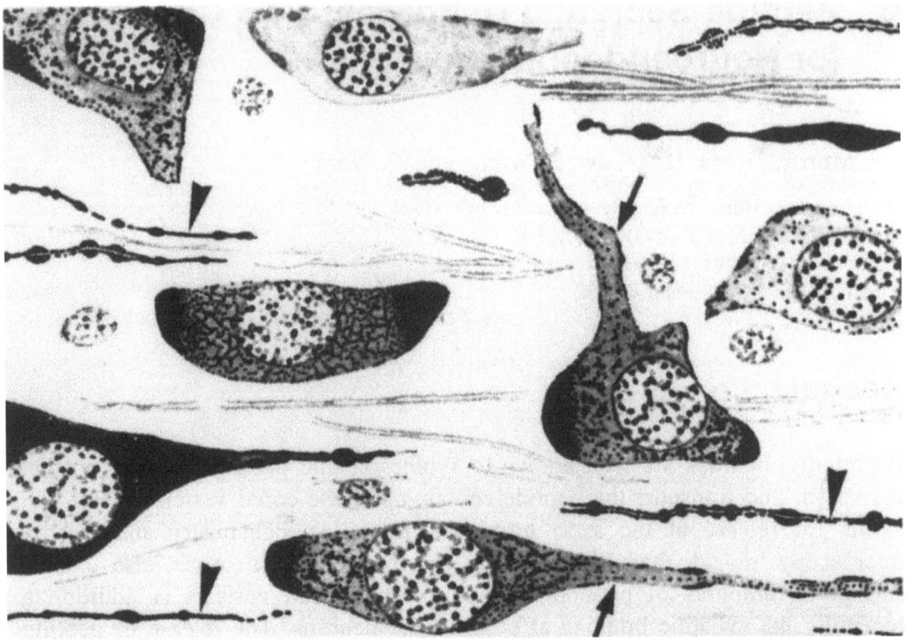

Fig. 1. Part of a drawing by Ernst Scharrer of neurosecretory neurons in the preoptic nucleus of *Scyllium stellare*. Note that whereas some of the processes containing neurosecretory material are thin and finely beaded (axons; *arrowheads*), there is also neurosecretory material in the thicker (dendritic) processes (*arrows*). x1110. (Adapted from Fig. 2, Scharrer 1951)

neurosecretory neurons (Fig. 1; Scharrer 1952). Understandably, the studies which followed and developed the concept of neurosecretion, focussed first on the transport of neurosecretory material through the axonal projections to the neurohaemal organs. However, some of the processes illustrated by their beautiful drawings may well show neurosecretory material in dendrites of the neurons (Fig. 1). This brief review is intended to show that, in addition to their well-known role as the receivers of afferent synapses, dendrites are active in protein synthesis and also release neurohormones which provide a new range of important local controls on the activity of the neuroendocrine neurons.

Dendrites of Magnocellular Neurosecretory Neurons Are Oriented in Specific Directions; They Contain and Can Release Neuropeptides

Most early immunocytochemical studies of the mammalian magnocellular neurosecretory system, which produces the hormones vasopressin (VP) and oxytocin (OT) and their associated neurophysin proteins, clearly stained the

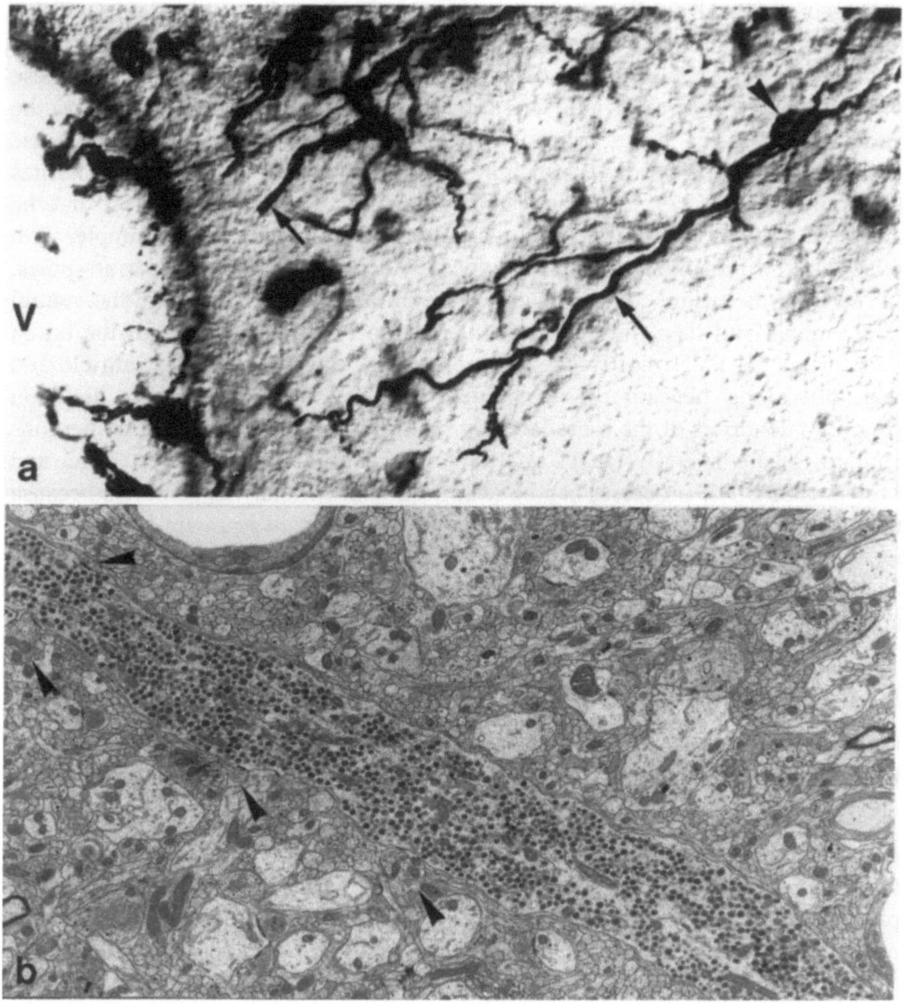

Fig. 2. Light microscopic (**a**) and low power electron microscopic (**b**) views of the dendrites of magnocellular neurons in the paraventricular nucleus of the rat hypothalamus. **a** The rather simple dendrites (*arrows*) are oriented ventromedially toward the third ventricle (*V*) and then run antero-posteriorly beneath the ventricle wall which is obliquely sectioned; they are filled with immunoreactive vasopressin neurophysin. x140. **b** The dendrite is filled with neurosecretory dense-cored vesicles and contacted by four synaptic boutons (*arrowheads*). x4200

neurosecretory cell bodies in the supraoptic (SON) and paraventricular (PVN) nuclei, and stained the neurohypophysis. However, for reasons that are not entirely clear, immunostained dendrites were not obvious in many of these preparations (Vandesande and Dierickx 1975). In many later preparations the axons could be seen as thin beaded processes and in some cases these axons arose from thicker, non-beaded branching processes which resembled dendrites (Sofroniew et al. 1979; Sofroniew and Glassmann 1981). It is now clear that the dendritic tree of magnocellular neurosecretory neurons is relatively simple, with relatively few branches and no unequivocal evidence for classical dendritic spines. In the SON the magnocellular dendrites are directed ventrally, toward the ventral glial lamina, and then run posteriorly or anteriorly along the floor of the brain; those from the PVN mostly pass ventromedially toward the third ventricle and then run along beneath the ventricular wall, often in an anterior direction (Fig. 2a); dendrites of the more laterally placed groups of magnocellular neurons tend to radiate laterally (Ju et al. 1986, 1992; Randle et al. 1986; Castel and Morris 1988). The processes can be identified with certainty as dendrites because they stain for the dendrite-restricted microtubule-associated protein MAP-2 (Pow et al. 1990) and are contacted by numerous synaptic boutons (Fig. 2b). Immunocytochemistry shows that the dendrites contain VP, OT and neurophysin apparently throughout their length (Fig. 2a).

Dendrites of Magnocellular Neurons Contain Neurosecretory Vesicles and Can Release Neurosecretory Peptides

Electron microscopy confirms the immunocytochemical appearances, showing that the magnocellular dendrites contain numerous hormone- and neurophysin-containing neurosecretory vesicles (NSV) (Fig. 3), though the proximal part of the dendrites often contains very few NSV (Morris et al. 1993), suggesting that they are actively transported away from the cell body into the dendrites. The presence of the NSV immediately raises the question of their function in the dendrites. At one time it was suggested that dendritic NSV represent either a reserve waiting to be transported into the axon or unwanted NSV destined for destruction. However, it is now clear that NSV in both the cell body and the dendrites can be released by exocytosis into the surrounding hypothalamus. In fact, the dendrites contain many more NSV than the cell bodies and would therefore contribute proportionally more to the amount of peptide secreted (Pow and Morris 1989). This accords well with demonstrations by in vitro incubations and microdialysis of local release of VP and OT (Moos et al. 1989; Landgraf 1992, 1997; Neumann et al. 1994) by a process which is insensitive to tetrodotoxin (DiScala Guenod et al. 1987).

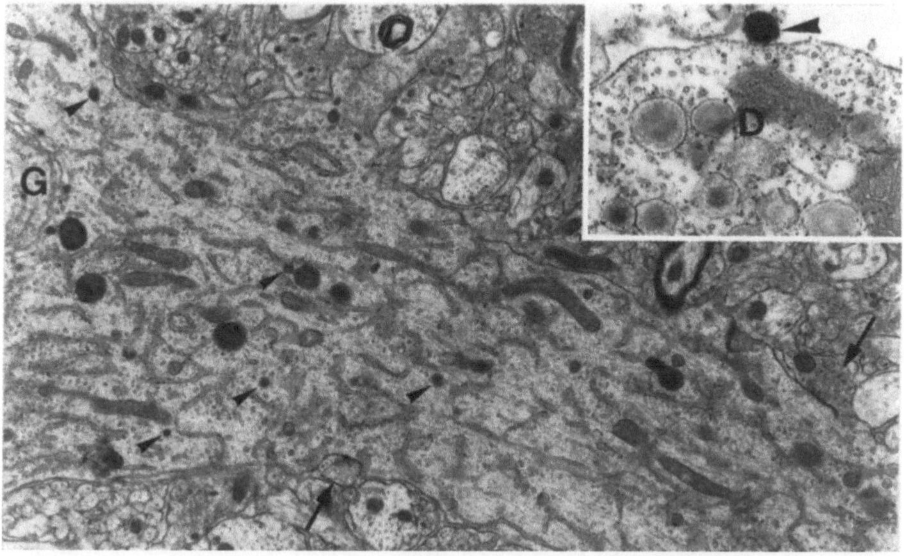

Fig. 3. Electron micrograph showing the origin of a dendrite from a magnocellular perikaryon (at *left*; *G* Golgi) in the paraventricular nucleus. Very few neurosecretory dense-cored vesicles (*arrowheads*) are found within this region. Rough endoplasmic cisterns continue into the dendrite which is contacted by two synaptic boutons (*arrows*). The insert shows exocytosis (*arrowhead*) of a dense-cored vesicle from a supraoptic magnocellular dendrite (*D*) in a hypothalamic slice stimulated with glutamate (10 mM) for 20 min in vitro. Magnification: x9200, insert: x30700

Dendrites of Magnocellular Neurons Can Synthesize Proteins Locally

In addition to their content of formed NSV, the dendrites of magnocellular neurons contain substantial amounts of the cellular machinery needed for the synthesis of peptides. Small cisterns of endoplasmic reticulum extend into the proximal part of the dendrites (Fig. 4); they become reduced in amount and size but are still present in the middle and more distal parts of the dendrites. Free polysomes, some associated with microtubules, can also be found in addition to the endoplasmic reticulum-associated ribosomes (Ma and Morris 1996). Immunohistochemistry and in situ hybridization reveals the presence of rough ER-associated enzymes such as protein disulphide isomerase (Ma and Morris, unpubl. obs.) and ribosomal and transfer RNAs (Ma and Morris 1997a). However, we could find no evidence for either classical Golgi cisterns or the packaging of NSV in magnocellular dendrites, and Golgi-specific proteins extend only into the proximal parts of dendrites of other neurons (Steward 1997). Furthermore, although we can detect OT precursor mRNA in the proximal part of the dendrites

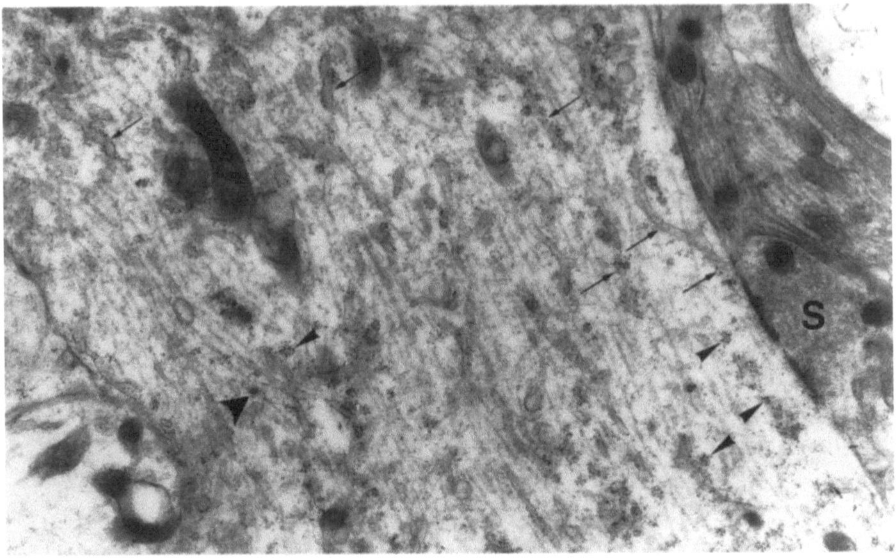

Fig. 4. High power electron micrograph to show protein-synthetic machinery in the hypertrophied dendrite of a magnocellular vasopressin neuron of a homozygous Brattleboro rat. No 160 nm dense-cored vesicles are present, only a 100 nm dense-cored vesicle (*large arrowhead*). Numerous strands of smooth and ribosome-studded endoplasmic reticulum (*arrows*) and many free polysomes (*arrowheads*) are present. A few of these are situated immediately beneath a synaptic bouton (*S*). x17000

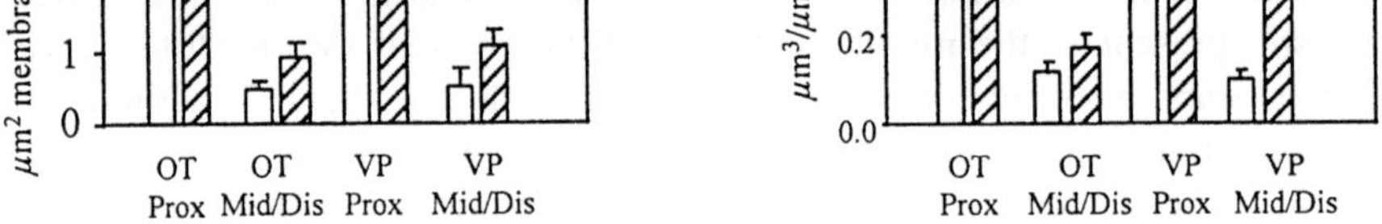

Fig. 5. Histograms showing data derived by morphometric analysis of (**a**) the amount of rough endoplasmic reticulum (*RER*) surface and (**b**) the amount of ribosomes in proximal (*Prox*), middle and distal (*Mid/Dis*) parts of the dendrites of oxytocin (*OT*) and vasopressin (*VP*) magnocellular neurons in the supraoptic nucleus of wild-type Long-Evans (*WT*) and homozygous Brattleboro (*BB*) rats. Data are mean + SEM; n = 5 animals

it cannot be detected very far along the dendrites. It therefore seems highly unlikely that the dendrites are normally involved in packaging hormone-containing NSV for exocytotic release. Indeed, when electron-dense material can be identified in cisterns of rough endoplasmic reticulum in the dendrites, it is frequently found in the form of aggregates of peptide much larger than typical NSV (Pow et al. 1991). However, studies on homozygous Brattleboro rats show that the increased activity of the neurosecretory neurons is associated with an enlargement, not only of their cell bodies (Morris 1987), but also of their dendrites. Associated with this enlargement is an increase in the amount of dendritic rough endoplasmic reticulum and ribosomes (Fig. 5).

Dendrites Contain a Variety of mRNAs

We have recently been able to demonstrate dendritic localisation of the mRNAs for $GABA_A$ and NMDA-R1 receptor proteins (Fig. 6). This, combined with a topographic association of a minor part (3%) of the polyribosomes with immediate post-synaptic cytoplasmic sites and of subsynaptic polyribosomes beneath 43% of sectioned synapse profiles, suggests that one function of dendritic peptide synthesis is the local generation of receptor proteins for insertion into the dendritic membrane (cf Steward 1997), though how this is achieved in the apparent absence of Golgi-type membranes is unclear. We can also demonstrate a dendritic location of the nNOS mRNA (Ma and Morris 1997b) which codes for

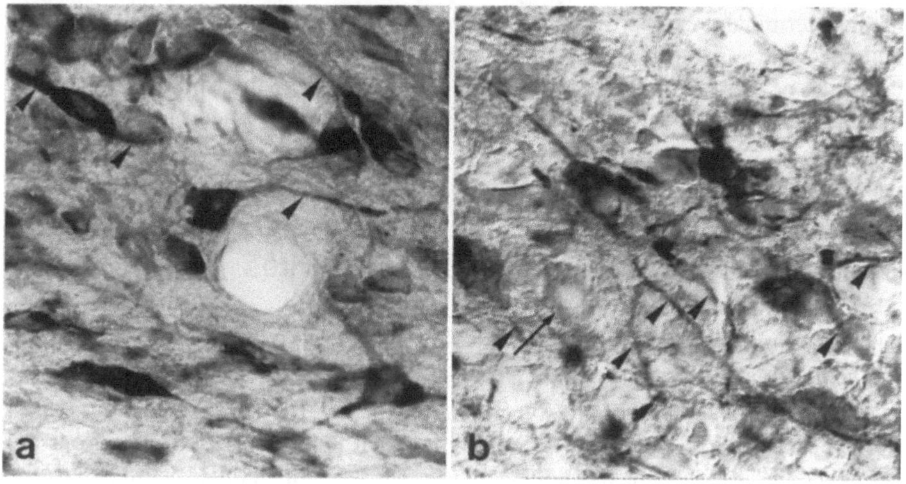

Fig. 6. Light micrographs of in situ hybridization preparations to show the presence in dendrites (*arrowheads*) of supraoptic magnocellular neurons of mRNAs for (**a**) $GABA_A$ and (**b**) NMDA R1 receptors. Note the variable intensity of the staining, especially for NMDA R1 in which some magnocellular neurons (*arrow*) appear unstained. x470

the neuronal nitric oxide synthase enzyme which appears to be distributed throughout the dendrites of magnocellular and some suprachiasmatic neurons (Wang and Morris 1996a, b); therefore, production of this activity modulator may also be locally controlled within the dendrites. The mRNA for the dendrite-specific MAP-2 can also be located in dendrites, which strongly suggests that local synthesis in the dendrites plays a role in the dendritic localization of MAP-2 which may be involved in the transport of NSV into the dendrites and thus, indirectly, contribute to the dendritic secretion of peptides.

Dendritic Exocytosis of Neurohormones is Stimulated by Neurotransmitters which Activate the Magnocellular Neurons

Our initial studies (Pow and Morris 1989) showed that depolarisation, barium, and calcium ionophores all caused exocytosis of the contents of NSV from both the dendrites and cell bodies of magnocellular neurons. We have since showed that release of both VP and OT from dendrites in hypothalamic slices can be stimulated by glutamate and acetylcholine (Fig. 7), two well known excitatory transmitters of the system, and that noradrenaline stimulates the dendritic release of VP but not OT (Ward 1994). Furthermore, both depolarisation and glutamate cause a transient rise of calcium in isolated dendrite preparations (Morris et al. 1993). It therefore appears that, when electrical activity of the neurosecretory neurons is stimulated by their synaptic input, this not only causes release from the axonal terminals in the neurohypophysis, but also causes release from the neurosecretory dendrites into the nuclei and surrounding areas. This almost certainly occurs because the dendrites are invaded by the spike of electrical activity generated in the axon hillocks of the neurosecretory neurons (see note on p. 85).

Oestradiol Can Stimulate the Release of Oxytocin and Vasopressin from Magnocellular Dendrites

The secretion of OT and VP has long been known to be influenced by circulating gonadal steroids, and OT-associated neurophysin has been referred to as oestrogen-stimulated neurophysin (Robinson et al. 1976; Perlow et al. 1982; Amico et al. 1986). However, the route by which the neurosecretory neurons are influenced by the steroids has not been clear. Most magnocellular neurons lack the well-documented classical nuclear oestrogen (α) receptors (Pfaff 1989; Herbison 1994) though whether or not they contain the newly discovered oestrogen β receptor (Kuiper et al. 1996; Mosselman et al. 1996) which is more highly expressed in rat brain than the α receptor (Kuiper et al. 1997) is not yet clear. Some actions of steroids in the brain occur too rapidly to be explained by a

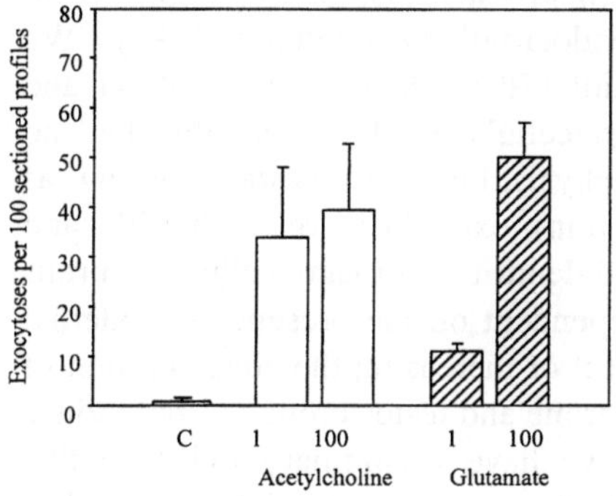

Fig. 7. Histogram showing the effect of glutamate and acetylcholine (1 nM, 100 nM) to cause exocytosis of neurosecretory vesicles from dendrites and cell bodies of supraoptic magnocellular neurons in hypothalamic slices from Long-Evans rats. *C* Saline control. Data are mean + SD

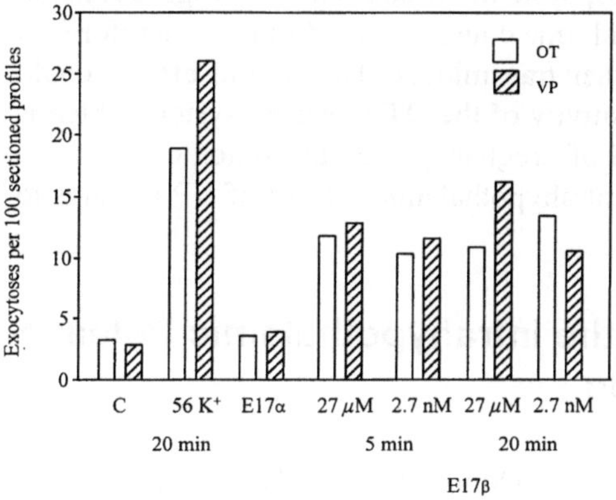

Fig. 8. Histogram showing the rapid (5 min, 20 min) effect of oestradiol-17β (*E17β*) to cause exocytosis of oxytocin, (*OT*) and vasopressin (*VP*) from dendrites of supraoptic magnocellular neurons in hypothalamic slices in vitro. *C* Saline control; oestradiol-17α (*E17α*, 27 μM) is ineffective; potassium depolarisation (56 mM K^+) provides a positive control

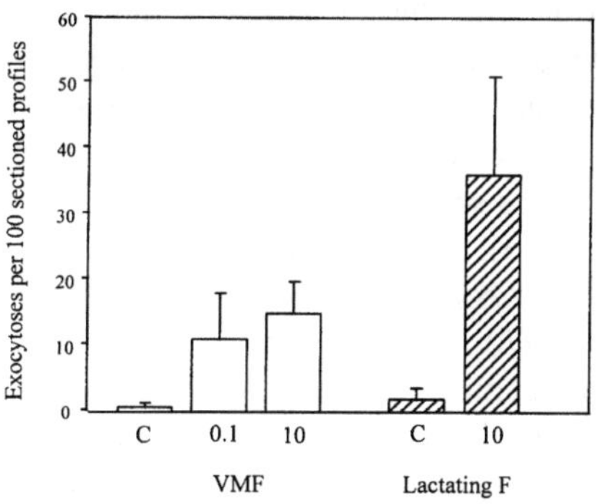

Fig. 9. Histogram to show the effect of oxytocin (0.1 or 10 μM) to stimulate the exocytosis of neurosecretory vesicles from dendrites and cell bodies of magnocellular neurons in the supraoptic nucleus of Long Evans rats. The effect is greater in lactating females (*lactating F*) than in virgin male or female (*VMF*) animals. *C* Saline control. Data are mean + SD

genomic effect (McEwen et al. 1990) and rapid non-genomic actions of steroids, some of which are produced in the brain (neuro-steroids) has emerged as an increasingly important concept in neuroendocrinology (Baulieu 1992). We therefore studied the acute effects of oestradiol-17β on the exocytosis of OT and VP in the neurohypophysis and in the magnocellular nuclei. Oestradiol had no exocytosis-inducing effect in the neurohypophysis, but concentrations as low as 2.7 nM caused the rapid (5 min or 20 min) exocytosis of both VP- and OT-containing NSVs from the cell bodies and dendrites of magnocellular neurons (Fig. 8). Interestingly, the effect was not dependent on the presence of external calcium and we were unable to detect an effect of increasing the concentration of oestradiol to 27 μM. Oestradiol-17α, progesterone and testosterone did not induce exocytosis (Wang and Morris 1996c) but we have preliminary evidence that pretreatment with progesterone can block the rapid oestradiol effect. The peptide-releasing effect of oestradiol is only about half that of depolarisation by potassium but is comparable to the effect of glutamate and acetylcholine applied to slices (compare Figs. 7, 8). Of course, systemic oestradiol and progesterone do not fluctuate markedly over such a rapid time course; oestradiol may therefore act primarily to sensitise the system to other transmitters. The steroid effects could play a part in the regulation of the sensitivity of the OT-releasing system which is being prepared for lactation at the end of pregnancy. The physiological function of an effect of oestradiol to cause the intrahypothalamic release of VP remains to be determined.

Physiological Functions of the Intrahypothalamic Release of Oxytocin and Vasopressin

Oxytocin. It has been known for many years (Moos et al. 1984) that oxytocin has a specific facilitatory effect on its own release from blocks of hypothalamus. In vivo, this locally released OT must be derived largely from the dendrites of magnocellular neurons. The physiological effects of OT centre on stimulation of the uterus during parturition and the ejection of milk during lactation, though it also has pronounced effects on various behaviours (Landgraf 1995). Lactation is associated with a plastic rearrangement of the glia surrounding OT neurons such that both the cell bodies and dendrites of the OT-secreting neurons are in greater contact, and there is an increase in synaptic boutons shared between two cells (see Theodosis et al. 1986; Theodosis and Poulain 1993). This structural plasticity can be induced by intraventricular infusion of OT. Oxytocin excites the electrical activity of OT-secreting neurons (Moss et al. 1972; Yamashita et al. 1987) and dendrites in hypothalamic slices derived from lactating rats are more sensitive to the hormone-releasing effects of oxytocin than are those from virgin rats (Fig. 9). Each NSV in an OT-secreting neuron contains ca. 85 000 molecules of OT so that, during lactation, when the glial processes between OT neurons and their dendrites are much reduced, the exocytosis of OT-containing NSV from dendrites must create a substantial local concentration of OT. Estimates from microdialysis measurements indicate that intranuclear concentrations reach micromolar levels (Landgraf 1995). Whereas it was at one time difficult to detect oxytocin receptors on magnocellular neurons (Tribollet et al. 1992) these have now been detected (Adan et al. 1995a,b) in the dorsal part of the SON where OT neurons are concentrated. It therefore appears that, at the end of pregnancy, the rising oestradiol and falling progesterone, together with other synaptic inputs causes the release of OT from dendrites of magnocellular neurons. This has an autoexcitatory effect on the cells' electrical activity and causes plastic changes in the relationships between the OT neurons and their surrounding glia which maintain a heightened state of sensitivity to suckling stimuli throughout lactation. When suckling stops, the stimulus to dendritic release of oxytocin is removed, and the system reverts to its pre-pregnant state.

Vasopressin. The vasopressin released from the dendrites seems likely to exert an auto- and paracrine inhibitory effect on VP-secreting neurons (Leng and Mason 1982) and can cause a rise in intracellular calcium in isolated rat VP-secreting supraoptic neurons (Dayanithi et al. 1996). However, the precise physiological function of this, apart from acting as an ultrashort loop negative feedback, is at present unclear.

Does Dendritic Release of Peptides Occur in Other Systems?

Immunocytochemical studies show that, in most systems and species investigated, immunoreactive peptide can be traced for considerable distances into the dendrites (e.g. Vigh-Teichmann et al. 1983; Glazer and Basbaum 1983; Lu and Ho 1992). Furthermore, the amount of immunoreactive peptide can be shown to vary with the physiological state (e.g. GnRH in a seasonal-breeding bird; Foster et al. 1988). In the mammalian suprachiasmatic nucleus, exocytosis of dendritic peptide can be demonstrated (Castel et al. 1996) and appears to be important for the pacemaker mechanisms which control the biological clock, because depressing synaptic activity by use of tetrodotoxin does not perturb the intrinsic time-keeping mechanisms of the nucleus (Schwartz et al. 1987; Shibata and Moore 1993).

Conclusions

At one time dendrites were viewed very largely as those parts of neurons which received synaptic signals from their afferent innervation. Studies on magnocellular neurosecretory neurons have now revealed that (1) they have the machinery for the local synthesis of proteins such as receptors and (2) they release in a controlled manner neuroactive substances which influence their own activity and sensitivity and also that of the neurons and glia which surround them. Important physiological functions can be attributed to this dendritic peptide release.

Magnocellular neurons have, for many years, provided a robust model system for understanding the function of peptidergic neurons in general. We believe that, since the presence of dendritic peptide is widespread among peptidergic neurons, this will continue to prove true, although the function of dendritic peptide in other systems is as yet largely unexplored.

Acknowledgments

We thank Lynne Scott, Sarah Rodgers, and Derek Hardiman for expert technical assistance. Studies in the authors' laboratory were funded in part by BBSRC, Wellcome Trust, The Royal Society, and The International Human Frontier Science Program.

References

Adan RA, Van Leeuwen FW, Sonnemans MA, Brouns M, Hoffman G, Verbalis JG, Burbach JP (1995a) Rat oxytocin receptor in brain, pituitary, mammary gland, and uterus: partial sequence and immunocytochemical localization. Endocrinology 136: 4022-4028

Adan RA, Van Leeuwen FW, Sonnemans MA, Hoffman G, Verbalis JG, Burbach JP (1995b) The rat oxytocin receptor. cDNA cloning and immunocytochemical localization in brain, pituitary, mammary gland and uterus. Adv Exp Med Biol 395: 345-346

Amico JA, Finn FM, Robinson AG (1986) Ampholyte displacement and high pressure liquid chromatographic separation of estrogen-responsive neurophysin from human plasma. J Clin Endocrinol Metab 63: 835-840

Baulieu EE (1992) Neurosteroids: an overview. Adv Biochem Psychopharmac 47: 1-16

Castel M, Morris JF (1988) The neurophysin-containing innervation of the forebrain of the mouse. Neuroscience 24: 937-966

Castel M, Morris JF, Belenky M (1996) Non-synaptic and dendritic exocytosis from dense-cored vesicles in the suprachiasmatic nucleus. NeuroReport 7: 543-547

DiScala Guenod D, Strosser MT, Richard P (1987) Electrical stimulations of perifused magnocellular nuclei in vitro elicit Ca^{2+}-dependent, tetrodotoxin-insensitive release of oxytocin and vasopressin. Neurosci Lett 76: 209-214

Dayanithi G, Widmer H, Richard Ph (1996) Vasopressin-induced intracellular Ca^{2+} increase in isolated rat supraoptic cells. J Physiol 490: 713-727

Foster RG, Plowman G, Goldsmith AR, Follett BK, Parry DM (1988) The LH-RH system of the male European starling: photoperiod induces changes to a possible multifunctional peptide system. Basic Appl Histochem 32: 95-102

Glazer EJ, Basbaum AI (1983) Opioid neurons and pain modulation: an ultrastructural analysis of enkephalin in cat superficial dorsal horn. Neuroscience 10: 357-76

Herbison AE (1994) Immunocytochemical evidence for oestrogen receptors within GABA neurons located in the perinuclear zone of the supraoptic nucleus and $GABA_A$ β_2/β_3 subunits on supraoptic oxytocin neurons. J Neuroendocrinol 6: 5-11

Ju G, Liu S, Tao J (1986) Projections from the hypothalamus and its adjacent areas to the posterior pituitary in the rat. Neuroscience 19: 803-828

Ju G, Ma D, Duan X-Q (1992) Third ventricular subependymal oxytocin-like immunoreactive neuronal plexus in the rat. Brain Res Bull 28: 887-896

Kuiper GGJM, Enmark E, Pelto-Huikko M, Nilsson S, Gustafsson J-A (1996) Cloning of a novel estrogen receptor expressed in rat prostate and ovary. Proc Natl Acad Sci USA 93: 5925-5930

Kuiper GGJM, Carlsson B, Grandien J, Enmark B, Haggblad J, Nilsson S, Gustafsson J-A (1997) Comparison of the ligand binding specificity and transcript tissue distribution of estrogen receptors α and β. Endocrinology 138: 863-870

Landgraf R (1992) Central release of vasopressin: stimuli, dynamics and consequences. Prog Brain Res 91: 29-39

Landgraf R (1995) Intracerebrally released vasopressin and oxytocin: measurement, mechanisms and behavioural consequences. J Neuroendocrinol 7: 243-253

Leng G, Mason WT (1982) Influence of vasopressin upon firing patterns of supraoptic neurons: a comparison of normal and Brattleboro rats. Ann NY Acad Sci 394: 153-158

Lu J, Ho RH (1992) Evidence for dorsal root projection to somatostatin-immunoreactive structures in laminae I-II of the spinal dorsal horn. Brain Res Bull 28: 17-26

Ma D, Morris JF (1996) Protein synthetic machinery in magnocellular dendrites of Brattleboro and wild-type rats. 16th Blankenese Conference, Abstract 16.

Ma D, Morris JF (1997a) Local protein synthesis in magnocellular dendrites: basic elements and their response to hyperosmotic stimuli. 1997 World Congress of Neurohypophysial Hormones, Montreal, Canada, August 1997

Ma D, Morris JF (1997b) Nitric oxide synthase gene expression is increased in magnocellular hypothalamic neurons of homozygous Brattleboro rats. J Endocrinol 152 (Suppl): OC40

McEwen BS, Coirini H, Schumacher M (1990) Steroid effects on neuronal activity: when is the genome involved? In: Chadwick D, Widdows K (eds) Steroids and neuronal activity. Ciba Found Symp 153: 3-12

Moos F, Freund-Mercier MJ, Guerne Y, Guerne JM, Stoeckel ME, Richard Ph (1984) Release of oxytocin and vasopressin by magnocellular nuclei in vitro: specific facilitatory effect of oxytocin on its own release. J Endocrinol 102: 63-72

Moos F, Freund-Mercier Y, Guerne Y, Vincent JD, Richard P (1989) Release of oxytocin within the supraoptic nucleus during the milk ejection reflex. Exp Brain Res 76: 593-602

Morris JF (1987) The Brattleboro magnocellular neurosecretory system: a model for the study of peptidergic neurons. Ann NY Acad Sci 394: 54-71

Morris JF, Pow DV, Sokol HW, Ward AR (1993) Dendritic release of peptides from magnocellular neurons in normal rats, Brattleboro rats and mice with hereditary nephrogenic diabetes insipidus. In: Gross P, Richter D, Robertson GL (eds) Vasopressin. John Libbey Eurotext, Paris, pp 171-182

Moss RL, Dyball REJ, Cross BA (1972) Excitation of antidromically identified neurosecretory cells of the paraventricular nucleus by oxytocin applied iontophoretically. Exp Neurol 34: 95-102

Mosselman S, Polman J, Dijkema (1996) ERβ: identification and characterization of a novel human estrogen receptor. FEBS Lett 392: 49-53

Neumann I, Koehler E, Landgraf R, Summy-Long JY (1994) An oxytocin receptor antagonist infused into the supraoptic nucleus attenuates intranuclear and peripheral release of oxytocin during suckling in conscious rats. Endocrinology 134: 141-148

Perlow MJ, Reppert SM, Artman HA, Fisher DA, Self SM, Robinson AG (1982) Oxytocin, vasopressin and estrogen-stimulated neurophysin: daily patterns of concentration in cerebrospinal fluid. Science 216: 1416-1418

Pfaff DW (1989) Patterns of steroid hormone effects on electrical and molecular events in hypothalamic neurons. Mol Neurobiol 3: 135-154

Pow DV, Morris JF (1989) Dendrites of hypothalamic magnocellular neurons release neurohypophysial peptides by exocytosis. Neuroscience 32: 435-439

Pow DV, Morris JF, Toescu EC (1990) Dendrosomes: a new preparation of isolated neurosecretory dendrites. J Neuroendocrinol 2: 103-106

Pow DV, Morris JF, Rodgers S (1991) Peptide accretions in the endoplasmic reticulum of magnocellular neurosecretory neurons in normal and experimentally manipulated rats. J Anat 178: 155-174

Randle JRC, Bourque CW, Renaud LP (1986) Serial reconstruction of Lucifer yellow-labelled supraoptic nucleus neurons in perfused rat hypothalamic explants. Neuroscience 17: 435-467

Robinson AG, Ferin M, Zimmerman EA (1976) Plasma neurophysin levels in monkeys: emphasis on the hypothalamic response to estrogen and ovarian events. Endocrinology 98: 468-475

Scharrer E, Scharrer B (1940) Secretory cells within the hypothalamus. Res Publ Assoc Res Nerv Ment Dis 20: 170-194

Scharrer E, Scharrer B (1945) Neurosecretion. Phys Revs 25: 171-181

Scharrer E (1951) Das Hypophysen-Zwischenhirnsystem von *Scyllium stellare*. Z Zellforsch 37: 196-204

Schwartz WJ, Gross RA, Morton MT (1987) The suprachiasmatic nuclei contain a tetrodotoxin-resistant circadian pacemaker. Proc Natl Acad Sci USA 84: 1694-1698

Shibata S, Moore RY (1993) Tetrodotoxin does not affect circadian rhythms in neuronal activity and metabolism in rodent suprachiasmatic nucleus in vitro. Brain Res 606: 259-266

Sofroniew MV, Weindl A, Schinko I, Wetzstein R (1979) The distribution of vasopressin-, oxytocin- and neurophysin-producing neurons in the guinea pig brain. Cell Tissue Res 196: 367-384

Sofroniew MV, Glassmann W (1981) Golgi-like immunoperoxidase staining of hypothalamic magnocellular neurons that contain vasopressin, oxytocin or neurophysin in the rat. Neuroscience 6: 619-643

Steward O (1997) Ultrastructural basis for gene expression at the synapse: synapse-associated polyribosome complexes. Neuron 18: 9-12

Theodosis DT, Montagnese C, Rodriguez F, Vincent JD, Poulain DA (1986) Oxytocin induces morphological plasticity in the adult hypothalamo-neurohypophysial system. Nature 322: 738-740

Theodosis DT, Poulain DA (1993) Activity-dependent neuronal-glial and synaptic plasticity in the adult mammalian hypothalamus. Neuroscience 57: 501-535

Tribollet E,, Barberis C, Dubois-Dauphin M, Dreifuss JJ (1992) Localization and characterization of binding sites for vasopressin and oxytocin in the brain of the guinea pig. Brain Res 589: 15-23

Vandesande F, Dierickx K (1975) Identification of the vasopressin-producing and the oxytocin-producing neurons in the hypothalamic magnocellular neurosecretory system of rat. Cell Tissue Res 164: 153-162

Vigh-Teichmann, Vigh B, Korf HW, Oksche A (1983) CSF-contacting and other somatostatin-immunoreactive neurons in the brains of *Anguilla anguilla*, *Phoxinus phoxinus*, and *Salmo gairdneri* (Teleostei) Cell Tissue Res 233: 319-34

Wang H, Morris JF (1996a) Constitutive nitric oxide synthase in hypothalami of normal and hereditary diabetes insipidus rats and mice: role of nitric oxide in osmotic regulation and its mechanism. Endocrinology 137: 1745-1751

Wang H, Morris (1996b) Presence of neuronal nitric oxide synthase in the suprachiasmatic nuclei of mouse and rat. Neuroscience 74: 1059-1068

Wang H, Morris JF (1996c) Oestradiol acutely stimulates exocytosis of oxytocin and vasopressin from dendrites and somata of hypothalamic magnocellular neurons. Neuroscience 68: 1179-1188

Ward AR (1994) The role of receptors in controlling the release of peptides from the dendrites of magnocellular neurons. PhD Thesis, University of Oxford

Yamashita H, Okuya S, Inenaga K, Kasai M, Uesugi S, Kannan H, Kaneko T (1987) Oxytocin predominantly excites putative oxytocin neurons in the rat supraoptic nucleus in vitro. Brain Res 416: 364-368

Note added in press: Although electrical impulses generated by axon hillock activation would be expected to invade the dendrites and cause release of dendritic peptide it has been shown recently that antidromic activation of the nuclei by stimulation of the neural lobe stalk does not cause intranuclear release of oxytocin and vasopressin (M. Ludwig, personal communication). This suggests that dendritic exocytosis may be quite separately controlled.

7 Intracerebral Release of Vasopressin and Oxytocin: New Aspects of the Old Concept of Neurosecretion

M. Engelmann and R. Landgraf

Max-Planck-Institut für Psychiatrie, Kraepelinstr. 2, D-80804 München, Germany;
Phone.: +49 89 30622 228, Fax: +49 89 30622 569, E-mail: engelm@mpipsykl.mpg.de

Abstract

The hypothalamic-neurohypophysial system belongs to the best investigated neurosecretory systems in the mammalian brain. The nonapeptides vasopressin and oxytocin are synthesized in the neurons of this system and are secreted either as hormones into the systemic circulation or as neuromodulators/neurotransmitters into the extracellular fluid of distinct brain areas. Recent findings of different lines of investigation provided new aspects in terms of stimuli, dynamics, mechanisms and consequences of the activation of the hypothalamic-neurohypophysial system and, consequently, the release of vasopressin and oxytocin into the different compartments. For instance, it is now evident that both nonapeptides are released not only from axon terminals but from virtually every part of the neuron's surface. Depending upon the quality and intensity of a given stimulus, the release patterns from dendrites, perikarya, axons en passant, and axon terminals can be dissociated from each other and, thus, seem to be differentially controlled. These regulatory patterns of the hypothalamic-neurohypophysial system are likely to coordinate appropriate endocrine and behavioral responses of the organism to challenging situations. It is tempting to speculate whether the new aspects of neuroendocrine regulation discussed here can also be extended to other neurosecretory systems.

Introduction

Almost 70 years ago, in his PhD thesis, Ernst Scharrer suggested that neurons of diencephalic brain structures are capable of releasing biologically active, chemical compounds into the systemic blood circulation (Scharrer 1928). During the following decades, he and his wife, Berta Scharrer, validated this concept of neurosecretion in numerous vertebrate and invertebrate species (Scharrer and Scharrer 1937). Twenty years after Ernst Scharrer's visionary suggestion, the hypothalamic-neurohypophysial system (HNS) was described as the first peptidergic neurosecretory pathway in the brain (Bargmann and Scharrer 1951). This finding became even more attractive after the chemical structures of its main peptide products, namely the nonapeptides vasopressin (AVP) and oxytocin (OXT), had been discovered (Acher and Chauvet 1953; du Vigneaud et al. 1953). Because of its clear and relatively simple structural organization, its easy experimental accessibility and the chemical stability of both AVP and OXT, the HNS became a model system for studying - inter alia - the synthesis, processing and release of neuropeptides (Hatton 1990).

The mammalian HNS consists of magnocellular neurons located within the hypothalamic paraventricular (PVN) and supraoptic (SON) nuclei. Axons of these neurons project via the internal zone of the median eminence to the posterior pituitary to terminate at fenestrated capillaries. AVP and OXT, which are synthesized as parts of larger precursor molecules, are released peripherally (i.e. from axon terminals into the blood stream) after appropriate stimulation (AVP: e.g. osmotic challenge and hemorrhage; OXT: e.g. parturition and lactation; Cunningham and Sawchenko 1991). Distributed by the general circulation, both nonapeptides reach their peripheral target organs to act as hormones (AVP particularly in kidney and liver; OXT particularly in uterus and milk glands).

For nearly 40 years the HNS has been regarded as consisting of simple "output" effector neurons, integrating inputs from a variety of afferent sources into spike discharge, resulting in peptide secretion from the posterior pituitary (Hatton 1990). There was no indication that magnocellular neurons communicate actively with other nerve or glia cells, either within the hypothalamus or outside. During the past 15 years, however, evidence has been accumulated requiring a revision of this concept. A number of in vitro investigations provided not only evidence that magnocellular neurons of the SON and PVN influence the firing pattern of both extrahypothalamic (Leng 1982; Mason et al. 1986) and intrahypothalamic neurons (Leng and Mason 1982) but suggested also AVP and/or OXT of intrahypothalamic origin to trigger these effects (Moos et al. 1984; Mason et al. 1986). To act in this way, AVP and OXT have to be released not only from nerve terminals in the posterior pituitary but also into the extracellular fluid of the hypothalamic nuclei of their origin. The results of various post mortem and in vitro experiments (Krisch 1980; Pow and Morris 1989) supported this hypothesis. However, only the development of microperfusion techniques in vivo such as push-pull perfusion and, in particular, microdialysis, enabled the monitoring of dynamic changes in the intrahypothalamic nonapeptide release patterns in freely moving animals (Landgraf 1995, Ludwig 1995).

The aim of the present article is to summarize the recent findings concerning mechanisms, dynamics and regulation of release patterns of both AVP and OXT from magnocellular neurons. In this way, we want to examine whether the concept of peptidergic neurosecretion in general and the HNS in particular still provides an attractive paradigm for interneuronal communication.

Intracerebral Release of AVP and OXT During Stress

Until recently, microdialysis was used mainly to monitor intrahypothalamic release of AVP and OXT in response to pharmacological stimulation. We extended these observations to more physiological conditions by investigating the release of both nonapeptides within the PVN and SON using a variety of ethologically relevant and naturally occurring stressors. As an emotional stress, the social defeat paradigm was applied which consists of placing the animal under ongoing microdialysis in the home cage of a dominant male resident that had been trained to become aggressive towards intruders. Immediately after the first attack, the intruder was physically separated from the resident by a wire mesh, still allowing olfactory and visual contact and, thus, emphasizing the emotional aspect of the stress. As control, microdialysates were collected from rats placed in cages with the inserted wire mesh, but without the resident. While AVP release under basal conditions was comparable in the two groups, social defeat caused a significant rise during and after emotional stress (Wotjak et al. 1996b). Exposure to the novel cage, in contrast, failed to alter AVP release within the PVN (Fig. 1A). This response is remarkably specific, as AVP release within the SON and that of OXT within the PVN (Fig. 1B) were not changed under identical experimental conditions.

Another group of animals was forced to swim for 10 min in 20 °C warm water under ongoing microdialysis. Forced swimming is considered to represent a combined physical and emotional stress. Under these experimental conditions, not

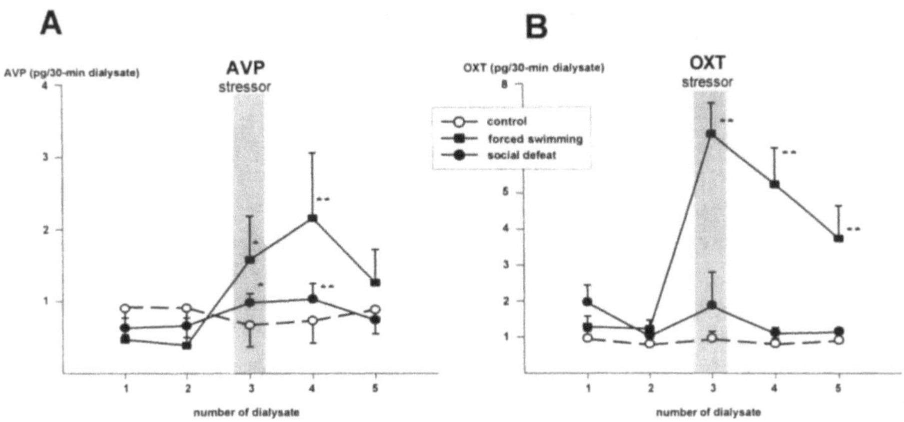

Fig. 1. AVP (**A**) and OXT (**B**) contents (means ± SEM) measured in consecutively collected 30-min microdialysates obtained from the PVN of adult male Wistar rats. At the beginning of the collection of dialysate number 3, the animals were either placed either into a novel cage (control) or forced to swim for 10 min or exposed to an aggressive resident in its home cage (n = 6-18 per group). ** $p < 0.01$; * $p < 0.05$ vs. dialysates 1 and 2 of the respective treatment group (ANOVA)

only AVP but also OXT release in both the PVN (Fig. 1A, B) and SON was significantly increased (Wotjak et al. 1997) indicating a stress-specific response of the HNS.

Independent Regulation of Central and Peripheral Nonapeptide Release

Typically, plasma hormone levels are used as indicators for the general secretory activity of the HNS. In other words, with respect to the intrahypothalamic release of AVP and OXT, one would expect that stressful stimuli known to trigger the secretion of the nonapeptides from the posterior pituitary also stimulate the release within the SON and/or PVN and vice versa. In fact, comparing intrahypothalamic and peripheral release patterns of OXT in response to forced swimming, we found nonapeptide release both within the hypothalamus and into the blood stream (Figs.1B, 2B). Parallel alterations in release patterns into both compartments have also recently been described in female animals during parturition and lactation (Neumann et al. 1993, 1996).

A first indication for a possible dissociation between central and peripheral release patterns was provided by studies investigating the effects of an osmotic challenge on the release of AVP and OXT. After intraperitoneal injection of hypertonic saline, an immediate increase in plasma AVP and OXT could be observed, whereas the intrahypothalamic release of the two neuropeptides was significantly delayed (Neumann et al. 1993). Further support for a dissociated release of AVP into different compartments was obtained from studies in which the intrahypothalamic and peripheral release patterns of this nonapeptide were

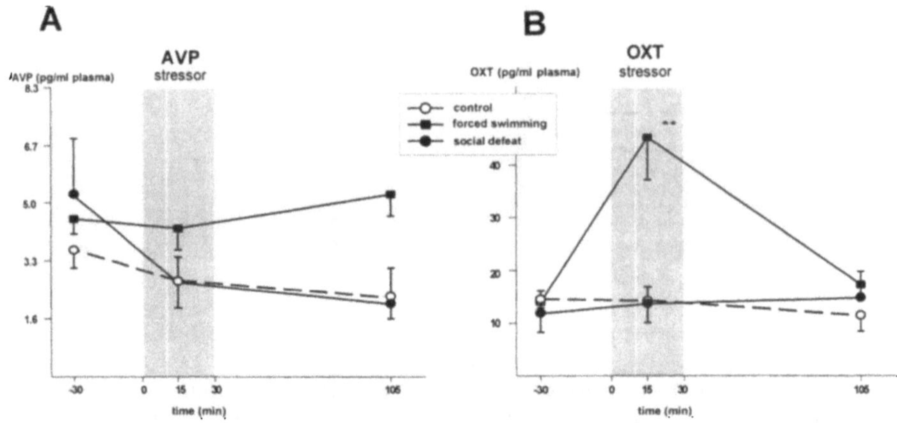

Fig. 2. AVP (**A**) and OXT (**B**) concentrations (means ± SEM) in plasma samples collected via chronically implanted jugular vein catheters in adult male rats. Samples were collected before, during and after the onset of the stressor (n = 6-10 per group). ** $p < 0.01$ vs. all other values (ANOVA)

monitored in response to social defeat. In defeated animals, the increase in the concentration of AVP in the extracellular fluid of the PVN was not accompanied by similar changes in plasma AVP (Wotjak et al. 1996b; Figs. 1A, 2A). However, the most striking evidence for a dissociated release was provided by studying release patterns of AVP in rats that were forced to swim. Whereas intrahypothalamic release of AVP reached very high levels in a range comparable to those after pharmacological stimulation (Fig. 1A), the peripheral secretion of the nonapeptide remained virtually unchanged (Fig. 2A). Interestingly, this stressor caused a marked increase in plasma osmolality (by 16 mOsm/kg H_2O). With regard to the close relationship between plasma osmolality and plasma AVP concentrations (Dunn et al. 1973; Windle et al. 1996), the failure to secrete AVP into the systemic circulation despite the marked rise in plasma osmolality appears to be even more remarkable and remains to be elucidated.

Regulatory Mechanisms Underlying Central and Peripheral Release Patterns

What might be the regulatory mechanisms underlying the observed dissociation between intrahypothalamic and peripheral release of AVP? As mentioned before, there is now good evidence indicating that the release of AVP and OXT into the extracellular fluid of the hypothalamic nuclei of their origin occurs predominantly from dendrites and perikarya (Pow and Morris 1989; Morris and Pow 1993). Recent studies suggest that, in contrast to the secretion from the posterior pituitary (Poulain and Theodosis 1988), this intrahypothalamic release does not require the generation and propagation of action potentials (di Scala-Guenot et al. 1987; Ludwig et al. 1995). Thus, it seems reasonable to suggest that an inhibition of the electrical activity of HNS neurons, e.g. by the local action of inhibitory neurotransmitters (e.g. taurine and GABA), blocks the secretion of AVP and OXT from axon terminals in the posterior pituitary, but fails to reduce intra-PVN and intra-SON release from dendrites and perikarya.

The intrahypothalamic release of AVP and OXT not only seems to be uncoupled from the electrical activity of the neuron but, moreover, the nonapeptides might be able to feedback in a positive manner on their own release patterns (Neumann et al. 1994, Wotjak et al. 1994). This mechanism of a self-amplified intrahypothalamic release of OXT probably also acts to potentiate the peripheral secretion of the same nonapeptide (Lambert et al. 1993; Neumann et al. 1996) (Fig. 3A). In the case of AVP, however, the physiological significance of the positive feedback action is less clear. We favor the hypothesis that, once AVP has reached a critical intrahypothalamic threshold by a positive feedback action, it suppresses its own peripheral release (Fig. 3B). This is supported by findings from electrophysiological studies providing evidence for a reduced electrical activity of vasopressinergic neurons in response to AVP administration (Leng and Mason 1982; Dayanithi et al. 1995; Hermes et al. 1996; Ludwig and Leng 1997). To

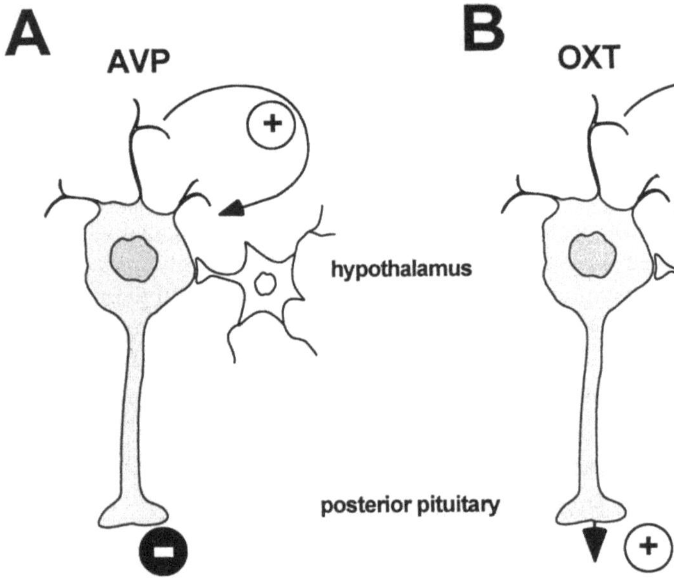

Fig. 3. Schematic diagram showing the effect of the positive feedback action of intrahypothalamically released AVP (**A**) and OXT (**B**) on the neurons of their origin to control their peripheral nonapeptide secretion

induce these effects, AVP released into the extracellular fluid of the SON or PVN could either directly inhibit the electrical activity of vasopressinergic neurons (Dayanithi et al. 1996) or act via a stimulation of the local release of inhibitory amino acids (see above) from astroglia, interneurons or neurons of the respective perinuclear zone (Decavel and Hatton 1995; Boudaba et al. 1996; Hermes et al. 1996).

The interaction between and coordination of positive and negative feedback actions at multiple levels seem to be a prerequisite for fine-tuned release patterns into different compartments.

Impact of the HNS on other Neuroendocrine Systems

AVP and OXT released within and from the HNS are thought to significantly influence the activity of numerous other neuroendocrine systems. Among them, the hypothalamic-pituitary-adrenal (HPA) axis is the prominent neuroendocrine system which is also activated during stress and originates within the PVN. Parvocellular neurons of this nucleus, which contain mainly AVP and the corticotropin releasing hormone (CRH), terminate in the external zone of the median eminence to secrete both peptides as releasing hormones into the long portal vessels. Transported via the portal blood to the anterior pituitary, AVP and CRH trigger the release of corticotropin (ACTH), which in turn controls the secretion of corticosteron (rat) or cortisol (human) from the adrenals (Antoni 1993). Although activated during stress, originally the HNS and the HPA axis were regarded as two relatively separate neurosecretory systems. However, different lines of recent investigations provided compelling evidence for a significant involvement of the HNS in HPA axis regulation at different levels. At the level of the PVN, for example, AVP released from magnocellular neurons into the extracellular fluid of this nucleus induces a negative tonus on the release of ACTH and, thus, on the activity of the HPA axis (Wotjak et al. 1996b). Moreover, a combined microdialysis/push-pull perfusion study in the rat revealed that, at the level of the median eminence, AVP is released en passant from axons of HNS neurons which project to the posterior pituitary (Wotjak et al. 1996a). AVP released en passant is thought to reach the portal blood and, thus, is likely to mediate an influence of the HNS on the HPA axis (Holmes et al. 1986; Buma and Nieuwenhuys 1987; Nordmann and Dayanithi 1988). Finally, at the level of the pituitary, AVP and OXT secreted from the posterior part may reach the anterior pituitary by short portal vessels to contribute to the regulation of ACTH release (Bergland and Page 1978). The physiological relevance of the regulatory control of the HPA axis by the HNS at different levels is likely to limit the activity of the stress axis to keep the homeostasis of the organism (Wotjak et al. 1997).

Intrahypothalamic nonapeptide release, however, might not only be a regulatory principle for neurosecretory systems, but might also have direct behavioral impact. The release of OXT into the extracellular fluid of the PVN, for instance, could participate in the initiation of self-grooming which the animals display under stressful conditions (Van Erp et al. 1993). Moreover, stimulation of the release of AVP within both the SON and PVN was shown to improve olfactory short-term memory in male rats (Engelmann et al. 1994, 1996). In this context it is tempting to speculate that intrahypothalamically released AVP and OXT might represent the neurochemical equivalent of emotions that color incoming information and, thus, contribute not only to an appropriate endocrine response, but also to an appropriate behavioral coping strategy.

Conclusion

Taken together, the data discussed here reveal that the spatial and temporal pattern of the release of AVP and OXT from and within the HNS is considerably more differentiated than originally thought. In addition to the release from axon terminals, vasopressinergic and oxytocinergic neurons are capable of releasing their nonapeptides from virtually every part of their surface, including dendrites, perikarya and axon en passant (Fig. 4). As a consequence, the resulting multiple levels at which a single peptidergic neuron is capable of communicating with other populations of nerve or glia cells requires subtle regulatory mechanisms which control the release of the nonapeptide into the different compartments. This might explain the ambivalent actions that intrahypothalamically released AVP

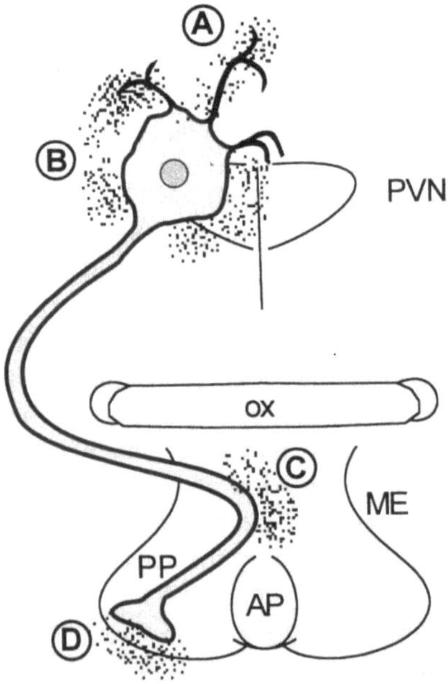

Fig. 4. Schematic drawing of nonapeptide release from different parts of the membrane of magnocellular neurons of the HNS (for example originating in the PVN). Release can occur within the hypothalamus from dendrites (**A**) and perikarya (**B**), at the level of the median eminence from the axon en passant (**C**), and from axon terminals within the posterior pituitary (**D**). *AP* Anterior pituitary; *ME* median eminence; *OX* optic chiasm; *PVN* paraventricular nucleus; *PP* posterior pituitary

exerts on the neurons of its origin (e.g. stimulating its intrahypothalamic release but inhibiting its own peripheral release).

The observed dissociation between intracerebral and peripheral release of AVP also has implications for the design and analysis of future animal studies. These findings indicate that one has to be careful in extrapolating from plasma hormone levels to the secretory activity of the respective neuropeptidergic system within the brain. Furthermore, it is important to note that the data and the hypotheses presented here argue strongly against the peripheral administration route in pharmacological experiments focusing on the role of endogenous, intracerebrally released nonapeptide in behavioral performance. Peripheral administration of AVP after a stressful learning and memory task, for example, can result in behavioral effects which are not due to the reinforcement of central actions of neuropeptidergic interneuronal communication (LeMoal et al. 1981). Thus, the concept of neurosecretion originally developed and established by Ernst and Berta Scharrer seems to become an even more attractive model for studying neurosecretory processes in a physiological context if the new aspects discussed above are considered.

Acknowledgments

The authors wish to thank Dr. Carsten T. Wotjak (ZNM Hamburg, Germany) for sharing unpublished data and inspirative discussions. Supported by Deutsche Forschungsgemeinschaft.

References

Acher R, Chauvet J (1953) La structure de la vasopressine de boef. Biochim Biophys Acta 12: 487-488

Antoni FA (1993) Vasopressinergic control of pituitary adrenocorticotropin secretion comes of age. Front Neuroendocrinol 14: 76-122

Bargmann W, Scharrer E (1951) The site of origin of the hormones of the posterior pituitary. Am Sci 39: 255-259

Bergland RM, Page RB (1978) Can the pituitary secrete directly to the brain? (Affirmative anatomical evidence). Endocrinology 102: 1325-1338

Boudaba C, Szabó K, Tasker JG (1996) Physiological mapping of local inhibitory inputs to the hypothalamic paraventricular nucleus. J Neurosci 16: 7151-7160

Buma P, Nieuwenhuys E (1987) Ultrastructural demonstration of oxytocin and vasopressin release sites in the neural lobe and median eminence of the rat by tannic acid and immunogold methods. Neurosci Lett 74: 151-157

Cunningham ET Jr, Sawchenko PE (1991) Reflex control of magnocellular vasopressin and oxytocin secretion. Trends Neurosci 14: 406-411

Dayanithi G, Moos F, Richard Ph (1995) Vasopressin controls magnocellular vasopressin neurones via V1-type receptors in the rat. J Physiol 489: 184-185

Dayanithi G, Widmer H, Richard P (1996) Vasopressin-induced intracellular Ca^{2+} increase in isolated rat supraoptic cells. J Physiol 490: 713-727

Decavel C, Hatton GI (1995) Taurine immunoreactivity in the rat supraoptic nucleus: prominent localization in glia cells. J Comp Neurol 354: 13-26

di Scala-Guenot D, Strosser MT, Richard P (1987) Electrical stimulations of perifused magnocellular nuclei in vitro elicit Ca^{2+}-dependent, tetrodotoxin-insensitive release of oxytocin and vasopressin. Neurosci Lett 76: 209-214

du Vigneaud V, Ressler C, Trippett S (1953) The sequence of amino acids in oxytocin, with a proposal for the structure of oxytocin. J Biol Chem 205: 949-957

Dunn FL, Brennan TJ, Nelson AE, Robertson GL (1973) The role of blood osmolality and volume in regulating vasopressin secretion in the rat. J Clin Invest 52: 3212-3219

Engelmann M, Ludwig M, Landgraf R (1994) Simultaneous monitoring of intracerebral release and behavior: endogenous vasopressin improves social recognition. J Neuroendocrinol 6: 391-395

Engelmann M, Wotjak CT, Ludwig M, Neumann I, Landgraf R (1996) Behavioral consequences of intracerebral vasopressin and oxytocin: focus on learning and memory. Neurosci Biobehav Rev 20: 341-358

Hatton GI (1990) Emerging concepts of structure-function dynamics in adult brain: the hypothalamo-neurohypophysial system. Prog Neurobiol 34: 437-504

Hermes MLHJ, Spanswick D, Renaud LP, Buijs RM (1996) Inhibitory action of vasopressin on neurons of the rat hypothalamic paraventricular nucleus. Soc Neurosci Abstr 22: 807.16

Holmes MC, Antoni FA, Aguilera G, Catt KJ (1986) Magnocellular axons in passage through the median eminence release vasopressin. Nature 319: 326-329

Krisch, B (1980) Immunocytochemistry of neuroendocrine systems: vasopressin, somatostatin, luliberin. Progr Histochem Cytochem 13: 1-160

Lambert RC, Moos FC, Richard Ph (1993) Action of endogenous oxytocin within the paraventricular or supraoptic nuclei: a powerful link in the regulation of the bursting pattern of oxytocin neurons during the milk-ejection reflex in rats. Neuroscience 57: 1027-1038

Landgraf R (1995) Intracerebrally released vasopressin and oxytocin: measurement, mechanisms and behavioural consequences. J Neuroendocrinol 7: 243-253

LeMoal M, Koob GF, Koda LY, Bloom FE, Manning M, Sawyer WH, Rivier J (1981) Vasopressor receptor antagonist prevents behavioural effects of vasopressin. Nature 291: 491-493

Leng G (1982) Lateral hypothalamic neurones: osmosensitivity and the influence of activating magnocellular neurosecretory neurones. J Physiol 326: 35-48

Leng G, Mason WT (1982) Influence of vasopressin upon firing patterns of supraoptic neurons: a comparison of normal and brattleboro rats. Ann NY Acad Sci 394: 153-158

Ludwig M (1995) Functional role of intrahypothalamic release of oxytocin and vasopressin: consequences and controversies. Am J Physiol 268: E537-E545

Ludwig M, Leng G (1997) Autoinhibition of supraoptic nucleus vasopressin neurones in vivo: a combined retrodialysis/electrophysiological study in rats. Eur J Neurosci (in press)

Ludwig M, Callahan MF, Morris M (1995) Effects of tetrodotoxin on osmotically stimulated central and peripheral vasopressin and oxytocin release. Neuroendocrinology 62: 619-627

Mason WT, Hatton GI, Ho YW, Chapman C, Robinson ICF (1986) Central release of oxytocin, vasopressin and neurophysin by magnocellular neurone depolarization: evidence in slices of guinea pig and rat hypothalamus. Neuroendocrinology 42: 311-322

Moos F, Freund-Mercier MJ, Guerne Y, Guerne JM, Stoeckel ME, Richard Ph (1984) Release of oxytocin and vasopressin by magnocellular nuclei in vitro: specific facilitatory effect of oxytocin on its own release. J Endocrinol 102: 63-72

Morris JF, Pow DV (1993) New anatomical insights into the inputs and outputs from hypothalamic magnocellular neurons. Ann NY Acad Sci 689: 16-33

Neumann I, Ludwig M, Engelmann M, Pittman QJ, Landgraf R (1993) Simultaneous microdialysis in blood and brain: oxytocin and vasopressin release in response to central and peripheral osmotic stimulation and suckling in the rat. Neuroendocrinology 58: 637-645

Neumann I, Koehler E, Landgraf R, Summy-Long J (1994) An oxytocin receptor antagonist infused into the supraoptic nucleus attenuates intranuclear and peripheral release of oxytocin during suckling in conscious rats. Endocrinology 134: 141-148

Neumann I, Douglas AJ, Pittman QJ, Russell JA, Landgraf R (1996) Oxytocin released within the supraoptic nucleus of the rat brain by positive feedback action is involved in parturition-related events. J Neuroendocrinol 8: 227-233

Nordmann JJ, Dayanithi G (1988) Release of neuropeptides does not only occur at nerve terminals. Biosci Rep 8: 471-483

Poulain DA, Theodosis DT (1988) Coupling of electrical activity and hormone release in mammalian neurosecretory neurons. Curr Top Neuroendocrinol 9: 73-104

Pow DV, Morris JF (1989) Dendrites of hypothalamic magnocellular neurons release neurohypophysial peptides by exocytosis. Neuroscience 32: 435-439

Scharrer, E (1928) Die Lichtempfindlichkeit blinder Elritzen; Untersuchungen über das Zwischenhirn der Fische I. Z vergl Physiol 7: 1-38

Scharrer E, Scharrer B (1937) Über Drüsen-Nervenzellen und neurosekretorische Organe bei Wirbellosen und Wirbeltieren. Biol Rev 12: 185-216

Van Erp AMM, Kruk MR, Semple DM, Verbeet DWP (1993) Initiation of self-grooming in resting rats by local infusion of oxytocin but not α-MSH. Brain Res 607: 108-112

Windle RJ, Luckman SM, Stoughton RP, Forsling ML (1996) The effect of pinealectomy on osmotically stimulated vasopressin and oxytocin release and Fos protein production within the hypothalamus of the rat. J Neuroendocrinol 8: 747-753

Wotjak CT, Ludwig M, Landgraf R (1994) Vasopressin facilitates its own release within the rat supraoptic nucleus. Neuroreport 5: 1181-1184

Wotjak CT, Kubota M, Kohl G, Landgraf R (1996a) Release of vasopressin from supraoptic neurons within the rat median eminence in vivo. A combined microdialysis and push-pull perfusion study in the rat. Brain Res 726: 237-241

Wotjak CT, Kubota M, Liebsch G, Montkowski A, Holsboer F, Neumann I, Landgraf R (1996b) Release of vasopressin within the rat paraventricular nucleus in response to emotional stress: a novel mechanism of regulating adrenocorticotropic hormone secretion? J Neurosci 16: 7725-7732

Wotjak CT, Ganster J, Kohl G, Holsboer F, Landgraf R, Engelmann M (1997) Dissociated central and peripheral release of vasopressin, but not oxytocin, in response to repeated swim stress: new insights into the regulatory capacities of peptidergic neurons. Neuroscience (submitted)

8 Effects of Neuropeptides on Brain Function in Humans

H.L. Fehm, R. Pietrowsky, B. Perras, C. Dodt, and J. Born

Medizinische Klinik I, Medizinische Universität zu Lübeck,
Klinische Forschergruppe, Ratzeburger Allee 160, D-23538 Lübeck, Germany;
Phone: +49 451 500 2305; Fax: +49 451 500 3339,
E-mail: fehm@gwsun.medinf.mu-luebeck.de

Abstract

The endocrine system can be viewed as a communication system between the brain and the body periphery with efferent and afferent pathways of equal importance. This implies that blood-borne hormones, including peptide hormones, can influence brain functions. This review summarizes a series of experiments designed to analyze effects of vasopressin after intravenous and intranasal administration on brain function in human subjects. Brain function was assessed by electroencephalographic methods, i.e. recording of "event-related brain potentials" and the sleep EEG. Vasopressin was shown to improve signs of attention and cognitive functions, but only with intranasal administration. Intranasal administration of peptides turned out to be a principal way to bypass the blood-brain barrier. Acute intranasal administration of vasopressin failed to restore impaired cognitive functions in elderly human subjects. However, with long-term intranasal vasopressin treatment, sleep disturbancies characteristic of aged people, were ameliorated. Time spent in slow-wave sleep as well as in rapid-eye-movement sleep substantially increased during a 3-month treatment period. Analyzing the actions of neuropeptides on brain function in humans could provide the basis for developing a new, specific, and "soft" neuropharmocology.

Introduction

The discovery of the principles of neurosecretion by Ernst und Berta Scharrer set the stage for a new subdiscipline of endocrinology: neuroendocrinology. Neuroendocrine research during the past few decades led to the discovery of a never ending stream of neuropeptides. Some of these, like the hypophysiotropic hormones of the hypothalamus found their way to clinical application and became important tools in the diagnosis and treatment of endocrine and non-endocrine diseases. For the vast majority of neuropeptides, physiological functions still have to be elucidated.

It is now well established that the functions of most if not all of the peripheral endocrine glands are controlled by the brain. The description of the mechanism and pathways through which this control is exerted continues to be a main topic of neuroendocrine research. Adequate functioning of the control mechanisms requires that the brain is fully informed about relevant humoral events occurring in the periphery. This implies that hormones produced by peripheral glands feed back to the brain. Since all parts of the brain, including the neocortex, are involved in the control of endocrine secretion it is reasonable to assume that the feedback effects of peripheral hormones pertain to all aspects of the CNS activity, including higher brain functions. In this vein, we would postulate that the endocrine system represents a communication system between brain and body periphery with equal importance for both the efferent and the afferent pathways (Fig. 1). The humoral system is thus viewed as a counterpart to another communication system, the autonomous nervous system where the mutual interdependence of efferent and afferent pathways has been acknowledged.

That peripheral hormones influence CNS functioning is well established for the hormones of the adrenal and thyroid glands. Peptide hormones were generally considered to be inactive in this respect because they are unable to cross the blood-brain barrier. Nevertheless, David de Wied's group could provide abundant evidence that ACTH and vasopressin influence higher brain functions as learning, memory and attention in experimental animals (de Wied et al. 1993). We have studied the effects of these peptides on brain function in humans. This review will focus on the actions of vasopressin, because this peptide played a major role in Ernst und Berta Scharrer's scientific work.

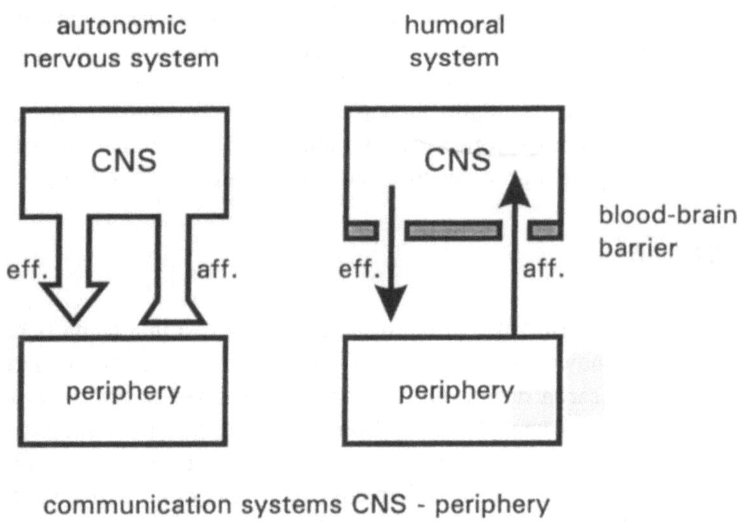

Fig. 1. Schematic representation of the two most important communication systems between the brain and the body periphery

Methods

To analyze the effects of vasopressin on brain function during wakefulness we employed electroencephalographic methods, especially the so-called event-related brain potentials (ERPs).

ERPs are typically evoked by repeatedly presented sensory stimuli. They include a series of negative and positive voltage deflections which are generated in different subcortical and cortical structures. The various components of the ERP can be linked to certain aspects of the brain's information processing at preattentive and attentive stages. In addition to ERP recording, we used somnopolygraphic recordings to examine effects of vasopressin on brain functions in the course of nocturnal sleep.

Results and Discussion

A first experiment was designed to determine the effects of vasopressin on attentional processes by analyzing auditory evoked potentials during a vigilance

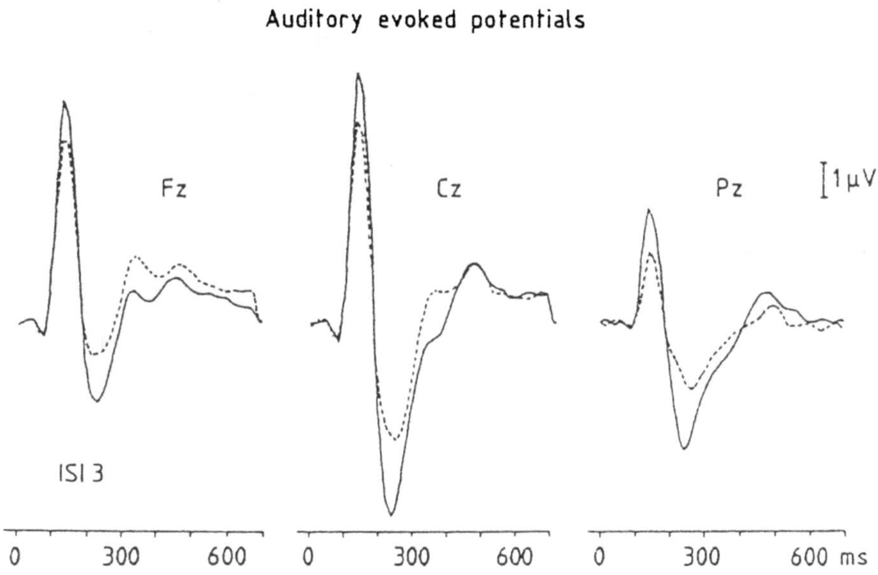

Fig. 2. The auditory evoked responses to tone pips recorded from electrode locations along the midline (*Fz* frontal, *Cz* central, *Pz* parietal). Vasopressin-treated subjects showed higher amplitudes of the vertex potential. —Vasopressin; ---- placebo

task (Fehm-Wolfsdorf et al. 1988). The task required subjects to covertly count series of monotonous tone pips. Averaged ERPs to these tones showed the expected decrease in amplitude during the course of the task (habituation). The decrease was most pronounced for the N1-P2-potential complex (also termed

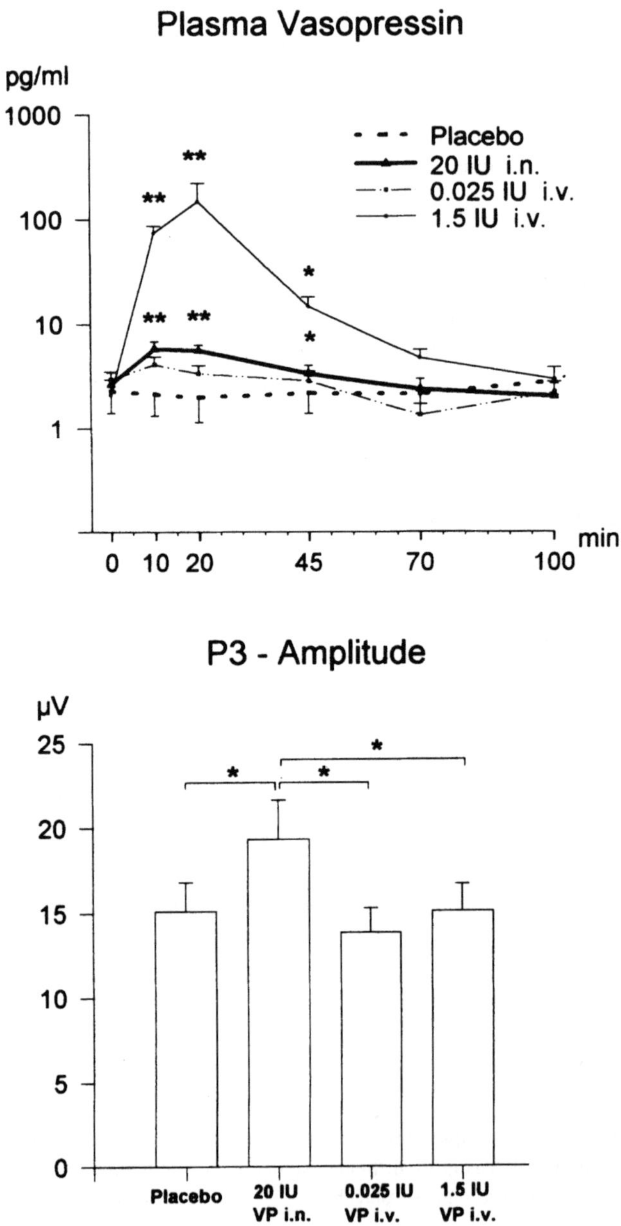

Fig. 3. Top Plasma vasopressin (VP) levels following 20 IU VP intranasal (i.n.), 0.025 IU AVP intravenous (i.v.), 1.5 IU AVP i.v., and placebo at 0, 10, 20, 30, 45, 70 and 100 min following administration. **Bottom** Peak amplitude of the P3 component of the event-related potential following placebo, i.n. and i.v. AVP administration

"vertex potential") which is a large biphasic (negative-positive) voltage deflection occurring around 100 msec after the tone. Subjects treated intranasally with 20 IU arginine vasopressin (AVP) one hour prior to the experimental session displayed markedly higher amplitudes of the vertex potential as compared to the placebo treated group (Fig. 2). The vertex potential reflects stimulus-related cortical arousal; an increase in amplitude indicates that more cortical energy, i.e. attentional capacity is provided for the processing of the stimulus.

In subsequent studies, we attempted to establish a dose-response curve for these ERP effects of vasopressin. For this purpose, different doses of AVP were administered intravenously (i.v.). Yet, contrary to our expection, intravenous AVP failed to affect the ERP even at the highest dose. This led us to suspect that the effects of AVP on ERPs can be obtained with intranasal administration only.

Therefore, in a further experiment, we compared the effect of AVP on ERPs after intranasal (i.n.) versus intravenous administration (Pietrowsky et al. 1996). In a double-blind cross-over study, subjects received AVP intranasally at a dose of 20 IU and intravenously at doses of 1.5 and 0.025 IU, and saline solution. ERPs were recorded while the subjects performed on an auditory attention task (oddball paradigm). In this paradigm, in a sequence of frequent standard tones some rare target tones (of slightly higher frequency) were interspersed, requiring a button press response. The target tones of this task elicit a pronounced P3 component in the ERP which reflects the attentive processing of the behaviorally relevant signal within short term memory. After intranasal administration, there was a minute increase in plasma AVP concentrations and a comparable small increase was observed with the intravenous dose of 0.025 IU of AVP. In contrast, intravenous administration of 1.5 IU AVP induced a strong increase in plasma AVP levels, substantially exceeding the increase after i.n. administration. Intranasal AVP increased the P3 component of the ERP significantly ($p < 0.01$). By contrast, intravenous administration had no consistent effect on the ERP responses, despite the higher plasma concentration achieved with this route of administration (Fig. 3).

We conclude from these results that intranasal administration of peptides represents a way to bypass the blood-brain barrier. This route of administration appears to provide direct access for peptides to brain sites. In fact, with intranasal but not intravenous administration we could demonstrate effects for several further peptides, including corticotropin releasing hormone, cholecystokinine, adrenocorticotropin and angiotensin II, which were comparable to those observed in animals after intracerebroventricular administration of the substances (Pietrowsky et al. 1996; Kern et al. 1997).

The clear-cut effects of AVP on signs of attention and cognitive stimulus processing led to the question of whether these effects can be exploited therapeutically in patients with impaired cognitive functions. Such impairment occurs in advanced age and is typically accompanied by a substantial decrease in the P3 amplitude and an increase in P3 latency. Therefore, we compared the effect of AVP on ERPs in old and young mentally healthy subjects (Dodt et al. 1994). A total of 22 old and 28 young subjects were tested. Again, ERPs were recorded while the subjects performed an auditory oddball task. As expected, in the elderly subjects the latency of the P3 wave was significantly prolonged and the amplitude was diminished. AVP treatment significantly enlarged the N2-P3 amplitude in the young subjects but failed to improve the deficient amplitude of the old subjects (Fig. 4).

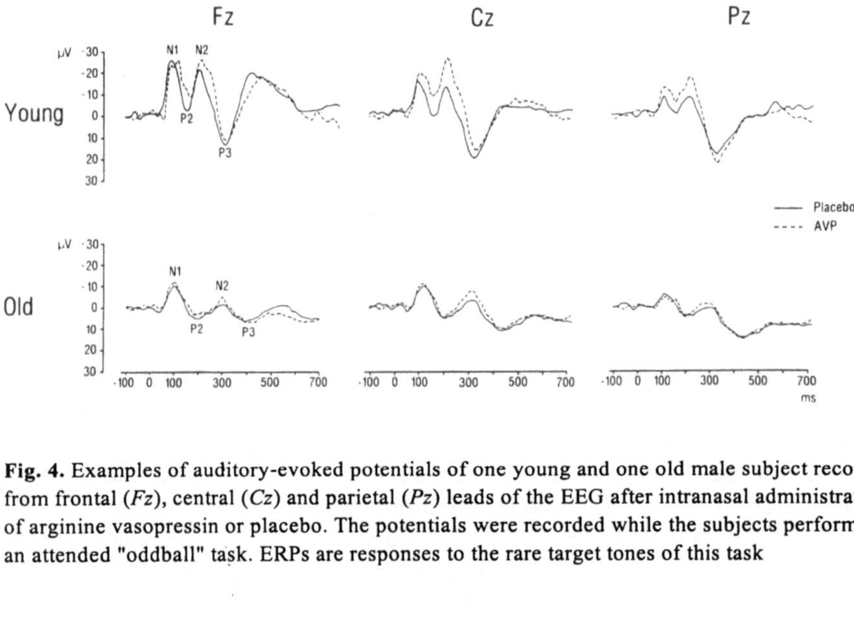

Fig. 4. Examples of auditory-evoked potentials of one young and one old male subject recorded from frontal (*Fz*), central (*Cz*) and parietal (*Pz*) leads of the EEG after intranasal administration of arginine vasopressin or placebo. The potentials were recorded while the subjects performed an attended "oddball" task. ERPs are responses to the rare target tones of this task

From the results presented so far, it appears that AVP is able to improve cognitive functions in young subjects, but is ineffective in the elderly, displaying significant deficits in these functions. This negative finding could be taken to indicate that in the elderly there is not only impaired availability of AVP in certain brain structures but also a resistance to the effects of AVP, induced presumably by a downregulation of VP-receptors. In this case, long-term administration could be expected to overcome this resistance. These considerations led us to study influences of a 3-month period of daily intranasal VP treatment (20 IU before bedtime and after awakening) on ERPs and nocturnal somnopolygraphical recordings in 26 healthy elderly human subjects (mean age 74.2 years; Perras et al. 1996; Perras et al., in press). Again, we failed to find any consistent effects of AVP on ERPs. However, subchronic intranasal treatment with VP improved dramatically the age-related impairment of sleep (Table 1). The total sleep time was increased by 45 minutes, time spent in slow wave sleep was increased on average by 21 minutes following VP, and also the time spent in REM sleep in the second half of the night was increased. Thus, VP promoted sleep and improved

Table 1. Effects of subchronic treatment with intranasal vasopressin on sleep parameters

	Placebo (n=13)		Vasopressin (n=13)		pairwise comparison	
	Mean	SEM	Mean	SEM	F value	*p* value
Sleep onset latency (min)	26.8	4.9	22.0	2.3	0.64	0.43
Sleep period time (min)	404.0	18.2	449.2	12.2	**12.34**	**0.002**
Sleep efficiacy (%)	86.9	2.1	86.9	1.9	0.00	0.99
Time of sleep stages (min)						
W	55.1	9.5	58.9	9.2	0.16	0.69
S1	81.2	12.9	75.1	13.4	0.86	0.36
S2	170.4	12.3	187.2	13.0	3.42	0.07
SWS (S3+S4)	36.5	6.0	58.0	9.4	**5.52**	**0.025**
REM	60.8	5.7	70.0	6.3	1.83	0.19
Sleep stages 1st half (min)						
W	27.1	6.0	26.4	6.5	0.00	0.95
S1	44.1	7.0	38.2	8.3	0.24	0.60
S2	77.8	7.5	94.8	6.6	4.20	0.055
SWS (S3+S4)	24.2	5.5	37.0	5.9	**4.87**	**0.038**
REM	30.1	3.3	28.2	3.8	0.01	0.92
Sleep stages 2nd half (min)						
W	27.4	4.3	33.0	5.2	1.41	0.30
S1	43.9	6.3	37.4	6.1	0.51	0.48
S2	86.4	5.7	91.3	6.5	0.23	0.64
SWS (S3+S4)	12.1	1.9	20.9	4.7	3.38	0.079
REM	31.7	3.6	42.0	3.0	**7.98**	**0.01**

Sleep onset latency: time between lights off and the onset of the first epoch of sleep stage 1 followed by sleep stage 2; sleep period time (SPT): time between sleep onset and final awakening, sleep efficacy defined by SPT minus awake (W) relative to total SPT (%); W: time awake; S1, S2: sleep stages 1 and 2; SWS: slow wave sleep (S3 plus S4); REM: rapid eye movement sleep. Time spent in the different sleep stages was determined for the total SPT and also separately for the first and second half of SPT. Means represent baseline adjusted averages across night 5 and 6, derived from ANCOVA. Right columns indicating F values and *p* values for pairwise comparisons (degrees of freedom 1, 23). Level of significance $p < 0.05$

sleep architecture in such a way that the sleep profile of the AVP-treated aged subjects was simular to those of the young control group. Considering the common impairment of sleep in senescence and the associated problems of treatment with presently available hypnotics, results suggest that the neuropeptide vasopressin is of potential value for the treatment of age-related insomnia.

References

De Wied D, Diamant M, Fodor M (1993) Central nervous system effects of the neurohypophyseal hormones and related peptides. Front Neuroendorinol 14: 251-302

Dodt C, Pietrowsky R, Sewing A, Zabel A, Fehm HL, Born J (1994) Effects of vasopressin on event-related potential indicators of cognitive stimulus processing in young and old humans. J Gerontol 49: 183-188

Fehm-Wolfsdorf G, Bachholz G, Born J, Voigt K, Fehm HL (1988) Vasopressin but not oxytocin enhances cortical arousal: an integrative hypothesis on behavioral effects of neurohypophyseal hormones. Psychopharmocology-Berl. 94: 496-500

Kern W, Schiefer B, Schwarzenburg J, Stange EF, Born J, Fehm HL (1997) Evidence for central nervous effects of corticotropin-releasing hormone on gastric acid secretion in humans. Neuroendocrinology 65: 291-298

Perras B, Mölle M, Born J, Fehm HL (1996) Sleep and signs of attention during 3 months of intranasal vasopressin: a pilot study in two elderly subjects. Peptides 17: 1253-1255

Perras B, Pannenborg H, Marshall L, Pietrowsky R, Born J, Fehm HL (1997) Beneficial treatment of age related sleep disturbancies by subchronic intranasal vasopressin. J Psychopharmacol (in press)

Pietrowsky R, Thiemann A, Kern W, Fehm HL, Born J (1996) A non-brain pathway for psychotropic peptides: evidence from a brain evoked potential study with cholecystokinin. Psychoneuroendocrinology 21: 559-572

Pietrowsky R, Strüben C, Mölle M, Fehm HL, Born J (1996) Brain potential changes after intranasal vs. intravenous administration of vasopressin: evidence for a direct nose-brain pathway for peptide effects in humans. Biol-Psychiatry 39: 332-340

9 Reduced Number of Nitric Oxide Synthase Immunoreactive Neurons in the Nucleus Paraventricularis Hypothalami of Depressed Patients and Schizophrenics

H.-G. Bernstein, A. Stanarius[1], D. Krell, G. Northoff, P. Danos, H. Henning, B. Baumann, and B. Bogerts

Klinik für Psychiatrie, Psychotherapie und Psychosomatische Medizin, Uniklinikum, Universität Magdeburg, Leipziger Str. 44, D-39120 Magdeburg, Germany;
Phone: +49 391 67 14249, Fax: + 49 391 67 15223
[1]Institut für Medizinische Neurobiologie, Universität Magdeburg

Abstract

Clinical studies have shown that affective disorders might causally be linked to a dysregulation of the hypothalamic-pituitary-adrenal (HPA) axis. In depressive patients, a strong increase in the hypothalamic expression of the HPA-triggering hormones corticotropin releasing factor, oxytocin and vasopressin have been described. Since these neuropeptides are thought to be under the control of the gas nitric oxide, we decided to look for the immunohistochemical distribution of the gas-generating enzyme, nitric oxide synthase, in hypothalamic neurons of depressed people in comparison to controls and schizophrenics. The paraventricular nucleus was the most prominent source of immunoreactive neurons in the hypothalamus. Fifteen to 20% of all paraventricular neurons were immunostained for nitric oxide synthase. Single enzyme-immunoreactive neurons were also found in other hypothalamic nuclei. The supraoptic nucleus was free of nitric oxide synthase. Only a small subset of paraventricular neurons co-expressed corticotropin releasing factor, whereas most of them were shown to co-express neurophysin. Quantitative morphology of nitric synthase oxide-containing paraventricular neurons revealed that their number was reduced both in depressive patients and in schizophrenics. Our morphological data point to a putative role of nitric oxide synthase in the deviant regulation of the HPA cascade in depression and, possibly, in schizophrenia.

Introduction

Nitric oxide (NO) is a relatively unstable gas which is increasingly recognized as a potential neurotransmitter or neuromodulator within the brain. NO formation is under the control of the enzyme NO synthase (NOS), which appears in distinct isoforms. Neurons contain constitutive forms (cNOSα, β and γ, according to Eliasson et al. 1997). In rat hypothalamic neurons NOS was shown to co-exist with neuropeptides involved in the regulation of the hypothalamic-pituitary-adrenal (HPA) axis, specifically those containing corticotropin releasing factor (CRF), vasopressin and oxytocin (Lee et al. 1995). These findings, together with pharmacological data on a putative role of NO in ACTH secretion through an inhibition of CRF release (Costa et al. 1993), found our interest as neuropsychiatrists, because (1) in humans a dysregulation of the HPA axis has frequently been observed in depressed patients (for review see Holsboer and Barden 1996), and (2) an overexpression of CRF, vasopressin and oxytocin in post mortem paraventricular neurons of depressed persons has been reported (Raadsheer et al. 1995; Purba et al. 1996). Unfortunately, knowledge about the cellular distribution of NOS within the normal and diseased human hypothalamus is very restricted (Sangruichi and Kowall 1991; Egberongbe et al. 1994). Therefore, we decided to immunolocalize the enzyme in the normal post mortem hypothalamus, to compare its cellular localization with those of CRF and neurophysin (the carrier protein of vasopressin and oxytocin) and to estimate quantitatively the amount of NOS-containing neurons within the paraventricular nucleus (PVN) in controls, schizophrenics and depressive patients.

Materials and Methods

Human Subjects

Controls. Thirteen patients (six women and seven men, mean age 50.7 years) without clinical signs of neurological or psychiatric disorders.
Depressed Patients. Eight patients (four women and four men, mean age 51.5 years) with a clinical diagnosis of mood disorder according to ICD-10. Four of them suffered from a monopolar depression, four had the bipolar form. All patients suffering from depressive syndrome received treatment with antidepressants over a rather long period of life. In addition, one of them received neuroleptic drugs. All patients but two received lithium.
Schizophrenics. Ten patients (five women and five men, mean age 51.7 years) with the clinical diagnosis of schizophrenia. Patients were treated with various neuroleptic drugs over a rather long period. Three schizophrenics received lithium and one was treated with antidepressants.

All cases were matched with regard to age, gender and time of autolysis. The matching procedure was done prior to all experiments.

Tissue Processing, Immunohistochemistry and Stereology

All brains were obtained from pathologists or medical examination offices. The brains were processed in a standard manner, including immersion fixation in formalin, embedding in paraplast and cutting with a microtome at 20 μm. Sections at intervals of 400 μm (according to the Cavalieri principle; Mayhev 1992) were stained for nitric oxide synthase using a well-characterized, monospecific polyclonal antiserum, raised in rabbits against the enzyme (Eurodiagnostica AB Stockholm). The working dilution was 1:2000 in phosphate-buffered saline. The avidin-biotin technique was used (Vectastain-peroxidase kit). 3,3′-diaminobenzidine served as a chromogen. For morphological orientation (Nieuwenhuys et al. 1988), every 50th section was stained for Nissl and Klüver-Barrera. Additionally, selected sections from control cases were counterstained with Nissl after NOS immunostaining, and the percentage of immunoreactive neurons was counted using a counting grid (Bernstein et al. 1996). Parallel sections to those immunostained for NOS from five control cases were subjected to immunohistochemical staining for CRF (monoclonal antibody from HyTest, working dilution 1:200) or neurophysin (monoclonal antibody from DAKO, working dilution 1:2000). The detection of hypothalamic peptides was performed using Sternberger′s unlabeled immunoenzyme technique (Sternberger 1979). Images of consecutive sections for either NOS, CRF or neurophysin were photographed using a video camera, digitalized, exactly superimposed by a computer-aided technique (Digitrace) and visualized on a monitor. Thus, the occurrence of cells double-labeled for NOS-CRH or NOS- neurophysin was easily detectable.

To estimate the amount of NOS-immunopositive paraventricular cells, we counted at higher magnification the number of cell profiles per slice using the optical dissector method (as recommended by Coggeshall and Lekan 1996). Left and right hemispheres were counted separately. Knowing the number of cell profiles between the two planes of the dissector, and the square of the counting area (as determined by the square of the PVN at the adjacent section) it was possible to calculate the amount of NOS-immunoreactive cells within a given tissue volume (cell density). The total number of NOS-containing neurons within the whole PVN was determined by multiplying the average numerical cell density by the volume of the population (Purba et al. 1996). Cell countings were carried out under blind conditions by two independent individuals. The data were statistically treated by the non-parametric, two-tailed U-test after Mann and Whitney.

Changes in NOS cell numbers in psychiatric patients were correlated to the duration of illness and medication using multivariant analysis of variation (MANOVA).

Results

Cellular Distribution of NOS-Positive Neurons in Normal Human Hypothalamus and Co-Localization of NOS with CRF and Neurophysin

The main locus of NOS-containing nerve cells was the PVN, where approximately 15 to 20% of the neurons expressed the antigen. NOS-immunoreactivity occurred both in magno- and parvocellular neurons (Figs. 1, 2). Outside the PVN, immunoreactive cells appeared less frequently. Single NOS-positive nerve cells were detected in the medial preoptic region (Fig. 3), the ventromedial nucleus, the infundibular nucleus and the suprachiasmatic nucleus. The supraoptic nucleus was completely devoid of immunopositive cells. CRF-immunoreactive cells were restricted to the PVN (Fig. 4) and a region near the median eminence. Neurophysin antiserum reacted with a majority of neurons situated in the paraventricular and supraoptic nuclei as well as with neurons in the suprachiasmatic nucleus. It was revealed that only a small subset of CRF-expressing neurons co-express NOS (5% or less), whereas the co-localization of NOS and neurophysin was observed for many PVN neurons (roughly 70% of all NOS-expressing nerve cells).

Quantitative Morphometry

Volumetric estimation of PVN size in controls, schizophrenics and depressive patients revealed that mean nuclear volume did not differ significantly between the groups. The difference in the total amount of NOS-containing PVN cells was significant between controls and schizophrenics (right side, $p \leq 0.02$) and between controls and depressive patients (left side, $p \leq 0.05$ and right side, $p \leq 0.002$). Differences in the density of PVN NOS neurons are shown in Fig. 5. Interestingly, only the right PVN of patients who had suffered from depression and schizophrenia showed a statistically significant decrease in the density of NOS immunopositive cells ($p \leq 0.018$ for both depressive patients and schizophrenics vs. controls, Fig. 5).

Discussion

The proposed important function of the messenger nitric oxide in the control of the release of various hypothalamic peptides, together with the observed elevated expression of CRF, vasopressin and oxytocin (Raadsheer et al. 1995; Purba et al. 1996), encouraged us to study the distribution of NOS in normal and diseased human hypothalamus. Although a certain role of NO in mood disorders has recently been supposed on the basis of theoretical considerations (Harvey 1996)

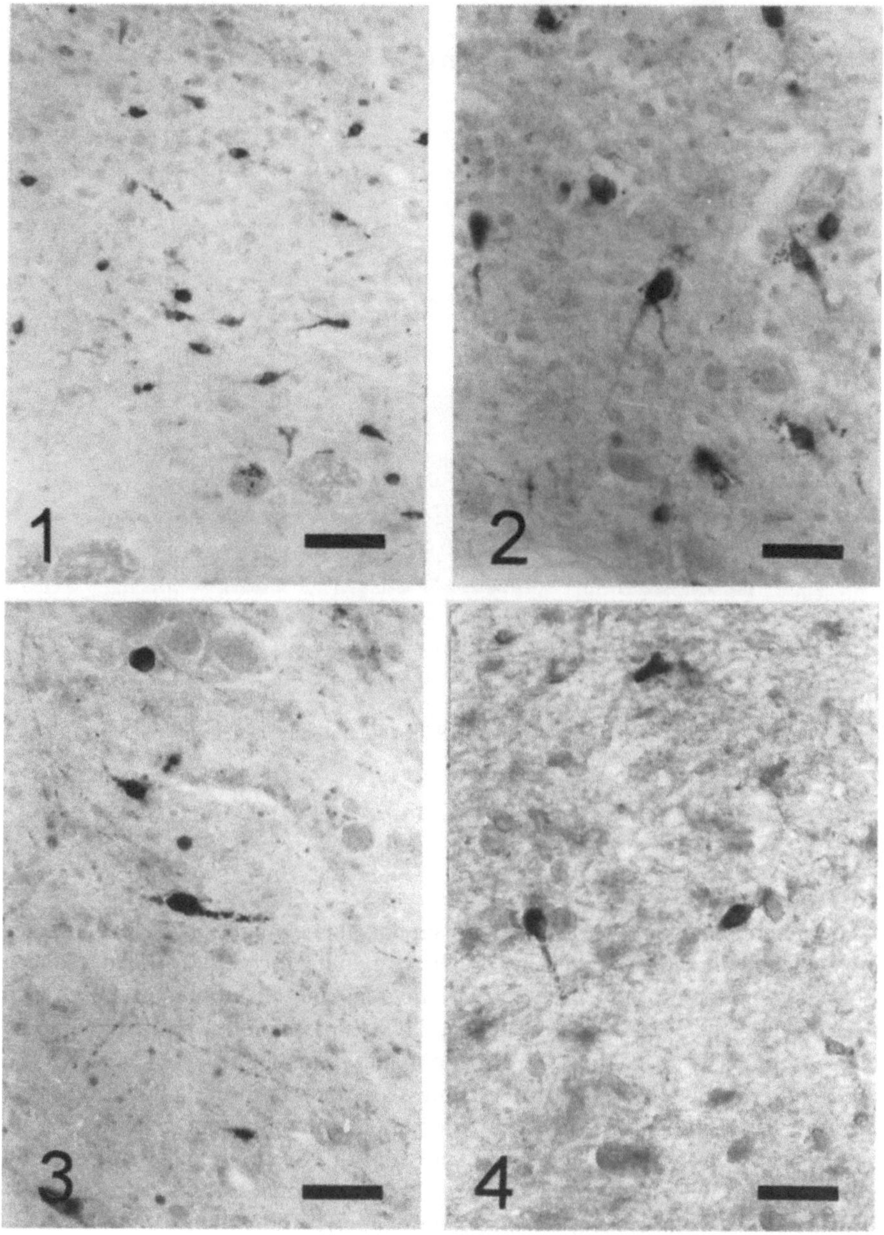

Fig. 1. Low power micrograph of NOS-containing neurons in the human paraventricular nucleus (PVN). *Bar:* 100 μm. **Fig. 2.** NOS-immunoreactive PVN nerve cells at higher magnification. *Bar:* 35 μm. **Fig. 3.** NOS-containing neurons in the preoptic area. Please note the appearance of the immunoreaction in cell bodies and nerve fibers. *Bar:* 35 μm. **Fig. 4.** CRF-immunoreactive neurons in the PVN. *Bar:* 35 μm

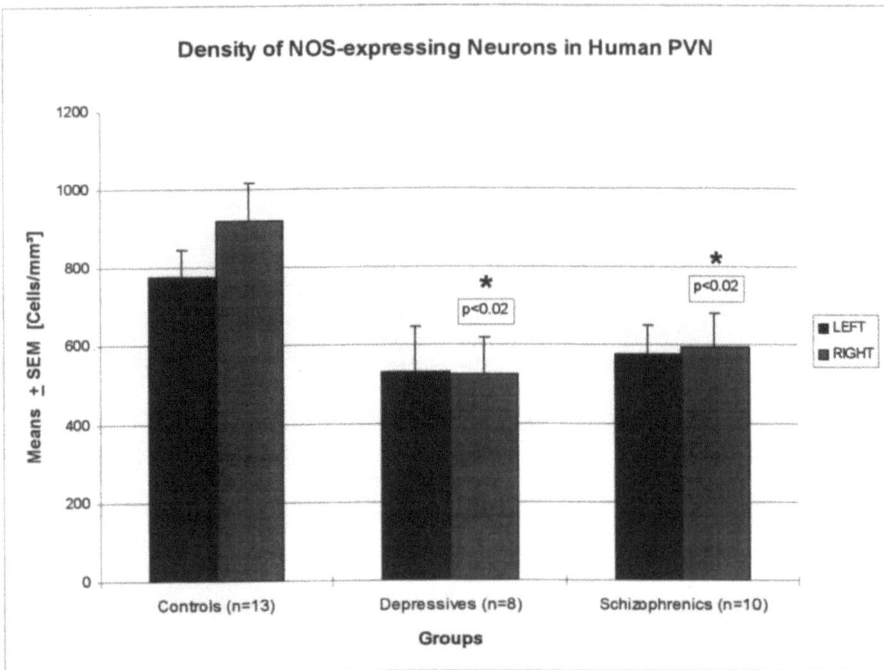

Fig. 5. Quantitative estimation of the density of NOS-containing PVN neurons in controls, depressive patients and schizophrenics

no experimental data are yet available to support this assumption. Schizophrenics were included in our study because there are hints for cell loss and reduced vasopressin/oxytocin-production in PVN neurons in schizophrenia (Lesch and Bogerts 1984; Mai et al. 1993). Moreover, NOS expression has been shown to be altered in other brain regions of schizophrenics (Gentleman et al. 1995). The interpretation of our results is difficult, since a reduction in the amount of NOS-expressing PVN in neurons of psychiatric patients might reflect either a discrete loss of PVN neurons or a decreased expression of NOS protein in still existing nerve cells. However, it is commonly thought that neuronal NOS is represented by a constitutive isoform. Hence, the first interpretation is more likely.

Another crucial point arising from our findings is the question of whether less NOS immunoreactive cells in the PVN means less hypothalamic NO production. It has been shown that the constitutive neuronal NOS is the dominant enzyme variant in the rat hypothalamus generating more than 90% of all NO in this brain area (Bhat et al. 1995). The assumption that there really is a decreased neuronal NO production in the PVN of psychiatric patients would help explain the overexpression of CRF, oxytocin and vasopressin in post mortem hypothalami of depressive patients (Raadsheer et al. 1995; Purba et al. 1996), since reduced levels of the gas might have a "disinhibiting" effect on peptide expression (Costa et al. 1993). The low degree of co-expression between NOS and CRF is not contradictory to this conjecture, since NO is diffusible and may act "par distance" (Wiklund et al. 1997). Alternatively, our findings may be interpreted as reflecting drug effects rather than disease-related changes. Indeed, there is evidence that the generation of NO is affected by neuroleptics, antidepressants and lithium (Chapleau and Aboud 1994; Hu et al. 1994; Coransaniti et al. 1995). However, these antipsychotics have been shown to stimulate NO production. Taken together, the reported results give reason to suppose that, in man, hypothalamic NO might play a regulatory role similar to that described for other mammalian species (Lee et al. 1995).

Acknowledgments

This research was supported by BMBF, DFG, Stanley Foundation and Alfried Krupp Stiftung.

References

Bernstein H-G, Henning H, Seliger N, Baumann B, Bogerts B (1996) Remarkable beta-endorphinergic innervation of human cerebral cortex as revealed by immunohistochemistry. Neurosci Letters 214: 1-4

Bhat GK, Mahesh VB, Lamar CA, Ping L, Aguan K, Brann DW (1995) Histochemical localization of nitric oxide neurons in the hypothalamus - association with gonadotropin-releasing hormone neurons and colocalization with N-methyl-D-aspartate receptors. Neuropharmacology 62: 187-197

Chapleau MW, Aboud FM (1994) Modulation of baroreceptor activity by ionic and paracrine mechanisms: an overview. Braz J Med Biol Res 27: 1001-1015

Coggeshall RE, Lekan HA (1996) Methods for determining numbers of cells ansd synapses: a case for more uniform standards of review. J Comp Neurol 364: 6-15

Coransaniti MT, Paoletti AM, Palma P, Granato T, Navarra M, Nistico G (1995) Systemic administration of pramiracetam increases nitric oxide synthase activity in the cerebral cortex of the rat. Funct Neurol 10: 151-155

Costa A, Trainer B, Besser M, Grossmann A (1993) Nitric oxide modulates the release of corticotropin-releasing hormone from rat hypophysis. Brain Res 605: 199-207

Egberongbe YL, Gentleman SM, Falkai P, Bogerts B, Polak JM, Roberts GW (1994) The distribution of nitric oxide synthase immunoreactivity in the human brain. Neuroscience 59: 561-578

Eliasson MJL, Blackshaw S, Schell MJ, Snyder SH (1997) Neuronal nitric oxide synthase alternatively spliced forms: prominent functional localizations in the brain. Proc Natl Acad Sci USA 94: 3396-3401

Gentleman SM, Leclerq P, von Bussmann KA, Garey LJ, Royston MC (1995) Differential distribution of nitric oxide synthase neurons in the cerebral cortex of schizophrenics. Schizophrenia Res 15: 28

Harvey BH (1996) Affective disorders and nitric oxide - a role in pathways to relapse and refractoriness. Human Psychopharmacol - Clin Exper 11: 309-319

Holsboer F, Barden N (1996) Antidepressants and hypothalamic-pituitary-adrenocortical (HPA) regulation. Endocrine Rev 17: 187-205

Hu JG, Lee JH, Elfakhany EE (1994) Inhibition of neuronal nitric oxide synthase by antipsychotic drugs. Brain Res 114: 161-166

Lee S, Barnabel G, Rivier C (1995) Systemic endotoxin increases steady-state gene expression of hypothalamic nitric oxide synthase: comparison with corticotropin-releasing factor and vasopressin gene transcripts. Brain Res 705: 136-148

Lesch A, Bogerts B (1984) The diencephalon in schizophrenia: evidence for reduced thickness of the periventricular grey matter. Eur Arch Psych Neurol Sci 234: 212-219

Mai JK, Berger, K, Sofroniew MV (1993) Morphometric evaluation of neurophysin-immunoreactivity in the human brain: pronounced inter-individual variability and evidence for altered staining patterns in schizophrenia. J Hirnforsch 34: 133-154

Mayhev TM (1992) A review of recent advances in stereology for quantifying neural structure. J Neurocytol 21: 313-328

Nieuwenhuys R, Voogd J, Van Huijzen C (1988) The human central nervous system. A synopsis and atlas. 3rd edition. Springer, Berlin-New York-Heidelberg

Purba JS, Hoogendijk WJG, Hofman MA, Swaab DF (1996) Increased number of vasopressin- and oxytocin-expressing neurons in the paraventricular nucleus of the hypothalamus in depression. Arch Gen Psychiatry 53: 137-143

Raadsheer FC, Van Heerikhuize JJ, Lucassen PJ, Hoogendijk WJG, Tilders FJ, Swaab D (1995) Increased corticotropin-releasing hormone (CRH) mRNA in paraventricular nuclei of depressed patients. Am J Psychiatry 152: 1372-1376

Sangruichi T, Kowall NW (1991) NADPH diaphorase in the human hypothalamus. Neuroscience 40: 713-724

Sternberger LA (1979) Immunocytochemistry, 2nd edn. Wiley, New York

Wiklund NP, Cellek S, Leone AM, Iversen HH, Gustafson LE, Brundin L, Furst VW, Flock A, Moncada S (1997) Visualisation of nitric oxide released by nerve stimulation. J Neurosci Res 47: 224-232

10 Neurofibrillary Degeneration in Hypophysiotrophic Nuclei of the Aging Human Hypothalamus

C. Schultz, D. Koppers, E. Braak, and H. Braak

Dr. Senckenbergische Anatomie, Institute for Anatomy I, Johann Wolfgang Goethe-University Frankfurt, Theodor-Stern-Kai 7, D-60590 Frankfurt am Main, Germany;
Phone: +49 69 6301 6912, Fax : +49 69 6301 6425, E-mail: Schultz@em.uni-frankfurt.de

Abstract

The aging human hypothalamus of 69 males (mean age: 72 ± 12 years) and 56 females (mean age: 78 ± 10 years) was investigated for neurofibrillary pathology associated with abnormally phosphorylated cytoskeletal protein tau. A conspicuous pathology was identified in the mediobasal portion of the tuber cinereum. It was characterized by neurofibrillary tangles, large dystrophic neurites and small terminal-like processes contacting the vasculature of the infundibular nucleus and the adjacent posterior median eminence. Remarkably, this pathology revealed a striking sex-difference: a marked or severe form of the lesions was exclusively noted in males, while it was completely absent in females. Moreover, the vessel-associated neurofibrillary lesions develop independently of Alzheimer's disease-related brain destruction. In a small subpopulation, a previously unreported pattern of neurofibrillary changes was also identified in the magnocellular paraventricular and supraoptic nuclei. These lesions are remarkable in that the magnocellular nuclei are considered to be extremely resistant to the development of neurofibrillary changes in Alzheimer's disease. Notably, the neurofibrillary changes of the magnocellular nuclei were identified even in the hypothalamus of non-demented controls. Taken together, these findings demonstrate a distinct entity of neurofibrillary degeneration selectively affecting major hypophysiotrophic nuclei of the aging human hypothalamus.

Introduction

In recent years, pathological alterations of the human hypothalamus have been noted in various diseases of different etiologies (for review: Swaab 1997). In Alzheimer's disease (AD), marked neurofibrillary degeneration has been reported in certain hypothalamic nuclei (Saper and German 1987; Braak and Braak 1992; Swaab et al. 1992). In the present study, the magnocellular and parvocellular neurosecretory system of the aging human hypothalamus was investigated for neurofibrillary changes associated with abnormally phosphorylated tau protein.

Materials and Methods

Brains of 69 males (mean age: 72 ± 12 years) and 56 females (mean age: 78 ± 10 years) were obtained at general autopsy, fixed in 4% aqueous formaldehyde, and embedded in polyethylene glycol (Braak and Braak 1991b). The severity of AD-related cortical destruction was classified using a procedure distinguishing six stages of neurofibrillary changes (stages I-VI) and three stages of amyloid depositing (stages A-C; Braak and Braak 1991a). The hypothalamus was dissected from one hemisphere and serial coronal sections were cut at 100 μm. Nuclear grays of the chiasmatic and tuberal hypothalamus were delineated by staining with aldehyde fuchsine-Darrow red (Fig. 1; Braak and Braak 1987; Braak and Braak 1992). The modified Gallyas silver iodide method was used to demonstrate neurofibrillary changes (Iqbal et al. 1993; Gallyas 1971; Braak and Braak 1991b), and Aβ-amyloid deposits were stained using the Campbell-Switzer silver pyridine method (Campbell et al. 1987). Immunocytochemistry was performed in 100 μm thick sections with the monoclonal antibodies AT8 (mouse, 1:1000; Mercken et al. 1992; Braak et al. 1994) and PHF-1 (mouse, 1:250; Greenberg et al. 1992) for abnormally phosphorylated tau protein.

Additional sections of the chiasmatic region of 33 males (mean age: 77 ± 10 years) and 31 females (mean age: 78 ± 10 years) were investigated for neurofibrillary changes of the paraventricular (PVN) and supraoptic nucleus (SON), using the techniques described above. In one case, 4-chloro-1-naphthol was used to visualize AT8-immunoreactive (ir) neurons in the PVN and SON. AT8-ir neurons were photographed and mapped. After destaining of the chromogen and restaining with aldehyde fuchsine/Darrow red, the neuronal type of the previously AT8-ir neurons was assessed according to previously published criteria (Braak and Braak 1987). For electron microscopy, AT8-immunostained sections were incubated with a secondary antibody linked to colloidal gold (Aurion GAM IgG/M GP-US, 5 nm, 1:1000, 5 h). Sections were postfixed in glutaraldehyde and OsO_4, silver enhanced (Aurion), dehydrated and embedded in Araldite. Ultrathin sections were placed on FORMVAR-coated one hole copper grids, counterstained with 1% lead citrate and saturated uranylacetate and examined using a Zeiss EM 109 electron microscope.

Alzheimer-Related Neurofibrillary Changes in the Human Hypothalamus

In the human hypothalamus, the basolateral tuber cinereum is particularly vulnerable to the formation of AD-related neurofibrillary changes. In this area, the

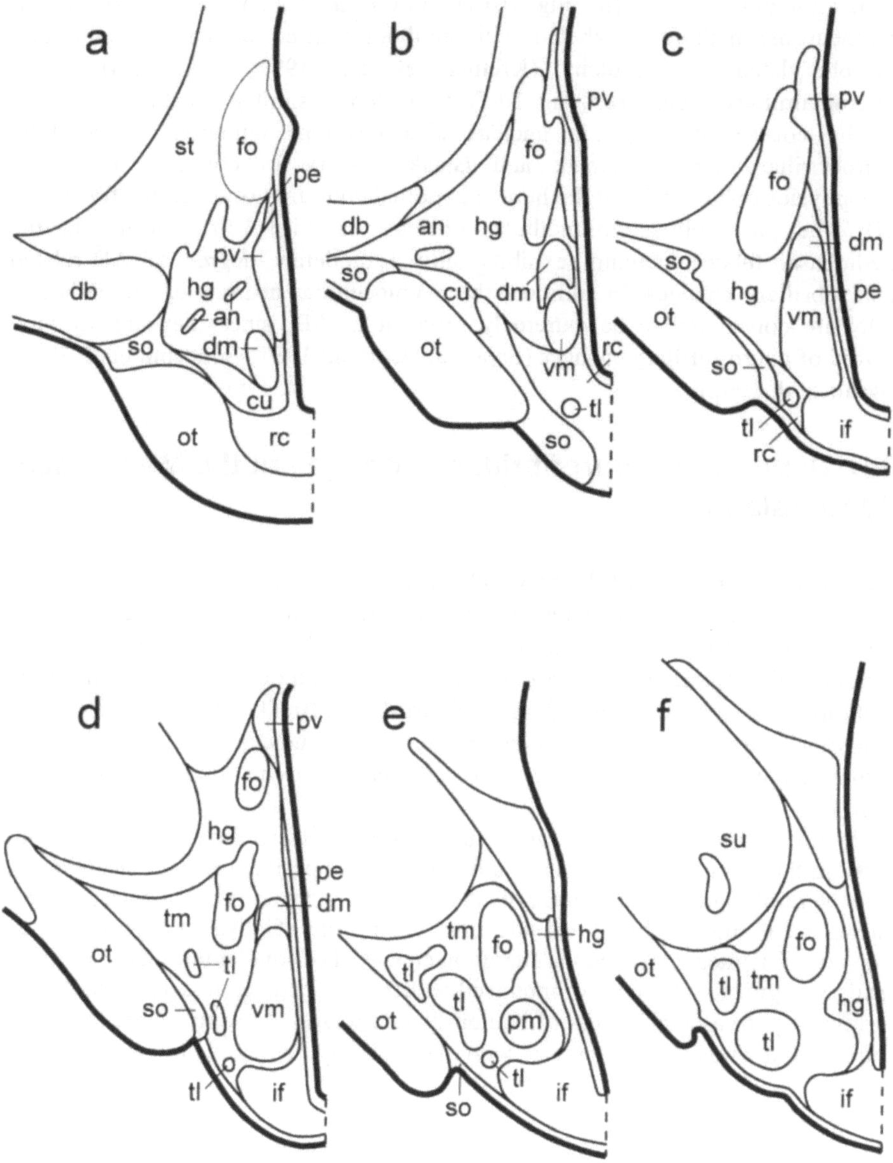

Fig. 1. Nuclear grays in coronal sections of the chiasmatic and tuberal region of the adult human hypothalamus. Each of the sections is spaced apart by 800 μm. *an* Accessory neurosecretory nucleus; *cu* cuneate nucleus; *db* nucleus of the diagonal band; *dm* dorsomedial nucleus; *fo* fornix; *hg* hypothalamic gray; *if* infundibular nucleus; *ot* optic tract; *pe* periventricular nucleus; *pv* paraventricular nucleus; *pm* posteromedial nucleus; *rc* retrochiasmatic nucleus; *so* supraoptic nucleus; *st* nucleus of the stria terminalis; *su* subthalamic nucleus; *tl* lateral tuberal nucleus; *tm* tuberomamillary nucleus; *vm* ventromedial nucleus. (Adapted from Braak and Braak 1992)

lateral tuberal nucleus (tl, Fig. 1b-f) shows an abundance of AD-related intraneuronal pathology, labeled with antibodies directed against abnormally phosphorylated tau protein (Kremer et al. 1991). The surrounding tuberomamillary nucleus (tm, Fig. 1d-f) is affected by similar cytoskeletal lesions. In the course of AD, this nucleus also develops numerous argyrophilic neurofibrillary tangles (Braak and Braak 1991a). In contrast, the major hypophysiotrophic nuclei of the human hypothalamus are not severely affected by AD. The parvocellular infundibular nucleus (if, Fig. 1c-f), located in the mediobasal tuber cinereum, exhibits only a moderate degree of AD-related neurofibrillary changes. In addition, the magnocellular neurons of the PVN and SON are considered to be extremely resistant to AD, since they are virtually devoid of neurofibrillary changes (Saper and German 1987; Goudsmit et al. 1990; Swaab et al. 1992).

Sex-Dependent Neurofibrillary Changes in the Mediobasal Hypothalamus

Two major components of the parvocellular neurosecretory system are comprised of the infundibular (or arcuate) and the periventricular nuclei, located in the mediobasal hypothalamus (MBH). The horseshoe-shaped infundibular nucleus surrounds the lateral and posterior entrance of the infundibulum, filling profound portions of the tuber cinereum (Fig. 1c-f; Braak and Braak 1992). Superiorly, the infundibular nucleus merges into the periventricular nucleus. The latter is located within a narrow zone underlying the anterior ependymal lining of the third ventricle.

In the present study, a dense accumulation of neurofibrillary changes was noted in the infundibular nucleus and in the adjacent periventricular zone (Fig. 2). A network of extremely distended, dystrophic neurites with neurofibrillary tangles interspersed among them was identified using the Gallyas silver technique (Fig. 2a). The most conspicuous alteration consisted of smaller terminal-like fibers easily recognizable by their pronounced tendency to contact vessels of the MBH, forming a perivascular plexus of bouton-like structures. In some cases, the entire course of a branching capillary loop was clearly outlined by such vessel-

Fig. 2. Severe degree of the sex-dependent neurofibrillary pathology in the mediobasal hypothalamus of a 76-year-old male. **a** A vascular network is labeled in the infundibular nucleus using the Gallyas silver technique. The vessels are outlined by their dense perivascular cuffing with small argyrophilic processes (*arrowhead* points to a branching vessel). *Small arrows* point to neurofibrillary tangles in the basolateral infundibular nucleus. **b** The pattern of cytoskeletal pathology labeled by PHF-1 resembles the pathology in the silver-stained section. **c** A particularly large number of structures is labeled by AT8. The perivascular pathology is partially concealed by the dense network of dystrophic neurites (not asscociated to vessels), which extend into the periventricular region (*arrowheads*). **d** Drawing summarizing the

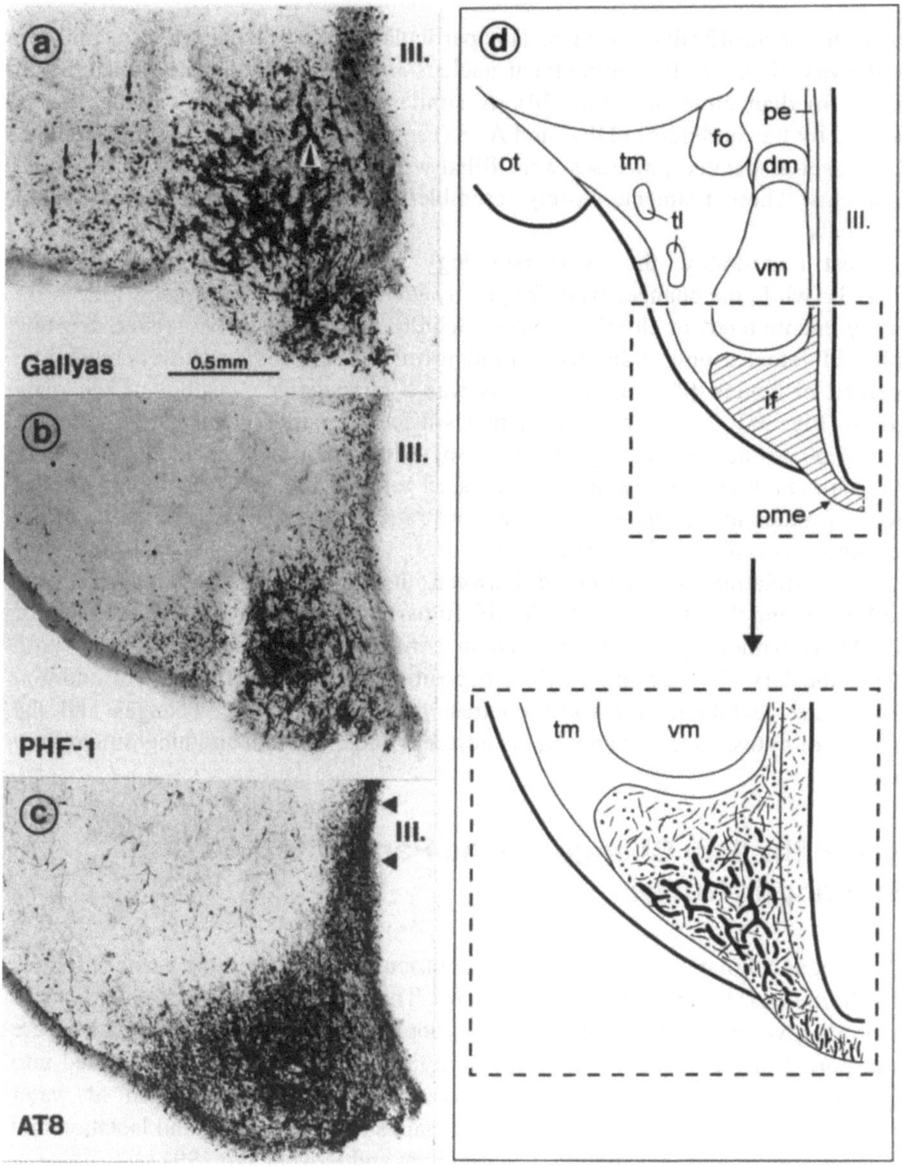

distribution of the MBH pathology. The pathology (*hatched area*) is confined to the infundibular nucleus (*if*), the adjoining periventricular region (*pe*), and the posterior median eminence (*pme*). *III* Third ventricle. **Insert:** Detailed drawing illustrating the major histopathological components of the cytoskeletal degeneration in the MBH. The perivascular pathology outlines a network of branched vessels (*bold lines*) in the posterior median eminence and in the adjacent infundibular nucleus. NFTs (*small dots*) and large dystrophic neurites (*small lines*) are found in the infundibular nucleus and the periventricular region. (With permission from Schultz et al. 1996). For other abbreviations see legend to Fig. 1

associated neurofibrillary lesions. This particular type of pathology was confined to the vasculature of the infundibular nucleus and the adjacent posterior extension of the median eminence (Fig. 2d). A similar pattern of MBH pathology was labeled by the antibodies PHF-1 and AT8 (Fig. 2b, c). Using electron microscopy, the vessel-associated processes were filled with bundles of AT8 immunopositive filaments. These filaments closely resembled the Alzheimer-type paired helical filaments.

Most remarkably, the MBH pathology revealed a striking sex-difference (Fig. 3): while the changes were frequently identified in elderly males, they were rarely encountered in elderly females (Schultz et al. 1996). Moreover, the few affected females only exhibited a mild form of the lesions, while a moderate, marked or severe form of the lesions was a characteristic feature of the male group (Fig. 3a). The percentage of males affected markedly increased from the age of 60 to the age of 90 years (Fig. 3b). More than 80% of males older than 75 years exhibited a more or less expressed form of the MBH pathology. About 30% of males in this age category demonstrated a severe or marked form of the lesions even visible to the unaided eye.

No correlation was observed between the severity of cortical Alzheimer pathology and the changes of the MBH. In particular, a marked or severe form of the MBH pathology was even noted in control cases virtually devoid of both neurofibrillary changes and amyloid depositing in neocortical areas. In addition these cases exhibited a marked contrast between the MBH changes and the paucity of cytoskeletal changes in the adjacent lateral tuberal and tuberomamillary nuclei.

Neurofibrillary Changes in the Paraventricular and Supraoptic Nuclei

The magnocellular neurons of the paraventricular and supraoptic nuclei produce the hormones vasopressin and oxytocin. The neurosecretory axons of these neurons give rise to the hypothalamo-hypophysial tract. The neuropeptides are transported to the neurohypophysis, where they are either stored or released into the blood stream. While vasopressin contributes to the regulation of water balance, oxytocin is involved in the maintenance of female labor and lactation and is implicated in male ejaculation (Cunningham and Sawchenko 1991).

A small subpopulation of brains investigated showed a severe neurofibrillary pathology in the magnocellular neurosecretory system (Schultz et al. 1997). Five of 66 cases (4 males, 1 female) exhibited the pathology in the PVN. Three of them displayed similar lesions in the SON. The neurofibrillary changes were labeled using specific antibodies directed against abnormally phosphorylated tau protein. They were also visualized by the modified Gallyas silver staining technique (Fig. 4).

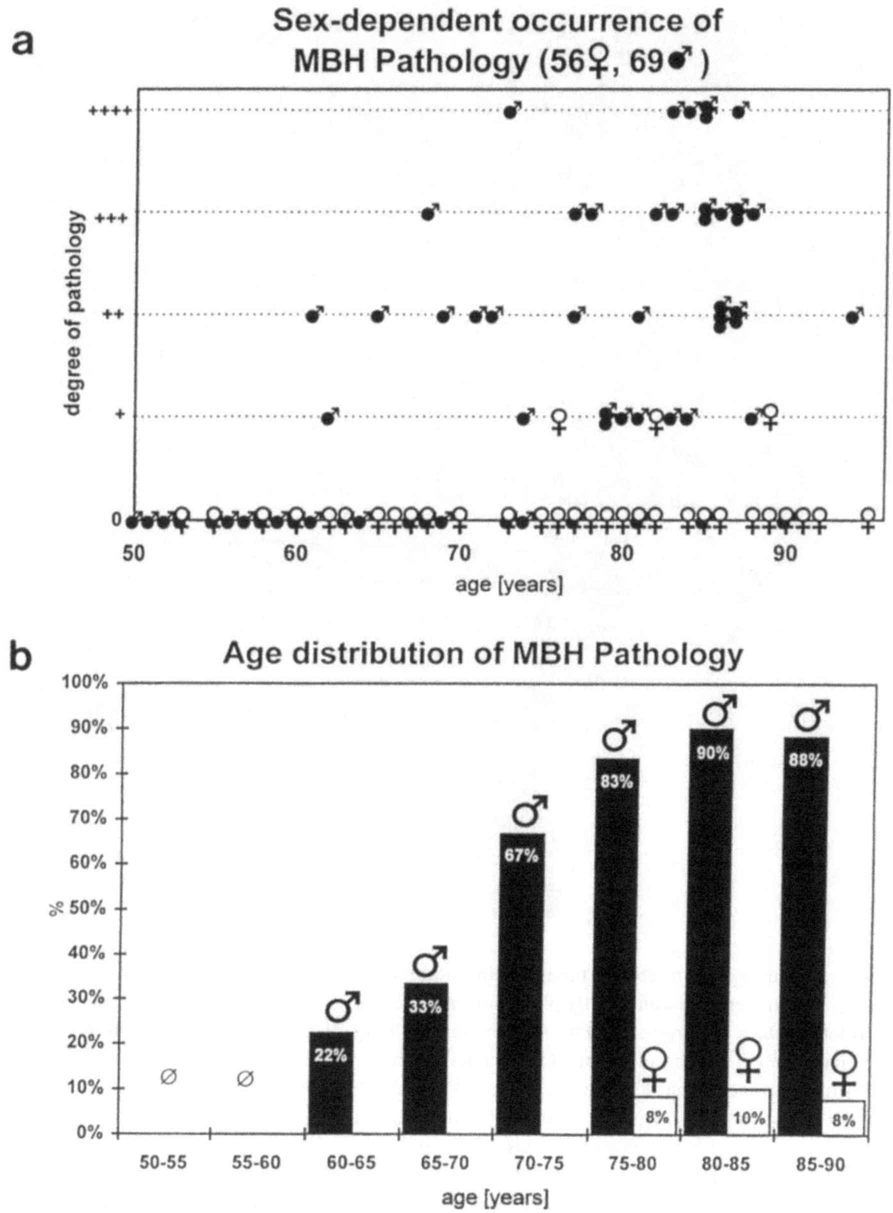

Fig. 3. **a** The degree of argyrophilic vessel-associated neurofibrillary lesions in the MBH was rated to be severe (++++), marked (+++), moderate (++), mild (+) or absent (0). An advanced form of the MBH pathology is exclusively identified in elderly males. In marked contrast, only three out of 56 females exhibit a mild form of the MBH pathology. **b** The percentage of male individuals affected by the MBH pathology markedly increases from the age of 60 to the age of 90 years. A marked or severe degree of the MBH pathology was identified in 30% of all males at this age (not shown). In contrast, only a small percentage of elderly females is affected

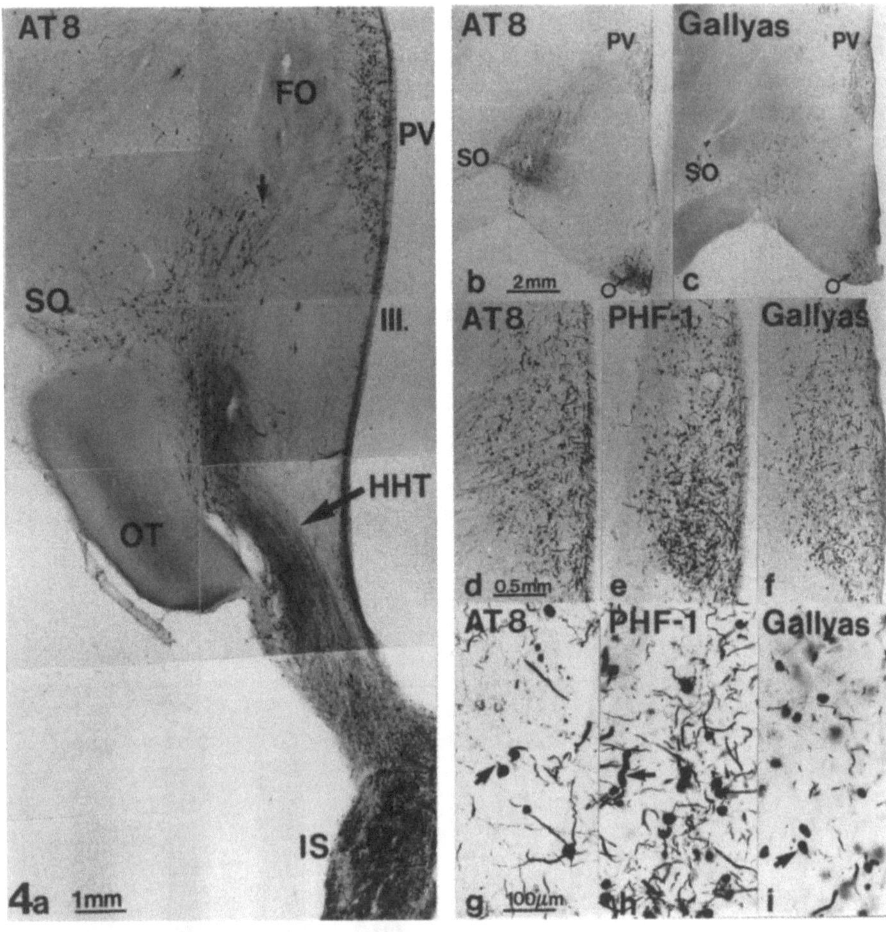

Fig. 4. a Frontal section through the tuberoinfundibular region of the hypothalamus (male, 77 years). The presence of abnormally phosphorylated tau protein in the magnocellular neurosecretory system is demonstrated by AT8 immunostaining. Neuronal perikarya of the paraventricular (*PV*) and the supraoptic (*SO*) nuclei are immunolabeled. Additional clusters of AT8-immunoreactive perikarya bridging the PVN and SON (*small arrow*) represent accessory neurosecretory neurons. The hypothalamo-hypophysial tract (*HHT, large arrow*), descending to the proximal infundibular stalk (*IS*), is also affected by the cytoskeletal changes. **b, c** Consecutive sections through the tuberal hypothalamus (male, 73 years, NFT stage: I), demonstrating the neurofibrillary pathology of the PVN and SON. Note the prominent labeling of the proximal hypothalamo-hypophysial tract by AT8 immunostaining (**c**). Both techniques, the AT8 immunostaining and the Gallyas silver technique, label a severe degree of the sex-dependent pattern of neurofibrillary pathology in the MBH (♂). Scale in **b** is also valid for **c**. **d-f** Neurofibrillary pathology of the paraventricular nucleus demonstrated in consecutive sections. The pathology is labeled by AT8 immunostaining (**d**), PHF-1 immunostaining (**e**), and by the Gallyas silver technique (**f**). Note the additional presence of AT8-immuolabeled neurosecretory axons in **d**. Scale in **d** is also valid for **e** and **f**. **g** Neurofibrillary pathology of the para-

ventricular nucleus shown at higher magnification. Neuronal perikarya with dystrophic neurites and axons are AT8-immunostained. *Arrow* neuron with a neurosecretory axon. **h** Neuronal perikarya and dystrophic neurites (*arrow*) are labeled by PHF-1 immunostaining. **i** Neurofibrillary tangles (*arrow*) and dystrophic neurites are also demonstrated by the Gallyas silver technique. Scale in **g** is also valid for **h** and **i**. (With permission from Schultz et al. 1997)

The neurofibrillary changes of the PVN and SON consisted of a dense accumulation of neurofibrillary tangles and large dystrophic neurites (Fig. 4d-i). An abundance of AT8-ir axons, generated by the PVN, arched ventro-laterally towards the SON (Fig. 4a, b). Small clusters of immunoreactive perikarya were scattered throughout the proximal portion of the fiber tract (Fig. 4a). The PVN axons finally merged with those of the SON, forming a well-defined fiber tract descending to the pituitary stalk. The majority of the AT8-ir neurons in the PVN and SON were large or medium-sized and contained faintly stained dust-like lipofuscin granules. According to their size and pigmentation pattern they could be classified as belonging to the magnocellular neurosecretory cell type (Braak and Braak 1987). At the ultrastructural level, bundles of Alzheimer-type paired helical filaments were identified in AT8-ir neurons of the PVN and SON.

The stages of AD-related neurofibrillary changes and amyloid deposition showed no association with the PVN/SON pathology.

Discussion

In this study, neurofibrillary changes were identified in human hypothalamic nuclei which are of pivotal importance for pituitary function. A unique feature of these hypothalamic alterations is the dramatic sex-difference in neurofibrillary degeneration of the MBH, which preferentially affects elderly males. This finding clearly demonstrates that the human hypothalamus is sexually dimorphic not only in relation to morphology (Swaab and Hofmann 1995) but also in relation to the development of age-related pathological alterations. Remarkably, the presence of advanced MBH lesions can be used as a reliable neuropathological indicator for an elderly male brain. In thick sections (100 μm) of cases affected by a severe form of the MBH pathology this indicator can be detected even by the unaided eye.

The sex-specific occurrence of argyrophilic neurofibrillary changes associated with abnormally phosphorylated tau protein is particularly interesting in that it does not occur in AD-related neurofibrillary degeneration. Moreover, even though the MBH pathology exhibits immunocytochemical and ultrastructural characteristics of Alzheimer-type cytoskeletal alterations, it is not related to the severity of cortical or hypothalamic Alzheimer pathology. This suggests that the sex-dependent MBH changes correspond to a distinct entity of neurofibrillary degeneration developing independently of the neocortical lesions leading to senile dementia of the Alzheimer type.

The distribution and the morphological appearance of the sex-dependent neurofibrillary pathology suggests an impairment of neurosecretory function in the male MBH. The neurosecretory terminals of the parvocellular infundibular and periventricular nuclei contact the capillaries of the median eminence, which penetrate the tissue of the human MBH (Green 1948; Duvernoy 1972). This neurovascular connection is crucial to the neuroendocrine regulation of posterior pituitary function (Barry 1977; Plant et al. 1978). In this context it is noteworthy that the infundibular nucleus is a major component of the hypothalamic circuitry regulating gonadal function. Experimental lesions of the infundibular nucleus and the adjacent posterior median eminence interfere with the hypothalamic stimulation of gonadotropin secretion in primates (Plant et al. 1978) and lead to testicular and prostate atrophy in dogs (Davidson and Ganong 1960). Interestingly, an impaired stimulation of gonadotropin secretion was recently suggested for the hypothalamus of elderly males (Kaufmann et al. 1990; Vermeulen et al. 1995). At the ultrastructural level, a severe disarrangement of the neuronal cytoskeleton was noted, probably leading to a functional impairment of the neuroendocrine processes in the MBH. It appears tempting to speculate that the neurofibrillary MBH lesions provide an explanation for such a functional decline in parvocellular neuroendocrine function in elderly males. However, clinical information necessary for a correlation between neuroendocrine function and the presence of neurofibrillary changes in the hypophysiotrophic nuclei is not available yet.

The majority of cases included in this study were virtually devoid of neurofibrillary changes in the PVN and SON. This corroborates previous reports classifying these nuclei as particularly resistant to the development of AD-related pathology (Saper and German 1987; Goudsmit et al. 1990; Swaab et al. 1992). Therefore, the occurrence of marked neurofibrillary alterations in the magnocellular neurosecretory system of a few individuals is an unexpected finding (Schultz et al. 1997). Although the relation between the PVN/SON changes and the sex-dependent MBH lesions remains to be determined, it is noteworthy that in elderly males the neurofibrillary changes in the PVN/SON are consistently accompanied by the sex-dependent lesions of the MBH (Fig. 4b, c). In this small number of cases, the distribution pattern of hypothalamic neurofibrillary alterations is suggestive of an impairment in both magnocellular and parvocellular neuroendocrine function.

Again, the PVN/SON pathology was noted in individuals whose clinical protocols did not mention overt cognitive decline and who were not diagnosed as suffering from a dementing disorder or other neurological illnesses. In these control cases, even the hypothalamic predilection sites for AD-related neurofibrillary pathology were spared. Similar to the MBH pathology, this unusual pattern of severe changes in regions which are particularly stable in AD, and absence of or only mild alterations in areas prone to development of AD-related changes, suggest that the PVN/SON pathology develops independently of senile dementia of the Alzheimer type.

In summary, the neurofibrillary degeneration of both magnocellular and parvocellular hypophysiotrophic nuclei represents a unique variant of neurofibrillary pathology developing in the aging human hypothalamus.

Acknowledgments

This study was supported by the Deutsche Forschungsgemeinschaft and the Bundesministerium für Forschung und Technologie. The generous gift of the antibody PHF-1 by S. G. Greenberg is gratefully acknowledged. The authors thank J. Bohl and H.H. Goebel, Institute for Neuropathology, Mainz, for providing autopsy material. The skillful technical assistance of A. Biczysko, I. Szasz, and U. Trautmann is gratefully acknowledged.

References

Barry J (1977) Immunofluorescence study of LRF neurons in man. Cell Tissue Res 181: 1-14

Braak E, Braak H, Mandelkow EM (1994) A sequence of cytoskeleton changes related to the formation of neurofibrillary tangles and neuropil threads. Acta Neuropathol 87: 554-567

Braak H, Braak E (1987) The hypothalamus of the human adult: chiasmatic region. Anat Embryol 176: 315-330

Braak H, Braak E (1991a) Neuropathological staging of Alzheimer-related changes. Acta Neuropathol 82: 239-259

Braak H, Braak E (1991b) Demonstration of amyloid deposits and neurofibrillary changes in whole brain sections. Brain Pathol 1: 213-216

Braak H, Braak E (1992) Anatomy of the human hypothalamus (chiasmatic and tuberal region): In: Swaab DF, Hofmann MA, Mirmiran M, Ravid R, van Leeuwen FW (eds) The human hypothalamus in health and disease. Progress in Brain Research, vol 93. Elsevier, Amsterdam, pp 3-16

Campbell SK, Switzer RC, Martin TL (1987) Alzheimer's plaques and tangles: a controlled and enhanced silver-staining method. Soc Neurosci Abstr 13: 678

Cunningham ET, Sawchenko PE (1991) Reflex control of magnocellular vasopressin and oxytocin secretion. Trends Neurosci 14: 406-411

Davidson JM, Ganong WF (1960) The effect of hypothalamic lesions on the testes and prostate of male dogs. Endocrinology 66: 480-488

Duvernoy H (1972) The vascular architecture of the median eminence. In: Knigge KN, Scott DE, Weindl A (eds) Brain-endocrine-interaction. Median eminence: structure and function. Karger, Basel, pp 79-108

Gallyas F (1971) Silver staining of Alzheimer's neurofibrillary changes by means of physical development. Acta Morphol Acad Sci Hung 19: 1-8

Goudsmit E, Fliers E, Swaab DF (1990) The supraoptic and paraventricular nuclei of the human hypothalamus in relation to sex, age and Alzheimer's disease. Neurobiol Aging 11: 529-536

Green JD (1948) The histology of the hypophyseal stalk and median eminence in man with special reference to blood vessels, nerve fibers and a peculiar neurovascular zone in this region. Anat Rec 100: 273-296

Greenberg SG, Davies P, Schein JD, Binder LI (1992) Hydrofluoric acid-treated tau/PHF proteins display the same biochemical properties as normal tau. J Biol Chem 267: 564-569

Iqbal K, Braak H, Braak E, Grundke-Iqbal I (1993) Silver labeling of Alzheimer neurofibrillary changes and brain β-amyloid. J Histotechnol 16: 335-342

Kaufman JM, Deslypere JP, Giri M, Vermeulen A (1990) Neuroendocrine regulation of pulsatile luteinizing hormone secretion in elderly men. J Steroid Biochem Mol Biol 37: 421-430

Kremer B, Swaab D, Bots G, Fisser B, Ravid R, Roos R (1991) The hypothalamic lateral tuberal nucleus in Alzheimer's disease. Ann Neurol 29: 279-284

Mercken M, Vandermeeren M, Lübke U, Six J, Boons J, Van de Voorde A, Martin JJ, Gheuens J (1992) Monoclonal antibodies with selective specificity for Alzheimer tau are directed against phosphatase-sensitive epitopes. Acta Neuropathol 84: 265-272

Plant TM, Krey LC, Moossy J, McCormack JT, Hess DL, Knobil E (1978) The arcuate nucleus and the control of gonadotropin and prolactin secretion in the female rhesus monkey (*Macaca mulatta*). Endocrinology 102: 52-62

Saper CB, German DC (1987) Hypothalamic pathology in Alzheimer's disease. Neurosci Lett 74: 364-370

Schultz C, Braak H, Braak E (1996) A sex difference in neurodegeneration of the human hypothalamus. Neurosci Lett 212: 103-106

Schultz C, Ghebremedhin E, Braak H, Braak E (1997) Neurofibrillary pathology in the human paraventricular and supraoptic nuclei. Acta Neuropathol 94: 99-102

Swaab DF, Grundke-Iqbal I, Iqbal K, Kremer HPH, Ravid R, Van de Nes JAP (1992) τ and ubiquitin in the human hypothalamus in aging and Alzheimer's disease. Brain Res 590: 239-249

Swaab DF, Hofman MA (1995) Sexual differentiation of the human hypothalamus in relation to gender and sexual orientation. Trends Neurosci 18: 264-270

Swaab DF (1997) Neurobiology and neuropathology of the human hypothalamus. In: Bloom FE, Björklund A, Hökfelt T (eds) Handbook of chemical neuroanatomy, vol 13. Elsevier, Amsterdam, pp 39-136

Vermeulen A, Kaufman JM (1995) Ageing of the hypothalamo-pituitary-testicular axis in men. Horm Res 43: 25-28

11 Sexual Dimorphism, Steroid-Induced Plasticity, and Behavioral Significance of the Vasotocinergic Innervation of the Avian Brain

G.C. Panzica[1], N. Aste[2], C. Castagna[1], J. Balthazart[3], and C. Viglietti-Panzica[1]

[1]Department of Anatomy, Pharmacology, and Forensic Medicine, University of Torino, c.so M. D'Azeglio 52, I-10126 Torino, Italy; Phone: +39 11 6707729, Fax: +39 11 6707732, E-mail: giancarlo.panzica@unito.it
[2]Institute of Molecular Neurobiology, Shiga University of Medical Science, Otsu, Shiga, Japan
[3]Lab. Biochemistry, University of Liège, Liège, Belgium

Abstract

This review focuses on the immunocytochemical demonstration of vasotocin (VT), the antidiuretic hormone in the avian brain. This peptide is synthetized by a well described system of diencephalic magnocellular neurons, but more recent studies have also demonstrated the presence of immunoreactive sexually dimorphic parvocellular groups or fibers in diencephalic and extradiencephalic regions of different avian species, including the Japanese quail. The main cluster has been detected in the male in a region considered as the avian homologue of the mammalian nucleus of the stria terminalis (nST). These cells are not visible in the female. This dimorphism has also been confirmed by in situ hybridization studies. Moreover, sexually dimorphic vasotocin-positive fibers are present in regions involved in the control of different aspects of reproduction, i.e. the nucleus preopticus medialis (copulatory behavior), the lateral septum (secretion of GnRH), and the nucleus intercollicularis (vocalization control). In the male, the vasotocin-immunoreactivity in these regions is strictly testosterone-dependent: castration, or exposure to a short-day photoperiod, decrease VT-immunoreactivity to female levels. Administration of estradiol-benzoate to embryos (a treatment that abolishes masculine sexual behavior) results in a dramatic decrease of the VT-immunoreactivity in all these brain regions of male quail. Behavioral experiments demonstrate that intracerebroventricular administration of vasotocin strongly inhibits male sexual behavior. These data suggest that in birds this peptide is not only the antidiuretic hormone, but also plays a central role in the control of diverse aspects of reproduction.

Introduction

Vasotocin (VT) is a non-mammalian neurohypophysial hormone (Acher et al. 1993) belonging to a large family of structurally and functionally related neuropeptides that includes arginin vasopressin (VP), oxytocin, mesotocin (MT), isotocin, and hydrins (Acher 1993). VT and VP were originally identified as neurohypophysial hormones produced by hypothalamic magnocellular elements (Leng et al. 1992) and regulating hydromineral balance (Acher and Chauvet 1995). Because of the role played by these neuropeptides in osmoregulation, they were called antidiuretic hormone, ADH. However, these two nonapeptides are also secreted at the level of the median eminence and probably, by a large amount of nerve terminals within several brain regions to regulate the adenohypophysis (Makara et al. 1996) and a variety of brain functions and behaviors (Dantzer and Bluthé 1993; Insel et al. 1993). In birds, the physiological functions classically attributed to VT are the regulation of water and electrolytic balance (see Simon Oppermann et al. 1988; Ramieri and Panzica 1989 for reviews) and the control of blood pressure (Szczepanska-Sadowska et al. 1985). Like VP (and oxytocin) in mammals, VT is also involved in the control of several aspects of avian reproduction, such as oviposition, by causing oviduct contractions (Rzasa and Ewy 1982; Shimada et al. 1986; Koike et al. 1988; Rice et al. 1985), and activating male sexual behavior (Bernroider and Leutgeb 1994; Goodson et al. 1996; Kihlström and Danninge 1972; Leutgeb 1995), as well as vocalizations (Voorhuis et al. 1991).

Distribution of Vasotocin in Birds

The Magnocellular System

The distribution of magnocellular neurosecretory neurons in the avian hypothalamus were first studied by means of histochemical techniques (Gomori reaction). These investigations demonstrated a scattered pattern, with local cluster-like aggregations of neurosecretory elements, but considerable species differences were observed in the pattern and cell size of the neurosecretory nuclei (for a review, see Oksche and Farner 1974). Later, the availability of antibodies to VT and MT (the analogue of mammalian oxytocin) and the development of immunocytochemical techniques, allowed the pinpointing of these nonapeptides and/or their respective carrier proteins (neurophysins) in the hypothalamo-neurohypophysial system of several avian species (for a full review of the less recent literature, see Viglietti-Panzica and Panzica 1991). In recent years, the development, the anatomical location and the response to osmoregulation challenges of VT-gene expressing neurons have been studied in the chicken and quail, by means of in situ hybridization (ISH) techniques. These studies demonstrated an excellent correspondence between the distribution of the VT-mRNA containing elements and of the VT immunoreactive (VT-ir) neurons (Milewski et al. 1989; Mühlbauer et al. 1993; Chaturvedi et al. 1994; Grossmann et al. 1995; Aste et al. 1996a; Jurkevich et al. 1997). The more precise identification of this system, obtained by means of these new techniques, suggested that the distribution of magnocellular neurosecretory elements follows a similar scheme in the species investigated (including both oscine and non-oscine birds) and that this distribution can be easily compared with the distribution observed in the hypothalamo-neurohypophysial system in mammals (Sánchez et al. 1991).

As a rule, it is very difficult to match immunocytochemically detectable peptidergic cell groups and cerebral nuclei defined by classical methods of neuroanatomy such as the Nissl stain. This explains why different terminologies have been used by different authors to describe the anatomical distribution of peptidergic systems and this confusion could result in inaccurate homologies with other vertebrates (for further discussion, see Fasolo et al. 1988; Viglietti-Panzica and Panzica 1991; Sánchez et al. 1991; Moore and Lowry 1997). For this reason, Berk et al. (1982) suggested a nomenclature for VT cell groups in the pigeon brain based exclusively on the topographic position of VT-immunoreactive (VT-ir) cell clusters. This nomenclature was subsequently used for the study of other avian species (Japanese quail, domestic mallard, domestic fowl) and has been proposed as a general rule for non-oscine birds (Viglietti-Panzica 1986; see Fig. 1).

According to these studies, three diencephalic regions contain VT-ir cells: the lateral preoptic area and lateral hypothalamus (L1-L5 groups), the periventricular preoptic area and periventricular hypothalamus (P1-P3 groups), and the dorsal diencephalon (DD1-DD2 groups). The L1 and L2 clusters are located close to the optic tract, lateral and medial respectively; the former is intermingled with the lateral forebrain bundle. The L3 group lies dorsal to L2 and medial to L1; these three groups correspond to the mammalian supraoptic nucleus (Sánchez et al. 1991). The periventricular system extends in the ventro-dorsal direction. The P2 and P3 groups roughly correspond to the mammalian paraventricular nucleus (PVN) pars parvocellularis, whereas the L4 and DD2 group correspond to the magnocellular portion of the PVN (Sánchez et al. 1991). In general, cells in the periventricular groups have a cell size slightly smaller than in the other groups (Sánchez et al. 1991). The hypothalamo-neurohypophysial system of oscine birds is organized with a similar distribution pattern (Goossens et al. 1977; Kiss et al. 1987; Voorhuis and De Kloet 1992; Deviche et al. 1996), even if no comparison has been performed yet among the VT-ir cell groups of these species and those identified in the pigeon.

The Parvocellular System

The combination of more sensitive immunocytochemistry (ICC) methods with newly developed ISH techniques, as well as the analysis of the system in various physiological conditions, contributed to the discovery of a broader distribution of VT-ir cell bodies (smaller and less intensely labeled than the so-called magnocellular neurons) and fibers that are not directly connected to the hypothalamo-neurohypophysial system. VT-ir neurons were observed around the nucleus rotundus (RT) and in the region between this nucleus and the lateral geniculate nucleus. Larger elements were also identified in the lateral reticular formation. A small but distinct cluster was located dorso-laterally to the RT near the isthmo-optic tract (Panzica et al. 1988; Aste et al. 1996a). A rather large group of parvocellular VT-ir elements was identified in the avian nucleus of the stria terminalis (nST). This group was at first identified in oscine birds (Kiss et al. 1987; Voorhuis et al. 1988; Voorhuis and De Kloet 1992; Deviche et al. 1996), and later on also in galliforms (Aste et al. 1995, 1996a, b, 1997a; Jurkevich et al. 1996, 1997).

VT-ir fibers were described in several brain regions of different species. However, a complete description is available only for a few species (e.g. canary: Kiss et al. 1987; zebra finch: Voorhuis, De Kloet 1992; *Junco hyemalis*: Deviche et al. 1996; quail: Viglietti-Panzica et al. 1997). In quail, VT-ir fiber endings were observed in the telencephalon (lateral septum, nST), in the whole diencephalon, in the mesencephalon (optic tectum, nucleus intercollicularis, substantia grisea centralis, area ventralis of Tsai, and substantia nigra), in the pons (raphe nuclei, locus coeruleus, and tegmentum), and in the medulla (nucleus of the solitary tract). Isolated VT-ir fibers were observed in several other regions all over the brain. A similar distribution has been reported in the canary and in *Junco hyemalis*. In addition, a large supply of VT-ir fibers was detected around the nucleus robustus archistriatalis in song birds (Kiss et al. 1987; Deviche et al. 1996). This nucleus and the ICo (which is also labeled by VT-ir fibers in both oscines and non-oscine birds) are both implicated in the control of several aspects of song.

Fig. 1. Distribution of VT-ir structures in the limbic, hypothalamic, and rostral mesencephalic regions of the male Japanese quail. The drawings correspond to six different levels taken from transverse sections through the rostro-caudal extent. The *left* side of the drawings shows the distribution of nuclei in Nissl-stained sections. The schematic drawings on the *right* side show the distribution of VT-ir structures. *Black dots* represent immunoreactive cell bodies, and *small crosses* indicate positive fibers. *Letters* indicate different VT-ir cell groups: *L1-L4* lateral hypothalamic groups; *P1-P3* periventricular groups; *DD1-DD2* dorsal diencephalic groups (Viglietti-Panzica 1986). The density of symbols indicates the relative differences in the number of cells or fibers but not their actual numbers. *Ac* Nucleus accumbens; *AM* nucleus anterior medialis hypothalami; *AVT* area ventralis of Tsai; *Ce* cerebellum; *CO* optic chiasma; *CP* commissura posterior; *DLAl* nucleus dorsolateralis anterior, pars lateralis; *DLAm* nucleus

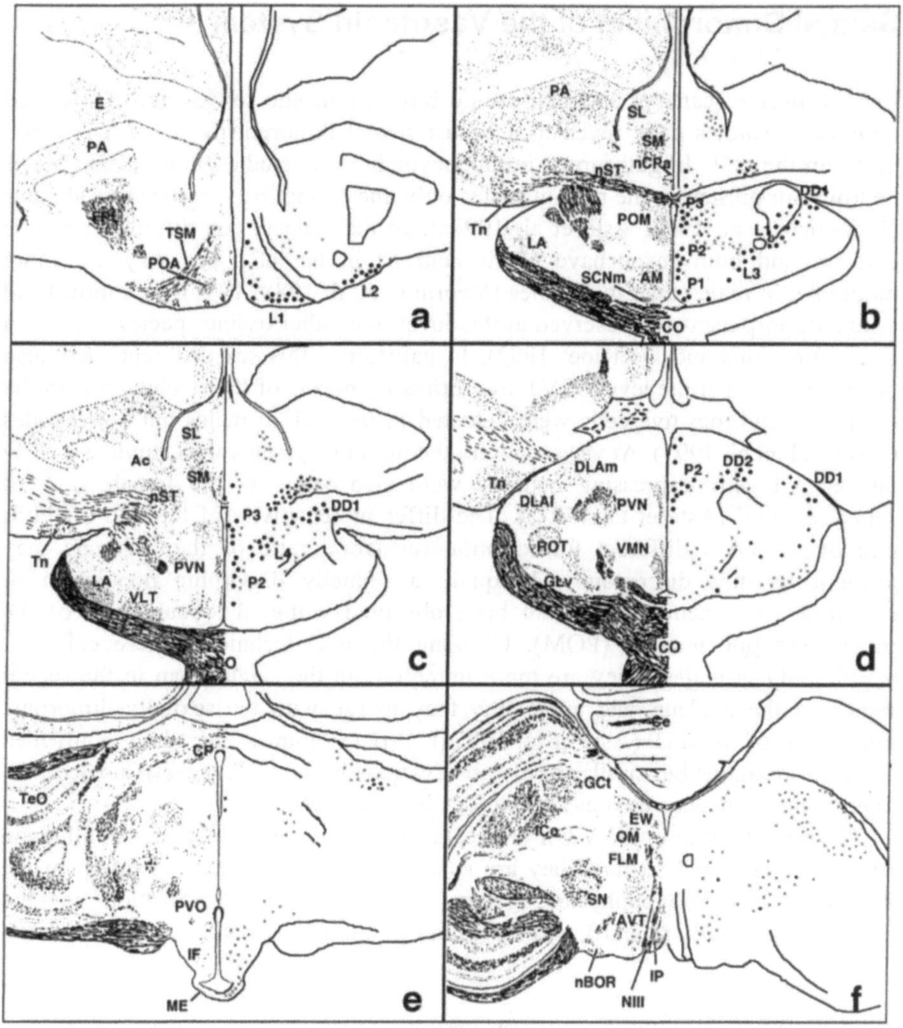

dorsolateralis anterior, pars medialis; *E* ectostriatum; *EW* nucleus of Edinger and Westphal; *FLM* fasciculus longitudinalis medialis; *FPL* fasciculus prosencephalicus lateralis; *GCt* substantia grisea centralis; *GLv* nucleus geniculatus lateralis, pars ventralis; *ICo* nucleus intercollicularis; *IF* nucleus infundibularis; *IP* nucleus interpeduncularis; *LA* nucleus lateralis anterior thalami; *ME* median eminence; *nBOR* nucleus of the optic basal root; *nCPA* nucleus commissurae pallii; *NIII* nervus oculomotorius; *nST* nucleus striae terminalis; *OM* nucleus nervi oculomotorii; *PA* paleostriatum augmentatum; *POA* nucleus praeopticus anterior; *POM* nucleus praeopticus medialis; *PVN* nucleus paraventricularis; *PVO* paraventricular organ; *ROT* nucleus rotundus; *SCNm* nucleus suprachiasmaticus, pars medialis; *SL* nucleus septalis lateralis; *SM* nucleus septalis medialis; *SN* substantia nigra; *TeO* optic tectum; *TSM* tractus septo-mesencephalicus; *Tn* nucleus taeniae; *VLT* nucleus ventrolateralis thalami; *VMN* nucleus ventromedialis

Sexual Dimorphism of the Vasotocin System

In both oscine (canary) and non-oscine birds (domestic fowl, quail), a limited number of studies have revealed a robust sexual dimorphism of the VT-ir cell group in the nST. In galliforms, immunocytochemical studies showed that VT-ir neurons are present in the nST of males only and cannot be visualized in females (Jurkevich et al. 1997; Aste et al. 1997a). In the canary, this difference is less extreme and both sexes have VT-ir neurons in the nST but they are more numerous in males than in females (Voorhuis et al. 1988, 1990). In contrast, no sexual dimorphism was observed at this level in another oscine species, the zebra finch (Voorhuis and De Kloet 1992). In galliforms, this sex difference has also been confirmed at the level of VT transcripts by means of ISH techniques. In the domestic fowl, positive cells were detected in the nST of males but not females (Jurkevich et al. 1997). At variance with the immunocytochemical studies, weakly labeled VT-gene expressing neurons were also found in the female nST of Japanese quail (Aste et al. 1997a). The different sensitivity of the two methods employed, and a different transcription/translation ratio of the VT-gene may account for this discrepancy. In quail, a sexually dimorphic population of scattered VT-ir neurons has also been observed within the boundaries of the medial preoptic nucleus (POM). Utilizing the ICC technique, these cells are visible only in males. They are more numerous in the caudal than in the rostral region of the nucleus, and apparently they merge with the sexually dimorphic population of the nST (Aste et al. 1997a). ISH techniques essentially confirmed these observations but similar to the observations in the nST, weakly labeled VT-expressing elements were detected in female quail POM.

The sexually dimorphic VT-ir elements have similar characteristics in all avian species investigated so far. They are smaller than the neurons belonging to the hypothalamo-neurohypophysial system, their dendritic arborization is less complex (generally bipolar), and the immunostaining is weaker than that in the

Fig. 2. Sex differences in the VT-ir cells and fibers in the quail brain. **A** Cross-sectional cell area of VT-ir magnocellular elements in adult quail of both sexes. Measurements were performed on Bouin fixed, paraffin embedded material. The values are the mean cell area (μm^2) ± standard error of the mean (number of birds in each group = 4). A minimum of 100 cells per bird were measured. A two-way ANOVA was performed with sex and position as factors. The results indicated a significant effect of position ($F=3.943, p=0.038$), whereas no effect was detected for sex ($F=0.866, p=0.3645$) or for the interaction of sex and position ($F=0.279, p=0.76$). A subsequent Fisher PLSD test confirmed that the cell sizes in the *L1-L2* groups are significantly larger than in the other two groups in both sexes (* = $p<0.05$). **B** Fractional area (FA) covered by VT-ir structures in the nucleus of the stria terminalis (*nST*) and in the lateral septum (*SL*). FA was measured on SUSA fixed, paraffin embedded material as described in Viglietti Panzica et al. 1994. In both *nST* and *SL* there is a significant (* = $p<0.01$) sexual difference of VT-ir structures that cover a larger area in male than in female quail. Data from Jurkevich et al. (1996)

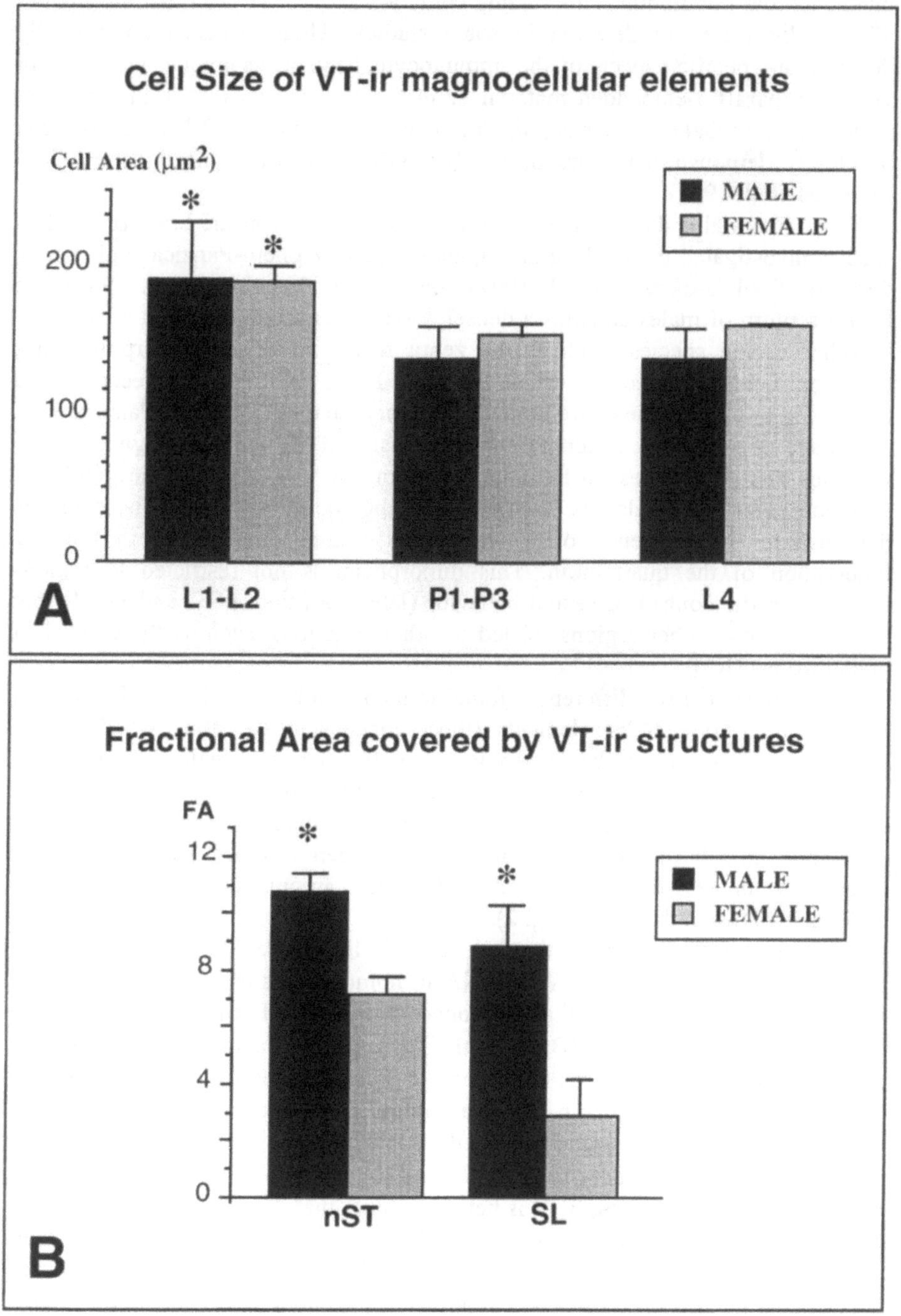
Cell Size of VT-ir magnocellular elements
Cell Area (μm²)
200
100
0
MALE
FEMALE
*
*
L1-L2
P1-P3
L4
A
Fractional Area covered by VT-ir structures
FA
12
8
4
0
MALE
FEMALE
*
*
nST
SL
B

magnocellular population. This finding could account for the fact that the smaller VT-ir cells were not detected in some studies. Their visualization critically depends on specific aspects of the immunocytochemical technique; they are not visible in paraffin-embedded material or in too thin sections. It is important to remember here that a similar population of sexually dimorphic VP-ir neurons can be clearly demonstrated in mammals, but only after colchicine injection (Van Leeuwen et al. 1985).

VT-ir fibers also display a sexually dimorphic pattern in the brain of the few species investigated in detail. In the Japanese quail (Viglietti-Panzica et al. 1992), domestic fowl (Jurkevich et al. 1997), and canary (Voorhuis et al. 1988), the lateral septum of males contains a denser VT-ir innervation than that of females. In other oscine species, such as the zebra finch, no sexual dimorphism has, however, been reported (Voorhuis and De Kloet 1992). More recently, we completed a study on the entire brain of the Japanese quail (Viglietti-Panzica et al. 1997) and observed sex differences in the density of VT-ir fibers (higher density in males than in females) in the lateral septum, the nST (Fig. 2), the POM, the periventricular hypothalamus, the ICo, and the locus coeruleus. These results demonstrate the existence of a widespread sexual dimorphism in the VT innervation of the quail brain. This dimorphism is not restricted to regions involved in the control of sexual behavior (lateral septum, nST, and POM), but involves several other regions related to other functions, such as the control of vocalizations (ICo).

In summary, the sex differences found in the avian brain so far involve at least part of the parvocellular elements (those located in the nST, which is the equivalent of the mammalian bed nucleus of the stria terminalis) and the fiber systems located mainly in the prosencephalon. Until now, no sexual dimorphism of the hypothalamo-neurohypophysial system has been described. No evident difference in cell distribution or number has been reported. Lastly, cell size measurements revealed significant differences among the groups, but no difference between the sexes (Fig. 2).

In contrast to these findings that indicate only limited sex differences in the VT circuits, the quantification of VT mRNA in homogenates of whole brains of the domestic fowl has revealed that the concentration of VT transcripts is twice as high in cockerels as in hens (Barth et al. 1995). Recent data have also confirmed the presence of a similar sex difference in homogenates of whole quail brain (Mühlbauer, Aste, Grossmann, Panzica, unpubl.). Taking into account the small number of sexually dimorphic parvocellular VT neurons, these data suggest that the observed sex differences in brain VT mRNA content cannot be related to the parvocellular cell groups only. It is hence possible that there is a less obvious sex difference in the number of magnocellular hypothalamic neurons or in their peptide content. Alternatively, it is also possible that the relatively large differences in VT mRNA concentrations in conjunction with only limited differences in the protein indicate sex specific controls in the translation of the VT transcripts or in the turnover rate of VT mRNA (Jurkevich et al. 1996).

Seasonal and Hormone-Dependent Plasticity of the Vasotocin System

Recent investigations on the avian vasotocin system have also adressed variations that could occur as a function of the season or the hormonal status of the bird. It is well known that many species of birds are highly photodependent (photoperiodic) and that their circulating testosterone (T) levels fall to very low values when the animals are exposed to short days (during the fall and the winter in natural environment). In these conditions, the amount of VT-ir fibers in the lateral septum, nST, and POM, as well as the number of cells in the nST, dramatically decreases both in oscines (canary: Voorhuis et al. 1988; junco: Deviche et al. 1996) and in non-oscine birds (quail: Viglietti-Panzica et al. 1992; Panzica et al. 1996a).

In male quail, changes in VT innervation of these brain regions are apparently controlled by T: the amount of ir fibers was strongly reduced or completely disappeared in gonadectomized birds, whereas T-treatment of castrated males restored the number of immunoreactive nerve fibers to a density that is typically seen in sexually mature males (Viglietti-Panzica et al. 1992, 1994) (Fig. 3).

The T-dependence of the VT-ir circuitries is also reflected by the diminution of the ir fibers in the POM and septum observed during aging (when circulating T levels spontaneously decrease to very low values) and by the restoration of VT innervation which was observed in aged male quail treated with exogenous T (Panzica et al. 1996a). Interestingly, when ovariectomized females are treated with doses of T that are sufficient to induce a full restoration of the innervation in castrated males, no obvious changes in the septal innervation and in the nST were detected (Viglietti-Panzica et al. 1992). The insensitivity of female vasotocinergic circuits to T has been recently confirmed by an independent experiment showing that T-treated gonadectomized female quail have very rare or no VT-ir fibers in the POM and the nST (Castagna et al. 1997b). Taken together, these data suggest that, in quail, the sexual dimorphism in VT innervation does not depend on a differential activation by T in the two sexes (activational effect) but develops during the embryonic period (organizational effect). These sex differences and controls by T are substantially different from those previously demonstrated in the canary brain (Voorhuis et al. 1988). In this species, T-treatment of females enhances the VT immunoreactivity in the lateral septum to a male-typical level. It therefore appears that the sex dimorphism observed in adult canaries reflects an activational role of T.

In male quail, a clear anatomical specificity was observed in the sexually dimorphic features and T sensitivity of the VT system which was affected at the level of the limbic and preoptic region, but not at the level of the magnocellular neurons or of the hypothalamo-neurohypophysial tract which were considered as controls (Viglietti-Panzica et al. 1992, 1994; Panzica et al. 1996a). The dimorphism of the quail vasotocinergic elements in the septum and in the nST is

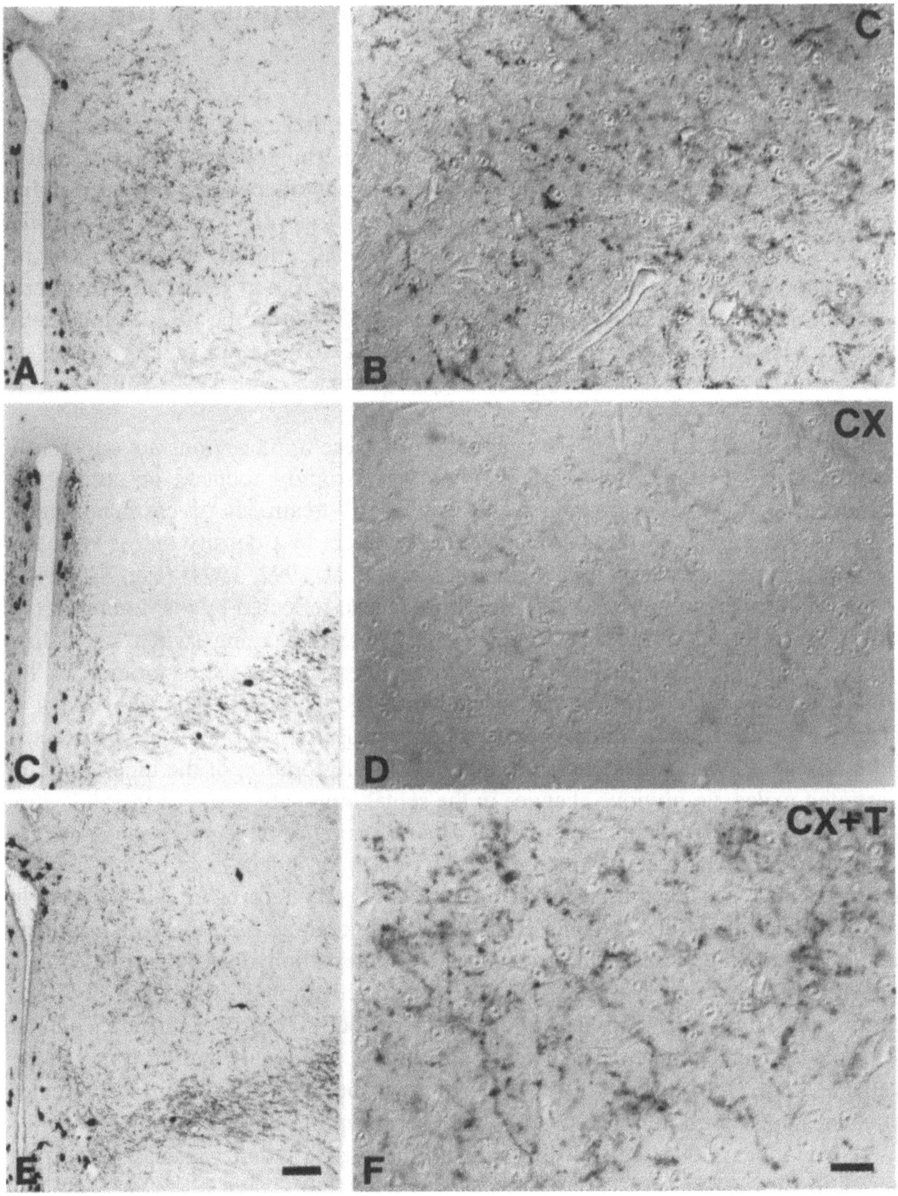

Fig. 3. Effects of testosterone on VT-ir structures in the POM of male quail. **A, B** Control sham-operated male (*C*); **C, D** castrated male (*CX*); **E, F** castrated male treated with testosterone (*CX+T*). Note the almost total loss of immunoreactivity in the CX male quail. *Bar* in **E** (for **A, C, E**): 100 µm, *bar* in **F** (for **B, D, F**): 25 µm. Also note the close contacts between VT-ir fibers and negative cell bodies in **B,D,F**

in some sense qualitatively similar to the situation observed in adult rats (De Vries and Al Shamma 1990). In this species there is also a sexual dimorphism in the induction by T of the VP innervation of the septum: a denser innervation is still observed in males than in females after a treatment with identical doses of T. Moreover, like in the rat (De Vries et al. 1985), the VT-ergic system of quail consists of elements that are sensitive and elements that are not sensitive to steroids.

There is no specific information on the origin of the steroid-sensitive VT-ir fibers in birds. Tracing studies in duck (Korf 1984) and quail (Balthazart et al. 1994, Balthazart and Absil 1997) have shown that the periventricular region projects to the septum and the POM, but there is no demonstration that the VT-ir cell bodies are implicated in these projections. Whether the VT-ir and VT mRNA expressing neurons in quail POM and nST (Aste et al. 1997a) are the origin of the immunoreactive fibers which outline these two nuclei and the lateral septum, remains to be experimentally analyzed. In rats, a sexually dimorphic T-sensitive population of VP-ir neurons has been found in the bed nucleus of the stria terminalis and in the medial amygdaloid nucleus (De Vries 1990). Recent studies testing the effects of various hormonal manipulations during rat development have demonstrated similar changes in the VP immunostaining of cells of the bed nucleus of the stria terminalis and the medial amygdaloid nucleus and of fibers in the lateral septum, suggesting that these two nuclei give rise to the sexually dimorphic VP-ir projection to the lateral septum (Wang et al. 1993). This conclusion has not been tested experimentally, however.

The results summarized above indicate that the sexually dimorphic innervation of the lateral septum and POM could be organizational in nature. To test this hypothesis, we injected quail embryos on day 8 of incubation either with estradiol benzoate (EB) or with the aromatase inhibitor, R76713, or with the solvent as a control. All subjects (males and females) were gonadectomized 4 weeks post-hatching and implanted with Silastic capsules filled with T, in order to place the adult birds in similar endocrine conditions. Major qualitative differences were still observed in the VT-ir innervation of the POM, lateral septum, and nST in the different experimental groups (see Table 1). In agreement with our previous studies, dense fiber networks were observed in control males but not in females. EB-treated males had completely lost this immunoreactivity and also the capacity to display copulatory behavior as previously shown in many experimental studies (see for a review Adkins Regan 1990). Conversely, R76713-treated females displayed a male-typical VT-ir innervation of these two brain regions and high levels of male-like copulatory behavior (Balthazart et al. 1992b).

No effect of the experimental treatments was detected at the level of the hypothalamo-hypophysial tract (Castagna et al. 1997b). These results indicate that the T-sensitivity of the male VT circuits terminating in the POM, lateral septum, and nST is organized during the embryonic period and becomes sexually differentiated following exposure of the female brain to estrogens. The sexual dimorphism observed in adults is hence truly organizational in its origin. Because this embryonic organization of the innervation of the POM by VT is controlled by mechanisms that are very similar to the mechanisms that organize the sex difference in male typical copulatory behavior (Balthazart and Ball 1995), these results also suggest a very close relationship between the VT innervation of POM and the control of male sexual behavior.

Table 1. Semiquantitative analysis of the distribution of VT-ir structures in the medial preoptic nucleus (POM), in the nucleus of the stria terminalis (nST), and in the lateral septum (SL) of male and female Japanese quail treated in ovo with estradiol benzoate (EB) or with the aromatase inhibitor, R76713. The last row indicates the presence or the absence of male copulatory behavior after each treatment. Details on the experimental procedure are given in the text

	MC	MR76	MEB	FC	FR76	FEB
POM	++++	++++	+	+	+++	+
NST	++++	++++	+	+	+++	+
SL	++++	++++	+	+	+++	+
Behavior	+	+	-	-	+	-

MC male controls; *FC* female controls; *MR76* male treated with aromatase inhibitor R76713; *FR76* female treated with aromatase inhibitor R76713; *MEB* male treated with EB; *FEB* female treated with EB. The number of + in each group is proportional to the area covered by VT-ir structures; + is very low and ++++ is a very high density of positive fibers

Effects of Vasotocin on Reproductive Behavior in Male Quail

The anatomical data reviewed above demonstrated a number of correlations between the brain vasotocinergic innervation and specific aspects of male reproductive behavior in quail. They established that (1) VT specifically innervates the sexually dimorphic medial preoptic nucleus, a structure that plays a key role in the activation of male sexual behavior (Panzica et al. 1996b), (2) this innervation is T-sensitive and (3) it is sexually dimorphic in the organizational sense (irreversibly differentiated by early estrogen action; Castagna et al. 1997b). Moreover, VT-ir fibers were found in close contact with aromatase-ir cells in the preoptic area and hypothalamus which supported the idea that VT could be implicated in the control of reproduction (aromatase is a limiting step in the activation of copulatory behavior by T; Balthazart et al. 1997a). VT and its mammalian homologue, VP were previously shown to have powerful effects on a variety of behaviors used in a reproductive context by amphibians, mammals and a few species of birds including the domestic fowl, pigeon, zebra finch and canary (see Castagna et al. 1997a for references). However, no data were available to assess the possible significance of VT for the control of reproductive behavior in quail.

In order to investigate this, castrated male quail that were chronically treated with exogenous T (subcutaneous Silastic implants), were injected with various doses of either [Arg^8]-vasotocin (VT; Sigma V0130) or the potent V1 receptor antagonist ([Deamino-Pen^1, Tyr(Me)^2, Arg^8]-Vasopressin or dPTyr(Me)AVP; Bachem H-5340), and the effects of these injections on appetitive sexual behavior were measured by a learned social proximity response (time spent near and frequency of looks through a window separating two adjacent cages, one containing the stimulus female and one containing the experimental male; see Balthazart et al. 1995, 1997b) and on consummatory sexual behavior (mount attempts and cloacal contact movements) as well as on crowing, a T-dependent vocalization.

VT inhibited, in a dose-dependent manner (doses ranging from 2 to 20 µg/bird), both the appetitive and consummatory aspects of male sexual behavior. All aspects of sexual behavior were inhibited by the highest dose of VT (20 µg) when tested immediately after the injection, but only measures of appetitive behavior were decreased if the test was delayed by 30 min. Mount attempts and cloacal contacts were not affected under these conditions (Castagna and Balthazart 1996). During independent experiments, crowing occurrences were counted by repeated 5 min-long observations of the birds in their home cage. The frequency of crowing was sharply decreased by the systemic injection of 2 or 10 µg VT per bird, and this effect lasted one to two hours. On the contrary, injections of the V1 receptor antagonist, dPTyr(Me)AVP (50-100 µg/bird), increased crowing frequency for periods of up to 4 hours suggesting that the behavioral effects of VT are receptor mediated and relatively specific (Fig. 4A-B).

These data suggested that VT plays an inhibitory role in many aspects of male sexual behavior in quail. To test the hypothesis that the behavioral effects of VT were centrally mediated (as suggested by the dense T-sensitive vasotocinergic innervation of the POM, a key center in the control of male reproductive behavior; Panzica et al. 1996b), VT was injected into the third ventricle (i.c.v.) of castrated male quail chronically treated with exogenous T. Increasing doses of VT in the range 2 to 200 ng per subject inhibited both appetitive and consummatory components of male sexual behavior in a dose-dependent manner. Appetitive behavior was, however, more sensitive to the treatment than consummatory behavior, the latter being significantly depressed only by the highest dose (Fig. 4C-D). At low doses, appetitive aspects of male sexual behavior were inhibited only. The highest VT dose (200 ng) that produced strong behavioral effects when injected i.c.v. had no effect when injected systemically (smallest effective dose was about 2 µg), indicating that the minimal effective dose was 10 to 100 times larger in systemic compared to central injection. This observation already provided some indication that the behavioral effects of VT were centrally mediated (Castagna et al. 1996).

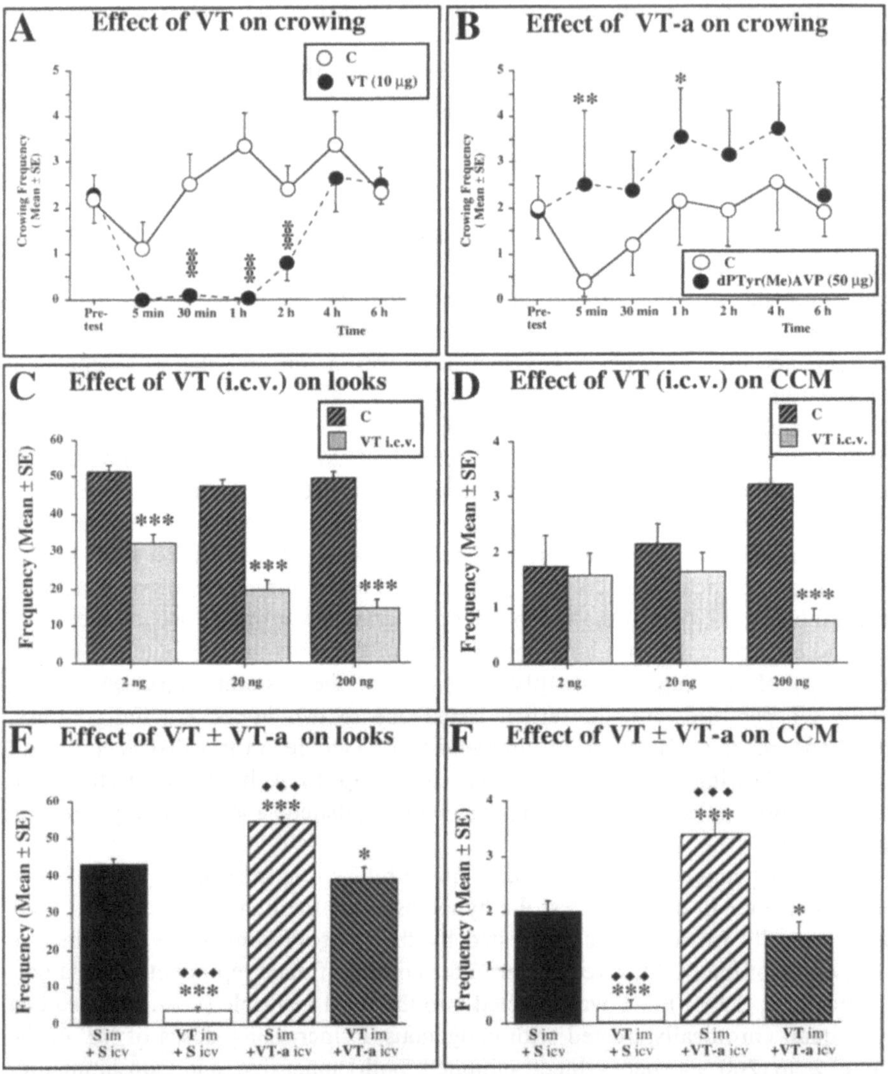

Fig. 4. Effects of *VT* and V1 receptor antagonist (*VT-a*) on crowing, appetitive (looks) and consummatory (cloacal contact movements; *CCM*) aspects of male sexual behavior in castrated male Japanese quail chronically treated with testosterone. **A, B** Effects of i.m. injections of *VT* (10 µg/bird) or of the *VT-a* (50 µg/bird) on crowing frequency. Birds were observed at different intervals after the injection (5 and 30 min, 1, 2, 4 and 6 hours; pretest is a test done as control before the injections). After the demonstration of a significant overall effect by analyses of variance (ANOVA), data were analyzed by post-hoc t tests for matched samples to compare control and *VT* or *VT-a*-injected birds at each time point. * = $p<0.05$; ** = $p<0.01$ and *** = $p<0.001$. **C, D** Effects of i.c.v. injections of *VT* (2-200 ng/bird) on appetitive and consummatory aspects of male sexual behavior. Data were first analyzed by ANOVA with two

repeated variables that demonstrated significant effects of the treatments. At each dose level, the control and *VT* data were compared then by post-hoc t-tests. * = $p<0.05$; ** = $p<0.01$ and *** = $p<0.001$. **E, F** Effects of i.m. injections of vasotocin (*VT*; 20 µg/bird), of i.c.v. injections of the *VT-a* (500 ng/bird), and of the combined treatment with *VT* and *VT-a* on appetitive and consummatory aspects of male sexual behavior. Injections of the vehicle (*S*) were always made to control for non specific effects of i.m. or i.c.v. injections. Data were analyzed by one way ANOVA that demonstrated significant overall effects of the treatments. The four treatments were then compared by post-hoc t-tests. * = $p<0.05$, ** = $p<0.01$ and *** = $p<0.001$ for comparisons with the control group (*S im* + *S icv*); ♦♦♦ = $p<0.001$ for comparisons with the *VT im* + *VT-a icv* group

This was further confirmed in subsequent experiments in which behavioral effects of a systemic VT injection could be blocked by a central injection of the V1 receptor antagonist dPTyr(Me)AVP. This study was carried out with one group of birds that had learned the social proximity response used to measure appetitive sexual behavior. Before each behavioral test, each bird received one i.c.v. injection followed 5 min later by one systemic (i.m) injection. The central injection contained either the VT-antagonist (VT-a) or 9 ‰ saline solution. The i.m. injection contained either VT or 9 ‰ saline. Behavioral tests were initiated immediately after the i.m. injection. The combination of the two central and two peripheral treatments therefore defined four separate experimental conditions. Quail were injected with (1) saline i.m. and saline i.c.v., or (2) VT i.m. and saline i.c.v., or (3) the VT-a i.c.v. and saline i.m., or finally (4) VT i.m. and the VT-a i.c.v. Each subject was tested for all four conditions and could therefore be used as its own control.

This experiment first confirmed that a central injection of VT very efficiently blocks both appetitive and consummatory aspects of male sexual behavior (Fig. 4E, F). In contrast, the VT antagonist, when injected alone, enhanced these aspects of the behavior, clearly indicating that under physiological conditions, endogenous VT exerts a tonic inhibition on these behaviors. The experiment also showed that the central injection of the antagonist was able to inhibit the behavioral effects of a systemic injection of VT, indicating that systemically injected VT really exerts its effects at the central level (Castagna et al. 1997a).

Conclusions

According to the literature reviewed above, four main groups of vasopressinergic neurons can be identified in the mammalian brain, namely the magnocellular and parvocellular neurons located in the PVN and SON, and the parvocellular elements identified (normally after colchicine injection) in the SCN, bed nucleus of the stria terminalis, and medial amygdala (De Vries et al. 1994). Magnocellular neurons of the PVN and SON are the origin of the fibers that project to the neurohypophysis through the hypothalamo-hypophysial tract. The parvocellular neurons of the PVN project to the hindbrain and spinal cord, probably participating in a pathway that controls neurovegetative functions. There is no evidence at present that these specific vasopressinergic projections are controlled by steroids.

The other three VP-ir cell groups could hence be the origin of the steroid-sensitive vasopressinergic projections that have been described in mammals (cf. De Vries 1995). However, available data suggest that the VP-ir cells of the SCN do not contain steroid receptors and do not have a sexually dimorphic distribution. It has therefore been suggested that all steroid-sensitive vasopressinergic fibers in the mammalian brain originate from the sexually dimorphic VP-ir cell populations located in the bed nucleus of the stria terminalis and medial amygdala (De Vries et al. 1992).

In birds, ICC techniques revealed a distribution pattern for the VT-ir cell bodies belonging to the magnocellular system that is comparable to the pattern observed in mammals (Sánchez et al. 1991). These neurons also appear to project into the neurohypophysis (Mikami et al. 1978). This projection is not modulated by steroids (Viglietti-Panzica et al. 1994) despite the fact that estrogen and androgen receptors are present in the paraventricular and supraoptic regions of the quail brain (Watson and Adkins-Regan 1988; Balthazart et al. 1989, 1992a; Foidart et al. 1995). No study has, however, analyzed the potential colocalization of steroid receptors with VT in these areas.

As in mammals, sexually dimorphic parvocellular VT-ir neurons are present in the avian brain; they are mainly clustered in a region above and caudal to the anterior commissure (nST, recently identified as the avian homologue of the mammalian bed nucleus of the stria terminalis, Aste et al. 1997a), and in the POM. The nST and the POM also contain both estrogen and androgen receptors (Watson and Adkins-Regan 1988; Balthazart et al. 1989, 1992a; Foidart et al. 1995) and in analogy to what has been proposed for mammals, these neurons could be the origin of the sexually dimorphic steroid-sensitive VT projections observed in the POM, the nST and the lateral septum. This is consistent with recent tracing studies that have identified connections between these areas (Balthazart et al. 1994; Balthazart and Absil 1997) but specific studies combining retrograde tracing with ICC for VT should be performed to confirm the anatomical organization of these projections. No VT-ir or VT mRNA-expressing cells have been identified in the areas of the quail brain that are believed to be homologous to the mammalian amygdala (archistriatum, and in particular the nucleus taeniae; Aste et al. 1996a). The organization of the parvocellular vasotocinergic system may therefore be quite different in quail and in rats.

In addition, small to medium VT-ir cells can be observed above the optic chiasma (Panzica 1985) in a position that could be considered homologous to the mammalian SCN, but this comparison is difficult to make given the uncertainty about the location of this nucleus in birds in general (Shimizu et al. 1994; Norgren and Silver 1990) and in quail in particular. No additional information is available at present for the other mesencephalic groups of parvocellular vasotocinergic neurons that have been described in quail both with ISH (Aste et al. 1996a) and ICC (Viglietti-Panzica et al. 1997).

The sexually dimorphic parvocellular VT neurons of the limbic/preoptic region are presumably the origin of the widespread innervation of brain areas implicated in the control of reproductive activities. These projections are steroid-sensitive and sexually differentiated. Taken together, the anatomical data suggest that VT may play a critical role in the control of reproductive behavior.

Because data in mammals (Smock et al. 1992; Wang et al. 1994; Ferris et al. 1994, 1996) and amphibians (Moore et al. 1992, 1994; Boyd 1997; Moore and Lowry 1997) indicated that indeed VP and VT control reproductive behavior in a variety of species, experiments were carried out to test this idea directly in quail. These studies showed that VT exerts a powerful inhibitory effect on both appetitive and consummatory components of male sexual behavior as well as on crowing. Additional studies demonstrated that the effect of VT is centrally mediated and can be dissociated from the general stress reaction induced by systemic injections of the peptide. Finally, studies using a powerful V1 antagonist also showed that endogenous VT is likely to exert tonic inhibitory effects on these behaviors, because blockade of these receptors leads to a significant stimulation of all these behavioral responses. The behavioral effects of VT injections mimic brain mechanisms that are part of the normal physiology of the species because the V1 receptor antagonist was able to produce opposite effects. In mammals, the V1 receptor subtype is broadly distributed in the brain where it is supposed to mediate most, if not all, behavioral effects of this neuropeptide. Our results suggest that a similar situation occurs in birds although additional studies would be required to test the receptor specificity of the behavioral effects described.

A number of studies on a variety of vertebrate species have previously demonstrated that VP or VT have a diversity of actions on behaviors ranging from the spawning reflex in some species of fishes to the mating call of amphibians and the flank marking response in mammals (Moore 1992; De Wied et al. 1988). Many of these studies suggest that VT or VP increase the occurrence frequency of the behaviors but some inhibitory effects have also been demonstrated (e.g. inhibition of release call in *Rana pipiens*, of spontaneous locomotion in *Rana catesbiana,* of lordosis in female rats: Diakow 1978; Södersten et al. 1983; Moore 1992). Little experimental work has unfortunately been carried out in birds. One early study showed that the injection of VT to intact sexually mature pigeons or cocks produces a short-term increase in the frequency of copulatory acts (Kihlström and Danninge 1972) but more recently, Bernroider and Leutgeb (1994) presented data in abstract form suggesting that VT decreases motivational aspects of sexual learning in quail. Two abstracts also recently reported that central VT injections stimulate courtship behavior and aggression in zebra finches, *Taeniopygia guttata* (Goodson et al. 1996) or singing in the female white-crowned sparrow, *Zonotrichia leucophrys gambelii* (Maney et al. 1997).

Contrary to the data presented here, some of these experiments point to a stimulatory role of VT in reproductive behavior in birds. It is, however, difficult to make direct comparisons between the different studies because they were carried out in different species and under very different experimental conditions (different seasons, endocrine conditions, etc). One study has, for example, suggested that seasonal variations might explain these changes in VT action because injections of a VT analog activated singing in canaries in the early fall but inhibited this behavior in the winter (Voorhuis et al. 1991). No general theory can be offered at present to explain why VT stimulates aspects of reproductive behavior in some species but inhibits the same or similar behaviors in other species. Additional studies on a variety of animal models will be needed to answer this question.

The identification of inhibitory effects of VT on sexual behavior in male quail was somewhat unexpected because the anatomical studies that have been reviewed above had demonstrated that VT-immunoreactive fibers in the POM (Viglietti-Panzica et al. 1994) and septal region/nST (Viglietti-Panzica et al. 1992; Aste et al. 1997b) increased after treatment of castrates with T which at the same time activates male sexual behavior (Panzica et al. 1996b; Balthazart et al. 1996). The most obvious explanation was therefore that the increase in VT production is part of a cascade of biochemical events triggered by T in the brain that result in the activation of male sexual behavior. The behavioral studies speak against this interpretation but alternative explanations can be offered.

Based on a number of studies carried out mostly in mammals (De Vries 1995), it is unlikely that increases in the density of VT-ir fibers after T treatment reflect a blockade of the peptide release, resulting in its accumulation in fibers and terminals (see Castagna et al. 1997a for further discussions). Alternatively, one could postulate that a single injection of exogenous VT decreases rather than increases the VT activity in specific brain areas because the injected compound has a short half-life and is quickly metabolized but at the same time strongly and persistently inhibits the secretion of the endogenous peptide. This interpretation, however, does not take into account the fact that inhibition of sexual behavior was observed almost immediately after the injection of VT and disappeared in approximately 30 min. We therefore do not favor this possibility and suggest that the apparent contradiction between anatomical and behavioral data indicates that the T-induced increase in VT immunoreactivity should not be considered as one of the central consequence of the T action that leads to the activation of male sexual behavior, but rather reflects the development of a mechanism that is implicated in the maintenance of behavioral homeostasis.

When male sexual behavior is activated (high levels of circulating T), a negative control mechanism may need to be established in order to organize the distribution of behavioral occurrences on a short-term basis (birds should not be sexually active all the time). This mechanism is obviously not needed in castrates that are behaviorally inactive. A steroid-sensitive neuropeptidergic innervation would provide adequate support for such a control that could be directly sensitive to environmental stimuli. On a long-term basis, changes in steroid levels would establish the anatomical substrate of this behavioral control (control of the synthesis of VT and possibly of the growth of VT-containing fibers) while on a short-term basis, environmental and social stimuli could regulate the release of VT from its terminals and in this way switch off behavior for limited periods of time. Additional work is obviously required to test this hypothesis, including the identification of the brain areas (POM, septum, nST) where VT exerts its behavioral effects, and the analysis of environmental stimuli that are potentially able to modulate VT release at these specific anatomical sites.

In conclusion, the data collected so far indicate that the steroid-sensitive vasotocinergic innervation of the medial preoptic area and/or of other regions such as the nST in the quail brain plays a significant role in the modulation of reproductive behavior. Steroids could therefore affect reproductive behavior both directly, by acting on steroid-sensitive neurons in the preoptic area, and indirectly, by modulating peptidergic (specifically vasotocinergic) inputs to this area. Additional work will be needed to evaluate the full significance of the indirect effects mediated by vasotocinergic inputs.

Acknowledgments

The collaboration between the Laboratories of Torino and Liège was supported by grants from the European Community (grant: CT94-0472), the Italian Ministero degli Esteri, and the European Science Foundation (RG95/203). This work was also supported by grants from the NIMH (R01 MH50388), the Belgian FRFC (Nbr. 9.4565.96F), the University of Liège (Fonds Spéciaux pour la Recherche) and Government of the French Community of Belgium (Action Concertée # 93/98-171) to JB, and by grants from MURST (60 and 40%) to GCP and CVP, and from CNR to CVP. The European Science Foundation supported the stay of C. Castagna in Liège and of N. Aste in Celle (Germany) with specific grants.

References

Acher R (1993) Neuroendocrine control of water homeostasis: the vasopressin/vasotocin/hydrin regulatory system. In: Mornex R, Jaffiol C, Leclere J (eds) Progress in endocrinology. Parthenon, Carnforth

Acher R, Chauvet J (1995) The neurohypophysial endocrine regulatory cascade: precursors, mediators, receptors, and effectors. Front Neuroendocrinol 16: 237-289

Acher R, Chauvet J, Chauvet MT, Michel G (1993) The avian neurohypophysial hormone-neurophysin precursors: structure, post-translational processing and evolution. In: Sharp PJ (ed) Avian endocrinology. Journal of Endocrinology Ltd, Bristol, pp 149-160

Adkins Regan EK (1990) Hormonal bases of sexual differentiation in birds. In: Balthazart J (ed) Hormones, brain and behaviour in vertebrates, vol 1. Sexual differentiation, neuroanatomical aspects, neuropeptides and neurotransmitters. Karger, Basel, pp 1-14

Aste N, Panzica GC, Viglietti-Panzica C, Absil P, Balthazart J, Mühlbauer E, Grossmann R (1995) The vasotocin system of the nucleus of the stria terminalis in the Japanese quail. Soc Neurosci Abstr 21: 357

Aste N, Mühlbauer E, Grossmann R (1996a) Distribution of AVT gene expressing neurons in the prosencephalon of Japanese quail and chicken. Cell Tissue Res 286: 365-373

Aste N, Viglietti-Panzica C, Mühlbauer E, Grossmann R, Panzica GC (1996b) Sexual dimorphism of vasotocinergic structures in quail nucleus of stria terminalis. Italian J Anat Embryol 101 (Suppl 1): 136-137

Aste N, Balthazart J, Absil P, Grossmann R, Mühlbauer E, Viglietti-Panzica C, Panzica GC (1997a) Anatomical, neurochemical, and hodological definition of the nucleus of the stria terminalis in Japanese quail (*Coturnix japonica*). J Comp Neurol (in press)

Aste N, Viglietti-Panzica C, Balthazart J, Panzica GC (1997b) Testosterone modulation of peptidergic pathways in the septo-preoptic region of male Japanese quail. Poultry Avian Biol Rev (in press)

Balthazart J, Absil P (1997) Identification of catecholaminergic inputs to and outputs from aromatase-containing brain areas of the Japanese quail by tract tracing combined with tyrosine hydroxylase immunocytochemistry. J Comp Neurol 382: 401-428

Balthazart J, Ball GF (1995) Sexual differentiation of brain and behavior in birds. Trends Endocrinol Metab 6: 21-29

Balthazart J, Gahr M, Surlemont C (1989) Distribution of estrogen receptors in the brain of the Japanese quail: an immunocytochemical study. Brain Res 501: 205-214

Balthazart J, De Clerck A, Foidart A (1992a) Behavioral demasculinization of female quail is induced by estrogens: studies with the new aromatase inhibitor, R76713. Horm Behav 26: 179-203

Balthazart J, Foidart A, Wilson EM, Ball GF (1992b) Immunocytochemical localization of androgen receptors in the male songbird and quail brain. J Comp Neurol 317: 407-420

Balthazart J, Dupiereux V, Aste N, Viglietti-Panzica C, Barrese M, Panzica GC (1994) Afferent and efferent connections of the sexually dimorphic medial preoptic nucleus of the male quail revealed by in vitro transport of DiI. Cell Tissue Res 276: 455-475

Balthazart J, Reid J, Absil P, Foidart A, Ball GF (1995) Appetitive as well as consummatory aspects of male sexual behavior in quail are activated by androgens and estrogens. Behav Neurosci 109: 485-501

Balthazart J, Tlemçani O, Ball GF (1996) Do sex differences in the brain explain sex differences in the hormonal induction of reproductive behavior? What 25 years of research on the Japanese quail tells us. Horm Behav 30: 627-661

Balthazart J, Absil P, Viglietti-Panzica C, Panzica GC (1997a) Vasotocinergic innervation of areas containing aromatase-immunoreactive cells in the quail forebrain. J Neurobiol 33: 45-60

Balthazart J, Castagna C, Ball GF (1997b) Aromatase inhibition blocks the activation and sexual differentiation of appetitive male sexual behavior in Japanese quail. Behav Neurosci 111: 381-397

Barth SW, Jurkevich A, Grossmann R (1995) Sexual dimorphism in the expression of Arg-vasotocin in the chicken brain. Poultry and Avian Biology Reviews 6: 279

Berk ML, Reaves TA, Hayward JN, Finkelstein JA (1982) The localization of vasotocin and neurophysin neurons in the diencephalon of the pigeon, *Columba livia*. J Comp Neurol 204: 392-406

Bernroider G, Leutgeb S (1994) V1-receptor mediated effects of vasotocin on motivational, mnemonic and aversive components of sexual learning in quail. Soc Neurosci Abstr 20: 1602

Boyd SK (1997) Brain vasotocin pathways and the control of sexual behaviors in the bullfrog. Brain Res Bull (in press)

Castagna C, Balthazart J (1996) Effects of vasotocin on sexual behavior and crowing in male Japanese quail. Italian J Anatomy Embryol 101 (Suppl 1): 148-149

Castagna C, Absil P, Balthazart J (1996) Central effects of vasotocin on appetitive and consummatory sexual behavior in male quail. Soc Neurosci Abstr 22: 2068

Castagna C, Absil P, Foidart A, Balthazart J (1997a) Systemic and intracerebroventricular injections of vasotocin inhibit appetitive and consummatory components of male sexual behavior in Japanese quail. Behav Neurosci (in press)

Castagna C, Panzica GC, Russo C, Tlemçani O, Balthazart J (1997b) Preoptic vasotocin: a neurochemical marker of the brain circuits mediating sexually differentiated copulatory behavior in quail. Soc Neurosci Abstr 23 (in press)

Chaturvedi CM, Newton BW, Cornett LE, Koike TI (1994) An in situ hybridization and immunohistochemical study of vasotocin neurons in the hypothalamus of water-deprived chickens. Peptides 15: 1179-1187

Dantzer R, Bluthé RM (1993) Vasopressin and behavior - from memory to olfaction. Regul Pept 45: 121-125

De Vries GJ (1990) Sex differences in neurotransmitter systems. J Neuroendocrinol 2: 1-13

De Vries GJ (1995) Studying neurotransmitter systems to understand the development and function of sex differences in the brain: the case of vasopressin. In: Micevich P, Hammer RP (eds) Neurobiological effects of sex steroid hormones. Cambridge Univ. Press, Cambridge, pp 254-278

De Vries GJ, Al Shamma HA (1990) Sex differences in hormonal responses of vasopressin pathways in the rat brain. J Neurobiol 21: 686-693

De Vries GJ, Buijs RM, Van Leeuwen FW, Caffé AR, Swaab DF (1985) The vasopressinergic innervation of the brain in normal and castrated rats. J Comp Neurol 233: 236-254

De Vries GJ, Crenshaw BJ, Al Shamma HA (1992) Gonadal steroid modulation of vasopressin pathways. Ann NY Acad Sci 652: 387-396

De Vries GJ, Al Shamma HA, Zhou L (1994) The sexually dimorphic vasopressin innervation of the brain as a model for steroid modulation of neuropeptide transmission. In: Luine VN, Harding CF (eds) Hormonal restructuring of the adult brain. Basic and clinical perspectives. Ann NY Acad Sci 743: 95-120

De Wied D, Joëls M, Burbach JPH, De Jong W, De Kloet ER, Gaffori HD, Urban IJA, Van Ree JM, Van Wimersma Greidanus TB, Veldhuis HD, Versteeg DHG, Wiegant VM (1988) Vasopressin effects on the central nervous system. In: Negro-Vilar A, Conn PM (eds) Peptide hormones: effects and mechanisms of action. CRC Press, Boca Raton, pp 97-140

Deviche P, García-Ojeda E, Plumari L, Panzica GC (1996) Vasotocinergic innervation in a male passerine bird (*Junco hyemalis*): effect of photoperiodic condition. Soc Neurosci Abstr 22: 1551

Diakow C (1978) Hormonal basis of breeding behavior in female frogs: vasotocin inhibits the release call of *Rana pipiens*. Science 199: 1456-1457

Fasolo A, Panzica GC, Viglietti-Panzica C, Renda T, D'Este L (1988) Comparative chemical anatomy of the brain: concepts and methods. Basic Appl Histochem 32: 15-30

Ferris CF, Delville Y, Irvin RW, Potegal M (1994) Septo-hypothalamic organization of a stereotyped behavior controlled by vasopressin in golden hamsters. Physiol Behav 55: 755-759

Ferris CF, Delville Y, Brewer JA, Mansour K, Yules B, Melloni RH (1996) Vasopressin and developmental onset of flank marking behavior in golden hamsters. J Neurobiol 30: 192-204

Foidart A, Houbart M, Prins GS, Balthazart J (1995) Distribution and neurochemical characterization of androgen receptor-containing cells in the quail brain. Soc Neurosci Abstr 21: 101

Goodson JL, Greenwood VR, Adkins-Regan E (1996) The central control of courtship and aggression in male zebra finches (*Taeniopygia guttata*): effect of vasotocin infusions. Soc Neurosci Abstr 22: 2068

Goossens N, Blähser S, Oksche A, Vandesande F, Dierickx K (1977) Immunocytochemical investigation of the hypothalamo-neurohypophysial system in birds. Cell Tissue Res 184: 1-13

Grossmann R, Kisliuk S, Xu B, Mühlbauer E (1995) The hypothalamo-neurohypophyseal system in birds. Adv Exp Med Biol 395: 657-666

Insel TR, Winslow JT, Williams JR, Hastings N, Shapiro LE, Carter CS (1993) The role of neurohypophyseal peptides in the central mediation of complex social processes - evidence from comparative studies. Regul Pept 45: 127-131

Jurkevich A, Barth SW, Aste N, Panzica GC, Grossmann R (1996) Intracerebral sex differences in the vasotocin system in birds: possible implication on behavioral and autonomic functions. Horm Behav 30: 673-681

Jurkevich A, Barth SW, Grossmann R (1997) Sexual dimorphism of arg-vasotocin gene expressing neurons in the telencephalon and dorsal diencephalon of the domestic fowl. An immunocytochemical and in situ hybridization study. Cell Tissue Res 287: 69-77

Kihlström JE, Danninge I (1972) Neurohypophysial hormones and sexual behavior in males of the domestic fowl (*Gallus domesticus*) and the pigeon (*Columba livia* Gmel.). Gen Comp Endocrinol 18: 115-120

Kiss JZ, Voorhuis TAM, Van Eekelen JAM, De Kloet ER, De Wied D (1987) Organization of vasotocin-immunoreactive cells and fibers in the canary brain. J Comp Neurol 263: 347-364

Koike TI, Shimada K, Cornett LE (1988) Plasma levels of immunoreactive mesotocin and vasotocin during oviposition in chickens: relationship to oxytocic action of the peptides in vitro and peptide interaction with myometrial membrane binding sites. Gen Comp Endocrinol 70: 119-126

Korf HW (1984) Neuronal organization of the avian paraventricular nucleus: intrinsic, afferent, and efferent connections. J Exp Zool 232: 387-395

Leng G, Dyball REJ, Luckman SM (1992) Mechanisms of vasopressin secretion. Horm Res 37: 33-38

Leutgeb S (1995) Social preference in the Japanese quail (*Coturnix coturnix japonica*): hormonal modulation. Thesis, Univ. Salzburg,

Makara GB, Kiss A, Lolait SJ, Aguilera G (1996) Hypothalamic-pituitary corticotroph function after shunting of magnocellular vasopressin and oxytocin to the hypophyseal portal circulation. Endocrinol 137: 580-586

Maney DL, Goode CT, Wingfield JC (1997) Intraventricular infusion of arginine vasotocin induces singing in a female songbird. SBN-CRB Meet, Baltimore, Abstr 91

Mikami SI, Tokado H, Farner DS (1978) The hypothalamic neurosecretory systems of the Japanese quail as revealed by retrograde transport of horseradish peroxidase. Cell Tissue Res 194: 1-15

Milewski N, Ivell R, Grossmann R, Ellendorff F (1989) Embryonal development of arginine vasotocin/mesotocin gene expression in the chicken brain. J Neuroendocrinol 1: 473-484

Moore FL (1992) Evolutionary precedents for behavioral actions of oxytocin and vasopressin. In: Pedersen CA, Caldwell JD (eds) Oxytocin in maternal, sexual, and social behaviors. Ann NY Acad Sci 652: 156-165

Moore FL, Lowry CA (1997) Comparative neuroanatomy of vasotocin and vasopressin in the vertebrate brain. In: Trudeau VL (ed) Comparative neuroendocrinology of vertebrate reproduction and growth: organismal, cellular and molecular aspects. Comp Biochem Physiol (in press)

Moore FL, Wood RE, Boyd SK (1992) Sex steroids and vasotocin interact in a female amphibian (*Taricha granulosa*) to elicit female-like egg-laying behavior or male-like courtship. Horm Behav 26: 156-166

Moore FL, Lowry CA, Rose JD (1994) Steroid-neuropeptide interactions that control reproductive behaviors in an amphibian. Psychoneuroendocrinology 19: 581-592

Mühlbauer E, Hamann D, Xu B, Ivell R, Udovic B, Ellendorff F, Grossmann R (1993) Arginine vasotocin gene expression and hormone synthesis during ontogeny of the chicken embryo and the newborn chick. J Neuroendocrinol 5: 281-288

Norgren RB, Silver R (1990) Distribution of vasoactive intestinal peptide-like and neurophysin-like immunoreactive neurons and acetylcholinesterase staining in the ring dove hypothalamus with emphasis on the question of an avian suprachiasmatic nucleus. Cell Tissue Res 259: 331-339

Oksche A, Farner DS (1974) Neurohistological studies of the hypothalamo-hypophysial system of *Zonotrichia leucophrys gambelii* (Aves, Passeriformes) with special attention to its role in the control of reproduction. Adv Anat Embryol Cell Biol 48: 1-136

Panzica GC (1985) Vasotocin-immunoreactive elements and neuronal typology in the suprachiasmatic nucleus of the chicken and Japanese quail. Cell Tissue Res 242: 371-376

Panzica GC, Calcagni M, Ramieri G, Viglietti-Panzica C (1988) Extrahypothalamic distribution of vasotocin-immunoreactive fibers and perikarya in the avian central nervous system. Basic Appl Histochem 32: 89-94

Panzica GC, García-Ojeda E, Viglietti Panzica C, Thompson NE, Ottinger MA (1996a) Testosterone effects on vasotocinergic innervation of sexually dimorphic medial preoptic nucleus and lateral septum during aging in male quail. Brain Res 712: 190-198

Panzica GC, Viglietti-Panzica C, Balthazart J (1996b) The sexually dimorphic medial preoptic nucleus of quail: a key brain area mediating steroid action on male sexual behavior. Front Neuroendocrinol 17: 1-75

Ramieri G, Panzica GC (1989) Comparative neuroanatomical aspects of the salt and water balance in birds and mammals. J Endocrinol Invest 12: 59-74

Rice GE, Arnason SS, Arad Z, Skadhauge E (1985) Plasma concentrations of arginine vasotocin, prolactin, aldosterone, and corticosterone in relation to ovoposition and dietary NaCl in the domestic fowl. Comp Biochem Physiol 81A: 769-777

Rzasa J, Ewy Z (1982) The effect of ovarian steroids in the response of the hen uterus to the neurohypophysial hormones. Acta Physiol Pol 33: 249-255

Sánchez F, Panzica GC, Viglietti-Panzica C, Aste N, Carretero J, Vázquez R (1991) A comparative analysis of the vasotocin and vasopressin systems in the chicken and rat hypothalamus. An immunocytochemical study. J Hirnforsch 32: 27-37

Shimada K, Neldon H, Koike TI (1986) Arginine vasotocin (AVT) release in relation to uterine contractility in the hen. Gen Comp Endocrinol 64: 362-367

Shimizu T, Cox K, Karten HJ, Britto LR (1994) Cholera toxin mapping of retinal projections in pigeon (*Columba livia*), with emphasis on retinohypothalamic connections. Visual Neurosci 11: 441-446

Simon Oppermann C, Simon E, Gray DA (1988) Central and systemic antidiuretic hormone and angiotensin II in salt and fluid balance of birds as compared to mammals. Comp Biochem Physiol 90A: 789-803

Smock T, Arnold S, Albeck D, Emerson P, Garritano J, Burrows K, Derber W, Sanson C, Marrs K, Weatherly H, Kruse K (1992) A peptidergic circuit for reproductive behavior. Brain Res 598: 138-142

Södersten P, Henning M, Melin P, Ludin S (1983) Vasopressin alters female sexual behaviour by acting on the brain independently of alterations in blood pressure. Nature 301: 608-610

Szczepanska-Sadowska E, Simon-Oppermann C, Gray DA, Simon E (1985) Blood pressure and arginine vasotocin in normonatremic and hypernatremic ducks. Basic Res Cardiol 80: 116-125

Van Leeuwen FW, Caffé AR, De Vries GJ (1985) Vasopressin cells in the bed nucleus of the stria terminalis of the rat: sex differences and the influence of androgens. Brain Res 325: 391-394

Viglietti-Panzica C (1986) Immunohistochemical study of the distribution of vasotocin reacting neurons in avian diencephalon. J Hirnforsch 27: 559-566

Viglietti-Panzica C, Panzica GC (1991) Peptidergic neurons in the avian brain. Ann Sci Nat Zool Paris 12: 137-155

Viglietti-Panzica C, Anselmetti GC, Balthazart J, Aste N, Panzica GC (1992) Vasotocinergic innervation of the septal region in the Japanese quail: sexual differences and the influence of testosterone. Cell Tissue Res 267: 261-265

Viglietti-Panzica C, Aste N, Balthazart J, Panzica GC (1994) Vasotocinergic innervation of sexually dimorphic medial preoptic nucleus of the male Japanese quail: influence of testosterone. Brain Res 657: 171-184

Viglietti-Panzica C, García-Ojeda E, Aste N, Castagna C, Panzica GC (1997) Sexual dimorphism of vasotocin system in the Japanese quail. Soc Neurosci Abstr 23 (in press)

Voorhuis TAM, De Kloet ER (1992) Immunoreactive vasotocin in the zebra finch brain (*Taeniopygia guttata*). Dev Brain Res 69: 1-10

Voorhuis TAM, Kiss JZ, De Kloet ER, De Wied D (1988) Testosterone-sensitive vasotocin-immunoreactive cells and fibers in the canary brain. Brain Res 442: 139-146

Voorhuis TAM, De Kloet ER, De Wied D (1990) The vasotocin system in the canary brain. In: Balthazart J (ed) Hormones, brain and behaviour in vertebrates. Vol.1 Sexual differentiation, neuroanatomical aspects, neurotransmitters and neuropeptides. Karger, Basel, New York, pp 168-179

Voorhuis TAM, De Kloet ER, De Wied D (1991) Effect of a vasotocin analog on singing behavior in the canary. Horm Behav 25: 549-559

Wang Z, Bullock NA, De Vries GJ (1993) Sexual differentiation of vasopressin projections of the bed nucleus of the stria terminals and medial amygdaloid nucleus in rats. Endocrinology 132: 2299-2306

Wang Z, Ferris CF, De Vries GJ (1994) Role of septal vasopressin innervation in paternal behavior in prairie voles (*Microtus ochrogaster*). Proc Natl Acad Sci USA 91: 400-404

Watson JT, Adkins Regan EK (1988) Neuroanatomical localization of sex steroid-concentrating cells in the Japanese quail (*Coturnix japonica*): autoradiography with (^{3}H)-testosterone, (^{3}H)-estradiol, and (^{3}H)-dihydrotestosterone. Neuroendocrinology 49: 51-64

12 The Stimulatory Action of the Glutamate Agonist, N-Methyl-Aspartate, on Luteinizing Hormone Release in the Cockerel with Immunocytochemical Observations on Its Mode of Action

R. Józsa[1,2], A. Mess[2], R. Gladwell[3], F.J. Cunningham[3], and P.J. Sharp[1]

[1]Division of Development and Reproduction, Roslin Institute (Edinburgh), Roslin, Midlothian, EH25 9PS, UK
[2]Department of Anatomy, University Medical School of Pécs, Hungary; Fax: +36 72 326 244, E-mail: jozsar@apacs.pote.hu
[3]School of Animal and Microbial Sciences, University of Reading, Whiteknights, PO Box 228, Reading RG6 2AJ, UK

Abstract

Bolus intravenous injections of 5-40 mg/kg body weight of the glutamate agonist, N-methlyl-aspartate (NMA), induced an immediate increase in the concentration of plasma luteinizing hormone (LH) in juvenile (8-10 week old) cockerels. The concentration of plasma LH returned to baseline values within 30 minutes. The NMA treatment resulted in the appearance of the product of an immediate-early gene, *c-fos,* localized immunocytochemically in cell nuclei in the preoptic, periventricular and basal hypothalamus. Although fos-immunoreactivity occurred in cell bodies close to gonadotropin releasing hormone-I (GnRH-I) perikarya, double immunocytochemical labeling demonstrated that the NMA treatment did not induce fos expression in GnRH-I neurons. Potential sites of action of NMA in releasing GnRH-I and LH were explored immunocytochemically. Perikarya, immunoreactive for N-methyl-D-aspartate (NMDA) receptor, were detected throughout the hypothalamus and were particularly prominent in the supraoptic and paraventricular nuclei, but none were observed in the the vicinity of GnRH-I neurons. Immunoreactive fibers were abundant throughout the hypothalamus, including the internal zone of the median eminence. Many were found in the vicinity of GnRH-I cell bodies, suggesting the possibility of direct contact. The presence of NMDA receptor was also demonstrated in the cephalic lobe of the anterior pituitary. These observations provide evidence for the first time in a bird, that LH release is stimulated by an excitatory amino acid. Apparently, the action of NMA does not involve expression of *c-fos* or NMDA-like receptors in GnRH-I neurons.

Introduction

Excitatory amino acids (EAAs), such as glutamate and aspartate, play an important role in the neuroendocrine control of gonadotropin secretion in mammals (Brann and Mahesh 1994). Their actions may be mediated by several types of receptors, but the most intensively investigated is the ionotropic glutamate receptor of the N-methyl-D-aspartate (NMDA) subtype (Brann and Mahesh 1994). Treatment with NMDA stimulates luteinizing hormone (LH) or gonadotropin releasing hormone (GnRH) release in several mammalian species including the rat (Price et al. 1978; Ondo el al. 1988; Bourguignon et al. 1989), hamster (Ebling et al. 1993, 1995; Urbanski et al. 1993), sheep (Lincoln et al. 1991), monkey (Wilson and Knobil 1982; Gay and Plant 1987), horse (Fitzgerald 1996), pig (Sesti and Britt 1992) and mouse (Saitoh et al. 1991). The action of NMDA on gonadotropin release is not exclusively stimulatory. Inhibitory effects have been reported, for example, in the guinea pig (Giri and Kaufman 1995) and gonadectomized monkey (Reyes et al. 1990). The action of NMDA on gonadotropin secretion is dependent on the steroid environment (Estienne et al. 1990; Reyes et al. 1990; Brann and Mahesh 1994), sexual maturity (e.g. Plant et al. 1989), oestrous cycle (e.g. Luderer et al. 1993) or the stage of a photo-induced breeding cycle (e.g. Urbanski 1990; Hui et al. 1992).

The mechanism through which NMDA regulates GnRH release is not fully established. Immortalized mouse GnRH neurons (GT-1) contain NMDA type 1 receptor (NMDAR1) and respond to NMDA with an increased release of GnRH (Mahachoklertwattana et al. 1994). However, observations in the rat suggest that only 8% (Eyigor and Jennes 1996) or between 2 and 19% (Gore et al. 1996) of GnRH neurons contain NMDAR1. There are four other NMDA receptor subtypes and of these only one, NMDAR2a, has been observed in a low percentage (19%) of GnRH neurons in the rat brain (Eyigor and Jennes 1996). It is therefore possible that the effects of NMDA on GnRH release are mediated indirectly. Attempts have been made to identify the mode of action of NMDA in the control of GnRH release in mammals using fos immunocytochemistry (Saitoh et al. 1991; Lee et al. 1993; MacDonald et al. 1993, Ebling et al. 1993, 1995) which provides a marker of the expression of the immediate early gene *c-fos* (Hoffman et al. 1993). These studies show that *c-fos* expression is induced in many cells in the hypothalamus after NMDA treatment, suggesting that some of them may mediate the action of NMDA on GnRH release. However, the absence of *c-fos* expression in GnRH neurons after NMDA treatment suggests that NMDA does not act to induce GnRH gene transcription.

The aim of this study was to establish whether, as in mammals, gonadotropin secretion is controlled by NMDA receptors in the domestic chicken, and if so, to determine, using fos- and NMDAR-immunocytochemistry, whether NMDA acts directly on the GnRH-I neurons. In the domestic chicken GnRH-I, rather than GnRH-II, directly controls the release of luteinizing hormone from the anterior pituitary gland (Sharp et al. 1990).

Materials and Methods

Birds and Treatments

Juvenile cockerels (8-10 weeks old) from a commercial brown egg laying breed (ISA Poultry Services, Peterborough, Cambridgeshire, UK) were reared and maintained under a short day photoschedule of 8 h light and 16 h darkness with free access to food and water. The birds were caged in groups until 7 weeks of age, after which they were caged individually. Blood samples (0.5 ml) were collected using a heparinized syringe from a wing vein, and after centrifugation, the plasma was separated and stored frozen until radioimmunoassay for LH.

N-Methyl-Aspartate Administration and LH Measurement

N-methyl-D,L-aspartate (NMA, Sigma) was dissolved at 40 mg/ml in 0.1M phosphate-buffered-saline (PBS), and the pH was adjusted to neutral with 5M NaOH. The drug was injected into a wing vein at doses of 5, 10, 20, and 40 mg/kg body weight. Control birds were injected with neutralized PBS vehicle. For immunocytochemical studies, NMA was administered at a dose of 10 mg/kg body weight one hour before sacrifice by perfusion under terminal anaesthesia. Blood samples were taken immediately before and 2 min after injection of NMA or vehicle. For the time course study, additional samples were taken at 15 and 30 min. Plasma LH was measured by radioimmunoassay as described by Sharp et al. (1987). The intra-and inter-assay coefficients of variation for the assay were 8.4 and 10.7%, respectively.

Preparation of Brain Sections for Immunocytochemistry

Cockerels were deeply anaesthetized with sodium pentobarbitone (Sigma) and perfused via cannulae inserted into the carotid arteries with 0.1 M PBS, pH 7.4. After the blood had been washed out of the brain, the perfusion was completed with Zamboni's fixative containing 4% paraformaldehyde in 0.1 M phosphate buffer, with 15% picric acid, pH 7.4. Serial coronal sections (70 µm, 30-40/bird) were cut through the hypothalamus in the plane of the Kuenzel and Masson (1988) chicken stereotaxic atlas, on a shaking microtome and processed for immunocytochemistry. Fos-immunolabeling was localized in the vehicle (n = 4), or NMA-injected birds (n = 4) and NMDAR-immunolabeling was localized in NMA injected birds (n = 4) only. Fos/GnRH-I, fos/NMDAR and GnRH-I/NMDAR double labeling were observed in different groups (n = 3) of birds after treatment with NMA.

Immunocytochemical Procedures

Free-floating sections were processed for immunocytochemistry using the peroxidase-anti-peroxidase method of Sternberger (1986). In brief, after washing in 0.2M PBS, the sections were immersed in 0.3% hydrogen peroxide to block endogenuos peroxidase activity. After this, they were washed in PBS and incubated for 30 min at 4 °C in 10% non-immune donkey serum in PBS to block non-specific labeling. For single labeling studies, the sections were then sequentially incubated in rabbit chicken-fos antibody (code 9/3, 1:10 000), or rabbit NMDAR1-receptor antibody (1:5000), diluted in PBS containing 1% non-immune donkey serum (48 h at 4 °C), anti-rabbit IgG (1:100) and rabbit peroxidase-antiperoxidase (1:100, DAKO). Peroxidase was visualized using diaminobenzidine tetrahydrochloride (DAB, Sigma). After dehydration, sections were mounted on glass slides and coverslipped. For fos/GnRH-I double labeling, the first primary antibody (rabbit anti-chicken fos) was visualized using nickel-DAB substrate (20 mg DAB, 80 mg nickel ammonium sulphate $6H_2O$ in 100 ml 0.05 M TRIS buffer, pH 7.6, with 7 μl 30% H_2O_2). After rinsing in PBS, the sections were incubated in the second primary antibody, sheep anti-chicken GnRH-I (1:10000), followed sequentially by donkey anti-sheep IgG (1:100) and sheep-PAP (1:100, DAKO). The reaction product was visualized using DAB. For NMDAR1/GnRH-I double labeling, the same procedure was adopted, but the fos primary antibody was replaced with the NMDAR1-antibody. For Fos/NMDAR1 double labeling, the first primary antibody (rabbit anti-chicken fos) was visualized using nickel-DAB and the second primary antibody (rabbit anti-NMDAR1) was visualized using DAB. Extensive washing between the application of the two primary antibodies ensured that fos was not visualized with DAB.

Antisera for Immunocytochemistry

All the antisera used in the study were polyclonal. The fos antibody was raised in rabbit against the 20 amino-acid C-terminal sequence of chicken fos as described and validated for immunocytochemistry by Sharp et al. (1995). The anti-chicken GnRH-I antibody was raised in sheep using the method described by Sharp et al. (1990). The antibody showed no cross-reactivity with GnRH-II, and immunolabeling was completely abolished by preincubation with chicken GnRH-I. The antibody to rat NMDAR1 was provided by Dr. F.J.P. Ebling (Department of Anatomy, University of Cambridge, UK). It was raised against a C-terminus peptide sequence of the rat NMDAR1 (Petralia et al. 1994) and would be expected to recognize the 1a, 1b, 2a, and 2b splice variants (Hollman et al. 1993).

Results

Effect of Intravenous Injection of NMA on LH Release

Bolus intravenous injections of 5, 10, 20, and 40 mg NMA/kg body weight resulted in immediate increases in the concentration of plasma LH which decreased to baseline values within 30 minutes (Fig. 1). In all cockerels taken for immunocytochemistry and injected with 10 mg NMA/kg body weight (bw), it was confirmed that the treatment induced an acute release of LH. No change in the concentration of LH was observed in vehicle-injected birds.

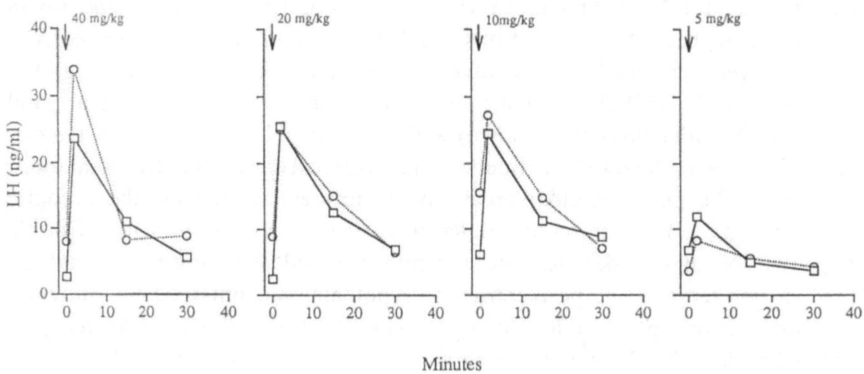

Fig. 1. Changes in the concentration of plasma LH after intravenous injections of 5-40 mg NMA/kg body weight in 8- week-old juvenile cockerels (n = 2 for each dose). *Open squares* and *open circles* represent two different animals

Fos-Immunolabeling in the Hypothalamus After NMA Injection

Fos-immunoreactivity was observed exclusively in cell nuclei and was not seen in cytoplasm or nerve fibers (Figs. 2A-F, 3E). One hour after an intravenous injection of 10 mg NMA/kg body weight, fos-labeling appeared within cells in the hypothalamus which were not labeled in vehicle-injected control birds. Prominent clusters of fos-labeled cells appeared in the ventral area of the tractus corticohabenularis et corticoseptalis (CHCS; Fig. 2A) and nucleus habenularis. These cell groups extended rostro-ventrally, with a particularly prominant group associated with the nucleus commissurae pallii (nCPa; Fig. 2B), passing above the anterior commissure into the dorsal preoptic region surrounding the organum vasculosum laminae terminalis (Fig. 2C). Other groups of fos-labeled cells occurred in the basal tuberal hypothalamus close to the median eminence (Fig. 2D) and in the periventricular areas.

Double Fos- and GnRH-I - Immunolabeling After NMA Injection

Immunolabeling for GnRH-I was restricted to the cytoplasm (Fig. 2E, F). Immunolabeled GnRH-I perikarya were observed in the hypothalamus extending from the preoptic-OVLT area caudo-dorsally above the anterior commissure into the medial septal-CHCS region. This continuum of GnRH-I perikarya was associated with cells containing fos-immunolabeled nuclei but fos-immunoreactivity was not observed in GnRH-I cells (Fig. 2E, F).

Immunolocalization of Hypothalamic NMDAR1-Like Receptor

Immunolabeling for NMDAR1-like receptor was found in cytoplasm and fibers (Fig. 3A-F). The greatest concentrations of NMDAR1-immunoreactive (NMDAR-ir) perikarya were observed in the paraventricular and supraoptic nuclei (Fig. 3A, B, D). Scattered NMDAR-ir cells were seen in the periventricular and ventromedial hypothalamus (not shown) and in the cephalic lobe of the anterior pituitary (Fig. 3C). NMDAR-ir fibers were widespread in the hypothalamus, particularly in the periventricular areas, the ventromedial nucleus, the preoptic area, septum, the CHCS region and inner zone of the median eminence. Double labeling for fos and NMDAR1-like receptor showed that some, but not all NMDAR-ir perikarya in the periventricular hypothalamus contained fos-immunoreactive nuclei (Fig. 3E). Double labeling for GnRH-I and NMDAR1-like receptor revealed abundant NMDAR1-ir fibers associated with GnRH-I perikarya and processes (Fig. 3F).

Discussion

To our knowledge, this is the first report in a bird demonstrating a stimulatory effect of NMA on LH release. It is therefore likely that, as in several mammals (see Introduction), endogenous glutamate plays a stimulatory role in the regulation of the avian GnRH and LH release. The present observations are not able to distinguish between possible sites of action at the level of the hypothalamus to release GnRH-I or at the level of the anterior pituitary gland to stimulate LH release directly. Evidence from the mammalian literature suggests that NMA has no direct stimulatory action at the level of the pituitary gland to release LH (Ondo et al. 1976; Tal et al. 1983). This observation remains to be confirmed in the cockerel. The time course of the LH response to an intravenous injection of NMA is the same as that observed after an intravenous injection of GnRH-I (Sharp et al.

Fig. 2. Representative photomicrographs of coronal sections through the brains of 9-week-old cockerels 1 h after an intravenous injection of NMA (10 mg/kg bw).The distribution of fos-immunoreactive neurons is seen within the medial subdivision of the tractus corticohabenularis

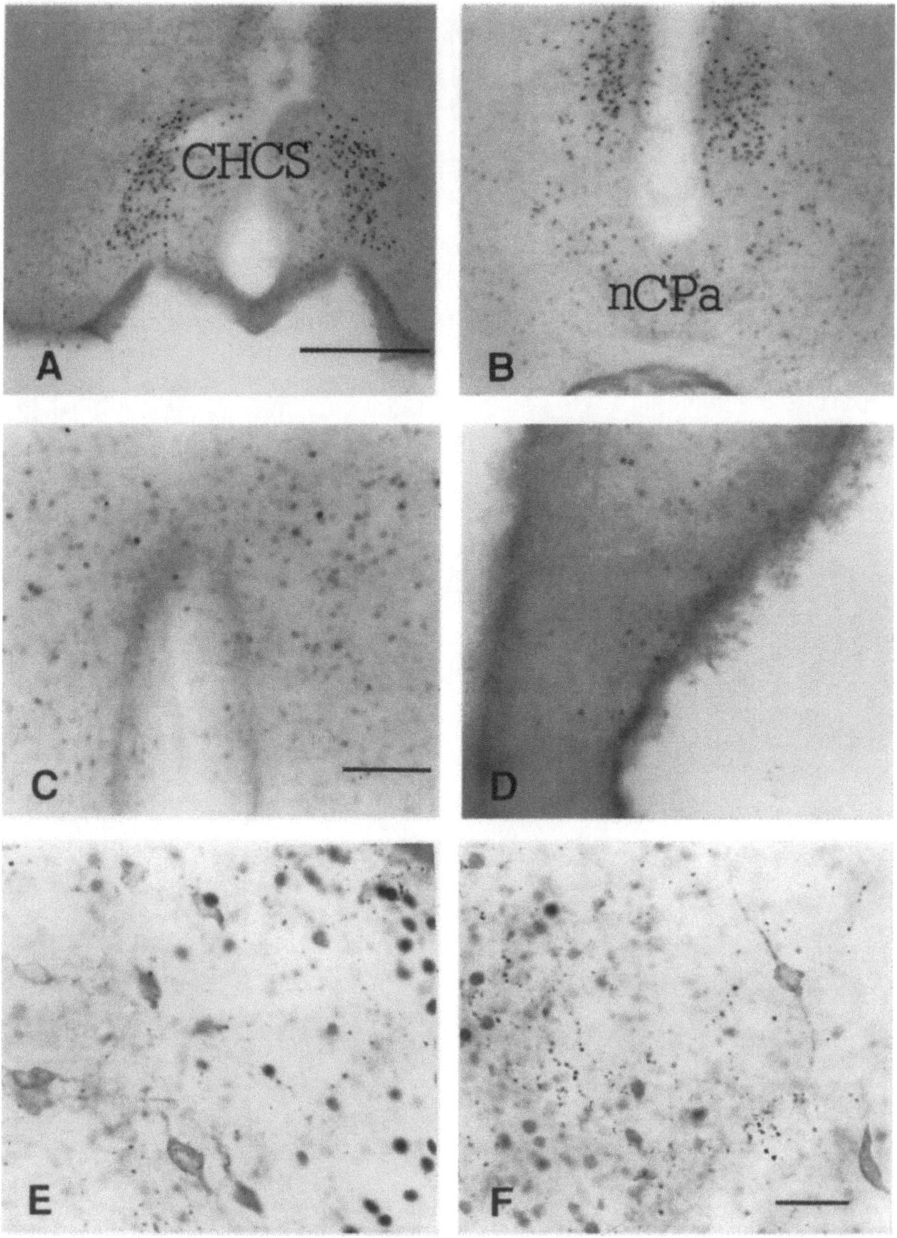

et corticoseptalis (**A**, *CHCS*), within and around the nucleus of the commissura pallii (**B**, *nCPa*), in the dorsal preoptic area at the level of the organum vasculosum laminae terminalis (**C**) and in the infundibular hypothalamic nucleus (**D**). Relationships between GnRH-I-immunoreactive perikarya (cytoplasmic labeling) and fos-immunoreactive cells (nuclear labeling) are seen in the tractus corticohabenularis et corticoseptalis (**E**) and in the preoptic area (**F**). *Bar* in **A** (for **A** and **B**): 100 μm; *bar* in **C** (for **C** and **D**): 50 μm; *bar* in **F** (for **E** and **F**): 50 μm

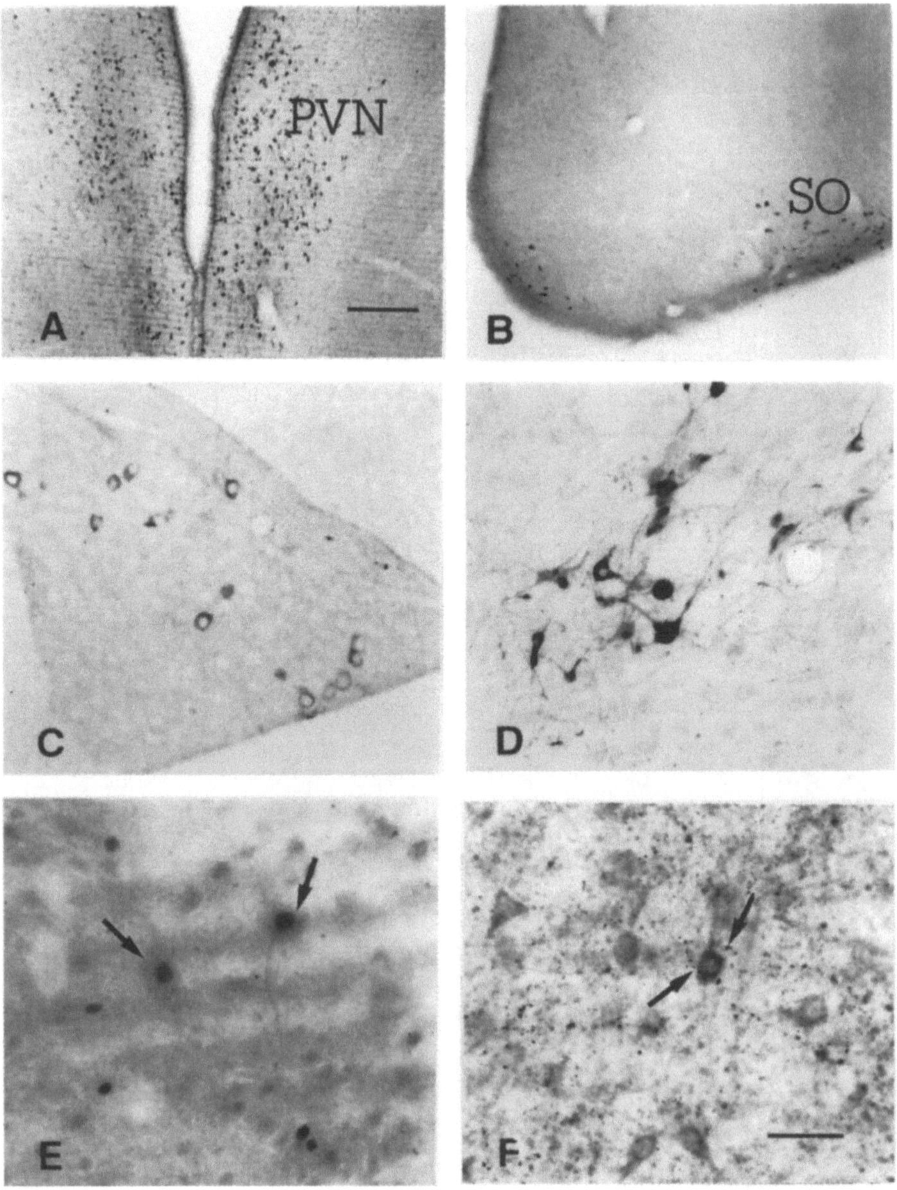

Fig. 3. Distribution of NMDAR1-immunoreactive (NMDAR1-ir) cells in coronal sections within the paraventricular nucleus (**A**, *PVN*), the supraoptic nucleus (**B**, *SO*) and in the cephalic lobe of the anterior pituitary (**C**). **D** High power view of NMDAR1-ir cells in the supraoptic nucleus. **E** Coexpression of fos- and NMDAR1-immunoreactivity within cells in the the periventricular hypothalamus. Some fos-positive cells (*arrows*), but not all, are immunoreactive for NMDAR1. **F** Double immunolabeling shows NMDAR1-ir fibers surrounding GnRH perikarya in the CHCS area (*arrows*). *Bar* in **A** (for **A** and **B**): 100 µm, *bar* in **F** (for **C**-**F**): 50 µm

1987). If a pituitary site of action is excluded, it seems that bolus intravenous injection of NMA in the juvenile cockerel acts to induce a pulse of GnRH-I release as observed in several mammals (e.g. Price et al. 1978; Gay and Plant 1987; Lincoln et al. 1991; Fitzgerald 1996). The mechanism of action of NMA in the induction of GnRH release has been explored in rodents through the use of fos and NMDA receptor in situ hybridization/immunocytochemistry (Fos: Saitoh at al. 1991; Lee et al. 1993; Ebling et al. 1993, 1995; NMDA receptor: Ebling et al. 1995; Eyigor and Jennes 1996; Gore at al. 1996). Consistent with these studies, the administration of NMA induced the appearance of fos-immunoreactivity in nuclei of cells in the preoptic-OVLT region and in the tuberal hypothalamus in birds. In the cockerel, fos-immunoreactivity also occurred in association with the full continuum of GnRH-I neurons extending from the preoptic-OVLT region caudo-dorsally into the septal-CHCS region (Józsa and Mess 1982; Sterling and Sharp 1982). In agreement with the data from rodents, NMA treatment did not induce the expression of *c-fos* in GnRH-I neurons. In the rat, evidence has been presented that NMA stimulation of GnRH-I release does not involve a change in the rate of GnRH gene transcription but may increase the stability of the GnRH mRNA (Gore and Roberts 1994). This suggestion may also explain why NMA failed to induce fos-immunoreactivity in cockerel GnRH-I neurons.

Interpretation of the possible functional significance of the immunocyto-chemical observations on localization of NMDAR1-like receptors in the glutamatergic control of LH secretion is hampered by a lack of knowledge of the amino acid sequence of the chicken NMDA receptor. The NMDAR1 receptor antibody used was raised against a C-terminal amino-acid sequence from the rat NMDAR1 receptor (Petralia et al. 1994) and would be expected to recognize 1a, 1b and 2a, 2b splice variants (Hollman et al. 1993). For the purposes of this discussion it will be assumed that evolutionary conservation is such that the antibody recognizes the chicken NMDA receptor.

The rapidity of the LH response to NMA injection suggests either a direct effect on the anterior pituitary gland and/or the instant release of GnRH-I stored in GnRH-I neurons. A direct effect on the anterior pituitary gland cannot be excluded since NMDAR1-like receptor immunolabeling occurred in the anterior pituitary gland, which contains gonadotrophs (Mikami 1986). However, in view of the consensus opinion in the mammalian literature (Brann and Mahesh 1994) that NMA exerts a direct or indirect effect on the GnRH neuron, it is probable that this also applies for the cockerel. The possibility that NMA acts directly at level the GnRH-I cell bodies and/or the GnRH-I terminals in the median eminence is unlikely, since NMDAR1-like receptor immunolabeling was not observed at either of these sites. This observation is consistent with findings in rats showing that NMDA receptor gene-expression in GnRH neurons is rare (Eyigor and Jennes 1996; Gore et al. 1996) or absent (Ebling et al. 1995) and that NMDA receptors play a minor role in the regulation of GnRH release at the level of the median eminence (Donoso et al. 1990; Lopez et al. 1992).

The possibility that NMA exerts a stimulatory effect on GnRH-I neurons indirectly is suggested by the observation that fibers exhibiting NMDAR1-like receptor immmunoreactivity occur in close proximity to GnRH-I cell bodies. The identity of the neurotransmitter in these fibers is unknown but the cell bodies, from which they originate, may be those identified in the periventicular hypothalamus as containing fos-immunoreactivity after NMA treament. In contrast with observation in the hamster (Ebling et al. 1995), only scattered NMDAR1-like immunoreactive fibers were seen in the external layer of the median eminence. This appears to rule out the possibility that NMDA receptor mediates the release of GnRH from GnRH neuronal terminals in the median eminence.

In conclusion, it is established for the first time in a bird that NMA treatment stimulates LH release, indicating a role for excitatory amino acids in avian reproductive neuroendocrine function. The mechanism involved does not appear to be mediated by the expression of *c-fos* or NMDAR1 receptor subtype in GnRH-I neurons.

Acknowledgments

This work was supported by a BBSRC Linked Research Group Grant LR45/602, a Roslin Institute BBSRC Core Strategic Grant, and partly by OTKA 016356 Grant. We thank Mr. R.T. Talbot and Mr. P. Wilson for help with the NMA injections and the LH assays, and Dr. F.J.P. Ebling for the gift of the NMDAR1-antibody.

References

Brann WD, Mahesh VB (1994) Excitatory amino acids: function and significance in reproduction and neuroendocrine regulation. Front Neuroendocrinol 15: 3-49

Bourguignon J, Gerard A, Mathieu J, Simons J, Franchimont P (1989) Pulsatile release of gonadotropin-releasing hormone from hypothalamic explants is restrained by blockade of N-methyl-D, L-aspartate receptors. Endocrinology 125: 1090-1096

Donoso A, Lopez F, Negro-Vilar A (1990) Glutamate receptors of the non N-methyl-D-aspartic acid type mediate the increase in luteinizing-releasing hormone release by excitatory amino acids. Endocrinology 126: 414-420

Ebling FJ, Hui Y, Mirakhur A, Maywood ES, Hastings MH (1993) Photoperiod regulates the LH response to central glutamergic stimulation in the male Syrian hamster. J Neuroendocrinol 5: 609-618

Ebling FJP, Alexander IHA, Urbanski HF, Hastings MH (1995) Effects of N-methyl-D-aspartate (NMDA) on seasonal cycles of reproduction, body weight and pelage colour in the male Siberian hamster. J Neuroendocrinol 7: 555-566

Eyigor O, Jennes L (1996) Identification of glutamate receptor subtype mRNAs in gonadotropin-releasing hormone neurons in rat brain. Endocrine 4: 133-139

Estienne MJ, Shillo KK, Hileman SM, Green MA, Hayes SH (1990) Effect of N-methyl-D, L-aspartate on luteinizing hormone secretion in ovariectomized ewes in the absence and presence of estradiol. Biol Reprod 42: 126-130

Fitzgerald, BP (1996) Effects of administration of N-methyl-D, L-aspartate (NMA) on gonadotrophin secretion in untreated and steroid-treated ovariectomized mares during the breeding season and in intact and ovariectomized mares during anestrus. Dom Anim Endocrinol 13: 211-218

Gay VL, Plant TM (1987) N-methyl-D,L-aspartate elicits hypothalamic gonadotropin releasing hormone release in prepubertal male Rhesus monkeys (*Macaca mulatta*). Endocrinology 120: 177-179

Giri M, Kaufman JM (1995) Involvement of neuroexcitatory amino acids in the control of gonadotropin releasing hormone release from the hypothalamus of the adult male guinea pig: predominantly inhibitory action of N-methyl-D, L-aspartate-mediated neurotransmission and its reversal after ovariectomy. Endocrinology 136: 2404-2411

Gore AC, Roberts JL (1994) Regulation of gonadotrophin-releasing hormone gene expression by the excitatory amino acids kainic acid and N-methyl-D,L-aspartate in the male rat. Endocrinology 134: 2026-2031

Gore AC, Wu TJ, Rosenburg JJ, Roberts JL (1996) Gonadotropin releasing hormone and NMDA receptor gene expression and colocalization change during puberty in female rats. J Neurosci 16: 5281-5289

Hollman M, Boulter J, Maron C, Beasley L, Sullivan J, Pecht G, Heinemann S (1993) Zinc potentiates agonist-induced currents at certain splice variants of the NMDA receptor. Neuron 10: 943-954

Hoffman, GE, Smith MS, Verbalis JG (1993) C-fos and related immediate early gene products as markers of activity in neuroendocrine systems. Front Neuroendocrinol 3: 173-213

Hui, Y, Hastings MH, Maywood ES, Ebling FJP (1992) Photoperiodic regulation of glutamatergic stimulation of secretion of luteinizing hormone in male Syrian hamsters. J Reprod Fert 95: 935-946

Józsa R, Mess B (1982) Immunohistochemical localization of luteinizing hormone releasing hormone in the central nervous system of the domestic fowl. Cell Tissue Res 227: 451-458

Kuenzel WJ, Masson M (1988) A stereotaxic atlas of the brain of the chicken (*Gallus domesticus*). The Johns Hopkins University Press, Baltimore

Lee W-S, Abbud R, Hoffman G, Smith S (1993) Effects of N-methyl-D-aspartate receptor activation on c-fos expression in luteinizing-hormone releasing hormone neurons in female rats. Endocrinology 133: 2248-2254

Lincoln GA, Wu FCW (1991) Luteinizing hormone responses to N-methyl-D,L-aspartate during a photoperiodically-induced reproductive cycle in the ram. J Neuroendocrinol 3:309-317

Lopez F, Donoso A, Negro-Volar A (1992) Endogenous excitatory amino acids and glutaminergic receptor subtypes involved in the control of hypothalamic luteinizing hormone-releasing hormone secretion. Endocrinology 130: 1986-1992

Luderer U, Strobl F, Levine J, Schwartz N (1993) Differential gonadotrophin responses to N-methyl-D, L-aspartic acid in metestrous, proestrous and ovariectomized rats. Biol Reprod 48: 857-866

MacDonald MC, Robertson HA, Wilkinson M (1993) Age- and dose-related NMDA induction of Fos-like immunoreactivity and c-fos mRNA in the arcuate nucleus of immature female rats. Dev Brain Res 73: 193-198

Mahachoklertwattana P, Sanchez J, Kaplan SL, Grumbach MM (1994) N-methyl-D-aspartate (NMDA) receptors mediate the release of gonadotrophin-releasing hormone (GnRH) by NMDA in a hypothalamic GnRH neuronal cell line (GT-1). Endocrinology 134: 1023-1030

Mikami S (1986) Immunocytochemistry of the avian hypothalamus and adenohypophysis. Int Rev Cytol 105: 189-247

Ondo J, Pass K, Baldwin R (1976) The effects of neurally active amino acids on pituitary gonadotrophin secretion. Neuroendocrinology 21: 79-87

Ondo J, Wheeler D, Dom R (1988) Hypothalamic site of action for N-methyl-D-aspartate (NMDA) on LH secretion. Life Sci 43: 2283-2286

Petralia RS, Yokotani N, Wenthold RJ (1994) Light and electron microscopic distribution of the NMDA receptor subunit NMDAR1 in the rat nervous system using a selective anti-peptide antibody. J Neurosci 14: 667-696

Plant TM, Gay VL, Marshall GR, Arslan M (1989) Puberty in monkeys is triggered by chemical stimulation of the hypothalamus. Proc Natl Acad Sci USA 86: 2506-2510

Price MT, Olney JW, Cicero TJ (1978) Acute elevations of serum luteinizing hormone induced by kainic acid, N-methyl-aspartic acid or homocystic acid. Neuroendocrinology 26: 342 -358

Reyes A, Luckhaus J, Ferin M (1990) Unexpected inhibitory action of N-methyl-D, L-aspartate on luteinizing hormone release in adult ovariectomized rhesus monkeys: a role of the hypothalamic-adrenal axis. Endocrinology 127: 724-729

Saitoh Y, Silverman AJ, Gibson MJ (1991) Norepinephrine neurons in the mouse locus coeruleus express c-fos protein after N-methyl-D,L-aspartate acid (NMDA) treatment: relation to LH release. Brain Res 561: 11-19

Sesti LA, Britt JH (1992) Elicitation of release of luteinizing hormone by N-methyl-D,L-aspartic acid during three paradigms of suppressed secretion of luteinizing hormone in the female pig. Anim Endocrinol 9: 105-114

Sharp PJ, Dunn IC, Talbot RT (1987) Sex differences in the response to chicken LHRH-I and -II in the domestic fowl. J Endocrinol 115: 323-331

Sharp PJ, Talbot RT, Main GM, Dunn IC, Fraser HM, Huskisson NS (1990) Physiological roles of chicken LHRH-I and -II in the control of gonadotrophin release in the domestic chicken. J Endocrinol 1124: 291-299

Sharp PJ, Li Q, Talbot RT, Barker P, Huskinsson N, Lea RW (1995) Identification of hypothalamic nuclei involved in osmoregulation using fos immunocytochemistry in the domestic hen (*Gallus domesticus*), ring dove (*Streptopelia risoria*), Japanese quail (*Coturnix japonica*) and zebra finch (*Taenopygia guttata*). Cell Tissue Res 285: 351-361

Sterling RJ, Sharp PJ (1982) The localization of LH-RH neurones in the diencephalon of the domestic hen. Cell Tissue Res 222: 283-298

Sternberger LA (1986) Immunocytochemistry, 3rd edn.Wiley, New York

Tal J, Price M, Olney J (1983) Neuroactive amino acids influence gonadotrophin suprapituitary mechanisms in either rodents or primates. Brain Res 273: 179-182

Urbanski H A (1990) A role for N-methyl-D,L-aspartate receptors in the control of seasonal breeding. Endocrinology 127: 2223-2228

Urbanski HF, Fahy MM, Collins PM (1993) Influence of N-methly-D,L-aspartate on the reproductive axis of male Syrian hamsters. J Endocrinol 137: 247-252

Wilson RC, Knobil E (1982) Acute effect of N-methyl-D,L-aspartate on the release of pituitary gonadotrophins and prolactin in the adult female rhesus monkey. Brain Res 248: 177-179

13 Neuroendocrine Role of Receptors for Insulin and IGF-1 in the Pituitary Gland: Morphological Considerations

J.W. Unger, M. Betz, and W. Lange

Department of Anatomy, University of Munich,
Pettenkoferstr. 11, D-80336 München, Germany;
Phone: +49 89 5160 4851, Fax: +49 89 5160 4857,
E-mail: unger@anat.med.uni-muenchen.de

Abstract

Previous investigations have demonstrated that interaction of peripheral insulin with neuroendocrine centers of the brain may play a key role in the regulation of body metabolism. Since the various levels of interaction are only incompletely known, the present study focused on the adenohypophysis of the rat, where the localization of the insulin receptor and the structurally and functionally closely related insulin-like growth factor-1 (IGF-1) receptor was investigated by light- and electronmicroscopic immunocytochemistry. The two receptors were found on separate subpopulations of secretory cells of the pars distalis, the IGF-1 receptors preferentially occurred in the postero-lateral region, whereas insulin receptors were dispersed throughout the pars distalis. Almost 90% of cells containing insulin receptors were also shown to be immunoreactive for β-endorphin. In contrast, IGF-1 receptors were almost exclusively located on follicle stimulating hormone secreting cells, suggesting a regulatory role of IGF-1 in the pituitary gonadotropin system. In addition, key proteins of receptor tyrosine kinase signal transduction, such as insulin receptor substrate-1 and phosphatidyl inositol-3 kinase, were demonstrated in the anterior lobe. Both markers were highly colocalized in two populations of morphologically distinct secretory cells, i.e. large, smooth-shaped cells and small-sized cells with irregular cytoplasmic protrusions. These data indicate the ability of insulin and IGF-1 receptors to transduce biological signals in the pituitary in vivo and further support the hypothesis that peripheral insulin acts as a regulatory hormone in the control of body homeostasis via various steps in the neuroendocrine axis, including the pituitary gland.

Introduction

As early as 1953, G.C. Kennedy presented a paper about "the role of depot fat in the hypothalamic control of food intake in the rat". Thus, the idea that body weight is regulated around a given set point and that the brain may be a sensor for overall body energy balance was developed decades ago (Kennedy 1953). However, the ways in which signals are transmitted from the periphery to the central nervous system in the control of food intake are still a matter of intense research. The hormonal regulatory system of feeding behavior consists of a number of mechanisms, both in the periphery (i.e., acute increase or decrease in appetite as a direct result of changes in blood glucose) as well as in neuroendocrine and autonomic centers of the brain, i.e. direct influence on neuronal activity by corticosteroids and neuropeptides, such as neuropeptide Y, galanin or opioid peptides (Grossmann 1986; Stanley et al. 1986; Schick et al. 1991, 1993; Leibowitz 1992). In addition, there is evidence that insulin plays a key role as a mediator in the communication between the periphery and central neuroendocrine cell systems in the long-term control of body metabolism and weight (Woods et al. 1985; Baskin et al. 1987, 1993; Unger et al. 1991; Schwartz et al. 1992a).

The brain has been regarded for a long time to be insulin-independent, a dogma which is still true concerning the neuronal glucose metabolism in most parts of the brain. However, investigations from several groups have collected data that indicate an endocrine role of insulin in the adult central nervous system. First of all, insulin receptors are present in abundance in a variety of specific brain regions involved in the regulation of central autonomic activity, including feeding behavior. Earlier binding assays and autoradiographic investigations (Werther et al. 1987, 1989) as well as more detailed histological mapping studies, carried out in our lab with immunocytochemistry using specific receptor antibodies (Unger et al. 1989), demonstrated dense receptor staining in neurons in many brain areas including hypothalamic nuclei, such as the arcuate and paraventricular nuclei, and limbic regions, including amygdala, hippocampus and autonomic brainstem areas, i.e. nucleus of the solitary tract. Subsequent work by Baskin's group showed that one mechanism of insulin's long-term ability to reduce food intake is the inhibition of neuropeptide Y release from the hypothalamic arcuate-paraventricular pathway (Schwartz et al. 1992b; Cusin et al. 1995). An important morphological correlate for immediate insulin uptake from the peripheral blood into these strategic neuroendocrine centers may be the presence of insulin receptors in nerve terminals in the external zone of the median eminence (Van Houten et al. 1980; Unger et al. 1993).

Since some of the relay nuclei of the hypothalamo-hypophysial system are obviously under influence of peripheral insulin, it is of interest to expand the "neuroendocrine aspects" of insulin from the brain to the pituitary, especially to the adenohypophysis as the next lower level of endocrine regulation. In the past,

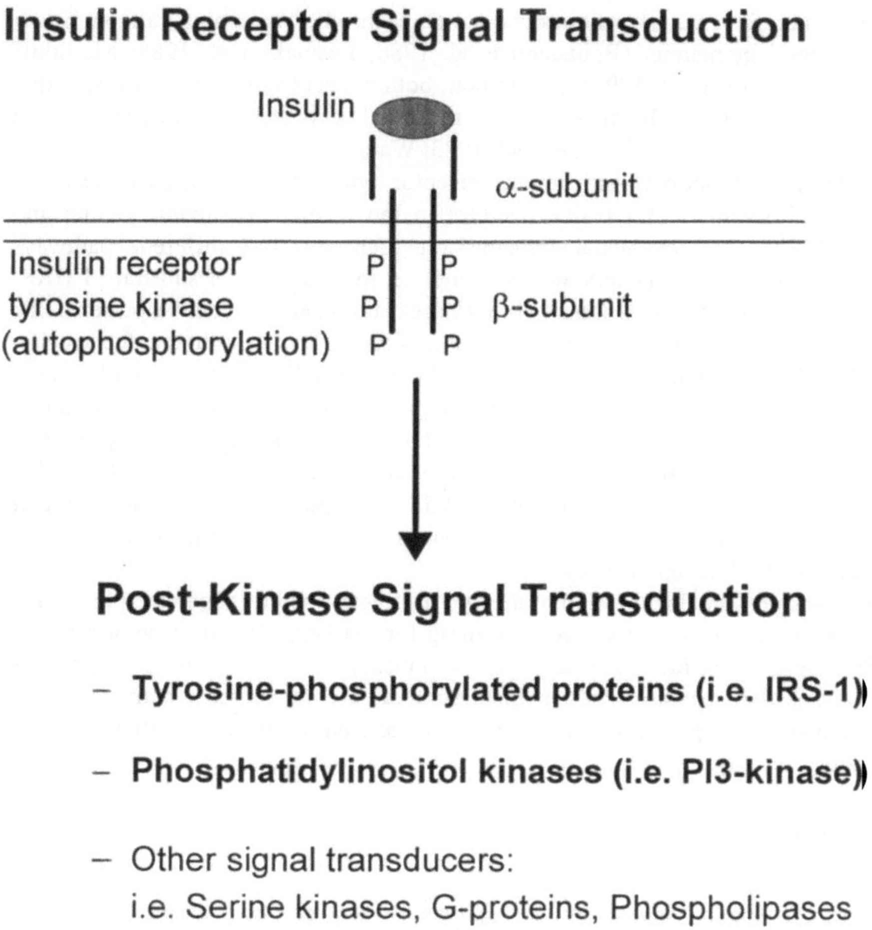

Fig. 1. Model of insulin receptor signal transduction via tyrosine kinases. The activated and autophosphorylated receptor phosphorylates cytoplasmic proteins, i.e. insulin receptor substrate-1 (*IRS-1*) which in turn mediates the postreceptor signaling cascade by activation of phosphatidylinositol 3-kinase (PI3-kinase) and recruitment of a number of other substrates, such as src homology 2 (SH2) containing proteins

only a few studies have tested ligand binding of insulin and IGF-1 in rat and human pituitary. Results from these investigations indicated significant amounts of binding activity predominantly in the anterior lobe (Rosenfeld et al. 1984; Werther et al. 1987, 1989; Kanematsu et al. 1991; De Keyser et al. 1994). However, no data on the precise cellular localization have been available so far. It is important to note that the insulin gene family consists of several members and that the structures of the receptors for the two most prominent members, namely insulin and insulin-like growth factor-1 (IGF-1) are very similar (Ullrich et al.

1986). Therefore, both hormones can act on each others' receptor - although in a dose-dependent manner (Bohannon et al. 1988; Lesniak et al. 1988; McElduff 1988; De Keyser et al. 1994). In addition, both are receptor tyrosine kinases that use the same intracellular substrates in initial key steps of receptor signal transduction pathways (Shia and Pilch 1983; Waters and Pessin 1996).

After ligand-binding, endogenous receptor tyrosine kinase activity leads to autophosphorylation of tyrosine residues in the receptor β-subunit (White and Kahn 1989). Further signal transmission branches into different pathways involving tyrosine phosphorylated proteins, i.e. insulin receptor substrate-1 (IRS-1), phosphatidyl inositol-3 kinase (PI3-kinase) and other intracellular proteins like serine kinases, G-proteins or phospholipases (Fig. 1). It is thought that the specific combination and temporal activation of these intracellular pathways ultimately leads to specific biological effects, i.e. cell growth or metabolic change, although the precise cellular steps are only incompletely known (Häring 1991; Myers et al. 1994; Folli et al. 1996; Waters and Pessin 1996). Because of the close relationship between the structure of the insulin and IGF-1 receptor and in their intracellular steps of signal transmission, it is necessary to correlate and compare the cellular presence of both receptor types.

In our study, the localization of the two receptors was investigated by light- and electron microscopy. Furthermore, staining for markers of post-receptor tyrosine kinase signal transduction (IRS-1 and PI3-kinase) was carried out to support the hypothesis that receptors present in these areas are functionally active. Finally, receptor-positive pituitary cells were characterized in regard to their specific secretory activity.

Materials and Methods

For morphological investigations, perfusion-fixed pituitary tissue (phosphate-buffered 4% paraformaldehyde for light microscopy or 4% paraform-aldehyde/0.2% glutaraldehyde for electron microscopy) from male, adult Sprague-Dawley rats (Charles River, Kisslegg im Allgäu, Germany) was either embedded in paraffin or immersed in phosphate-buffered sucrose (10-30%) for cryostat sectioning. Series of consecutive paraffin sections were cut either at a thickness of 1-2 μm for colocalization studies or at a thickness of 7 μm for routine histology. Frozen sections were cut after cryoprotection with sucrose at a thickness of 30-40 μm and processed immediately for pre-embedding immunocytochemistry,

Fig. 2. Insulin receptors in the pars distalis of the rat pituitary gland. **a** Survey, depicting the presence of insulin receptor immunoreactivity in a morphologically distinct subpopulation of secretory cells. **b, c** Colocalization of insulin receptors with β-endorphin. Immunostaining on consecutive 1 μm sections demonstrates that almost all insulin receptor positive cells (**b**) are also immunoreactive for β-endorphin (**c**). Examples are labeled by *arrows*. *Asterisks* in **b** and **d** mark an identical blood vessel. **d, e** The ultrastructural analysis of insulin receptor

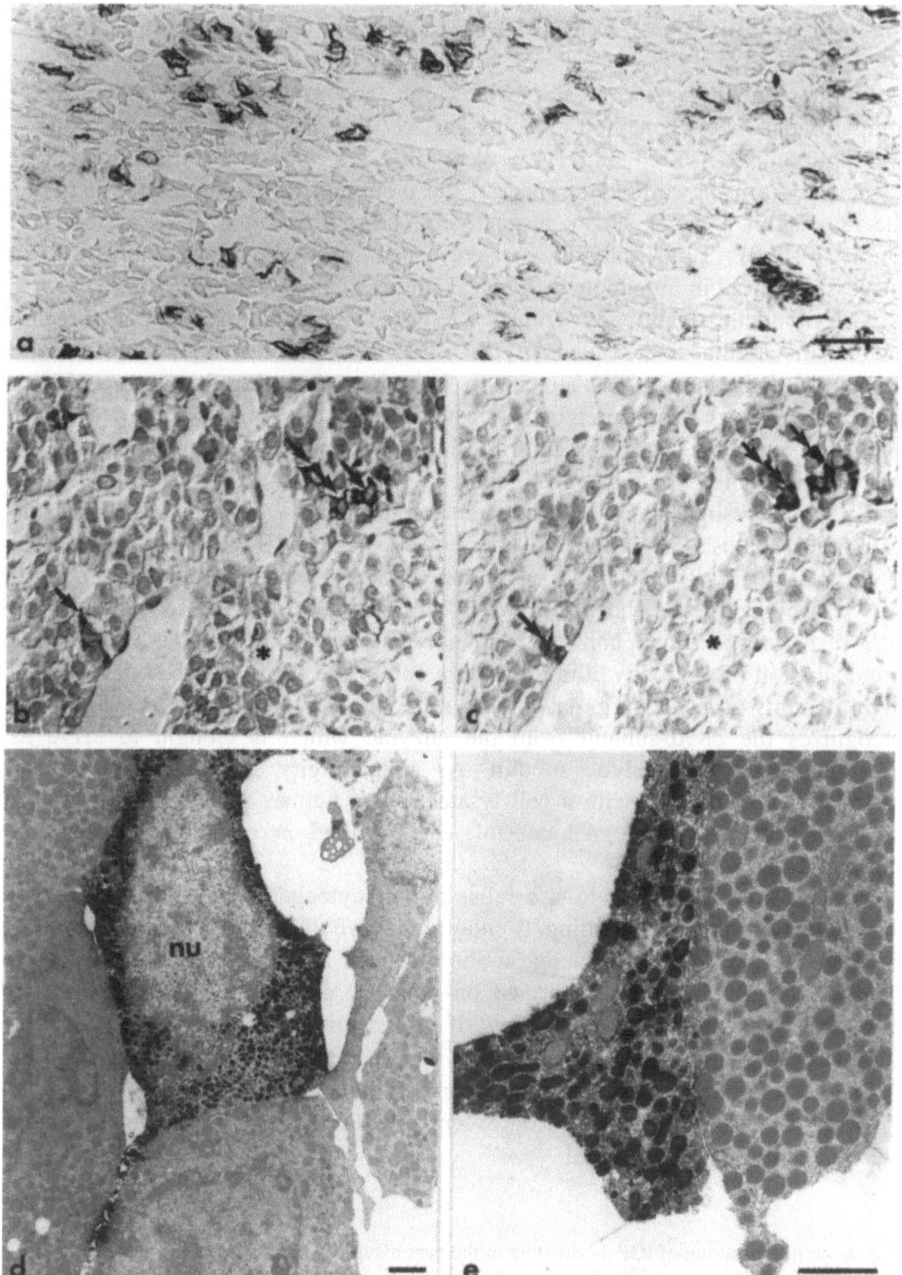

positive secretory cells demonstrates their irregular shape with small processes and cytoplasmic protrusions and their content of numerous secretory vesicles. Immunoreaction product is found freely in the cytosol as well as associated with secretory granules, whereas mitochondria and the nucleus (*nu*) are spared. **a-c** x880; **d** x4800; **e** x10700. *Bar* in **a** 10 µm; *bars* in **d, e** 1 µm followed by processing for electron microscopic analysis (Livingston and Unger 1992; Unger et al. 1993).

Immunocytochemistry was carried out according to the peroxidase-antiperoxidase (PAP) method (Sternberger 1979) with the following antibodies: two antibodies (A-1314 and R49) were directed against specific amino acid sequences from the carboxyterminus of the β-subunit of the human (A-1314) or the rat insulin receptor (R49), respectively (Ullrich et al. 1985; Goldstein and Dudley 1990). Both antibodies have been previously characterized and successfully used in immunocytochemical studies (Unger et al. 1989, 1993).

The monoclonal antibody against the IGF-1 receptor (aIGF-1r) is directed against the amino acid sequence 1284-1295 from the carboxyterminus of the β-subunit of the human IGF-1 receptor (Ullrich et al. 1986). Routine specificity tests were carried out by preincubation of antibody-dilutions with the respective antigenic peptides (10-20 µg/ml of diluted antibody) or replacement of the specific antibodies with nonspecific IgG.

For the investigation of receptor signal transduction, commercially available antibodies against IRS-1 and PI3-kinase were used (IRS-1: UBI, Lake Placid, USA; PI3-kinase: Transduction Laboratories, Lexington, Kentucky, USA). Anti-IRS-1 recognizes the carboxy-terminal 14 amino acid peptide of 160-185 kDa rat IRS-1. As a specificity control, immunoreactivity for IRS-1 was blocked by preincubation of antibody dilutions with a cell lysate from 3T3 cells that contain high amounts of IRS-1 (Fig. 4a, b). The monoclonal antibody against PI3-kinase recognizes the p85α-subunit from the enzyme which is thought to serve as a regulator for the catalytic subunit. As a specificity control for PI3-kinase antibody, preincubation with a cell lysate from a human epidermoid carcinoma cell line was performed. All control tests resulted in complete blockage of immunostaining.

For characterization of pituitary cells, immunocytochemical demonstration of a variety of hormones, including β-endorphin (β-END) and follicle-stimulating hormone (FSH) (both monoclonal antibodies from Boehringer Mannheim Corp., Penzberg, Germany) was performed on adjacent paraffin sections through the pituitary. To increase the sensitivity of the staining, sections were placed in 0.01 M citrate buffer, pH 6.0 and treated with a microwave for two times 5 min. All monoclonal or polyclonal, affinity-purified antibodies were used in a specific IgG-concentration of 10 - 20 µg/ml, except anti-IRS-1 which gave optimal results at a concentration of 1-3 µg/ml.

Fig. 3. Immunostaining of IGF-1 receptors in the pars distalis. **a** Immunoreactive secretory cells are relatively large and characterized by a smooth ovoid shape. Occasionally, large intracellular vacuoles are present. **b** The electron microscope analysis reveals numerous secretory granules and clusters of immunoreaction product that is not homogeneously distributed throughout the cytosol, but more concentrated along the cell surface (*nu* nucleus). **c, d** Colocalization of IGF-1 receptors with FSH. Immunostaining on adjacent sections through the pars distalis demonstrates an almost complete overlap between the cell population containing IGF-1

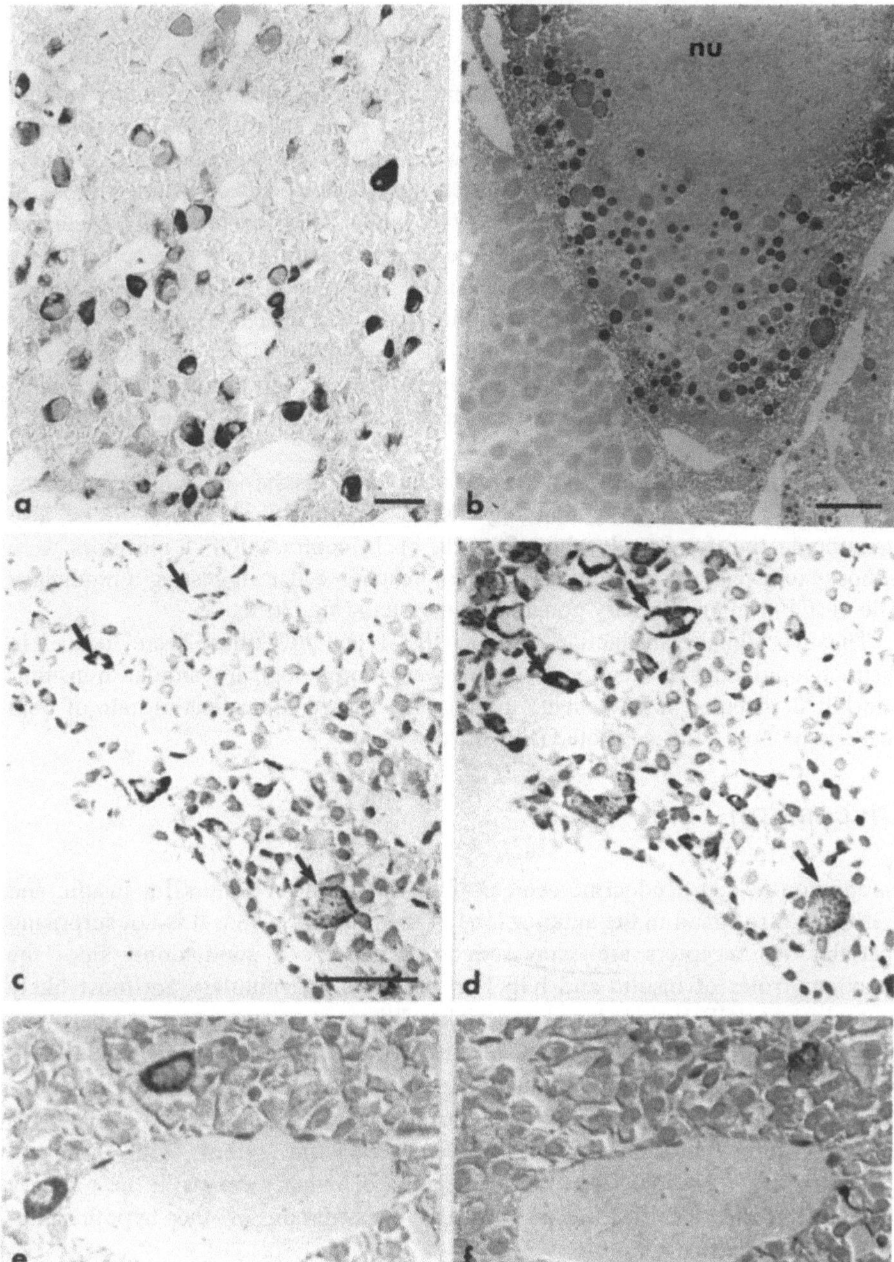

receptors (**c**) and FSH (**d**). Examples are labeled by *arrows*. **e, f** Lack of colocalization between IGF-1 and insulin receptors. Immunostaining of IGF-1 receptors (**e**) and insulin receptors (**f**) on consecutive sections demonstrates that the two receptors are located on different cell populations. **a** x650; **b** x8300; **c-f** x1300. *Bars* in **a, c** 10 μm; *bar* in **b** 1 μm

Results

Receptors for insulin and IGF-1 were found in subpopulations of secretory cells of the pars distalis of the adenohypophysis with a preference of IGF-1 receptors to the postero-lateral region, whereas insulin receptors were dispersed throughout the pars distalis (Figs. 2a, 3a). Further investigation showed two different types of cells that were either immunoreactive for insulin receptors or IGF-1 receptors (Figs. 3e, f). Ultrastructurally, the cells containing insulin receptors were characterized by a relatively small size, numerous secretory granules and cytoplasmic processes. The reaction product was located freely in the cytosol, but also associated with granules (Figs. 2d, e). In contrast, IGF-1 receptor positive cells were relatively large and ovoid-shaped. The reaction product was found in clusters along the cell surface, as well as in association with dense core vesicles (Fig. 3b).

When further characterizing the cells carrying insulin- or IGF-1 receptors, almost 90% of the cells containing insulin receptors were found to be also immunoreactive for β-endorphin (Fig. 2b, c). In contrast, IGF-1 receptors were almost exclusively located on FSH-immunoreactive cells, suggesting a regulatory role of IGF-1 in the pituitary gonadotropin system (Fig. 3c, d).

The two signal transduction proteins IRS-1 and PI3-kinase were located in distinct populations of secretory cells that were morphologically similar to insulin- and IGF-1 receptor positive cells. In addition, a large colocalization rate of both markers in these cells was noted (Fig. 4c, d).

Discussion

In addition to neuroendocrine centers of the brain, the receptors for insulin and IGF-1 are also found in the anterior lobe of the pituitary gland. It is not surprising that the two receptors are expressed on different cell populations, since the functional roles of insulin and IGF-1 in the brain and pituitary are most likely quite distinct. IGF-1 is produced in peripheral tissues, i.e. liver, and also partly by endogenous synthesis in the brain, i.e. the hypothalamus (D'Ercole et al. 1984; Sandberg et al. 1988). Insulin, in contrast, circulates in the peripheral blood and is thought to play a key role in the control of feeding behavior and body weight via feedback communication of the peripheral endocrine system with the brain (Schwartz et al. 1992a). Thus, the localization of either receptor in the pituitary may reflect its specific function in the organization of the hypothalamo-hypophysial system.

In the hypothalamo-hypophysial system IGF-1 is thought to be involved in the regulation of various neuroendocrine functions, i.e. growth hormone release, where its negative feedback control is most likely exerted indirectly at the level of hypothalamic somatostatin release from nerve terminals into portal blood vessels in the median eminence (Berelowitz et al. 1981) and/or directly at the level of the

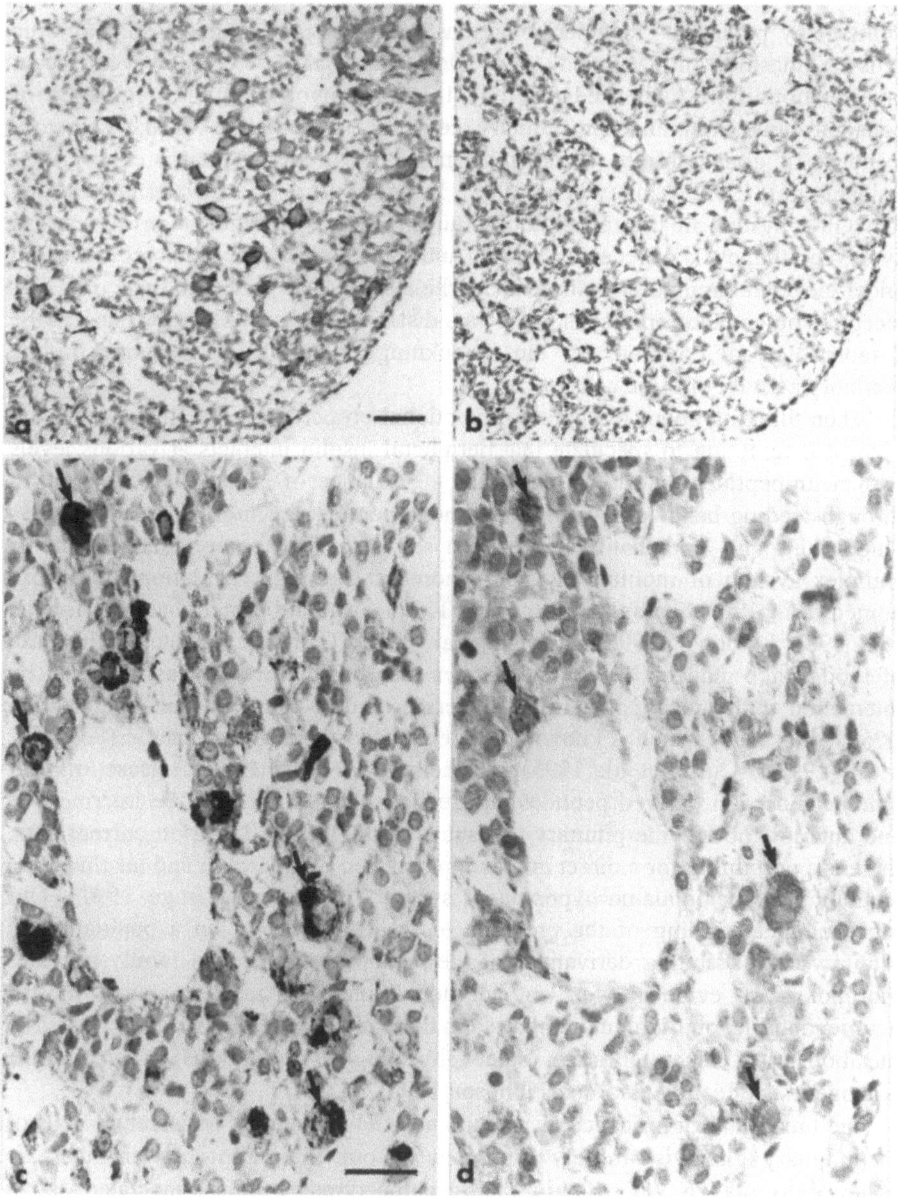

Fig. 4. Immunocytochemical detection of the intracellular signal transduction proteins IRS-1 and PI3-kinase in the pars distalis. **a** Survey, depicting a group of IRS-1 positive secretory cells. **b** Specificity control on a neighboring section, incubated with the IRS-1 antibody after blockage with lysate from 3T3 cells that contain a high amount of IRS-1. Note blockage of immunostaining. **c** Immunocytochemistry of PI3-kinase demonstrates two different sub-populations of secretory cells (small and irregular versus large and ovoid cells). **d** Comparison with IRS-1 on an adjacent 1 μm thin section reveals identical cell populations with a high degree of colocalization between the two markers (*arrows*). **a, b** x650; **c, d** x880. *Bars* 10 μm

pituitary by paracrine or autocrine actions, i.e. local production in somatotropic cells (Yamashita and Melmed 1986; Noguchi et al. 1989). However, recent investigations have suggested that IGF-1 is also involved in the regulation of gonadotropins from anterior pituitary cells, i.e. potentiation of the secretory response to gonadotropin releasing hormone in vitro (Kanematsu et al. 1991; Soldani et al. 1994; Whitley et al. 1995). In this context, it is also of interest that the IGF-1 system in the rat anterior pituitary lobe is dependent on the estrous cycle, i.e., the levels of receptor binding and IGF-1 binding proteins are increased under elevated circulating oestrogen (Michels et al. 1993). Our findings of IGF-1 receptors on gonadotropin cells in the pars distalis of the adenohypophysis may be a morphological correlate for the modulating effect of the hormone on the secretory activity of these cells.

When thinking about the putative functional properties of insulin receptors in vivo, it is tempting to speculate that peripheral insulin interacts at various levels with neuropeptides that are relevant for the balance of energy and metabolism through feeding behavior. A number of neuropeptides, including opioid peptides (endorphins and enkephalins) have been shown to play important roles in the complex system of monitoring energy homeostasis, e.g., regulation of glucose homeostasis (Ramabadran and Bansinath 1990) or fat and protein balance (Levine et al. 1985; Leibowitz 1992; Dalayeun et al. 1993). There is evidence that increase of food intake induced by opioid peptides may be caused - at least in part - by interaction with the hypothalamic NPYergic arcuate-paraventricular connection (Kotz et al. 1995) which is known to be regulated by peripheral insulin (Schwartz et al. 1992b; Cusin et al. 1995). Furthermore, NPY inhibits release of pro-opiomelanocortin derived peptides and reduces their mRNA in the intermediate and anterior lobe of the pituitary (Blasquez et al. 1995). Based on current data, there is a possibility for a direct interaction between β-endorphin and insulin at the level of the hypothalamo-hypophysial system (Unger and Lange 1997). Our morphological finding of the presence of insulin receptors on a population of pituitary cells secreting derivatives of the pro-opiomelanocortin family provides morphological evidence for an additional putative insulin action in the neuroendocrine, opioid-containing system that may also be significantly related to metabolism of macronutrients under physiological conditions or during the pathogenesis of eating disorders (Johnson 1995).

The functional significance of insulin and IGF-1 receptors in secretory cells of the pituitary gland is strongly supported by our finding of proteins that are believed to play a key role in the receptor tyrosine kinase-mediated signal transduction cascade. Earlier studies have demonstrated phosphotyrosine-containing proteins as general markers for substrates of tyrosine kinase activity (White et al. 1985). By immunoblotting, several phosphotyrosine-containing proteins were detected in the anterior pituitary. Two prominent protein bands were found at a molecular weight of 185 and 85 kDa (Unger et al. 1993), which are most likely identical with the two recently characterized signal transduction proteins pp185 or IRS-1 (Rothenberg et al. 1991; Sun et al. 1991) and pp85α, a subunit of PI3-kinase (Endemann et al. 1990; Hayashi et al. 1991; Lavan and Lienhard 1993).

Although the details on the signal transduction cascades of the insulin or IGF-1 receptors are still to be elucidated, recent studies have demonstrated that, upon ligand binding, IRS-1 is phosphorylated by the insulin- or IGF-1 receptor tyrosine kinase (Shemer et al. 1987; Shoelson et al. 1992) and subsequently serves as a docking protein for src homology 2 (SH2) containing substrates, i.e. the pp85α subunit of phosphatidylinositol-3 kinase which is activated in this way (Lamphere and Lienhard 1992; Waters et al. 1993; Rordorf-Nikolic et al. 1995; Waters and Pessin 1996). In turn, PI3-kinase phosphorylates phospho-inositides on the 3-position of the inositol ring. These mechanisms are thought to be initial steps in the downstream signaling pathway of insulin- and IGF-1 receptors.

Our morphological data showing the presence and colocalization of IRS-1 and PI3-kinase therefore strongly indicate the in vivo ability of cells in the anterior pituitary to transduce biological signals via tyrosine kinase pathways that originate at least in part from receptors for insulin and IGF-1. In summary, the central effects of insulin are most likely not restricted to brain areas, such as the hypothalamus, but may also involve the pituitary gland, in particular the opioid peptide axis. This mechanism may represent an additional connection between peripheral insulin and hormones of the brain-pituitary system.

Acknowledgment

This research was supported by a DFG grant (Un 59/2-3).

References

Baskin DG, Figlewicz DP, Woods SC, Porte D jr, Dorsa DM (1987) Insulin in the brain. Annu Rev Physiol 49: 335-347

Baskin DG, Sipols AJ, Schwartz MW, White MF (1993) Immunocytochemical detection of insulin receptor substrate-1 (IRS-1) in rat brain: colocalization with phosphotyrosine. Regul Peptides 48: 257-266

Berelowitz M, Szabo M, Frohman LA, Firestone S, Chu L, Hintz RL (1981) Somatomedin-C mediates growth hormone negative feedback by effects on both the hypothalamus and the pituitary. Science 212: 1279-1281

Blasquez C, Jégou S, Friard O, Tonon MC, Fournier A, Vaudry H (1995) Effect of centrally administered neuropeptide Y on hypothalamic and hypophyseal pro-opiomelanocortin-derived peptides in the rat. Neuroscience 68: 221-227

Bohannon NJ, Corp ES, Wilcox BJ, Figlewicz DP, Dorsa DM,Baskin DG (1988) Localization of binding sites for insulin-like growth factor-I (IGF-I) in the rat brain by quantitative autoradiography. Brain Res 444: 205-213

Cusin I, Dryden S, Wang Q, Rohner-Jeanrenaud F, Jeanrenaud B, Williams G (1995) Effects of sustained physiological hyperinsulinemia on hypothalamic neuropeptide Y and NPY mRNA levels in the rat. Neuroendocrinology 7: 193-197

Dalayeun JF, Norès JM, Bergal S (1993) Physiology of β-endorphins. A close-up view and a review of the literature. Biomed Pharmacother 47: 311-320

De Keyser J, Wilczak N, De Backer J-P, Herroelen L, Vauquelin G (1994) Insulin-like growth factor-I receptors in human brain and pituitary gland: an autoradiographic study. Synapse 17: 196-202

D'Ercole AJ, Stiles AD, Underwood LE (1984) Tissue concentrations of somatomedin C: further evidence for multiple sites of synthesis and paracrine and autocrine mechanisms of action. Proc Natl Acad Sci USA 81: 935-939

Endemann G, Yonezawa K, Roth RA (1990) Phosphatidylinositol kinase or an associated protein is a substrate for the insulin receptor tyrosine kinase. J Biol Chem 265: 396-400

Folli F, Saad MJA, Kahn CR (1996) Insulin receptor/IRS-1/PI 3-kinase signalling system in corticosteroid-induced insulin resistance. Acta Diabetologia 33: 185-192

Goldstein BJ, Dudley AL (1990) The rat insulin receptor: primary structure and conservation of tissue-specific alternative messenger RNA splicing. Mol Endocrinol 2: 235-244

Grossman SP (1986) The role of glucose, insulin and glucagon in the regulation of food intake and body weight. Neurosci Behav Rev 10: 295-315

Hayashi H, Miyake N, Kanai F, Shibasaki F, Takenawa T, Ebina Y (1991) Phosphorylation in vitro of the 85kDa subunit of phosphatidylinositol 3-kinase and its possible activation by insulin receptor tyrosine kinase. Biochem J 280: 769-775

Häring HU (1991) The insulin receptor: signalling mechanism and contribution to the pathogenesis of insulin resistance. Diabetologia 34: 848-861

Johnson RD (1995) Opioid involvement in feeding behaviour and the pathogenesis of certain eating disorders. Medical Hypothesis 45: 491-497

Kanematsu T, Irahara M, Miyake T, Shitsukawa K, Aono T (1991) Effect of insulin-like growth factor-I on gonadotropin release from the hypothalamus-pituitary axis in vitro. Acta Endocrinol 125: 227-233

Kennedy GC (1953) The role of depot fat in the hypothalamic control of food intake in the rat. Proc R Soc Lond (Biol) 140: 579-592

Kotz CM, Grace MK, Briggs J, Levine AS, Billington CJ (1995) Effects of opioid antagonists naloxone and naltrexone on neuropeptide Y-induced feeding and brown fat thermogenesis in the rat. J Clin Invest 96: 163-170

Lamphere L, Lienhard GE (1992) Components of signaling pathways for insulin and insulin-like growth factor-I in muscle myoblasts and myotubes. Endocrinology 131: 2196-2202

Lavan BE, Lienhard GE (1993) The insulin-elicited 60-kDa phosphotyrosine protein in rat adipocytes is associated with phosphatidylinositol 3-kinase. J Biol Chem 268: 5921-5928

Leibowitz SF (1992) Neurochemical-neuroendocrine systems in the brain controlling macronutrient intake and metabolism. TINS 15: 491-497

Lesniak MA, Hill JA, Kiess W, Rojeski M, Pert CB, Roth J (1988) Receptors for IGF-I and -II: autoradiographic localization in the rat brain and comparison to receptors for insulin. Endocrinology 123: 2089-2099

Levine AS, Morley JE, Gosnell BA, Billington CJ, Bartness TJ (1985) Opioids and consummatory behaviour. Brain Res Bull 14: 663-672

Livingston JN, Unger JW (1992) Immunohistochemical localization and biochemical analysis of tyrosine kinases in the central nervous system. Neuroprotocols 1: 193-200

McElduff A, Poronnik P, Baxter RC, Williams P (1988) A comparison of the insulin and insulin-like growth factor I receptors from rat brain and liver. Endocrinology 122: 1933-1939

Michels KM, Lee WH, Seltzer A, Saavedra JM, Bondy CA (1993) Up-regulation of pituitary [^{125}I]insulin-like growth factor-I (IGF-I) binding and IGF binding protein-2 and IGF-I gene expression by estrogen. Endocrinology 132: 23-29

Myers MG, Sun XJ, White MF (1994) The IRS-1 signalling system. TIBS 19: 289-293

Noguchi T, Sugisaki T, Kanamatsu T, Nishikawa N (1989) Presence of an insulin-like growth factor I (somatomedin C) immunoreactive substance in GH producing cells in the bovine anterior pituitary. Horm Metabol Res 12: 165-167

Ramabadran K, Bansinath M (1990) Glucose homeostasis and endogenous opioid peptides. Int J Clin Pharamacol Ther Toxicol 28: 89-98

Rordorf-Nikolic T, VanHorn DJ, Chen D, White MF, Backer JM (1995) Regulation of phosphatidylinositol 3′-kinase by tyrosyl phosphoproteins. J Biol Chem 270: 3662-3666

Rosenfeld RG, Ceda G, Wilson DM, Dollar LA, Hoffmann AR (1984) Characterization of high affinity receptors for insulin-like growth factors I and II on rat anterior pituitary cells. Endocrinology 114: 1571-1575

Rothenberg PL, Lane WS, Karasik A, Backer J, White MF, Kahn CR (1991) Purification and partial sequence analysis of pp185, the major cellular substrate of the insulin receptor tyrosine kinase. J Biol Chem 266: 8302-8311

Sandberg A-C, Engberg C, Lake M, von Holst H, Sara VR (1988) The expression of insulin-like growth factor I and insulin-like growth factor II genes in the human fetal and adult brain and in glioma. Neurosci Lett 93: 114-119

Schick RR, Schusdziarra V, Nussbaumer C, Classen M (1991) Neuropeptide Y and food intake in fasted rats: effect of naloxone and site of action. Brain Res 552: 232-239

Schick RR, Samsami S, Zimmermann JP, Eberl Th, Endres C, Schusdziarra V, Classen M (1993) Effect of galanin on food intake in rats: involvement of lateral and ventromedial hypothalamic sites. Am J Physiol 264: R355-R361

Schwartz MW, Figlewicz DP, Baskin DG, Woods SC, Porte D jr (1992a) Insulin in the brain: a hormonal regulator of energy balance. Endocr Rev 13: 387-414

Schwartz MW, Sipols AJ, Marks JL, Sanacora G, White JD, Scheurink A, Kahn SE, Baskin DG, Woods SC, Figlewicz DP, Porte D jr (1992b) Inhibition of hypothalamic neuropeptide Y gene expression by insulin. Endocrinology 130: 3608-3616

Shemer J, Adamo M, Wilson GL, Heffez D, Zick Y, LeRoith D (1987) Insulin and insulin-like growth factor-I stimulate a common endogenous phosphoprotein substrate (pp185) in intact neuroblastoma cells. J Biol Chem 262: 15476-15482

Shia MA, Pilch PF (1983) The β-subunit of the insulin receptor is an insulin-activated protein-kinase. Biochemistry 22: 717-721

Shoelson SE, Chatterjee S, Chaudhuri M, White MF (1992) YMXM motifs of IRS-1 define substrate specificity of the insulin receptor kinase. Proc. Natl. Acad. Sci. USA 89: 2027-2031

Soldani R, Cagnacci A, Yen SS (1994) Insulin, insulin-like growth factor I (IGF-I) and IGF-II enhance basal and gonadotrophin-releasing hormone-stimulated luteinizing hormone release from rat anterior pituitary cells in vitro. Eur J Endocrinol 131: 641-645

Stanley BG, Kyrkouli SE, Lampert S, Leibowitz SF (1986) Neuropeptide Y chronically injected into the hypothalamus: a powerful neurochemical inducer of hyperphagia and obesity. Peptides 7: 1189-1192

Sternberger LA (1979) Immunocytochemistry, 2nd edn. Wiley, New York

Sun XJ, Rothenberg P, Kahn CR, Backer JM, Araki E, Wilden PA, Cahill DA, Goldstein BJ, White MF (1991) Structure of the insulin receptor substrate IRS-1 defines a unique signal transduction protein. Nature 352: 73-77

Ullrich A, Bell JR, Chen E, Herrera R, Petruzzelli LM, Dull TJ, Gray A, Coussens L, Liao Y-C, Tsubokawa M, Mason A, Seeburg PH, Greenfeld C, Rosen OM, Ramachandran J (1985) Human insulin receptor and its relationship to the tyrosine kinase family of oncogenes. Nature 313: 756-761

Ullrich A, Gray A, Tam AW, Yang-Zeng T, Tsubokawa M, Collina C, Henzel W, LeBon T, Kathuria S, Chen E, Jacobs S, Frabcke U, Ramachandran J, Fujita-Yamaguchi Y (1986) Insulin-like growth factor I receptor primary structure: comparison with insulin receptor suggests structural determinants that define functional specificity. EMBO J 5: 2503-2512

Unger J, McNeill TH, Moxley III RT, White M, Moss AM, Livingston JN (1989) Distribution of insulin receptor-like immunoreactivity in the rat forebrain. Neuroscience 31: 143-157

Unger JW, Livingston JN, Moss AM (1991) Insulin receptors in the central nervous system: localization, signalling mechanisms and functional aspects. Progr Neurobiol 36: 343-362

Unger JW, Moss AM, Livingston JN (1993) The pars tuberalis is enriched with distinct phosphotyrosine-containing proteins not detected in other areas of the brain and pituitary. Cell Tissue Res 272: 499-507

Unger JW, Lange W (1997) Insulin receptors in the pituitary gland: morphological evidence for influence on opioid peptide-synthesizing cells. Cell Tissue Res 288: 471-483

Van Houten M, Posner BI, Kopriwer BM, Brawer JR (1980) Insulin binding sites localized to nerve terminals in rat median eminence and arcuate nucleus. Science 207: 1081-1083

Waters SB, Pessin JE (1996) Insulin receptor substrate 1 and 2 (IRS1 and IRS2): what a tangled web we weave. Trends Cell Biol 6: 1-4

Waters SB, Yamauchi K, Pessin JE (1993) Functional expression of insulin receptor substrate-1 is required for insulin-stimulated mitogenic signaling. J Biol Chem 268: 22231-22234

Werther GA, Hogg A, Oldfield BJ, McKinley MJ, Figdor R, Allen AM, Mendelsohn FAO (1987) Localization and characterization of insulin receptors in the rat brain and pituitary gland using in vitro autoradiography and computerized densitometry. Endocrinology 121: 1562-1570

Werther GA, Hogg A, Oldfield BJ, McKinley MJ, Figdor R, Mendelsohn FAO (1989) Localization and characterization of IGF-I receptors in rat brain and pituitary gland using in vitro autoradiography and computerized densitometry: a distinct distribution from insulin receptors. J Neuroendocrinol 1: 369-377

White MF, Maron R, Kahn CR (1985) Insulin rapidly stimulates tyrosine phosphorylation of a M_r-185,000 protein in intact cells. Nature 318: 183-186

White MF, Kahn CR (1989) Cascade of autophosphorylation in the beta-subunit of the insulin receptor. J Cell Biochem 39: 429-441

Whitley NC, Barb CR, Utley RV, Popvell JM, Kraeling RR, Rampacek GB (1995) Influence of stage of the estrous cycle on insulin-like growth factors-I modulation of luteinizing hormone secretion in the gilt. Bioi Reprod 53: 1359-1364

Woods SC, Porte D jr, Bobbioni E, Ionescu E, Sauter J-F, Rohner-Jeanrenaud F, Jeanrenaud B (1985) Insulin: its relationship to the central nervous system and control of food intake and body weight. Am J Clin Nutr 42: 1063-1071

Yamashita S, Melmed S (1986) IGF-I action on rat anterior pituitary cells: suppression of growth hormone secretion and mRNA levels. Endocrinology 118: 176-182

14 Insulin-Like Growth Factor I in Bony Fish

M. Reinecke

Division of Neuroendocrinology, Institute of Anatomy, University of Zürich,
Winterthurerstr. 190, CH-8057 Zürich, Switzerland;
Phone: +41 1 635 53 70, Fax: +41 1 635 57 02,
E-mail: reinecke@anatomie.unizh.ch

Abstract

Based on our own studies and those of other groups, the components of the insulin-like growth factor (IGF)-I system known to date in bony fish, i.e. the hormone IGF-I, the binding proteins and the type 1 IGF receptor, are presented. As deduced from cDNA cloning, bony fish prepro-IGF-I(s) are analogous to their mammalian counterparts. They also contain a signal peptide, B-, C-, A-, D- and E-domains and all residues neccessary for maintenance of tertiary structure. The primary structure of IGF-I, especially in the functionally important B- and A-domains, is highly conserved during phylogeny. As in mammals, the liver of bony fish produces the majority of circulating (endocrine) IGF-I, and the primary stimulus for the regulation of liver IGF-I is growth hormone. The IGF-I gene, however, is also expressed in various extrahepatic sites, such as the pancreas, gastro-intestinal tract, kidney, gills, head kidney, cartilage, male and female gonads, and brain. We identified parenchymal cells as potential local production sites by the use of immunohistochemistry, i.e. pancreatic islet cells, gastro-entero-endocrine cells, cells of renal proximal tubules, gill chondrocytes, chloride cells of the gill epithelium, interrenal cells of the head kidney, granulosa cells in the ovary, spermatocytes and Sertoli cells in testis, and neurons in retina and brain. Correspondingly, some in vitro and in vivo studies on bony fish indicate that local IGF-I systems most likely regulate organ-specific functions in an autocrine/paracrine manner.

Introduction

The insulin-like growth factors (IGFs) constitute a family of polypeptides that exert insulin-like metabolic and mitogenic/growth-promoting effects on a variety of target cells (see Sara and Hall 1990; Jones and Clemmons 1995; Reinecke and Collet 1997). This family consists of two major forms, IGF-I and IGF-II, and some molecular variants (Humbel 1990; Sara and Hall 1990). Mammalian IGF-I and IGF-II are 66% identical to each other. IGF-I is a basic peptide of 70 amino acid residues whereas IGF-II is a neutral peptide of 67 amino acid residues. The primary and tertiary structure of the IGFs and proinsulin show substantial similarities (see Humbel 1990; Sara and Hall 1990). Consequently, it has been presumed that in phylogeny the IGFs and insulin have been derived from a common ancestral molecule (Plisetskaya 1989; Chan et al. 1990).

The IGFs contain B- and A-domains which share 50% sequence identity with the B- and A-domains in insulin. Like proinsulin, the IGFs also show C-domains which connect the B- and A-domains. The C-domains of the IGFs, however, are considerably shorter than the C-peptide of proinsulin, exhibit quite different sequences, and are not removed during prohormone processing. Consequently, the mature IGF molecules are single chain polypeptides. In further contrast to insulin, the mature IGFs contain additional short C-terminal D-domains and the prohormones of both IGFs comprise further C-terminal extensions, the E-peptides, which are generally cleaved during posttranslational processing.

While insulin circulates in a free form in blood and body fluids, the majority of the IGFs are linked to specific binding proteins (IGFBPs). In fact, only less than 5% of serum IGF-I is not bound to BPs (see Humbel 1990). The IGFBPs transport the IGFs towards their target cells. To date, six different, but structurally related binding proteins, i.e. IGFBP-1 to -6, have been characterized in mammals (Shimazaki and Ling 1991). There is increasing evidence that the binding proteins not only stabilize the IGFs but also modulate several of their physiological activities (see Jones and Clemmons 1995; Zapf 1995).

Two different IGF receptors, the type 1 and type 2 receptor, have been identified. The type 1 receptor is a tyrosine kinase and, thus, is closely related to the insulin receptor. The type 1 receptor shows a high affinity for IGF-I, about three times lower affinity for IGF-II and a particularly low affinity for insulin. Similarly, the affinity of the insulin receptor for the IGFs is considerably low. The so-called type 2 receptor is also the mannose-6-phosphate (Man-6-P) receptor and, thus, is structurally different from the type 1 and insulin receptors. The Man-6-P receptor shows a high affinity for IGF-II, almost no affinity for IGF-I and does not bind insulin. The functional impact of the type 2 receptor that is present only in mammals is still enigmatic. However, binding of IGF-II to the Man-6-P receptor leads to internalization and degradation of IGF-II. Therefore, the only yet established function of the Man-6-P receptor is to remove IGF-II from the circulation (Jones and Clemmons 1995). For further considerations about the physiological impact of IGF-II and the Man-6-P receptor, see Reinecke and Collet (1997).

The mammalian liver exhibits the highest level of IGF-I expression but other organs also contain considerable amounts of IGF-I mRNA. Thus, IGF-synthesis occurs also in extrahepatic sites. As in the liver, the production of IGF-I in different tissues seems to be growth hormone (GH) dependent (see Sara and Hall 1990; Jones and Clemmons 1995; Reinecke and Collet 1997).

Molecular Features of IGF-I in Bony Fish

Fish are the oldest living vertebrates and have evolved some 500 million years ago (Powers 1989). The class of bony fish (Osteichthyes) is the most studied non-mammalian vertebrate group with respect to the insulin-like growth factors. Two obvious reasons apply for this focus of interest. First, bony fish larvae are the smallest independently functioning vertebrates and realize dramatic rates of growth during early life. Thereby they pass through gradual morphological and physiological transformations in order to develop from the larval to the juvenile phase (Segner et al. 1994). Thus, the functional status of endocrine systems that enables larval fish to regulate developmental and growth processes as well as to maintain homeostasis under varying environmental conditions is of basal importance (see Dickhoff et al. 1990). Secondly, bony fish have high commercial value and, thus, scientific results on growth regulation are of major interest for aquaculture industry (see Reinecke and Collet 1997). Consequently, most studies on the IGF system in bony fish deal with growth regulation and focus on the small group of salmonids.

In accordance, cDNA sequences encoding IGF-I peptides from several species of bony fish have been determined. Early studies dealt with salmonids, i.e. coho salmon *Oncorhynchus kisutch* (Cao et al. 1989), Atlantic salmon *Salmo salmar* (Duguay et al. 1992), rainbow trout *O. mykiss* (Shamblott and Chen 1992), and chinook salmon *O. tshawytschwa* (Wallis and Devlin 1993). In contrast to the salmonids which are tetraploid (Hinegardner and Rosen 1972), most other bony fish have a diploid genome. Recently, sequences have been determined for two perciform species, the gilthead seabream *Sparus aurata* (Duguay et al. 1996) and the tilapia *Oreochromis mossambicus* (Reinecke et al. 1997b), for catfish *Clarias macrocephalus* (McRory and Sherwood 1994) and for the daddy sculpin *Cottus scorpius* (Loffing-Cueni et al. 1997), a member of the scorpaeniforms, one of the most highly developed groups of the teleosts. The predicted amino acid sequences of salmonid and catfish mature IGF-I differ from those of the perciforms and *C. scorpius* which lack the two polar amino acids His^{C10} and Asn^{C12} in the C-domain. The homologies of bony fish A- and B-domain, however, are remarkably high when compared with humans (Table 1). As has been established in mammals, the particularly conservative phylogeny of the IGF-I A- and B-domains corresponds to their major physiological impact (see: Sara and Hall 1990; Jones and Clemmons 1995). This is also stressed by the finding that the B-domain residues Phe^{B23}-Tyr^{B24}-Phe^{B25} which are involved in the recognition of the type 1 receptor in mammals (Humbel 1990) are also found in bony fish IGF-I (Loffing-Cueni et al. 1997). In the following, the expression and localization of IGF-I in various organs and their physiological impact in fish will be presented.

Table 1. Extent of amino acid sequence identity of prepro-IGF-I from man and some bony fish species

Bony fish species	Signal-peptide	B-domain	C-domain	A-domain	D-domain	E-domain	Prepro-IGF-I
Rainbow	6/48	26/29	7/12	19/21	4/8	27/74	82/190
trout	12.5 %	90 %	58 %	90 %	50 %	36 %	43 %
Catfish	5/48	22/29	5/12	15/21	4/8	13/81	64/199
	10 %	76 %	42 %	71 %	50 %	16 %	32 %
Seabream	7/48	25/29	5/12	19/21	4/8	20/71	80/192
	14.5 %	86 %	42 %	90 %	50 %	27 %	42 %
Tilapia	7/48	27/29	6/12	18/21	3/8	21/70	82/188
	14.5 %	93 %	50 %	86 %	37.5 %	30 %	44 %
Sculpin	6/48	25/29	5/12	19/21	4/8	20/73	79/191
	12.5 %	86 %	42 %	90 %	50 %	27 %	41 %

Liver and the Growth Hormone (GH) - IGF-I Axis

As in mammals, the major site of IGF-I gene expression in bony fish is the liver (Duan et al. 1993; Duguay et al. 1996). In accordance with the presence of high-affinity binding sites for GH in salmon and tilapia liver (Gray et al. 1992; Shepherd et al. 1997), the expression of the liver IGF-I gene is GH dependent (Gray and Kelley 1991; Shepherd et al. 1997). Injections of GH induce significant enhancement of body growth in bony fish with the effect being more pronounced in juvenile individuals than in adults. Several in vitro and in vivo studies strongly indicate the presence of the GH-IGF-I axis in fish (see: Skyrud et al. 1989; Reinecke and Collet 1997).

The nutritional status also seems to influence the amount of circulating IGF-I in fish. Food deprivation or fasting for about 2 weeks significantly reduced IGF-I gene expression in coho salmon liver and led to growth retardation (Duan and Hirano 1992; Duan and Plisetskaya 1993). Nutrition has been shown to be a major regulator of hepatic GH receptors in coho salmon (Gray et al. 1992), and starving rainbow trout exhibited low plasma GH levels and IGF-like levels which were significantly lower than those of fed fish (Niu et al. 1993). In accordance, food deprivation for about 4 weeks decreased plasma GH levels in coho salmon and refeeding for about 2 weeks increased GH levels (Duan and Plisetskaya 1993). Thus, GH seems to be involved in the regulatory effect of nutrition on the level of circulating (endocrine) IGF-I in fish. However, recent evidence indicates that the effect of nutrition on IGF-I gene expression in fish maybe selectively exerted on the liver. Although prolonged fasting in coho salmon significantly decreased the IGF-I mRNA level in liver, it did not influence the IGF-I mRNA expression in several extrahepatic sites (Duan and Plisetskaya 1993).

Pancreas

The IGF-I gene is expressed in the pancreas of different groups of bony fish (Plisetskaya et al. 1993; Reinecke et al. 1997b). Immunohistochemical studies indicate that the endocrine islets are the potential production sites of pancreatic IGF-I (Reinecke et al. 1992, 1993b, 1997b; Berwert et al. 1995; Richardson et al. 1995). The likely production of IGF-I in bony fish endocrine pancreas is supported by the expression of IGF-I mRNA in the principal islets of the salmon *Oncorhynchus gorbusha* (Plisetskaya et al. 1993) and *Cottus scorpius* (Loffing-Cueni et al. 1997). In regard to the localization of IGF-I in the different islet cell subpopulations, pronounced species differences were found. IGF-I-immunoreactivity occurred in non-insulin cells in 5 out of 6 different bony fish species yet studied, i.e. in somatostatin cells in eel (*Anguilla anguilla*), tilapia (*Oreochromis mossambicus)* and *Carassius auratus* (Fig. 1a, b), and in pancreatic polypeptide, glucagon and somatostatin cells in turbot (*Scophthalmus maximus)* and carp (*Cyprinus carpio*) (Reinecke et al. 1993b, 1997a, b; Berwert et al. 1995). These distribution patterns are in line with results showing that IGF-I-immunoreactivity is restricted to non-insulin islet cells above the phylogenetic level of bony fish (Maake and Reinecke 1993; Reinecke et al. 1995).

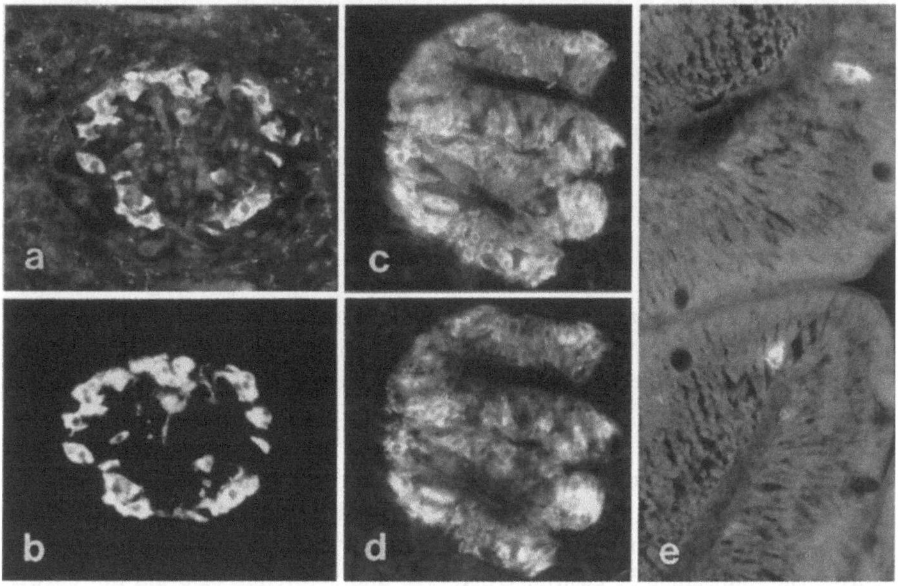

Fig. 1. IGF-I-immunoreactivity in the gastro-intestinal system of bony fish. **a, b** In a pancreatic islet of *Carassius auratus,* IGF-I-immunoreractivity (**a**) is confined to somatostatin-immunoreactive cells (**b**) as shown by double immunofluorescenc. x360. **c, d** In contrast, in the daddy sculpin (*Cottus scorpius*) IGF-I-immunoreactivity (**c**) occurs exclusively in insulin-immunoreactive (**d**) cells. x360. **e** IGF-I-immunoreactive entero-endocrine cells are present in the lower small intestine of the turbot (*Scophthalmus maximus*). x380

The exception found was *Cottus scorpius*, the daddy sculpin, where IGF-I-immunoreactivity occurred exclusively in insulin cells (Fig. 1c, d; Reinecke et al. 1993a). However, IGF-I-immunoreactive parenchymal cells in other sculpin organs, including neurons of the brain did not contain insulin-immunoreactivity (Loffing-Cueni et al. 1997) which corresponds to Northern Blot results in salmonids (Plisetskaya et al. 1993). Therefore, the sculpin B-cells probably produce both insulin and IGF-I. This hypothesis is supported by the structure of sculpin insulin (Cutfield et al. 1986) which differs considerably from sculpin IGF-I since the B- and A-domains of sculpin IGF-I show only about 7% and 57% homology, respectively, to those of sculpin insulin (Loffing-Cueni et al. 1997).

Islet IGF-I appears early during fish ontogeny as has been shown for turbot (Berwert et al. 1995), thus indicating important biological functions of islet-derived IGF-I already during larval development. On the one hand, islet-derived IGF-I may be a regulator of insulin secretion from the B cells as is likely in mammals (Leahy and Vandekerkhove 1990). In that case, this action would be exerted in a paracrine manner in most teleosts whereas in *C. scorpius* an autocrine mode of regulation would apply. On the other hand, recent studies indicate that islet-derived IGF-I may also act as an endocrine hormone. Hypophysectomy did not influence the amount of sulfation activity in eel pancreas (Duan and Hirano 1992) indicating that the production of islet IGF-I is not under the control of GH. In goby, ilectomy led to a decrease in $^{35}SO_4$-incorporation in cartilage (Bern et al. 1991) but hepatic GH binding was unchanged (Kelley et al. 1993). Thus, in addition to the GH-dependent IGF-I system in the liver, islet-derived IGF-I may constitute a second endocrine GH-independent IGF-I system that is also involved in fish growth promotion.

Gastro-Intestinal Tract

IGF-I gene expression has been observed in the stomach (Loffing-Cueni et al. 1997) and intestine (Shamblott and Chen 1993; Duan et al. 1993; Loffing-Cueni et al. 1997; Reinecke et al. 1997b) of bony fish and the potential IGF-I production sites have been identified as endocrine cells of the mucosal epithelium of daddy sculpin, turbot (Fig. 1e) and tilapia (Berwert et al. 1995; Reinecke et al. 1992, 1997b). Although the islet cells of the bony fish studied exhibited IGF-I-immunoreactivity in coexistence with classical islet hormones, the IGF-I-IR gastro-entero-endocrine cells did not (Berwert et al. 1995; Reinecke et al. 1997a). Because numerous other endocrine hormones, such as NPY, pancreastatin, neurotensin, substance P, gastrin, CCK, VIP and serotonin, also did not coexist with IGF-I-immunoreactivity (Reinecke et al. 1997a) the IGF-I-IR cells likely represent a particular class of gastro-entero-endocrine cells.

This is also stressed by the observation that the gastro-intestinal IGF-I-immunoreactive cells show remarkable inter-individual variation in frequency and distribution (Reinecke et al. 1997a, b). Several individuals exhibited infrequent small accumulations of IGF-I-immunoreactive cells whereas others contained frequent big clusters of IGF-I-immunoreactive cells. Depending on the individual, IGF-I-immunoreactive cells occurred either in the stomach, small or large instestine or in several portions of the gastro-intestinal tract. The physiological meaning of this variation is not yet clear. In pigs, the amounts of IGF-I-immunoreactivity and type 1 receptors coincided with villous growth and maturation (Schober et al. 1990). Furthermore, IGF-I improved mucosal structure and function in transplanted rat small intestine (Zhang et al. 1995). In analogy, IGF-I released from fish mucosal cells has been suggested to exert paracrine effects in response to varying local demands (Reinecke et al. 1997a). Because, in addition to gills and kidneys (see below), the intestinal tract is an important osmoregulatory organ in fish, the IGF-immunoreactive mucosal cells may also be involved in osmoregulation.

Osmoregulatory Organs

Kidneys and gills, the major osmoregulatory organs in fish, both express IGF-I mRNA as has been shown for several different species (Duan et al. 1993; Sakamoto and Hirano 1993; Shamblott and Chen 1993; Duguay et al. 1996; Reinecke et al. 1997b, Loffing-Cueni et al. 1997). Although transfer to seawater increased the IGF-I mRNA level in salmonid kidney (Sakamoto and Hirano 1993) while leaving the amount of liver IGF-I mRNA unchanged, nothing is yet known about possible effects of IGF-I on bony fish kidney function. However, cells of the proximal tubule in tilapia (Reinecke et al. 1997b) and rainbow trout (Fig. 2a) have been identified as the potential sites of renal IGF-I production. For the rat kidney, similar locations of IGF-I-immunoreactivity have been reported (see: Reinecke and Collet 1997). Thus, it is tempting to speculate that IGF-I in the fish kidney may be involved in several parameters of renal function as is likely in mammals (Hirschberg 1996).

In the gills, the potential production sites have been identified as chloride cells of the filament epithelium of tilapia (Reinecke et al. 1997b; Fig. 2b). Some physiological studies on salmonids indicate the role of the gill IGF-I system in the autocrine/paracrine regulation of serum osmolarity (McCormick et al. 1991; Sakamoto and Hirano 1993). The osmoregulatory effects of IGF-I seem to be exerted directly since with in vitro preparations of salmonid gill, mammalian IGF-I stimulated Na^+/K^+-ATPase (Madsen and Bern 1993) and possibly depend on priming by GH. The latter assumption is supported by a study in tilapia where hypophysectomy lowered the levels of Na^+/K^+-ATPase compared with sham-operated controls (Shepherd et al. 1997).

Cartilage and the GH-IGF-I Axis

As in mammals, cartilage has been mainly used to study the growth promoting activities of both GH and IGF-I and their interactions in bony fish. The stimulatory effect of GH on sulfate incorporation in eel cartilage turned out to be exerted indirectly and mediated by some IGF-like plasma factor (Duan and Inui 1990). In agreement with this finding, neither thyroxine nor GH exerted direct effects on the in vitro sulfate uptake in coho salmon cartilage (McCormick et al. 1992a, b). Similarly, short-term incubation of coho salmon or goby cartilage with GH did not influence basal and/or IGF-I stimulated sulfate incorporation, but injection with GH over 2-3 weeks stimulated both parameters (Bern et al. 1991; McCormick et al. 1992b). Studies of GH binding by the liver and the effects of hypophysectomy and IGF-I injections suggested that the GH status of the individual is important for the sensitivity of fish cartilage to IGF-I (Gray and Kelley 1991).

In several bony fish species, mammalian IGF-I has been shown to stimulate in vitro sulfate uptake into cartilage (Duan and Hirano 1990, 1992; Gray and Kelley 1991; Gray et al. 1992; McCormick et al. 1992b). IGF-I-immunoreactivity occurs in chondrocytes in tilapia (Reinecke et al. 1997b) and turbot as shown here (Fig. 2c). Therefore, the observed GH-dependent growth promoting effect of IGF-I on bony fish cartilage may be exerted in an autocrine/paracrine manner by IGF-I released from local chondrocytes. However, the possibility that endocrine, i.e., liver-derived IGF-I is also involved in these processes can not be excluded. In any case, it can be stated that the GH-IGF-I axis seems to play an important role in fish growth.

Head Kidney

The head kidney of adult bony fish constitutes a mass of lymphoid tissue. Within the head kidney intermingled interrenal and adrenal cells surround the cardinal veins. No IGF-I transcripts were detected in salmonid head kidney by the use of Northern blot (Sakamoto and Hirano 1993) but in tilapia head kidney, the expression of IGF-I mRNA was observed by reverse transcriptase polymerase chain reaction (RT-PCR) (Reinecke et al. 1997b). The different results are likely due to the higher sensitivity of the PCR. Interrenal cells have been identified as the potential production sites of IGF-I (Fig. 2d) in the head kidney (Reinecke et al. 1997b). In mammalian adrenal cortex, poor and generally negative evidence has been presented on the presence of IGF-I (Hansson et al. 1988; Penhoat et al. 1989; Mesiano et al. 1993). Thus, the presence of IGF-I in interrenal cells may be a peculiar feature of bony fish. Although there is yet little evidence for IGF-I in fish interrenals we would like to suggest that IGF-I derived from fish interrenal cells may be involved in the local regulation of steroidogenesis as has been assumed for mammalian gonads (Chatelain et al. 1994).

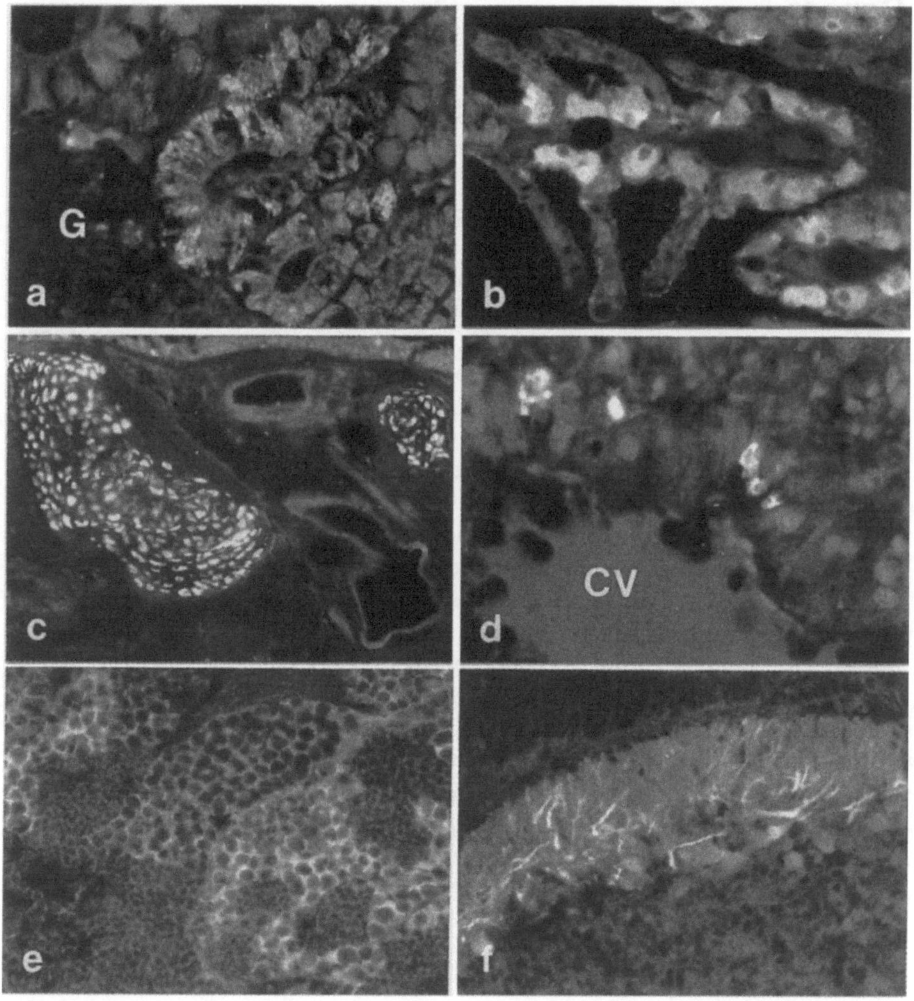

Fig. 2. IGF-I-immunoreactive parenchymal cells in organs of different bony fish species. **a** IGF-I-immunoreactivity occurs in some epithelial cells of the renal proximal tubules of rainbow trout. *G* Glomerulum. x390. **b** Chloride cells of the epithelium of the tilapia gill filaments contain IGF-I-immunoreactivity. x520. **c** Numerous chondrocytes in turbot gill cartilage exhibit IGF-I-immunoreactivity. x140. **d** In tilapia head kidney, IGF-I-immunoreactivity is present only in interrenal cells. *CV* Cardinal vein. x580. **e** Primary spermatocytes in testis of daddy sculpin are IGF-I-immunoreactive. x470. **f** Purkinje cell dendrites in turbot cerebellum are strongly IGF-I-immunoreactive. x390

Gonads

Recent evidence indicates that local IGF-I systems are present in both female and male fish gonads. In the ovary, the expression of IGF-I transcripts has been described for salmonids (Duan et al. 1993; Duguay et al. 1994), tilapia (Reinecke et al. 1997b) and daddy sculpin (Loffing-Cueni et al. 1997). Studies on red seabream (Kagawa et al. 1995) and tilapia (Reinecke et al. 1997b) revealed that IGF-I-immunoreactivity is found mainly in granulosa cells around developing and mature oocytes. There are also specific binding sites for IGF-I in the fish ovary as has been shown for carp. The IGF-I binding achieved depended on the season and reached a maximum at the end of pre-spawning (Gutierrez et al. 1993). Furthermore, IGF-I has been shown to induce final oocyte maturation in red seabream under in vitro conditions (Kagawa et al. 1994). The above results indicate that in fish an intraovarian IGF-I system seems to be present that regulates follicular growth and oocyte maturation in an autocrine/paracrine manner.

IGF-I mRNA is also expressed in the male gonad of bony fish as has been shown for the testes of rainbow trout (LeGac et al. 1996), tilapia (Reinecke et al. 1997b) and daddy sculpin (Loffing-Cueni et al. 1997). The likely production sites are spermatogenic cells and Sertoli cells which have been shown to express IGF-I mRNA in rainbow trout (LeGac et al. 1996) and to contain IGF-I-immunoreactivity in tilapia (Reinecke et al. 1997b) and sculpin (Fig. 2e). The expression of testicular IGF-I seems to be controlled by GH because the amount of IGF-I mRNA in juvenile rainbow trout testis was increased after long-term GH-treatment (LeGac et al. 1996). Since in trout testis the IGF type 1 receptor has also been identified (LeGac et al. 1996), IGF-I in fish testes may serve as paracrine/autocrine regulator of fish spermatogenesis and is likely to mediate the effects of GH (Reinecke et al. 1997b).

Central Nervous System

IGF-I expression in the brain has been demonstrated for representatives of quite different bony fish groups like salmonids (Duan et al. 1993; Duguay et al. 1994), catfish (McRory and Sherwood 1994), perciforms (Reinecke et al. 1997b), and scorpaeniforms (Loffing-Cueni et al. 1997). The potential sites of IGF-I synthesis have been identified in in the perciform species barramundi and tilapia using immunohistochemical techniques (Richardson et al. 1995; Reinecke et al. 1997b). IGF-I-immunoreactive neurons were present at all levels of the brain but were most prominent in the cerebellum. IGF-I-immunoreactive Purkinje cells (Fig. 2f) were consistently found but both frequency and distribution patterns of the IGF-I-immunoreactive neurons in other brain areas showed pronounced variation among different individuals (Reinecke et al. 1997b). IGF-I receptors have been identified in the brain of daddy sculpin (Drakenberg et al. 1993), carp and trout (Leibush et al. 1996). Therefore, a local IGF-I system seems to exist, indicating a physiological role for IGF-I in fish brain. The distribution patterns of IGF-I-immunoreactive neurons in the fish brain resemble the overall pattern of IGF-I mRNA expression as described in the rat brain (Bondy et al. 1992; Bondy and Lee 1993) and, thus, argue against a neurotransmitter or -modulator function of IGF-I (Reinecke and Collet 1997). Rather, IGF-I in the adult fish brain may support survival of neurons and glia cells, as is likely for mammals (Hepler and Lund 1990; Werther et al. 1994). Fish retina also expresses IGF-I mRNA (Mack et al. 1995; Reinecke et al. 1997b) and contains IGF-I-immunoreactive ganglionic, bipolar and receptor neurons (Richardson et al. 1995; Reinecke et al. 1997b). In retina, it has been suggested that IGF-I is additionally involved in the production of new receptor neurons (Reinecke et al. 1997b).

Binding Proteins

Studies on IFGBPs in bony fish are just appearing. However, at least three different IGFBPs have been identified in the sera of different species of bony fish, such as coho salmon, rainbow trout, striped bass, goby, tilapia and barramundi (Kelley et al. 1992; Niu and Le Bail 1993; Anderson and Matthews 1995). Based on their molecular weights and regulation it has been proposed that these fish binding proteins may be related to mammalian IGFBP-1,-2 and-3, respectively (Kelley et al. 1992). In goby, the levels of IGFBP-1 and IGFBP-2 changed with the insulin status. In coho salmon, IGFBP-3 serum level increased after GH injection and thus seems to be GH dependent. The levels of IGFBP-1 decreased during fasting in the striped bass but the levels of IGFBP-3 remained unchanged. These results indicate functional similarities between fish IGFBP-1 and IGFBP-3 and their mammalian counterparts. A physiological impact of IGFBPs in bony fish is also indicated by earlier studies with exogenous treatment of IGF-I. Injection of mammalian IGF-I in high doses led to severe hypoglycemia, growth inhibition and increased mortality (Skyrud et al. 1989; McCormick et al. 1991).

Although relatively few results on IGFBPs in bony fish have yet been obtained, these suggest that IGFBPs with molecular weights comparable to their mammalian counterparts are present. Furthermore, there is evidence that the expression of IGFBPs in fish may be hormonally regulated in manner that is similar to that of the mammalian IGFBPs. Finally, functional relationships are also suggested between the fish and mammalian IGFBPs.

Conclusions

The so far characterized constituents of the IGF-I system in bony fish are the hormone IGF-I, at least three different binding proteins and the type 1 IGF receptor. Thus, there is increasing evidence that bony fish contain all components of the IGF-I system as characterized in mammals (see Jones and Clemmons 1995).

As in representatives of other vertebrate classes investigated for sites and quantity of IGF gene expression, the liver of bony fish represents the main source of IGF-I and, thus, produces the majority of circulating (endocrine) IGF-I (see Reinecke and Collet 1997). In further agreement with results obtained in other vertebrates, the primary stimulus for the regulation of bony fish liver IGF-I is GH. As in mammals, in some bony fish evidence has been obtained that the nutritional status of the individual also modulates the amount of circulating IGF-I.

In addition to the liver, the production of IGF-I in extrahepatic sites is well established in bony fish. Numerous organs have been shown to express IGF-I genes and some also the type 1 receptor gene. Therefore, the production of IGF-I in various sites has a long phylogenetic history. However, research on the IGF-I system in bony fish has not progressed as far as it has in mammals. In particular, the presence and location of the type 1 receptor and the mode of its regulation in several extraheptaic sites must be clarified before an overall picture of the IGF-I system and its physiological impact in bony fish can be drawn. Likewise, research on the potential interactions between IGF-I and other hormones, such as prolactin, insulin, thyroxine and steroids, as well as on the regulation of organ-specific expression of IGF-I during development, is just beginning.

Among the extrahepatic sites that express the IGF-I gene are the endocrine pancreas, gastro-intestinal tract, kidney, head kidney, cartilage, male and female gonads, brain and gills. Our immunohistochemical studies strongly indicate that IGF-I is produced in parenchymal cells. Correspondingly, some in vitro and in vivo studies indicate that local IGF-I systems most likely regulate organ-specific functions in an autocrine/paracrine manner. Most of the observed local actions of IGF-I are obviously based on the primary effect of IGF-I; that is to cause mitogenesis and to promote differentiation and growth. However, some biological effects of IGF-I such as its involvement in osmoregulation in gills (and kidneys?) are probably based on different, yet largely unknown, modes of action.

In mammals and bony fish, production and/or secretion of IGF-I in several extrahepatic sites probably also depend on the GH status. The results so far obtained in these two distinct vertebrate classes which are separated by some millions of years of evolution are quite similar. Although few comparable results have been reported for amphibia and none are yet available for reptiles (see Reinecke and Collet 1997) the findings may suggest a straight evolutionary line. However, variant degrees of GH sensitivity seem to occur in bony fish extrahepatic sites which still await clarification.

Finally, some IFGBPs have yet been identified and characterized from sera of bony fish. There is preliminary evidence that they represent the bony fish counterparts to the mammalian IGFBPs. However, the influence of bony fish IGFBPs on the physiological effects of IGF-I as well as their own regulation are largely unknown.

Acknowledgments

Supported by the Swiss National Foundation (grants 32-33349 and 32-045351) and the Foundation for Scientific Research at the University of Zürich.

References

Bern HA, McCormick SD, Kelley KM, Gray ES, Nishioka RS, Madsen SS, Tsai PL (1991) Insulin-like growth factors "under water": role in growth and function of fish and other poikilothermic vertebrates. In: Spencer EM (ed) Modern concepts of insulin-like growth factors. Elsevier, New York, pp 85-96

Berwert L, Segner H, Reinecke M (1995) Ontogeny of IGF-1 and the classical islet hormones in the turbot, *Scophthalmus maximus*. Peptides 16: 113-122

Bondy CA, Lee W-H (1993) Patterns of insulin-like growth factor and IGF receptor gene expression in the brain. Ann NY Acad Sci 692: 33-43

Bondy CA, Werner H, Roberts C, LeRoith D (1992) Cellular patterns of type-I insulin-like growth factor receptor gene expression during maturation of the rat brain: comparison with insulin-like growth factors I and II. Neuroscience 46: 909-923

Cao QP, Duguay SJ, Plisetskaya E, Steiner DF, Chan SJ (1989) Nucleotide sequence and growth hormone-regulated expression of salmon insulin-like growth factor 1 mRNA. Mol Endocrinol 3: 2005-2010

Chan SJ, Cao Q-P, Steiner DF (1990) Evolution of the insulin superfamily: Cloning of a hybrid insulin/insulin-like growth factor cDNA from amphioxus. Proc Natl Acad Sci USA 87: 9319-9323

Chatelain PG, Avallet M O, Nicolino, M, Lejeune H, Clark A, Chuzel F, Penhoat A, Saez J (1994) Insulin-like growth factor I actions on steroidogenesis. Acta Pædiatr Scand 399: 176-177

Cutfield JF, Cutfield SM, Carne A, Emdin SO, Falkmer S (1986) The isolation, purification and amino-acid sequence of insulin from the teleost fish *Cottus scorpius* (daddy sculpin). Eur J Biochem 158: 117-123

Dickhoff WW, Brown CJ, Sullivan CV, Bern HA (1990) Fish and amphibian models for developmental endocrinology. J Exp Zool (Suppl) 4: 90-97

Drakenberg K, Sara VR, Falkmer S, Gammeltoft S, Maake C, Reinecke M (1993) Identification of IGF-1 receptors in primitive vertebrates. Regul Pept 43: 73-82

Duan C, Inui Y (1990) Effects of recombinant eel growth hormone on the uptake of (^{33}S) sulfate by ceratobranchial cartilages of the Japanese eel, *Anguilla japonica*. Gen Comp Endocrinol 79: 320-325

Duan C, Hirano T (1990) Stimulation of ^{35}S-sulphate uptake by mammalian insulin-like growth factors I and II in cultured cartilages of the japanese eel, *Anguilla japonica*. J Exp Zool 256: 347-350

Duan C, Hirano T (1992) Effects of insulin-like growth factor 1 and insulin on the in-vitro uptake of sulphate by eel branchial cartilage: evidence for the presence of independant hepatic and pancreatic sulphation factors. J Endocrinol 133: 211-219

Duan C, Plisetskaya EM (1993) Nutritional regulation of insulin-like growth factor-I mRNA expression in salmon tissues. J Endocrinol 139: 243-252

Duan C, Duguay SJ, Plisetskaya EM (1993) Insulin-like growth factor (IGF-I) mRNA expression in coho salmon, *Oncorhynchus kisutch*: tissue distribution and effects of growth hormone/prolactin family proteins. Fish Physiol Biochem 11: 371-379

Duguay SJ, Park LK, Samadpour M, Dickhoff WW (1992) Nucleotide sequence and tissue distribution of three insulin-like growth factor I prohormones in salmon. Mol Endocrinol 6: 1202-1210

Duguay SJ, Swanson P, Dickhoff WW (1994) Differential expression and hormonal regulation of alternatively spliced IGF-I mRNA transcripts in salmon. J Mol Endocrinol 12: 25-37

Duguay SJ, Lai-Zhang J, Steiner DF, Funkenstein B, Chan SJ (1996) Developmental and tissue-regulated expression of IGF-I and IGF-II mRNAs in *Sparus aurata*. J Mol Endocrinol 16: 123-132

Gray ES, Kelley KM (1991) Growth regulation in the gobiid teleost, *Gillichthys mirabilis*: roles of growth hormone, hepatic growth hormone receptors and insulin-like growth factor I. J Endocrinol 131: 57-66

Gray ES, Kelley KM, Law S, Tsai R, Young G, Bern H (1992) Regulation of hepatic growth hormone receptors in coho salmon *(Oncorhynchus kisutch)*. Gen Comp Endocrinol 88: 243-252

Guttiérrez J, Parrizas M, Carneiro N, Maestro JL, Maestro MA, Planas J (1993) Insulin and IGF-I receptors and tyrosine kinase activity in carp ovaries: changes with reproductive cycle. Fish Physiol Biochem 11: 247-254

Hansson H-A, Nilsson A, Isgaard J, Billig H, Isaksson O, Skottner A, Andersson IK, Rozell B (1988) Immunohistochemical localization of insulin-like growth factor I in the adult rat. Histochemistry 89: 403-410

Hepler JE, Lund PK (1990) Molecular biology of the insulin-like growth factors. Relevance to nervous system function. Mol Neurobiol 4: 93-127

Hinegardner R, Rosen DE (1972) Cellular DNA content and the evolution of teleostean fishes. Am Nat 106: 621-644

Hirschberg R (1996) Insulin-like growth factor I in the kidney. Miner Electrolyte Metab 22: 128-132

Humbel RE (1990) Insulin-like growth factors I and II. Eur J Biochem 190: 445-462

Jones JI, Clemmons DR (1995) Insulin-like growth factors and their binding proteins: biological actions. Endocrine Rev 16: 3-34

Kagawa H, Kobayashi M, Hasegawa Y, Aida K (1994) Insulin and insulin-like growth factors I and II induce final maturation of oocytes of red seabream, *Pagrus major*, in vitro. Gen Comp Endocrinol 95: 293-300

Kagawa H, Moriyama S, Kawauchi H (1995) Immunocytochemical localization of IGF-I in the ovary of the red seabream, *Pagrus major*. Gen Comp Endocrinol 99: 307-315

Kelley KM, Siharath K, Bern HA (1992) Identification of insulin-like growth factor-binding proteins in the circulation of four teleost fish species. J Exp Zool 263: 220-224

Kelley KM, Gray ES, Siharath K, Bern HA (1993) Experimental diabetes mellitus in a teleost fish. II. Roles of insulin, growth hormone (GH), insulin-like growth factor-I, and hepatic GH receptors in diabetic growth inhibition in the goby, *Gillichthys mirabilis*. Endocrinology 132: 2696-2702

Leahy JL, Vandekerkhove KM (1990) Insulin-like growth factor-I at physiological concentrations is a potent inhibitor of insulin secretion. Endocrinology 126:1593-1598.

Le Gac F, Loir M, Le Bail, P-Y, Ollitrault M (1996) Insulin-like growth factor I (IGF-I) mRNA and IGF-I receptor in trout testis and in isolated spermatogenic and Sertoli cells. Mol Reprod Dev 44: 23-35

Leibush B, Parrizas M, Navarro I, Lappova Y, Maestro MA, Encinas M, Plisetskaya EM, Gutiérrez J (1996) Insulin and insulin-like growth factor-I receptors in fish brain. Regul Pept 61: 155-161

Loffing-Cueni D, Schmid A, Graf H, Reinecke M (1997) IGF-I in the bony fish *Cottus scorpius*: cDNA, expression and differential localization in brain and islets. Mol Cell Endocrinol (submitted)

Maake C, Reinecke M (1993) Immunohistochemical localization of insulin-like growth factor 1 and 2 in the endocrine pancreas of rat, dog and man and their coexistence with classical islet hormones. Cell Tissue Res 273: 249-259

Mack AF, Balt SI, Fernald RD (1995) Localization and expression of insulin-like growth factor in the teleost retina. Vis Neurosci 12: 457-461

Madsen SS, Bern HA (1993) In-vitro effects of insulin-like growth factor-I on gill Na^+, K^+-ATPase in coho salmon, *Oncorhynchus kisutch*. J Endocrinol 138: 23-30.

Mesiano S, Mellon SH, Jaffe RB (1993) Mitogenic action, regulation, and localization of insulin-like growth factors in the human fetal adrenal gland. J Clin Endocrinol Metab 76: 968-976

McCormick SD, Sakamoto T, Hasegawa S, Hirano, T (1991) Osmoregulatory actions of insulin-like growth factor-I in rainbow trout (*Oncorhynchus mykiss*). J Endocrinol 130: 87-92

McCormick SD, Kelly KM, Young G, Nishioka RS, Bern HA (1992a) Stimulation of coho salmon growth by insulin-like growth factor I. Gen Comp Endocrinol 86: 398-406

McCormick SD, Tsai PI, Kelly KM, Nishioka RS, Bern HA (1992b) Hormonal control of sulfate uptake by branchial cartilage of coho salmon: Role of IGF-I. J Exp Zool 262: 166-171

McRory JE, Sherwood NM (1994) Catfish expresses two forms of insulin-like growth factor-I (IGF-I) in the brain. J Biol Chem 269: 18588-18592

Niu P-D, Le Bail P-Y (1993) Presence of insulin-like growth factor binding protein (IGF-BP) in rainbow trout (*Oncorhynchus mykiss*) serum. J Exp Zool 265: 627-636

Niu P-D, Perez-Sanchez J, Le Bail P-Y (1993) Development of a protein binding assay for teleost insulin-like growth factor (IGF)-like: relationship between growth hormone (GH) and IGF-like in the blood of the rainbow trout (*Oncorhynchus mykiss*). Fish Physiol Biochem 11, 1-6

Penhoat A, Naville D, Jaillard C, Chatelain PG, Saez JM (1989) Hormonal regulation of insulin-like growth factor I secretion by bovine adrenal cells. J Biol Chem 264: 6858-6862

Plisetskaya EM (1989) Pancreatic peptides. In: Holmgren S (ed) The Comparative Physiology of Regulatory Peptides. Chapman and Hall, London, pp 174- 202

Plisetskaya EM, Bondareva VM, Duan C, Duguay SJ (1993) Does salmon brain produce insulin? Gen Comp Endocrinol 91: 74-80

Powers DA (1989) Fish as model systems. Science 246: 352-358

Reinecke M, Drakenberg K, Falkmer S, Sara VR (1992) Peptides related to insulin-like growth factor 1 in the gastro-entero-pancreatic system of bony fish and cartilaginous fish. Regul Pept 37: 155-165

Reinecke M, Betzler D, Drakenberg K, Falkmer S, Sara VR (1993a) Occurrence of members of the insulin superfamily in central nervous system and digestive tract of protochordates. Histochemistry 99: 277-285

Reinecke M, Maake C, Falkmer S, Sara VR (1993b) The branching of insulin-like growth factor 1 and insulin: an immunohistochemical analysis during phylogeny. Regul Pept 48: 65-76

Reinecke M, Broger I, Brun R, Zapf J, Maake C (1995) Immunohistochemical localization of insulin-like growth factor I and II in the endocrine pancreas of birds, reptiles and amphibia. Gen Comp Endocrinol 100: 385-396

Reinecke M, Müller C, Segner H (1997a) An immunohistochemical analysis of the ontogeny, distribution and coexistence of 12 regulatory peptides and serotonin in endocrine cells and nerve fibers of the digestive tract of the turbot, *Scophthalmus maximus* (Teleostei). Anat Embryol 195: 87-101

Reinecke M, Schmid A, Ermatinger R, Loffing-Cueni D (1997b) Insulin-like growth factor I in the teleost *Oreochromis mossambicus*, the tilapia: gene sequence, tissue expression and cellular localization. Endocrinology 138: 3613-3619

Reinecke M, Collet C (1997) The phylogeny of the insulin-like growth factors. Int Review Cytol (in press)

Richardson NA, Anderson AJ, Rimmer MA, Sara VR (1995) Localization of insulin-like growth factor-I immunoreactivity in larval and juvenile Barramundi (*Lates calcarifer*). Gen Comp Endocrinol 100: 282-292

Sakamoto T, Hirano T (1993) Expression of insulin-like growth factor I gene in osmoregulatory organs during seawater adaption of the salmonid fish: possible mode of osmoregulatory action of growth hormone. Proc Natl Acad Sci USA 90: 1912-1916

Segner H, Storch V, Reinecke M, Kloas W, Hanke W (1994) The development of functional digestive and metabolic organs in turbot, *Scophthalmus maximus*. Mar Biol 119: 471-486

Shamblott MJ, Chen TT (1992) Identification of a second insulin-like growth factor in a fish species. Proc Natl Acad Sci USA 89: 8913-8917

Shamblott MJ, Chen TT (1993) Age-related and tissue-specific levels of five forms of insulin-like growth factor mRNA in a teleost. Mol Marine Biol Biotech 2: 351-361

Shepherd BS, Sakamoto T, Nishioka RS, Richman III NH, Mori I, Madsen SS, Chen TT, Hirano T, Bern HA, Grau EG (1997) Somatotrophic actions of homologous growth hormone and prolactins in the euryhyaline teleost, the tilapia, *Oreochromis mossambicus*. Proc Natl Acad Sci USA 94: 2068-2072

Shimasaki S, Ling N (1991) Identification and molecular characterization of insulin-like growth factor binding proteins (IGFBP-1, -2, -3, -4, -5 and -6). Growth Factor Research 3: 243-266

Schober DA, Simmen FA, Hadsell DL, Baumrucker, CR (1990) Perinatal expression of type I IGF receptors in porcine small intestine. Endocrinology 126: 1125-1132

Skyrud T, Andersen Ø, Aleström P, Gautvik KM (1989) Effects of recombinant growth hormone and insulin-like growth factor 1 on body growth and blood metabolites in brook trout (*Salvelinus fontinalis*). Gen Comp Endocrinol 75: 247-255.

Wallis AE, Devlin RH (1993) Duplicate insulin-like growth factor I genes in salmon display alternative splicing pathways. Mol Endocrinol 7: 409-422

Werther GA, Russo V, Cheesma H, Edmondson S (1994) IGF-I as a paracrine growth factor in developing rat brain. In: Baxter RC, Gluckman PD, Rosenfeld RG (eds) The insulin-like growth factors and their regulatory proteins. Elsevier Science B.V., pp 291-294

Zhang W, Frankel WL, Adamson WT, Roth JA, Mantell MP, Bain A, Ziegler TR, Smith RJ, Rombeau JL (1995) Insulin-like growth factor-I improves mucosal structure and function in transplanted rat small intestine. Transplantation 5: 755-761

Zapf J (1995) Physiological role of the insulin-like growth factor binding proteins. Eur J Endocrinol 132: 645-654

15 Corticosterone Upregulates Hippocampal Calbindin-D_{28k} in a Region Specific Manner and Exacerbates Ischemic Neuronal Damage Exclusively in Non-Calbindinergic Neurons

A. Rami, J. Winckler, and A. Rabié[1]

Dr. Senckenbergische Anatomie, Institute for Anatomy III, Johann Wolfgang Goethe-University Frankfurt, Theodor-Stern-Kai 7, D-60590 Frankfurt am Main, Germany;
Phone: +49 69 6301 6929, Fax: +49 69 6301 6017, E-mail: Rami@em.uni-frankfurt.de
[1]Centre de Pharmacologie, Rue de la Cardonille, 34095 Montpellier, France

Abstract

Administration of corticosterone to intact animals induces calbindin-D_{28k} protein in the hippocampal CA1-CA2 subfields. The fact that this effect on calbindin-D_{28k} was shown to be specific for the hippocampus argues for a receptor-mediated effect on gene expression. In addition, chronic pretreatment with corticosterone aggravates ischemia-induced neuronal damage in the CA3-CA4 subfields. This effect is similar to that of preischemic hyperglycemia, which also induces postischemic seizures and aggravates brain damage, since corticosterone raises the blood glucose level and enhances tissue lactic acidosis during ischemia. The energetically compromising qualities of corticosterone indicate that it is a key factor in hippocampal vulnerability. We assume that the increase of calbindin-D_{28k} expression in the CA1-CA2 subfields in corticosterone treated animals is an adaptive response to the exogenous stress. The lack of adaptive response in CA3-CA4 neurons endangers them by impairing the ability of these neurons to counteract the deleterious effects of calcium. This finding supports: the hypotheses (1) that corticosterone treatment, when paired with an ischemic insult, causes a prolonged elevation of intracellular concentration of free calcium ions $[Ca^{2+}]_i$ in neurons in an energy dependent manner, probably through the reduction of calcium efflux and (2) that neurons which do contain calbindin-D_{28k} are particularly predisposed to ischemic insults. The CA1-CA2 neurons express high amounts of calbindin-D_{28k} under stress conditions because their activity may involve a high rate of calcium buffering.

Introduction

In recent years, a number of neuronal high-affinity calcium-binding proteins have been identified in various animal species. At least 19 of these proteins have been detected in avian brain subfractions (Anthony and Babitch 1985). These calcium-binding proteins might be involved in the control of cellular calcium homeostasis. One of these proteins, calbindin-D_{28k} (CB), is present in the dentate granule cells and certain pyramidal cells of the CA1 and CA2 subfields of the rat hippocampal formation (Baimbridge and Miller 1982; Rami et al. 1987a, b, 1989, 1990, 1992). CB appears during development only when well defined input or output pathways to and from the hippocampal neurons become functional (Rami et al. 1987a). Mammalian CBs are highly homologous and share a 75% homology with chick CB. CB is also expressed in fish, squid and octopus. Phylogenetic surveys have shown that CB was originally neuronal and that its expression in other areas has occurred more recently (Rogers 1989). The discrete distribution of CB and the fact that it is phylogenetically old and has been conserved have led to the assumption that CB serves a crucial function in cellular calcium homeostasis (Feldman et al. 1990; Heizmann and Hunziker 1991). Although the precise physiological role of CB in the brain is still unknown, its high conservation during phylogeny, its high affinity for calcium, and its selective loss from some cortical neurons in Alzheimer´s disease, and from dentate granule cells after epileptic seizures, strongly suggests that this protein could be implicated in the regulation of $[Ca^{2+}]_i$ (Ichimiya et al. 1988; Baimbridge and Miller 1984; Rami et al. 1990, 1992; Heizmann and Braun 1992).

Beyond their role in the normal development and plasticity of neural circuitry, systems for calcium homeostasis are likely to be involved in neurodegenerative conditions. A great deal of evidence supports the "calcium-hypothesis" for neurodegeneration. While cell-specific differences in the types and density of excitatory amino acid (EAA) receptors contribute to selective neuronal vulnerability, differences in cell calcium-buffering systems may also be important. It will certainly be of interest to identify signals that directly influence neuronal CB.

Unlike CB present in the kidney and intestine, its expression in the brain is not regulated by vitamin D. Although there is a strict positive correlation between the CB-immunoreactivity and the vitamin D receptor distribution throughout the brain, the biological significance of this striking correlation is still unclear. It can be postulated that these vitamin D receptors are sensitive to steroids other than vitamin D. Iacopino and Christakos (1990) presented the first evidence that glucocorticoids like corticosterone (CORT) act as regulators of CB gene expression in the brain. The rat brain contains two corticosteroid-receptor types, the mineralocorticoid receptor (MR) with a high affinity for CORT (Kd = 0.5 nM) and the glucocorticoid receptor (GR) with a 10-fold lower affinity for CORT (Kd = 5 nM). The hippocampal formation contains high concentrations of MR and GR and is likely to be the principal target for corticosteroids.

In our previous studies on the role of CB in delayed neuronal death after cerebral ischemia (Rami et al. 1990, 1992), a correlation could be established between the presence of CB, the ability of the neuron to reduce $[Ca^{2+}]_i$, and the resistance to cell death. From these findings it seems that neuronal systems which reduce $[Ca^{2+}]_i$ may prove to be important therapeutic targets for preventing neuronal death in an array of degenerative disorders. Several in vitro studies have indicated that cells expressing CB exhibit lower Ca^{2+} entry through voltage-dependent Ca^{2+} channels and are better able to reduce $[Ca^{2+}]_i$ transients evoked by voltage depolarizations. The idea here is that CB-containing neurons will have an increased ability to buffer Ca^{2+} and to cope with the potentially destructive Ca^{2+} loads that are associated with the glutamate receptor overactivation which occurs under ischemic conditions (Schmidt-Kastner and Freund 1991; Rami et al. 1990, 1992; Hara et al. 1993).

To date, CORT has been identified as an in vivo regulator of CB in the nervous system, whereas adrenalectomy decreased hippocampal CB mRNA and CB-immunoreactivity (CB-IR) in the dentate gyrus (Krugers et al. 1995). Otherwise, it is well documented that glucocorticoids endanger the hippocampus during neurological insults by exacerbating numerous steps in the excitatory amino acid/calcium cascade of damage. Therefore, if the CORT-induced elevation in CB reflects increased protein levels, one might therefore expect that the Ca^{2+} buffering capacity of these cells would be increased. The goal of this study is to determine whether CB upregulation by chronic CORT treatment is able to rescue neurons from damage induced by transient cerebral ischemia. First, we investigated whether chronic CORT treatment could increase CB-IR in the hippocampus. Second, we compared the damage induced by cerebral ischemia in CORT treated animals with controls in various experimental schedules.

Materials and Methods

Cerebral Ischemia

Transient forebrain ischemia was performed in male Wistar rats (250-300 g) according to Smith et al. (1984). Cerebral ischemia was induced by clamping both common carotid arteries and lowering the mean arterial blood pressure to 40 mmHg. After 10 min the blood pressure was restored by the infusion of blood and the removal of the clamps. Arterial pH, pCO_2, pO_2, arterial blood pressure and plasma glucose concentration were determined 15 min before and 15 min after the induction of ischemia. Animals (50 rats) were divided into 5 groups (n = 10): control ischemic rats (Gr. 1), CORT treated rats which were subjected to cerebral ischemia immediately after the last injection of CORT (Gr. 2 or CORT-I) and CORT treated rats which were subjected to ischemia 2 days after terminating the CORT administration (Gr. 3 or CORT-II). Groups 4 and 5 served as controls and sham-operated rats for radioimmunoassay (RIA) and immunocytochemistry (for details on experimental groups see Table 1). Over a period of 2 weeks all treated animals received daily injections with CORT dissolved in sesame oil (10 mg/day, s.c.). Some animals were killed immediately after the last injection of CORT and their brains were prepared for radioimmunoassay and immunocytochemical staining of CB.

Table 1. Experimental protocol

Animal Groups	Treatment	Recovery (days)	Insult	Recovery (days)	Methods
Gr. 1	0	0	Ischemia	7	HIS / IHC
Gr. 2 (CORT-I)	CORT, 14 days 10 mg, s.c.	0	Ischemia	7	HIS / IHC
Gr. 3 (CORT-II)	CORT, 14 days 10 mg, s.c.	2	Ischemia	7	HIS / IHC
Gr. 4	CORT, 14 days 10 mg, s.c.	0	- -	0	RIA
Gr. 5	CORT, 14 days 10 mg, s.c.	0	Sham-operation	7	HIS /IHC

HIS histology for neuronal damage assessment; *IHC* immunohistochemistry; *RIA* radio-immunoassay; *CORT* corticosterone; *Gr.* group; *s.c.* subcutaneously

Assessment of Neuronal Damage

Seven days post-ischemia rats were anesthetized with 3.5% halothane and perfused transcardially with phosphate buffer (pH 7.4) containing 4% form-aldehyde. For histological evaluation of degenerating neurons, horizontal brain sections (5 μM) were stained with 1% celestine blue and 1% acid fuchsin. Neuronal injury was evaluated as the percentage of damaged neurons in the four hippocampal subfields (CA1, CA2, CA3 and CA4). In the staining procedure used, necrotic cells appeared bright red on examination under the light microscope, showed extensive cytorrhexis and/or karyorrhexis, and were readily distinguished from surviving neurons.

Determination of the CB and Protein Contents of the Hippocampus

The animals were killed by decapitation and their hippocampus rapidly dissected out, immediately frozen in liquid nitrogen, weighed and stored at -20 °C until processed. The samples were homogenized (20% wt./vol.) in buffer (TRIS-HCl 15 mM, NaCl 120 mM, KCl 3 mM, pH 7.4) and centrifuged 38 000 x *g* for 30 min. The resulting supernatant was diluted for protein determination and CB assay. CB levels were determined by radioimmunoassay (see Thomasset et al. 1982, Rami et al. 1989). The sensitivity of the assay was in the range of 1 ng CB per tube. CB was calculated as mean μg CB/mg total protein ± SEM. Group means were tested for significant differences by Student´s t-test.

Immunocytochemical Procedures

Rats were anesthetized with 3.5% halothane and perfused transcardially with phosphate buffer (pH 7.4) containing 4% formaldehyde. The sections were washed in several changes of phosphate buffered saline (PBS), containing 0.3% Triton X-100 at pH 7.4, treated with 1% hydrogen peroxide in PBS to eliminate endogenous peroxidase activity. A monoclonal antibody raised against calbindin-D_{28k} (working dilution 1:1000 in PBS, 18 h at 4 °C) was used. The sections were incubated in the secondary antibody, biotinylated sheep anti-mouse IgG (1:200), rinsed in PBS and subsequently incubated in streptavidin-biotin-peroxidase complex for 30 min. Following this, the sections were rinsed in PBS and reacted with a solution containing 3,3′-diaminobenzidine tetrahydrochloride, hydrogen peroxide (0.006%) and TRIS-HCl buffer (pH 7.6) for 10 min. To ensure comparable immunostaining, sections were processed together at the same time under the same conditions. The intensity of immunolabeling for CB was determined by measuring the optical density (OD), using an image analysis system (Zeiss, IBAS). The OD was expressed in arbitrary units corresponding to gray levels. The OD of the CA1-CA2 cells and dentate gyrus granule cells was calculated. For background staining we measured the CA3-CA4 areas which were immunonegative for CB. The OD of the corresponding area (CA1-CA2 areas or dentate gyrus) was related to the background value by the formula: OD area - OD background/ OD background x100. Statistical analysis was performed using the Mann-Whitney *U*-test.

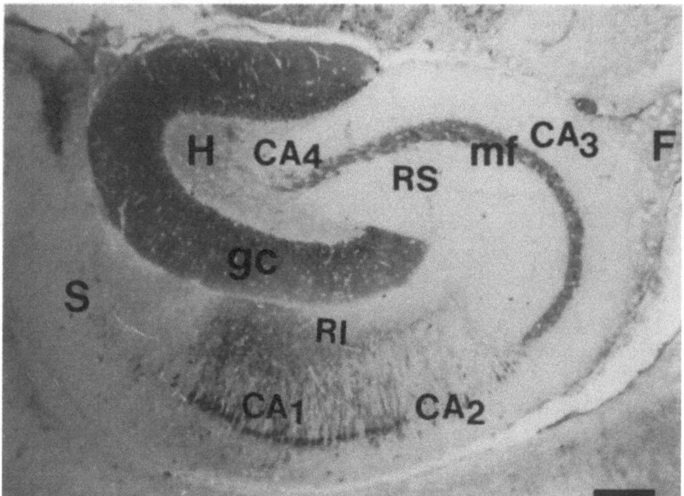

Fig. 1. Horizontal section of midseptotemporal hippocampus from sham-operated rat immunostained for Calbindin-D_{28k}. Immunoreactivity is located in granule cells (*gc*), mossy fibers (*mf*), and in the *CA1-CA2* subfields. *CA1-CA4* Pyramidal cell subfields; *F* fimbria; *H* hilus; *RI* regio inferior; *RS* regio superior; *S* subiculum. *Bar* 200 μm

Results

Immunocytochemistry and Radioimmunoassay (RIA)

The most intense immunoreactivity for CB was localized in a subpopulation of interneurons, which mainly occur in the strata radiatum and oriens of Ammon's horn. Granule cells, their dendritic trees and axons projecting to CA3-CA4 were also strongly immunoreactive for CB. No immunolabeling was found in more hilar neurons, and pyramidal cells in CA3-CA4 areas were also negative for CB. The majority of pyramidal cells in CA1-CA2 subfields were immunostained

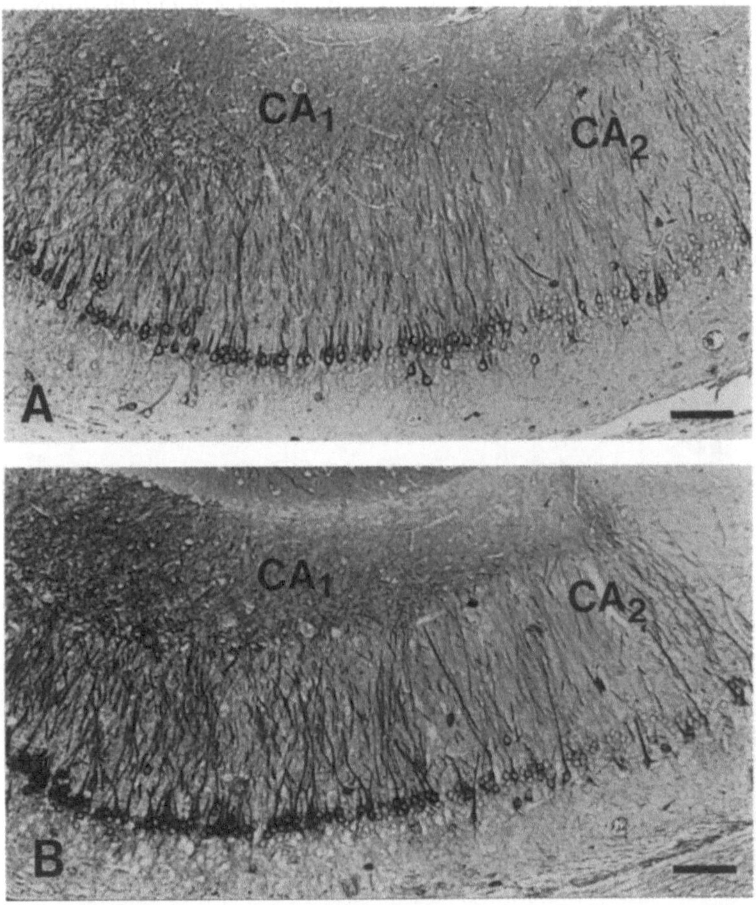

Fig. 2. Immunocytochemical characterization of Calbindin-D_{28k} neurons in the *CA1-CA2* subfields of rat hippocampus. Representative photomicrographs showing Calbindin-D_{28k}-immunoreactivity in controls (**A**) and in corticosterone treated animals (**B**). Note the relative dark staining in **B** (for details see Fig. 3B). *Bars* 100 µm

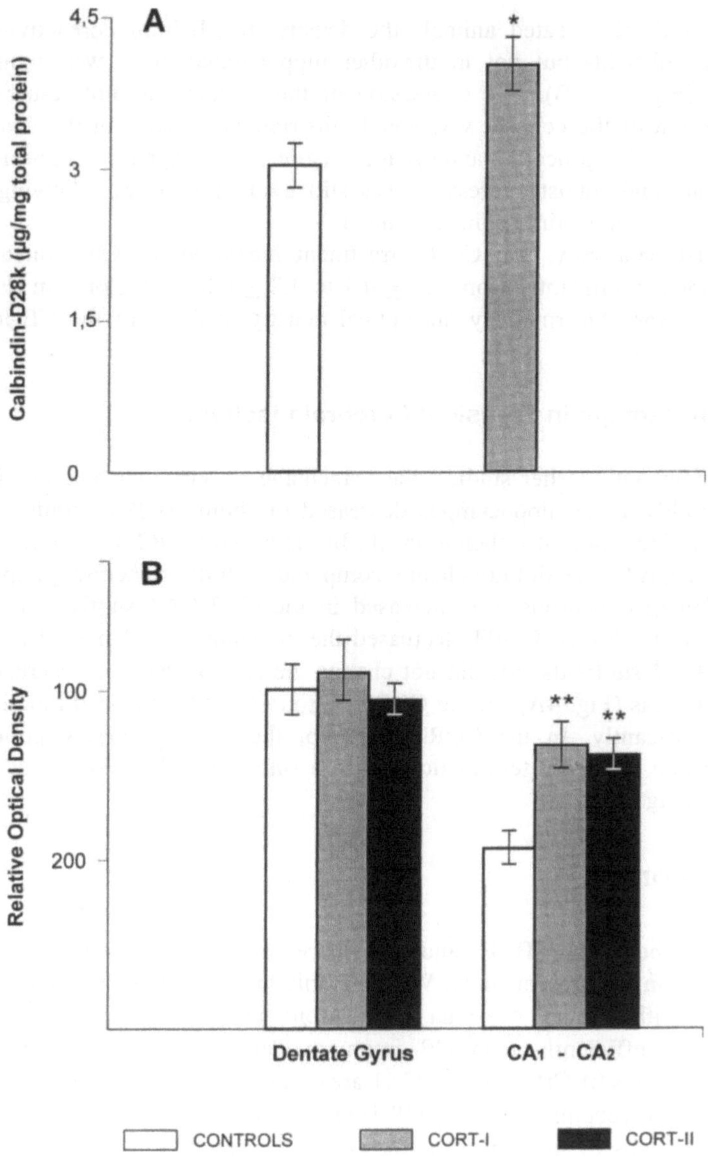

Fig. 3. A Calbindin-D_{28k} levels in rat hippocampus (n = 5 animals per group) after 14 days of s.c. administration of corticosterone or controls. Calbindin-D_{28k} was calculated as mean µg Calbindin-D_{28k}/mg total protein. Radioimmunoassay (RIA) values are reported as mean ± SEM. (n = 5 animals per goup). Group means were tested for significant differences by Student's t-test. * = $p < 0.05$. **B** Optical density of Calbindin-D_{28k} immunoreactivity in rat hippocampus of controls and corticosterone treated animals. Calbindin-D_{28k} immunoreactivity was significantly enhanced in the CA1-CA2 subfields of corticosterone treated rats when compared with controls. Note that corticosterone treatment did not change the density of Calbindin-D_{28k} immunoreactivity in dentate gyrus. ** = $p < 0.01$ by Mann-Whitney *U*-test

(Fig. 1). In CORT treated animals the density of CB-immunoreactivity in the CA1-CA2 subfields, but not in the other hippocampal areas, was significantly increased (Figs. 2, 3B). The expression of the protein was not restricted to a certain region of the cell but was widely distributed throughout the neuron and even filled the fine processes and synaptic regions. The expression appeared to be quite stable, and robust expression was still evident 10 days following CORT treatment, the farthest time point examined.

The RIA data show that CORT treatment increased the CB content in the hippocampus significantly from 3.7 ± 0.3 to 4.2 ± 0.8 μg/mg protein (Fig. 3A). This result was confirmed by the optical density analysis of the CB-immunoreaction (Fig. 3B).

Neuronal Damage in Transient Forebrain Ischemia

In agreement with earlier studies, the percentage of surviving neurons in CA1-CA2 subfields of rat hippocampus decreased to about 20 % in control animals 7 days after induction of ischemic insult. In the group CORT-I, neuronal damage in the CA1-CA2 areas did not change compared with the ischemic group, but the rate of damaged neurons was increased in the CA3-CA4 subfields (Fig. 4A). In the group CORT-II, CORT decreased the percentage of damaged neurons in the CA1-CA2 subfields, but did not change the rate of neuronal necrosis in the CA3-CA4 areas (Fig. 4A). In the CORT-I group CORT enhanced blood glucose levels significantly. In the CORT-II group, the 2-day recovery period after corticosterone treatment termination led to a stabilization of blood glucose concentration (Fig. 4B).

Discussion

A relation between CB-IR and resistance to ischemic neurotoxicity was documented in the present study. We were able to show indirectly that, following challenges with cerebral ischemia, CB neurons were able to reduce intracellular $[Ca^{2+}]_i$ more efficiently than CB immunonegative cells. The corticosterone-induced increases of CB-IR in the CA1 area may be mediated by the low affinity glucocorticoid receptor (GR). The GR is present in CA1 and CA3 pyramidal cells and granule cells of the dentate gyrus (DG). It remains to be clarified why the effect was exclusively seen in the CA1-CA2 areas and not in the DG. The antibody used in our study recognizes the calcium bound form of CB. The changes might be caused by altered Ca^{2+} entry or de novo synthesis of CB. This might have consequences for $[Ca^{2+}]_i$, since in vitro studies have indicated that cells expressing CB exhibit lower Ca^{2+} entry through voltage-dependent Ca^{2+} channels and reduce Ca^{2+} transients evoked by voltage depolarizations more effectively.

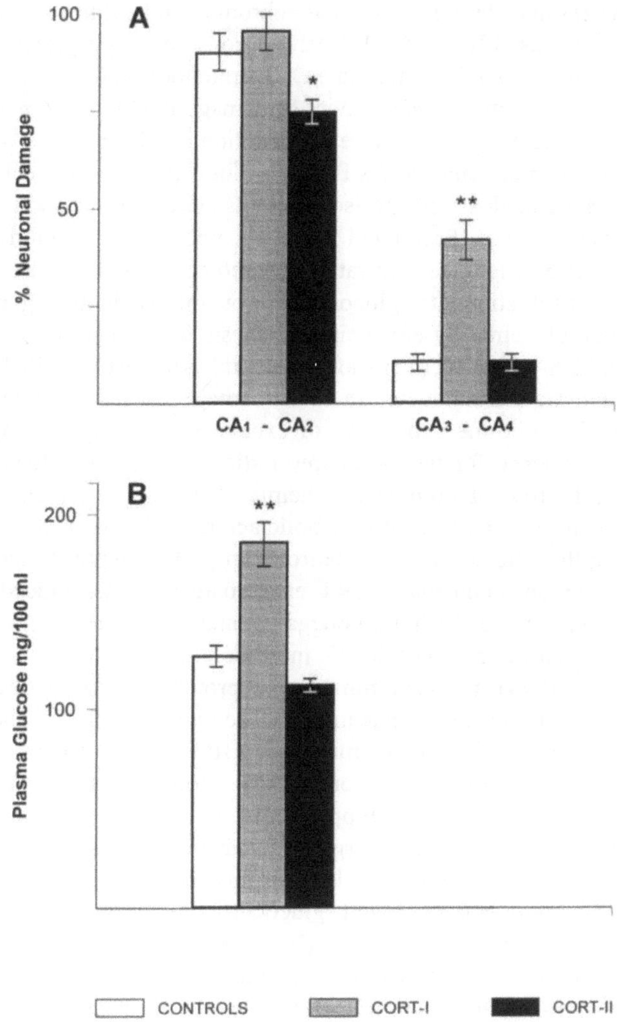

Fig. 4. A Quantification of neuronal necrosis (acidophilic cells) in rat hippocampus. Experimental groups (see Material and Methods): ischemic controls (*open bars*); corticosterone treated animals which were subjected to ischemia immediately after the last corticosterone injection (*gray bars*) and corticosterone treated animals which were subjected to ischemia after 2 days recovery (*filled bars*). Note that in CORT-I animals ischemic damage was exacerbated in *CA3-CA4* neurons only. In CORT-II animals the ischemic damage was attenuated in *CA1-CA2* subfields. Data are means ± S.D. Different from control: * = $p < 0.05$; ** = $p < 0.01$ by Mann-Whitney *U*-test. **B** Effects of corticosterone treatment on plasma glucose levels. Controls (*open bars*); corticosterone treated animals (*CORT-I* and *CORT-II*). Note that in *CORT-I* plasma glucose levels are significantly elevated. In *CORT-II*, the 2 day-recovery period after the last injection of corticosterone was sufficient to normalize plasma glucose levels. Data are means ± SD. ** = $p < 0.01$ by Mann-Whitney *U*-test

The present results demonstrate that chronic pretreatment with CORT upregulates CB exclusively in CA1-CA2 neurons and aggravates the brain damage induced by reversible ischemia in CB-immunonegative neurons (CA3-CA4 subfields). The selective enhancement of damage in CA3-CA4 neurons only seems to be mediated through the failure of induction of CB in CA3-CA4 neurons and/or by hyperglycemia. Since CORT raises blood glucose concentration and enhances tissue lactic acidosis during ischemia, its adverse effects are, at least in part, due to hyperglycemia. Prolonged treatment with glucocorticoids has been found to elevate the glucagon concentration (Marco et al. 1973; Wise et al. 1973). Since as glucagon itself stimulates gluconeogenesis, the rise in glucagon may also contribute to hyperglycemia by enhancing synthesis of glucose. Excessive lactic acidosis is thought to affect recovery after cerebral ischemia (Siesjö 1981). One other possible and significant factor in brain damage caused by hyperglycemic ischemia is thought to be the generation of excess free radicals. Lundgren et al. (1991) provided evidence for increased free radical formation in hyperglycemic brains following 15 min of complete ischemia. This suggests that free radical production is enhanced as a result of metabolic acidosis. Similar results have been reported by using the kainate-model of neurotoxicity. As a matter of fact, Elliot et al. (1993) have demonstrated that CORT exacerbates kainate-induced abnormal tau immunoreactivity in the CB-immunonegative neurons of the CA3 region.

It is known that elevated CORT levels increase voltage-gated Ca^{2+} currents in vitro (Karst et al. 1994) and that nimodipine protects against CORT-induced morphological changes in the hippocampus (Dachir et al. 1997). These results strongly support the calcium involvement in CORT related effects. Moreover, several studies point to the energetically compromising qualities of glucocorticoids as a key factor in hippocampal vulnerability to stressors. The steroid seems to place cells at an energetic disadvantage to such an extent that metabolically demanding insults are less readily survived. As a mechanism underlying this energetic endangerment, glucocorticoids decrease glucose uptake in hippocampal neurons and glia by 30 % (Virgin et al. 1991; Horner et al. 1990). The second interpretation is based on the fact that the CB antiserum used preferentially recognized the calcium loaded form of CB. Therefore, the increase in CB-IR may reflect a CORT induced calcium loading of CB. This could result from increased $[Ca^{2+}]_i$ after CORT treatment. Further studies using monoclonal antibodies recognizing the two different forms of CB are needed to clarify this issue.

In rats investigated 4, 6 and 12 h after forebrain ischemia, Andiné et al. (1992) observed an enhanced uptake of Ca^{2+} into the CA1 neurons. The Ca^{2+} uptake correlated with the extent of neuronal damage. Since our antibodies preferentially recognize the calcium loaded form of CB, one would expect an increase in CB-immunostaining after the ischemic insult. However, ischemia or treatment with Ca^{2+} ionophores induced decreases of CB and parvalbumin contents in CA1 pyramidal cells (Johansen et al. 1990; Rami et al. 1990, 1992), indicating Ca^{2+} induced destruction of these calcium binding proteins.

It should be kept in mind that under ischemic conditions protein synthesis is initially inhibited in the CA1 pyramidal cells. This inhibition is followed by a partial and then by permanent loss of protein synthesis which never recovers in the CA1 area after the ischemic challenge (Johansen and Diemer 1990). Moreover, the protein synthesis is also impaired in the resistent region for up to 12 h (Thilmann et al. 1986). However, if a persistently impaired synthesis of calcium binding proteins contributed to cell death, one would expect that neurons requiring calcium binding proteins would suffer the most. In other words, the lack of CB expression in CA3 neurons make them more susceptible to the deleterious effects of ischemia and CORT. Thus, CORT and the ischemic stress seem to worsen every step in the excitatory-amino acid/calcium cascade of neurodegeneration. The energy-state dependency of this effect is shown by the fact that normalization of glycemia partially blunts the effects of CORT on ischemia induced damage in hippocampal cells.

It seems clear from this study that an increase of CB expression in CA1-CA2 subfields under CORT treatment is an adaptive response to exogenous stress. The lack of adaptive response in CA3-CA4 subfields endangers them by impairing the ability of these neurons to counteract the deleterious effects of calcium. This finding supports the hypothesis that CORT treatment, when combined with an ischemic insult, causes a prolonged elevation of neuronal $[Ca^{2+}]_i$ in an energy-dependent manner, probably through the reduction of Ca^{2+} efflux. Since corticosterone enhancement of ischemia induced damage in CA3-CA4 neurons was prevented by normalization of blood glucose level, it appears possible that excessive lactic acidosis adversely affects recovery from cerebral ischemia. This underlines the multitude of factors implicated in neuronal rescue. Obviously, intracellular Ca^{2+} homeostasis is strictly related to energy state and neuronal activity. In summary, the results support our hypothesis that neurons which contain CB are particularly susceptible to ischemic insults. The CA1-CA2 neurons express high amounts of CB under stress conditions because their activity may involve a high rate of calcium buffering.

Acknowledgments

The authors would like to render their sincere thanks to Prof. Dr. H.-W. Korf for organizing the Scharrer Symposium and for giving us the opportunity to present these data. We thank Mrs. A. Langhagen for technical assistance and Mr. B. vom Schemm for photographic work.

References

Andiné P, Jacobson I, Hagberg H (1992) Enhanced calcium uptake by CA1 pyramidal cell dendrites in the postischemic phase despite subnormal evoked field potentials: excitatory amino acid receptor dependency and relationship to neuronal damage. J Cereb Blood Flow Metab 12: 773-783

Anthony FA, Babitch JA (1985) Calcium-binding proteins in whole brain and synaptic subfractions. Brain Res 345: 124-131

Baimbridge KG, Miller JJ (1982) Immunohistochemical localisation of calcium binding protein in the cerebellum, hippocampal formation and olfactory bulb of the rat. Brain Res 245: 223-229

Baimbridge KG, Miller JJ (1984) Hippocampal calcium-binding protein during commissural kindling-induced epileptogenesis: progressive decline and effects of anticonvulsants. Brain Res 324: 85-90

Dachir S, Kadar T, Robinzon B, Levy A (1997) Nimodipine's protection against CORT-induced morphological changes in the hippocampus of young rats. Brain Res 748: 175-183

Elliot EM, Mattson MP, Vanderklish P, Lynch G, Chang I, Sapolsky RM (1993) Corticosterone exacerbates kainate-induced alterations in hippocampal tau immunoreactivity and spectrin proteolysis in vivo. J Neurochem 61: 57-67

Feldman SC, Iacopino AM, Christakos S (1990) Vitamin D-dependent calcium binding protein (calbindin-D_{28k}) in the central nervous system: distribution and functional significance. In: Smith VL, Dedman JR (eds) Stimulus response coupling. CRC Press, Boca Raton, pp 133-157

Hara H, Sukamoto T, Kogure K (1993) Mechanism and pathogenesis of ischemia-induced neuronal damage. Prog Neurobiol 40: 645-670

Heizmann CW, Braun K (1992) Changes in calcium-binding proteins in human neurodegenerative disorders. TINS 15: 303-308

Heizmann CW, Hunziker W (1991) Intracellular calcium-binding proteins: more sights than insights.Trends Biochem Sci 16: 98-103

Horner H, Packan D, Sapolsky R (1990) Glucocorticoids inhibit glucose transport in cultured hippocampal neurons and glia. Neuroendocrinology 52: 57-64

Iacopino AM, Christakos S (1990) Corticosterone regulates calbindin-D_{28k} mRNA and protein levels in rat hippocampus. J Biol Chem 265: 10177-10180

Iacopino AM, Christakos, S Modi P, Alta CA (1992) Nerve growth factor increases calcium binding protein in rat olfactory bulb. Brain Res 578: 305-310

Ichimiya Y, Emson PC, Mounjoy CQ, Lawson DEM, Heizmann CW (1988) Loss of calbindin-D28k immunoreactive neurons from the cortex in Alzheimer-type dementia. Brain Res 475: 156-159

Johansen FF, NH Diemer (1990) Temporal profile of interneuron and pyramidal cell protein synthesis in rat hippocampus following cerebral ischemia. Acta Neuropathol 81: 14-19

Johansen FF, Tonder N, Zimmer J, Baimbridge KG, Diemer NH (1990) Short-term changes of parvalbumin and calbindin immunoreactivity in the rat hippocampus following cerebral ischemia. Neurosci Lett 120: 171-174

Karst H, Wadman WJ, Joels M (1994) Corticosteroid-dependent modulation of calcium currents in rat hippocampal CA1 neurons. Brain Res 649: 234-242

Krugers HJ, Medema RM, Postema F, Korf J (1995) Region-specific alterations of calbindin-D_{28k} immunoreactivity in the rat hippocampus following adrenalectomy and corticosterone treatment. Brain Res 696: 89-96

Lundgren S, Zhang H, Agardh CD, Smith ML, Evons PJ, Halliwell B, Siesjö BK (1991) Acidosis-induced ischemic brain damage: are free radicals involved? J Cereb Blood Flow Metab 11: 587-596

Marco J, Calle C, Roman D, Diaz-Ferros M, Villanueva ML, Valverde I (1973) Hyperglucagonism induced by glucocorticoids treatment in man. N Engl J Med 288: 128-131

Rami A, Bréhier A, Thomasset M, Rabié A (1987a) Cholecalcin (28-kDa calcium binding protein) in the rat hippocampus: development in normal animals and in altered thyroid states. An immunocytochemical study. Dev Biol 124: 228-238

Rami A, Bréhier A, Thomasset M, Rabié A (1987b) The comparative immunocytochemical distribution of 28 kDa cholecalcin (CaBP) in the hippocampus of rat, guinea pig and hedgehog. Brain Res 422: 149-153

Rami A, Lomri N, Bréhier A, Thomasset M, Rabié A (1989) Effects of altered thyroid states and undernutrition on the calbindin-D_{28k} (calcium-binding protein) content of the hippocampal formation in the developing rat. Brain Res 485: 20-28

Rami A, Rabié A, Thomasset M, Krieglstein J (1990) Calbindin-D_{28k} and delayed neuronal death in the rat hippocampus after cerebral ischemia. In: Krieglstein J, Oberpichler H (eds) Pharmacology of cerebral ischemia. Wissenschaftliche Verlagsgesellschaft Stuttgart, pp 123-128

Rami A, Rabié A, Thomasset M, Krieglstein J (1992) Calbindin-D_{28k} and ischemic damage of pyramidal cells in rat hippocampus. J. Neurosci Res 31: 89-95

Rogers JH (1989) Two calcium-binding proteins mark many chick sensory neurons. Neuroscience 31: 697-709

Schmidt-Kastner R, Freund TF (1991) Selective vulnerability of the hippocampus in brain ischemia. Neuroscience 40: 599-636

Siesjö BK (1981) Cell damage in the brain: a speculative synthesis. J Cereb Blood Flow Metab 1: 155-185

Smith ML, Bendek G, Dahgren N, Rosén, I, Wieloch T, Siesjö BK (1984) Models for studying long-term recovery following forebrain ischemia in the rat. A 2-vessel occlusion model. Acta Neurol Scand 69: 385-401

Thilmann R, Xie Y, Kleihues P, Kiessling M (1986) Persistent inhibition of protein synthesis precedes delayed neuronal death in postischemic hippocampus. Acta Neuropathol 71: 88-93

Thomasset M, Parkes CO, Cuisinier-Gleizes P (1982) Rat calcium-binding proteins: distribution, ontogeny and vitamin-D-dependence. Am J Physiol 243: E483-E488

Virgin C, Ha T, Packan D, Tombaugh G, Yang S, Horner H, Sapolsky R (1991) Glucocorticoids inhibit glucose transport and glutamate uptake in hippocampal astrocytes: implications for glucocorticoid neurotoxicity. J Neurochem 57: 1422-1428

Wise JK, Hendler R, Felig P (1973) Influence of glucocorticoids on glucagon secretion and plasma amino acid concentrations in man. J Clin Invest 52: 2774-2782

16 Structural Evidence for Direct and Indirect Neuro-Endocrine Interactions in the Human Adrenal Cortex

C. Heym

Institut für Anatomie und Zellbiologie, Universität Heidelberg,
Im Neuenheimer Feld 307, D-69120 Heidelberg, Germany;
Phone: +49 6221 548681, Fax: +49 6221 544951,
E-mail: heym@novsr1.pio1.uni-heidelberg.de

Abstract

Recent studies provide cytochemical and functional evidence for significant involvement of extrapituitary factors in adrenocortical functions in man. Among these factors, neural messengers play a crucial role in adrenocortical regulation. They appear to arise from postganglionic sympathetic and vagal neurons, with both extrinsic and intrinsic locations, as well as from sensory neurons, each population exhibiting characteristic chemical codes. The close association of neural messenger containing segments with steroid hormone synthesizing cells and, on the other hand, their apposition on arteries and sinusoid capillaries, are indicative of a dual regulatory involvement of nerve fibers in cortical functions. Multiple intraadrenal factors from innervated cortical and medullary endocrine cells are further partners in cortical regulation. Moreover, cells of the immune system which have immigrated into the adrenal cortex and are equipped with binding sites for neuromediators, contribute to the complex regulatory mechanisms controlling adrenocortical functions. The summarized data suggest that the basal endocrine influence of the hypothalamo-pituitary axis on adrenocortical growth and functions in man is controlled by the nervous system which also regulates fine tuning of human adrenocortical activity at organ and tissue levels.

Introduction

It is generally accepted that, in addition to the hypothalamo-pituitary-adrenal (HPA) axis, neural mechanisms participate in the regulation of adrenocortical functions (Engeland and Dallman 1975; Fehm et al. 1984; Edwards and Jones 1987; Holzwarth et al. 1987). There is compelling evidence for a high degree of differentiation in the human efferent adrenocortical innervation which provides the structural correlate for an efficient autonomic control of cortical functions (cf. Unsicker and Heym 1996; Heym 1997). Afferent adrenal nerve fibers may play an important role in local reflex control and its general homeostatic mechanisms, involving chemo- and mechanoreceptors (Niijima and Winter 1968; Ehrhart-Bornstein et al. 1991).

The human adrenal gland is a highly vascularized organ. Responses of the adrenal vasculature to pharmacological agents are different from responses observed in other vascular beds (Breslow 1992), indicating that the adrenal has developed unique vasoregulatory mechanisms. As an ultrastructural correlate of their function, axon varicosities are distributed in the wall of arteries and abut upon small capillaries (Dorovini-Zis and Zis 1991). A close relation between blood flow rate through the adrenal gland and hormone secretion was demonstrated in the rat (Hinson et al. 1986; Vinson and Hinson 1992). Therefore, messengers released by autonomic nerve fibers, most likely play a crucial role in blood flow regulation, both to the cortex and the medulla.

Interactions of both adrenal compartments, cortex and medulla, serve to adapt the body to increased physical or psycho-biological demands by releasing glucocorticoids and catecholamines (see Axelrod and Reisine 1984). Adrenomedullary chromaffin cells innervated by preganglionic autonomic neurons, and equipped with neuroactive modulators, constitute the paracrine link between medullary and cortical activity (Ehrhart-Bornstein et al. 1995). In support of a suggested ultrashort autocrine regulatory loop in cortical cell proliferation and steroid metabolism, peptides and growth factors have been immunochemically revealed in steroid cells of the human adrenal cortex (Ravid et al. 1986; Guillon et al. 1995).

Furthermore, the immunohistochemical presence of neuropeptides and cytokines in intraadrenal immunocytes, as well as the demonstration of respective binding sites on steroidogenic cells (see Unsicker and Heym 1996; Bornstein et al. 1997) indicate the modulatory role of additional paracrine pathways. Immune cells are equipped with receptors for neural mediators (Whalley et al. 1986; Maisel et al. 1989), again underlining the regulatory key role of the autonomic system.

Although the functional significance of the complex interaction of multiple neuro-endocrine factors in the human adrenal cortex has not been precisely deciphered, the regulatory effect of this interaction on adrenal functions can be considered prominent. This review attempts to summarize available structural and histochemical data on the sources of factors that are presumably involved in the extrapituitary regulation of the human adrenal cortex (Fig. 1). Some relevant findings in other mammals will be cited, when functional data on the human adrenal gland are lacking. However, considerable interspecies variations of messengers and modulators should be taken into consideration, when trying to transfer such data to man. Moreover, demonstration of one single component of the multifold regulatory feedback loops implies the danger of simplification, particularly misleading in the interpretion of pathological conditions.

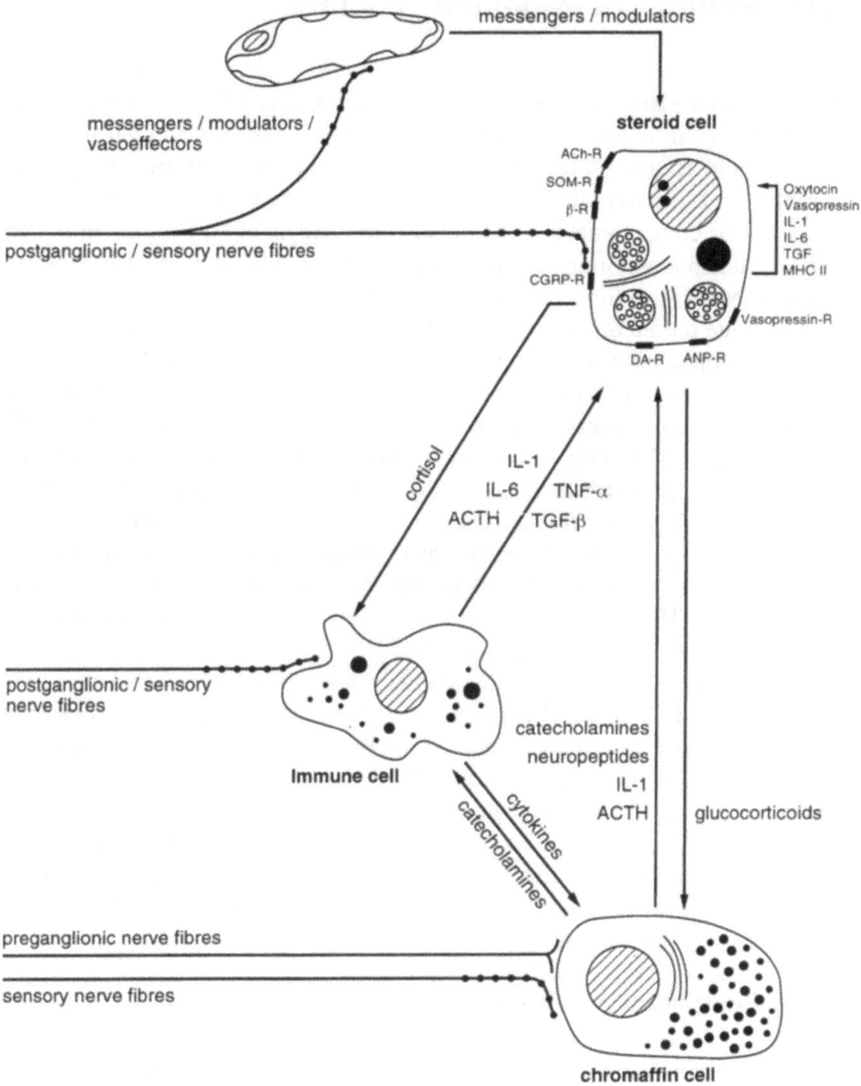

Fig. 1. Direct and indirect pathways in nervous regulation of the adrenal cortex. *R* receptor. For other abbreviations, see text. (Modified after Bornstein et al. 1997)

Extrinsic Sources of Regulatory Factors

The human adrenal gland is innervated by a complex network of varicose nerve fibers, that arise from extraadrenal sources as well as from intraadrenal neurons (see Unsicker and Heym 1996). Extrinsic nerve fibers branch from subcapsular fiber bundles and intraseptal nerve trunks; they contact endocrine cells and approach vascular components (Carlson et al. 1990; Heym et al. 1995b). Both the paracrine signal and autonomic control of blood flow are potential mechanisms by which even a sparse local innervation is able to affect many cells.

Our knowledge on the origin of adrenal nerve fibers has been greatly advanced by intraadrenal application of marker substances that are retrogradely transported in axons, thus allowing demonstration of their perikarya. With this technique, an extrinsic postganglionic innervation of the adrenal cortex was disclosed in small laboratory animals. This originates from thoracic ganglia of the sympathetic paravertebral chain (Holzwarth et al. 1987) and from the suprarenal ganglion (Kesse et al. 1988). Neuronal tracing also showed that sensory neurons in thoracic dorsal root (Heym et al. 1995a) and in vagal (nodose) ganglia (Coupland et al. 1989; Parker et al. 1990) contribute to the mammalian adrenal innervation. As recently shown (Heym et al. 1995b), the human adrenal cortex may also be innervated by neurons of the locations described.

Immunohistochemical colocalization studies with highly specific and sensitive antibodies provided important insights into both architecture and possible functional significance of adrenal innervation (Bacon and Smith 1988; Heym et al. 1994; 1995a, b; Colombo-Benkmann et al. 1995). As the sources of the adrenal innervation cannot be traced in man, correlation of the messenger content in intraadrenal nerve fibers and in neurons in autonomic and sensory ganglia and subsequent comparison with the messengers of adrenoprojecting neurons in laboratory animals may help to identify origin and targets of the human adrenal innervation, thereby unravelling specific nervous pathways.

Aminergic Postganglionic Innervation

The comparison between the plurichemical coding of adrenocortical nerve fibers and postganglionic nerve cell bodies in human sympathetic ganglia at paravertebral and suprarenal locations suggests that adrenocortical nerve fibers are in part derived from these ganglia also in man. According to their immunoreactivity for the enzymes of catecholamine synthesis tyrosine hydroxylase (TH) and dopamine β-hydroxylase (DBH) the vast majority of sympathetic postganglionic neurons contain **noradrenaline** (cf. Heym and Lang 1986). Therefore, the demonstration of human adrenocortical nerve fibers with TH- and DBH-immunoreactivity (Oomori et al. 1989; Baffi et al. 1992; Charlton et al. 1992) and of adrenergic receptors on corticocytes (Shima et al. 1984) indicates the presence of noradrenergic sympathetic efferents. **Neuropeptide tyrosin** (NPY)-

immunoreactive nerve fibers in the human adrenal cortex (Unsicker and Heym 1996) were shown to possess intimate nonsynaptic contacts with endrocrine cells (Unsicker and Heym 1996). As NPY is frequently present in noradrenergic postganglionic neurons of man (Baffi et al. 1992), NPY/TH-positive cortical nerve fibers (Baffi et al. 1992) are of presumably sympathetic origin (Lundberg et al. 1988; Gibbins 1989).

Nonaminergic Postganglionic Innervation

Applying acetylcholine esterase (AChE) histochemistry, varicose, presumably **acetylcholine** (ACh)-containing nerve fibers were described in all cortical layers (Coupland and Holmes 1958; Charlton et al. 1991). ACh is the classical transmitter of the postganglionic cholinergic adrenal innervation, provided by sympathetic and also by vagal efferents (see Lundberg et al. 1988).

Nitric oxide (NO), an important small molecule mediator in autonomic transmission was shown in human adrenocortical nerves by immunohistochemistry for nitric oxide synthase (NOS), the NO generating enzyme (Heym et al. 1994). Accordingly, we observed NOS-positive postganglionic neurons in human paravertebral ganglia (unpubl. obs.). In subprimates, NOS-positive fibers are TH-immunonegative; also in man such efferents may derive from a non-catecholaminergic neuron population described in human paravertebral ganglia (Schmitt et al. 1988; Järvi et al. 1989).

Nerve fibers immunolabeled for **vasoactive intestinal polypeptide** (VIP) supply subcapsular blood vessels and surround endocrine cells in the human adrenal cortex (Linnoila et al. 1980; Heym et al. 1995b; Unsicker and Heym 1996). Ultrastructurally, VIP-terminals are apposed to glomerulosa cells (Unsicker and Heym 1996). The observations that TH and VIP are not colocalized in a minor postganglionic neuron population of human paravertebral ganglia (Järvi et al. 1989), and that in these neurons VIP is colocalized with AChE (Lundberg et al. 1988), suggest a postganglionic sympathetic origin of adrenocortical VIP-labeled fibers (Lundberg et al. 1988). In some human adrenocortical fibers, NOS-immunoreactivity is colocalized with VIP-immunolabeling (Unsicker and Heym 1996).

Recently, colocalization of VIP and **NPY** was demonstrated in cholinergic nerve cells of peripheral vagal ganglia (Talmage et al. 1996). Likewise, in a substantial proportion of adrenocortical fibers NPY, is colocalized with VIP (Unsicker and Heym 1996). This combination is not present in human paravertebral symathetic neurons (Schmitt et al. 1988). NPY, therefore, appears to occur in both the catecholaminergic and the non-catecholaminergic (presumably cholinergic) adrenal innervation; the latter is probably of vagal origin.

Somatostatin (SOM)-immunolabeled varicose fibers occur mainly in the human zona glomerulosa and partly belong to the VIP-ergic fiber population (Heym et al. 1995b; Unsicker and Heym 1996). As an identical coding has been reported for ganglion cells in the human thoracic sympathetic chain (Schmitt et al.

1988), VIP/SOM-labeled nerve fibers likely derive from the nonaminergic fiber population in paravertebral ganglia.

Pharmacological findings presented evidence for the participation of **calcitonin gene-related peptide** (CGRP) in efferent autonomic functions. CGRP/SOM-positive adrenocortical nerve fibers (Heym et al. 1995b) likely originate from CGRP/SOM-containing sympathetic postganglionic neurons (Schmitt et al. 1988), as CGRP is absent from human vagal postganglionic neurons (Talmage et al. 1996).

Immunohistochemical approaches have revealed a dense network of **substance P** (SP)-immunostained nerve fibers supplying the human adrenal cortex (Linnoila et al. 1980; Helen et al. 1984; Heym et al. 1995b). Some of the SP-fibers were nonreactive for CGRP, in contrast to the general colocalization pattern of SP/CGRP in sensory nerves. SP has been described in vagal postganglionic neurons of other human visceral organs (Talmage et al. 1996), but is absent from postganglionic neurons in sympathetic ganglia (Schmitt et al. 1988; Colombo-Benkmann et al. 1996), suggesting that SP-positive but CGRP-nonreactive fibers may derive from vagal ganglia of an as yet unknown location.

The recent description of the messenger combination SP/VIP/NPY in cholinergic neurons of vagal ganglia in the human gallbladder (Talmage et al. 1996), along with the co-occurrence of SP/VIP (Heym et al. 1995b) or SP/SOM in human intracortical nerve fibers that lacked CGRP-labeling (Heym et al. 1995b) substantiate the supposition of a postganglionic vagal participation in the adrenocortical innervation of extrinsic (Parker et al. 1993) and/or intrinsic origin (Colombo-Benkmann et al. 1996). Provided a basically similar chemical coding of vagal postganglionic neurons in human abdominal organs (which remains to be proven), the differentiation between a non-catecholaminergic innervation connected either with the sympathetic or with the vagal system may be feasable with regard to the absence of substance P in sympathetic perikarya and, on the other hand, by the lack of CGRP in vagal postganglionic neurons.

Sensory Adrenocortical Innervation

The human adrenal gland receives a significant sensory innervation, as deduced from the apparent congruency of a specifically coded fiber population with the chemical equipment of human dorsal root ganglion cells at thoracic levels (Heym et al. 1995b).

Opioid peptides, i.e. **leu-enkephalin** and **dynorphin** (1-8) appear to contribute to the human sensory adrenocortical innervation (Pelto-Huikko et al. 1985; Heym et al. 1995b), since dynorphin is present in dorsal root ganglia (Heym et al. 1995b) but does not occur in human paravertebral neurons (Baffi et al. 1992). This latter observation corresponds to the absence of enkephalins from such neurons (Helen et al. 1984; Baffi et al. 1992).

In thoracic human sensory ganglia, **SP**-labeled perikarya occur early in ontogenesis (Charnay et al. 1983). In a considerable proportion of adrenocortical nerve fibers, SP is colocalized with immunostaining for **CGRP** (Heym et al. 1995b), and, less frequently, for dynorphin. In addition, the C-terminal octapeptide **cholecystokinin**-8 (CCK) and also **NOS** have been distinguished in some of the SP-positive cortical nerve fibers (Heym et al. 1995b). Cell bodies with identical immunolabelings (SP/CGRP, SP/dynorphin, SP/CCK, SP/NOS) are present in thoracic dorsal root ganglia of man (Heym et al. 1995b), underscoring the sensory character of such fibers. This assumption is consistent with the finding of capsaicin-sensitive adrenoprojecting dorsal root neurons (Parker et al. 1990) in guinea pig containing SP/CGRP, SP/dynorphin, SP/CCK, or SP/NOS (Heym et al. 1995a). The absence of CCK in efferent postganglionic neurons in man (Baffi et al. 1992) is in line with this view.

Some **galanin** (GAL)-immunolabeled nerve fibers are visible in the human adrenal cortex (Bauer et al. 1986). Galanin was described in animals to participate in effectory (Kummer 1987, Lindh et al. 1989) and sensory functions (Ju et al. 1987). However, galanin appears to be absent from human paravertebral sympathetic (Baffi et al. 1992) and vagal ganglia (Talmage et al. 1996). Furthermore, galanin-containing sensory neurons were shown in the guinea pig not to project to the adrenal (Heym et al. 1995a). Colocalization experiments will help to elucidate the origin of adrenocortical galanin-positive nerve fibers.

Functional Implications of Messengers and Modulators in Adrenocortical Nerves

While basal hormone production in the adrenal cortex is induced by secretion of ACTH (Axelrod and Reisine 1984; Chrousos et al. 1985), nerve-related effects on glucocorticoid levels have been reported to be in the range of 50% (Edwards and Jones 1987, Engeland and Gann 1989; Charlton 1990). The morning peak of plasma cortisol in man is ACTH-independent (Fehm et al. 1984); nerve activity not only influences diurnal variations in steroid synthesis (Dijkstra et al. 1996), but also regulates cortical functions in response to stress situations (Khalil et al. 1986). Moreover, adrenocortical growth (Dallman et al. 1977) and volume changes related to functional activity (Engeland and Dallman 1975) are governed by neural mechanisms. Electrical stimulation of the splanchnic nerves increases adrenocortical sensitivity to ACTH in hypophysectomized calves (Edwards and Jones 1987) and dogs (Engeland and Gann 1989). In addition, nerve stimulation causes an increased conversion of 11-deoxycortisol to cortisol by enhanced 11β-hydroxylase activity (Engeland and Gann 1989). Vice versa transection of these nerves causes a reduction of cortical responsiveness (Edwards and Jones 1987).

Apart from classical transmitters, numerous comodulators are involved in this differential control. From morphological and functional evidence, it is assumed that the messengers are released from the terminal boutons, diffuse into the extracellular space, and act on receptor sites of steroidogenic cells. The demonstration of receptors for some of these mediators in the human adrenal cortex underscores their functional relevance (see Unsicker 1984; Holzwarth et al. 1987; Charlton 1990; Unsicker and Heym 1996).

ACh was shown to act on steroid secretion (Hadjian et al. 1982), both directly by affecting muscarinic receptors (Hadjian et al. 1981), and indirectly by increased ACTH delivery to the cells via vasodilation (Edwards and Jones 1990; Charlton 1990). A third action of ACh on intraadrenal neurons is likely by causing release of comodulators (see Heym 1997). **Noradrenaline** (and also adrenaline) directly stimulate adrenal steroidogenesis (Bornstein et al. 1990; Walker et al. 1991; Charlton 1995). Apart from the putative regulatory influence of **NO** on blood flow, the close association of NOS-positive axons with cortical cells (Unsicker and Heym 1996) suggests a direct but hitherto unknown role of NO in the regulation of steroid secretion. The significant role of **SP** has been shown in the rat, i.e., modulation of adrenocortical innervation (Konishi et al. 1983) and ACh-action (Livett et al. 1990), as well as direct inhibition of aldosterone and potentiation of corticosterone release (Neri et al. 1990). **CGRP** has inhibitory effects on aldosterone secretion (Murakami et al. 1989) but stimulates the zona fasciculata in laboratory animals (Kuramoto et al. 1987; Hinson and Vinson 1990). In animals, exogeneously administered **VIP** causes hypertrophy of the zona glomerulosa (Mazzocchi et al. 1987), and stimulates aldosterone and corticosterone secretion (Cunningham and Holzwarth 1988), along with a considerable rise in blood levels of aldosterone. VIP further potentiates the action of ACTH by stimulating intracellular cAMP (Bloom et al. 1987). Hence, VIP present in the innervation of the adrenal cortex may be a potent stimulatory element, independent from the HPA-axis. Consistent with effects of **NPY** on glomerulosa and fasciculata cells (Rebuffat et al. 1988; Malendovicz et al. 1990, 1996), and on plasma aldosterone levels in laboratory animals (Mazzocchi and Nussdorfer 1987), NPY-binding sites were distinguished on corticocytes of cattle (Torda et al. 1988). In animals, **SOM** decreases the aldosterone response to angiotensin II (Aguilera et al. 1981) and inhibits growth (Boscaro et al. 1982) and steroidogenic capacity of glomerulosa cells (Rebuffat et al. 1984). The modulatory action of SOM on SP by competing with an SP-endopeptidase (Le Grèves et al. 1985) explains the functional significane of SOM/SP-coexistence. Regulation of steroidogenesis is also a likely function of **opioids,** as dynorphin in vivo and in vitro inhibits adrenocortical secretion (Mazzocchi et al. 1990; Neri et al. 1990; Unsworth and Viveros 1987). In the rat, **CCK** was shown to affect adrenocortical secretion (Itoh et al. 1982). Corticosteroid release is inhibited in rats not only by hypothalamic administration (Crawley et al. 1990), but also by direct effects of **galanin** (Mazzocchi et al. 1992). Finally, **nerve growth factors** may play a hitherto unknown role in adrenocortical growth and metabolism.

Neural influence on steroidogenic cells is in part related to changes in blood flow. It is known that adrenocortical vascular resistance is reduced by splanchnic nerve stimulation (Engeland et al. 1985; Edwards et al. 1986; Engeland and Gann 1989). In this regulation, some peptides, e.g. VIP, act as vasodilators (Said and

Mutt 1970; Linton and Lowry 1989), while others, e.g. NPY induce vasoconstriction (see Malendovicz 1993).

Intrinsic Sources of Regulatory Factors

Intraadrenal Neurons

The human adrenal cortex receives a considerable intrinsic innervation by postganglionic neurons at subcapsular, cortical and medullary locations (Mikhail and Amin 1969; Heym et al. 1994, 1995b; Unsicker and Heym 1996), with approximately 40 cell bodies/mm^3 medullary tissue (Colombo-Benkmann et al. 1996).

Intraadrenal neurons were suggested to have arisen from the sympathoadrenal cell lineage (Le Douarin 1982). This suggestion is substantiated by the immunohistochemical presence of **TH** in human adrenal nerve cells (Colombo-Benkmann et al. 1996). In a second TH-negative adrenal nerve cell population, AChE-activity was demonstrated (Charlton et al. 1991). Although sympathetic ganglia contain a cholinergic cell population (see Lundberg et al. 1988), evidence from chick-quail chimera (Le Douarin 1982) and from tracing experiments in laboratory animals (Coupland et al. 1989) points to a second origin of non-adrenergic adrenal neurons from the brainstem neural crest at the site of the vagus anlage.

By demonstration of mRNAs for ChAT or DBH, the two neuron populations in rat were shown to be equipped with specific modulator combinations (Dagerlind and Hökfelt 1991). Several pieces of evidence suggest that this applies also for man: while in aminergic neurons TH is associated with NPY-labeling (Järvi et al. 1989), SP-positive neurons (Heym et al. 1995b) that make up approximately 20% of the human intramedullary neuron population may be counted to the cholinergic neuron subset due to their non-reactivity for TH (Colombo-Benkmann et al. 1996). Recently, immunolabeling for VIP (Bryant et al. 1976) or SOM was found in human medullary ganglion cells that were TH-negative (Heym et al. 1995b). Likewise, NOS-positive nerve cell bodies were demonstrated to belong to the non-amingeric neuron population (Heym et al. 1994). Some of the non-aminergic nerve cell bodies contained SP/VIP (Colombo-Benkmann et al. 1996), substantiating the assumption of a vagal participation in the intrinsic adrenal innervation. The absence of CGRP from human intraadrenal neurons (Colombo-Benkmann et al. 1995) is in line with this supposition. The origin of a remaining intraadrenal neuron population cannot be traced yet; these neurons display surprisingly similar codes as the sensory adrenal innervation described above (SP/dynorphin or SP/cholecystokinin), but they do not reveal CGRP-immunoreactivity (Heym et al. 1995b).

Cortical Steroidogenic Cells

Cortical cells produce and secrete, under nervous control, factors that directly and indirectly adjust the fine tuning of cortical functions (see Bornstein et al. 1997). In support of a suggested short autocrine loop (Lupka and Szczundlik 1985), locally produced steroids, e.g. **ouabain** (see Hamlyn et al. 1996) were shown to affect aldosterone production (Szalay 1971). An indirect feedback is provided by stimulatory effects of **glucocorticoids** on adrenaline synthesis in medullary chromaffin cells (Wurtman and Axelrod 1966) which, in return, stimulate steroidogenesis through catecholamine release (Bornstein et al. 1990).

Regulatory peptides have been immunocytochemically revealed in steroid cells of the human adrenal cortex: the hypothalamic peptides **oxytocin** and **vasopressin** are present in droplets of the cytoplasm in all human fetal cortical zones (Ravid et al. 1986). In rat, oxytocin stimulates aldosterone secretion (Hinson et al. 1987), but ACh-induced aldosterone release is inhibited by oxytocin (Porter et al. 1988). Proliferatory effects of vasopressin on glomerulosa cells are still visible in the hypophysectomized rat (Payet and Lehoux 1980). The demonstration of binding sites for vasopressin in human cortical cells (Guillon et al. 1995) as well as the stimulatory influence of vasopressin on steroid secretion in vitro (Perraudis et al. 1993) also underline the direct involvement of this hormone in human adrenocortical functions.

Cytokines, produced and secreted by cortical cells, are thought to be involved in the fine regulation of adrenal functions under normal and stressful conditions (Abraham 1991). Human reticularis cells produce two members of the **interleukin** (IL) family (IL-1 and IL6), with potentiating effects on glucocorticoid and androgen secretion (González-Hernández et al. 1994b, 1995; Bornstein et al. 1996), in part by acting upon adrenergic mechanisms (Gwosdow et al. 1992). Members of the the α-subgroup of **tumor necrosis factor** (TNF, González-Hernández et al. 1996), have been shown to inhibit human adrenocortical steroidogenesis in vitro and to hinder growth of human fetal adrenocortical cells. Expression of **major histocompatibility complex (MHC)** molecules class II, present in the human zona reticularis (Jackson and McNicol 1988) is influenced by age and hormonal status (Khoury et al. 1987). Modulatory effects of IL-1 and TGF-β on MHC class II expression emphasize the complex intra- and intercellular cross talk.

Adrenomedullary Chromaffin Cells

Chromaffin cells in medullary and cortical positions were electronmicroscopically shown to be in direct contact with cortical cells (Bornstein et al. 1994). Cortisol and aldosterone release is induced by the sympatho-adrenal system, which stimulates cortical cells in a paracrine manner (Bornstein et al. 1994; Bornstein et al. 1997).

Noradrenaline and **adrenaline** are the main constituents of human adrenal chromaffin cells (Benchimol and Cantin 1977) and directly stimulate steroid production (Bornstein et al. 1990). Also, **dopamine** (DA) may act as a modulator of aldosterone secretion in man (Bevilacqua et al. 1982). Accordingly, dopamine binding sites have been demonstrated in the human adrenal cortex (Stern et al. 1986).

Again, neuroactive peptides act as comediators. Regulation of corticogenesis is a likely function of opioids provided by **enkephalin**-containing human chromaffin cells (Racz et al. 1980; Hervonen et al. 1989) and interacting with dopamine in the control of aldosterone production (Bevilacqua et al. 1982, Andreis et al. 1988). Moreover, **SP** (Colombo-Benkmann et al. 1996), **CGRP** (Pelto-Huikko and Salminen 1987), **NPY** (Malendovicz et al. 1996), **SOM** (Heym et al. 1994), and **atrial natriuretic peptide** (ANP; Suda et al. 1984) are constituents of human adrenal chromaffin cells (see Parker 1996). Injections of ANP in man cause a marked decrease in the plasma aldosterone concentration (Ishii et al. 1985). ANP lowers aldosterone, cortisol, and dehydroepiandrosterone secretion in human adrenal cell cultures, but increases the accumulation of intracellular cGMP. Therefore, an involvement of ANP in the early pathway of human steroidogenesis has been was supposed (Higuchi et al. 1986). In the primate adrenal cortex, two types of binding sites for ANP have been reported and a third type of receptor was found in the adrenal medulla (Fuchs et al. 1986). This latter finding raises the possibility of an indirect ANP-action on cortical functions via the regulation of catecholamines.

The presence of hypothalamic/pituitary peptides in human adrenal chromaffin cells again marks their relevance in adrenocortical functions. **Oxytocin** and **vasopressin** (Ang and Jenkins 1984) have already been described in the adrenal cortex. Moreover, medullary cells in man are a possible source of POMC-derived peptides, as they were shown to contain **melanocyte-stimulating hormone**, β-**endorphin** (Evans et al. 1983) and **ACTH** (Lloyd et al. 1984). Accordingly, mRNA from the human adrenal medulla hybridizes with the cDNA probe for the ACTH precursor (Jingami et al. 1984). The functional role of **transforming growth factor** (TGF)-β, recently discovered in chromaffin cells (Krieglstein and Unsicker 1996) is not yet understood.

Chromaffin cells in return, are influenced by glucocorticoids through the induction of phenylethanolamine N-methyl transferase (PNMT) synthesis, the enzyme catalyzing the conversion of noradrenaline to adrenaline (Wurtman and Axelrod 1966). In correspondence with this finding, injections of glucocorticoids in rats during development resulted in an increase of adrenaline-containing chromaffin cells (Eränkö et al. 1966).

Cells of the Immune System

Macrophages and lymphocytes occur frequently in the human adrenal gland (González-Hernández et al. 1994a). A direct influence on cortical steroid cells by intra-adrenal immunocytes has recently been demonstrated (Ehrhart-Bornstein et al. 1996; Bornstein et al. 1997). The presence of β-receptors (Maisel et al. 1989) and binding sites for neuropeptides (Hartung et al. 1986; Spengler et al. 1990) on human macrophages suggest a regulatory influence of the sympathetic nervous system. In return, an exchange of information between the immune system and adjacent autonomic nerve fibers may take place through the action of immune-derived products (Besedovsky and delRey 1992), because T-lymphocytes were shown to modulate peripheral sympathetic actions (Besedovsky et al. 1987). Cortisol via glucocorticoid receptors downregulates liberation of cytokines in monocytes (Jones and Kennedy 1993) that affect cortical cells. A reduced number of glucocorticoid receptors on peripheral lymphocytes in depressed patients may be the reason why adrenal cortex activation in these patients does not result in features matching with Cushing's syndrome (Whalley et al. 1986).

Synthesis of **ACTH** in human peripheral leukocytes (Smith and Blalock 1981) and mouse spleen macrophages (Lolait et al. 1984) is suppressed by dexamethasone (Smith et al. 1986). This finding suggests the expression and control of the POMC-gene in both, human leukocytes and macrophages. Moreover, macrophages appear to produce some cytokines, which affect adrenocortical functions: the expression of **IL-1** (González-Hernández et al. 1995) and of **IL-6** (González-Hernández et al. 1994b; Bornstein et al. 1996) has been demonstrated on both mRNA and protein levels. The activating effects of these interleukins on corticocytes have been described above. An opposite effect is provided by **TNF-α**, which inhibits basal and ACTH-stimulated cortisol synthesis in human fetal adrenals (González-Hernández et al. 1996). Another inhibitory cytokine in human macrophages is a **TGF-β** (Assoian et al. 1987), which also occurs in medullary cells (Krieglstein and Unsicker 1996). Mast cells produce **histamin**, inducing hypertrophy of the zona glomerulosa and stimulating aldosterone and cortisol secretion (Mikolajczyk 1965; Hirose et al. 1978). On the other hand, mast cell **heparin** decreases human aldosterone release, presumably by inhibition of the renin-angiotensin system (Schlatmann et al. 1964). In the rat, heparin is a powerful depressor of adrenocortical actions (Boulouard 1963; Mikolajczyk 1965).

References

Abraham E (1991) Effects of stress on cytokine production. Methods Arch Exp Pathol 14: 45-62

Aguilera G, Harwood JP, Catt KJ (1981) Somatostatin modulates the effects of angiotensin II in adrenal glomerulosa zone. Nature 292: 262-263

Andreis PG, Belloni AS, Cavallini L, Mazzocchi G, Nussdorfer GG (1988) Evidence that long-term administration of a methionine-enkephalin analogue stimulates the growth and steroidogenic capacity of rat inner adrenocortical cells. Neuropeptides 12: 165-170

Ang VT, Jenkins JS (1984) Neurohypophysial hormones in the adrenal medulla. J Clin Endocrinol Metab 58: 688-691

Assoian RK, Fleurdelys BE, Stevenson HC, Miller PJ, Madtes DK, Raines EW, Ross, R, Sporn MB (1987) Expression and secretion of type beta transforming growth factor by activated human macrophages. Proc Natl Acad Sci USA 84: 6020-6024

Axelrod J, Reisine TD (1984) Stress hormones: their interaction and regulation. Science 224: 452-459

Bacon SJ, Smith AD (1988) Preganglionic sympathetic neurones innervating the rat adrenal medulla: immunocytochemical evidence of synaptic input for nerve terminals containing substance P, GABA or 5-hydroxytryptamine. J Auton Nerv Syst 24: 97-122

Baffi J, Görcs T, Slowik F, Horváth M, Lekka N, Pásztor E, Palkovits M (1992) Neuropeptides in the human superior cervical ganglion. Brain Res 570: 272-278

Bauer FE, Hacker GW, Terenghi G, Adrian TE Polak JM, Bloom SR (1986) Localization and molecular forms of galanin in human adrenals: elevated levels in pheochromocytomas. J Clin Endocrinol Metab 63: 1372-1378

Benchimol S, Cantin M (1977) Ultrastructural cytochemistry of the human adrenal medulla. Histochemistry 54: 9-26

Besedovsky HO, delRey A (1992) Immune-neuroendocrine circuits: integrative role of cytokines. Front Neuroendocrinol 13: 61-94

Besedovsky HO, delRey, A Sorkin E, Burri R, Honegger CG, Schlumpf M, Lichtensteiger W (1987) T lymphocytes affect the development of sympathetic innervation of mouse spleen. Brain Behav Immun 1: 185-193

Bevilacqua M, Vago T, Raggi U, Scorza D, Proverbio M, Malacco E, Norbiato G (1982) In vitro steroidogenic properties of FK 33824, a stable analog of methionine-enkephalin. Opiate-dopamine interaction in the control of aldosterone production. J Endocrinol Invest 5: 277-280

Bloom SR, Edwards AV, Jones CT (1987) Adrenal cortical responses to vasoactive intestinal peptide in conscious hypophysectomised calves. J Physiol 391: 441-450

Bornstein SR, Ehrhart-Bornstein M, Scherbaum WA, Pfeiffer EF, Holst JJ (1990) Effects of splanchnic nerve stimulation on the adrenal cortex may be mediated by chromaffin cells in a paracrine manner. Endocrinology 127: 900-906

Bornstein SR, Ehrhart-Bornstein M, Scherbaum WA (1997) Morphological and functional studies of the paracrine interaction between cortex and medulla in the adrenal gland. Micr Res Tech 36: 520-533.

Bornstein SR, González-Hernández JA, Erhart-Bornstein M, Adler G, Scherbaum WA (1994) Intimate contact of chromaffin and cortical cells within the human adrenal gland forms the cellular basis for important intra-adrenal interactions. J Clin Endocrinol Metab 78: 225-232

Bornstein SR, Späth-Schwalbe E, Scherbaum WA (1996) Direct effects of interleukin-6 on human adrenal cells. Endocrine Res 22: 867-873

Boscaro M, Scaroni C, Edwards CRW, Mantero F (1982) Inhibitory effect of somatostatin on the aldosterone response to angiotensin II: in vitro studies. J Endocrin Invest 5: 173-178

Boulouard R (1963) Effects of cold and starvation on adrenocortical activity of rats. Fed Proc 22: 750-754

Breslow MJ (1992) Regulation of adrenal medullary and cortical blood flow. Am J Physiol 262: 1317-1330

Bryant MG, Polak MM, Modlin I, Bloom SR, Albuquerque R, Pearse AG (1976) Possible dual role for vasoactive intestinal peptide as gastrointestinal hormone and neurotransmitter substance. Lancet 1: 991-993

Carlson S, Skarphedinsson JO, Jennische E, Delle M, Thoren P (1990) Neurophysiological evidence for and characterization of the postganglionic innervation of the adrenal gland in the rat. Acta Physiol Scand 140: 491-499

Charlton BG (1990) Adrenal cortical innervation and glucocorticoid secretion. J Endocrin 126: 5-8

Charlton BG (1995) Noradrenergic innervation to the adrenal cortex is responsible for control of basal glucocorticoid section: a model. Med Hypotheses 44: 214-216

Charlton BG, McGadey JM, Russell D, Neal DE (1992) Noradrenergic innervation of the human adrenal cortex as revealed by dopamine-beta-hydroxylase immunohistochemistry. J Anat 180: 501-506

Charlton BG, Nkomazana OF, McGadey J, Neal DE (1991) A preliminary study of acetylcholinesterase-positive innervation in the human adrenal cortex. J Anat 176: 99-104

Charnay Y, Paulin C, Chayvialle JA, Dubois PM (1983) Distribution of substance P-like immunoreactivity in the spinal cord and dorsal root ganglia of the human foetus and infant. Neuroscience 10: 41-55

Chrousos G, Schnermeyer T, Doppman J, Oldfield E, Schulte H, Gold P, Loriaux D (1985) Clinical application of corticotropin releasing factor. Ann Intern Med 102: 344

Colombo-Benkmann M, Heym C, Schemann M (1995) Preganglionic sympathetic and intrinsic autonomic neurons, innervating the guinea pig adrenal gland, immunohistochemically contain acetyltransferase and in part also leu-enkephalin. Neurosci Lett 190: 155

Colombo-Benkmann M, Klimaschewski L, Heym C (1996) Immunohistochemical heterogeneity of nerve cells in the human adrenal gland with special reference to substance P. J Histochem Cytochem 44: 369-375

Coupland RE, Holmes RL (1958) The distribution of cholinesterase in the adrenal glands of the rat, cat and rabbit. J Physiol 141: 97-106

Coupland RE, Parker TL, Kesse WK, Mohamed AA (1989) The innervation of the adrenal gland. III. Vagal innervation. J Anat 163: 173-181

Crawley JN, Austin MC, Fiske SM, Martin B, Consolo S, Berthold M, Langel U, Fisone G, Bartfai T (1990) Activity of centrally administered galanin fragments on stimulation of feeding behavior and on galanin receptor binding in the rat hypothalamus. J Neurosci 10: 3695-3700

Cunningham LA, Holzwarth MA (1988) Vasoactive intestinal peptide stimulates adrenal aldosterone and corticosterone secretion. Endocrinology 122: 2090-2097

Dagerlind A, Hökfelt T (1991) Intrinsic ganglion cells and chromaffin cells in the rat adrenal gland. In: Genes, products, signals, receptors. 6th Int Symp on chromaffin cell biology, Marburg, Germany, August 18-23, p 58

Dallman MF, Engeland WC, McBride MH (1977) The neural regulation of compensatory adrenal growth. Ann NY Acad Sci 297: 373-392

Dijkstra I, Binnekade R, Tilders FJH (1996) Diurnal variation in resting levels of corticosterone is not mediated by variation in adrenal responsiveness to adrenocorticotrophin but involves splanchnic nerve integrity. Endocrinology 137: 540-547

Dorovini-Zis K, Zis AP (1991) Innervation of the zona fasciculata of the adult human adrenal cortex: a light and electron microscopic study. J Neural Transm Gen Sect 84: 75-84

Edwards AV, Jones CT, Bloom, SR (1986) Reduced adrenal cortical sensitivity to ACTH in lambs with cut splanchnic nerves. J Endocrinol 110: 81-85

Edwards AV, Jones CT (1987) The effect of splanchnic nerve stimulation on adrenocortical activity in conscious calves. J Physiol 382: 385-396

Edwards AV, Jones CT (1990) Adrenal response to intra-aortic infusions of acetylcholine in conscious functionally hypophysectomised calves. J Physiol 423: 1-13

Ehrhart-Bornstein M, Bornstein SR, Scherbaum WA, Pfeiffer EF, Holst JJ (1991) Role of the vasoactive intestinal peptide in a neuroendocrine regulation of the adrenal cortex Neuroendocrinology 54: 623-628

Ehrhart-Bornstein M, Bornstein SR, González-Hernández JA, Holst JJ, Waterman MR, Scherbaum WA (1995) Sympathoadrenal regulation of adrenocortical steroidogenesis. Endocrine Res 21: 13-24

Ehrhart-Bornstein M, Bornstein SR, Scherbaum WA (1996) Sympathoadrenal system and immune system in the regulation of adrenocortical function. Eur J Endocrinol 135: 19-26

Engeland WC, Dallman MF (1975) Compensatory adrenal growth is neurally mediated. Neuroendocrinology 19: 352-362

Engeland WC, Gann DS (1989) Splanchnic nerve stimulation modulates steroid secretion in hypophysectomised dogs. Neuroendocrinology 50: 124-131

Engeland WC, Lilly MP, Gann DS (1985) Sympathetic adrenal denervation decreases adrenal blood flow without altering the cortisol response to hemorrhage. Endocrinology 117: 1000-1010

Eränkö O, Lempinen M, Raisanen L (1966) Adrenaline and noradrenaline in the organ of Zuckerkandl and adrenals of newborn rats treated with hydrocortisone. Acta Physiol Scand 66: 253-254

Evans CJ, Erdelyi E, Weber E, Barchas JD (1983) Identification of pro-opiomelanocortin-derived peptides in the human adrenal medulla. Science 221: 957-960

Fehm, HL, Klein E, Holl R, Voigt KH (1984) Evidence for extrapituitary mechanisms mediating the morning peak of plasma cortisol in man. J Clin Endocrinol Metab 58: 410-414

Fuchs E, Shigematsu K, Saavedra JM (1986) Binding sites of atrial natriuretic peptide in tree shrew adrenal gland. Peptides 7: 873-876

Gibbins IL (1989) Co-existence and co-function. In: Holmgren S (ed) Comparative Physiology of Regulatory Peptides. Chapman and Hall, London, pp 308-343

González-Hernández JA, Bornstein SR, Erhart-Bornstein M, Geschwend JE, Adler G, Scherbaum WA (1994a) Macrophages within the human adrenal gland. Cell Tissue Res 278: 201-205

González-Hernández JA, Bornstein SR, Erhart-Bornstein M, Späth-Schwalbe E, Jirikowski GF, Scherbaum WA (1994b) Interleukin-6 mRNA expression in human adrenal gland in vivo: New clue to a paracrine or autocrine regulation of adrenal function. J Clin Endocrinol Metab 79: 1492-1497

González-Hernández JA, Bornstein SR, Erhart-Bornstein M., Geschwend JE, Gwosdow AR, Jirikowski GF, Scherbaum WA (1995) Interleukin-1 is expressed in human adrenal gland in vivo. Clin Exp Immunol 99: 137-141

González-Hernández JA, Bornstein SR, Erhart-Bornstein M, Späth-Schwalbe E, Scherbaum WA (1996) Human adrenal cells express TNFα-mRNA: evidence for a paracrine control of adrenal function. J Clin Endocrinol Metab 81: 807-813

Guillon G, Trueba M, Joubert D, Grazzini E, Chouinard L, Cote M, Payet MD, Manzoni O, Barberis C, Robert M, Gallo-Payet N (1995) Vasopressin stimulates steroid secretion in human adrenal glands: comparison with angiotensin-II effect. Endocrinology 136: 1285-1295

Gwosdow AR, O'Connell NA, Spencer JA, Kumar MSA (1992) Interleukin-1-induced corticosterone release occurs by an adrenergic mechanism from rat adrenal gland. Am J Physiol Endocrinol Metab 263: E461-E466

Hadjian AJ, Ventre R, Chambaz EM (1981) Cholinergic muscarinic receptors in bovine adrenal cortex. Biochem Biophys Res Comm 98: 892-900

Hadjian AJ, Guidicelli C, Chambaz EM (1982) Cholinergic muscarinic stimulation of steroidogenesis in bovine adrenal cortex fasciculata cell suspensions. Biochim Biophy Act 714: 157-163

Hamlyn JM, Hamilton BP, Mantua P (1996) Endogenous ouabain, sodium balance and blood pressure: a review and a hypothesis. J Hypertension 14: 151-167

Hartung HP, Wolter K, Joyka KV (1986) Substance P: binding properties and studies on cellular responses in guinea pig macrophages. J Immunol 136: 3856-3863

Hawthorn J, Nussey SS, Hartung HP, Wolter K, Joyka KV (1986) Substance P: binding properties and studies on cellular responses in guinea pig macrophages. J Immunol 136: 3856-3863

Helen P, Panula P, Yang HY, Hervonen A, Rapoport I (1984) Location of substance P, bombesin-gastrin-releasing peptide, [Met^5] enkephalin- and [Met^5] enkephalin-Arg^6-Phe^7-like immunoreactivities in adult human sympathetic ganglia. Neuroscience 12: 907-916

Hervonen A, Linnoila I, Vaalasti A, Alho H, Pelto-Huikko M (1989) Electron microscopic localization of enkephalin-like immunoreactivity in the human adrenal medulla. J Electr Microsc Techn 12: 380-388

Heym, C (1997) Immunocytochemical correlates of an extrapituitary adrenocortical regulation in man. Histol Histopathol 12: 567-581

Heym C, Braun B, Klimaschewski L, Kummer W (1995a) Chemical codes of sensory neurons innervating the guinea-pig adrenal gland. Cell Tissue Res 279: 169-181

Heym C, Braun B, Yin S, Klimaschewski L, Colombo-Benkmann M (1995b) Immunohistochemical correlation of human adrenal nerve fibres and thoracic dorsal root neurons with special reference to substance P. Histochem Cell Biol 104: 233-243

Heym C, Colombo-Benkmann M, Meyer B (1994) Immunohistochemical demonstration of the synthesis enzyme for nitric oxide and of cọmediators in neurons and chromaffin cells of the human adrenal medulla. Ann Anat 176: 11

Heym C, Lang R (1986) Transmitters in sympathetic postganglionic neurons. In: Unsicker K (ed) Neurohistochemistry: Modern Methods and Applicyations. Alan Liss, New York, pp 493-525

Higuchi K, Nawata H, Kato KI, Ibayashi H, Matsuo H (1986) Human atrial natriuretic polypeptide inhibits steroidogenesis in cultured human adrenal cells. J Clin Endocrinol Metab 62: 941-944

Hinson JP, Vinson GP (1990) Calcitonin gene-related peptide stimulates adrenal cortical function in the isolated perfused rat adrenal gland in situ. Neuropeptides 16: 129-133

Hinson JP, Vinson GP, Whitehouse BJ (1986) The relationship between perfusion medium flow rate and steroid secretion in the isolated perfused rat adrenal gland in situ. J Endocrinol 111: 391-396

Hinson JP, Vinson GP, Porter ID, Whitehouse BJ (1987) Oxytocin and arginine vasopressin stimulate steroid secretion by the isolated perfused rat adrenal gland. Neuropeptides 10: 1-9

Hirose T, Matsumoto I, Aikawa T (1978) Direct effect of histamine on cortisol and corticosterone production by isolated dog adrenal cells. J Endocrinol 76: 371-372

Holzwarth MA, Cunningham LA, Kleitman N (1987) The role of adrenal nerves in the regulation of adrenocortical functions. Ann NY Acad Sci 512: 449-464

Ishii T, Sugimoto T, Matsuoka H (1985) Clinical investigation of human atrial natriuretic polypeptide (hANP) on blood pressure, renal function and blood pressure body fluid regulatory hormones. Folia Endocrinol Jpn 61: 323

Itoh S, Hirota R, Katsuura G (1982) Effect of cholecystokinin octapeptide and vasoactive intestinal polypeptide on adrenocortical secretion in the rat. Jap J Physiol 32: 553-560

Jackson R, McNicol AM (1988) Class II MHC expression in normal adrenal cortex and cortical cells in autoimmune Addison's disease. J Pathol 155 113-120

Järvi R, Helen P, Hervonen A, Pelto-Huikko M (1989) Vasoactive intestinal peptide (VIP)-like immunoreactivity in the human sympathetic ganglia. Histochemistry. 90: 347-351

Jingami H, Nakanishi S, Imura H, Numa S (1984) Tissue distribution of messenger RNAs coding for opioid peptide precursors and related RNA. Eur J Biochem 142: 441-417

Jones TH, Kennedy RL (1993) Cytokines and hypothalamic-pituitary function. Cytokine 5: 531-538

Ju G, Hökfelt T, Brodin E, Fahrenkrug J, Fischer JA, Frey P, Elde RP, Brown JC (1987) Primary sensory neurons of the rat showing calcitonin gene-related peptide immunoreactivity and their relation to substance P-, somatostatin-, galanin-, vasoactive intestinal polypeptide- and cholecystokinin-immunoreactive ganglion cells. Cell Tissue Res 247: 417-431

Kesse WK, Parker TL, Coupland RE (1988) The innervation of the adrenal gland I. The source of pre- and postganglionic nerve fibres to the rat adrenal gland. J Anat 157: 33-41

Khalil Z, Livett BG, Marley PD (1986) The role of sensory fibres in the rat splanchnic nerve in the regulation of adrenal medullary secretion during stress. J Physiol 370: 201-215

Khoury EL, Greenspan JS, Greenspan FS (1987) Adrenocortical cells of the zona reticularis normally express HLA-DR antigenic determinants. Am J Pathol 127: 580-591

Konishi S, Otsuka M, Folkers K, Rosell S (1983) A substance P antagonist produces non-cholinergic slow excitatory postsynaptic potential in guinea pig sympathetic ganglia. Acta Physiol Scand 117: 157

Krieglstein K, Unsicker K (1996) Bovine chromaffin cells release a transforming growth factor-β-like molecule contained within chromaffin granules. J Neurochem 65: 1423-1426

Kummer W (1987) Galanin- and neuropeptide Y-like immunoreactivities coexist in paravertebral sympathetic neurons of the cat. Neurosci Lett 78: 127-131

Kuramoto H, Kondo H, Fujita T (1987) Calcitonin gene-related peptide (CGRP)-like immunoreactivity in scattered chromaffin cells and nerve fibers in the adrenal gland of rats. Cell Tissue Res 247: 309-315

Le Douarin NM (1982) The neural crest. University Press, Cambridge

Le Grèves P, Nyberg F, Terenius L, Hökfelt T (1985) Calcitonin gene-related peptide is a potent inhibitor of substance P degeneration. Eur J Pharmacol 115: 309-311

Lindh B, Lundberg JM, Hökfelt T (1989) NPY-, galanin-, VIP-, PHI-, CGRP-, and substance P-immunoreactive neuronal subpopulations in cat autonomic and sensory ganglia, and their projections. Cell Tissue Res 256: 259-273

Linnoila RI, Diagustine RP, Hervonen A, Miller RJ (1980) Distribution of (Met^5)- and (Leu^5)-enkephalin-, vasoactive intestinal polypeptide- and substance P-like immunoreactivities in human adrenal glands. Neuroscience 5: 2247-2259

Linton EA, Lowry PJ (1989) Corticotrophin-releasing factor in man and its measurement: a review. Clin Endocrinol 31: 225-249

Livett BG, Marley PD, Wan DCC, Zhou XF (1990) Peptide regulation of adrenal medullary function. J Neural Transm (Suppl) 29: 77-89

Lloyd RV, Shapiro B, Sisson JC, Kalff V, Thompson, NW, Beierwaltes WA (1984) An immunohistochemical study of pheochromocytomas. Arch Pathol Lab Med 108: 541-544

Lolait SJ, Lim ATW, Toh BH, Funder JW (1984) Immunoreactive beta-endorphin in a subpopulation of mouse spleen macrophages. J Clin Invest 73: 277-280

Lundberg JM, Martling CR, Hökfelt T (1988) Airways, oral cavity and salivary glands: classical transmitters and peptides in sensory and autonomic motor neurons. In: Björklund A, Hökfelt T, Owman C (eds) Handbook of chemical neuroanatomy, vol 6. Elsevier, Amsterdam, pp 391-445

Lupka A, Szczundlik S (1985) Dexamethasone suppresses cortisol but not ACTH and beta-endorphin plasma concentrations in healthy men. Hormone Metab Res 17: 547

Maisel AS, Fowler P, Rearden A, Motulsky HJ, Michel MC (1989) A new method for isolation of human lymphocyte subsets reveals differential regulation of β-adrenergic receptors by terbutaline treatment. Clin Pharmacol Therapy 46: 429-439

Malendowicz LK (1993) Involvement of neuropeptides in the regulation of growth, structure and function of the adrenal cortex. Histol Histopathol 8: 173-186

Malendovicz LK, Lesniewska B, Miskowiak B (1990) Neuropeptide Y inhibits corticosterone secretion by isolated rat adrenocortical cells. Experientia 46: 721-722

Malendowicz LK, Markowska A, Zabel M (1996) Neuropeptide Y-related peptides and hypothalamo-pituitary adrenal axis function. Histol Histopathol 11: 485-494

Mazzocchi G, Nussdorfer GG (1987) Neuropeptide-Y acutely stimulates rat zona glomerulosa in vivo. Neuropeptides 9: 257-262

Mazzocchi G, Malendowicz LK, Meneghelli V, Nussdorfer GG (1990) Acute effect of dynorphin on rat adrenocortical secretion in vivo. Neuroendocrinol Lett 12: 95-99

Mazzocchi G, Malendowicz LK, Rebuffat P, Nussdorfer GG (1992) Effects of galanin on the secretory activity of the rat adrenal cortex: in vivo and in vitro studies. Res Exp Med 192: 373-381

Mazzocchi G, Robba C, Malendowicz LK, Nussdorfer GG (1987) Stimulatory effect of vasoactive intestinal peptide (VIP) on the growth and steroidogenic capacity of rat adrenal zona glomerulosa. Biomed Res 8: 19-23

Mikhail Y, Amin F (1969) Intrinsic innervation of the human adrenal gland. Acta Anat 194: 105-112

Mikolajczyk H (1965) Urinary output of 17-ks and microscopic changes in the adrenal cortex of rats treated with heparin, histamine or serotonin. Endokrynol Polska: 347-358

Murakami M, Hiromichi S, Nakajima S, Nakamoto H, Kageyama Y, Saruta T (1989) Calcitonin gene-related peptide is an inhibitor of aldosterone secretion. Endocrinology 125: 2227-2234

Neri G, Andreis PG, Nussdorfer GG (1990) Inhibitory effect of dynorphin on the secretory activity of isolated rat adrenocortical cells. Biomed Res 11: 277-281

Niijima A, Winter DL (1968) Baroreceptors in the adrenal gland. Science 159: 434-435

Oomori Y, Okuno S, Fujisawa H, Ishikawa K, Satoh Y, Matsuda M, Yamano M, Ono K (1989) Tyrosine-hydroxylase-immunoreactive nerve fibers in the separated capsule of the rat adrenal gland. Acta Anat 136: 49-54

Parker TL (1996) The innervation of the adrenal medulla. In: Unsicker K (ed) Neuro-endocrine interactions. In: Burnstock G (ed) Handbook of the autonomic nervous system. Harwood, Amsterdam, pp 289-314

Parker TL, Afework M, Coupland RE (1990) Sensory innervation of the rat adrenal gland. Neurosci Lett 38: 62

Parker TL, Kesse WK, Mohamed AA, Afework M (1993) The innervation of the mammalian adrenal gland. J Anat 183: 265-276

Payet N, Lehoux JP (1980) A comparative study of the role of vasopressin and ACTH in the regulation of growth and function of rat adrenal glands. J Steroid Biochem 12: 461-467

Pelto-Huikko M, Salminen T (1987) Localization of calcitonin gene-related peptide (CGRP) in chromaffin cells and nerve terminals in adrenal medulla. Exp Brain Res Series 16: 73-77

Pelto-Huikko M, Salminen T, Hervonen A (1985) Localization of enkephalins in adrenaline cells and the nerves innervating adrenaline cells in rat adrenal medulla. Histochemistry 82: 377-383

Perraudis V, Delarue C, Lefebre H, Contesse V, Kuhn JM, Vaudry H (1993) Vasopressin stimulates cortisol secretion from human adrenocortical tissue through activation of VI receptors. J Clin Endocrinol Metab 76: 1522-1528

Porter ID, Whitehouse BJ, Taylor AH, Nussey SS (1988) Effect of arginine vasopressin and oxytocin on acetylcholine-stimulation of corticosteroid and catecholamine secretion from the rat adrenal gland perfused in situ. Neuropeptides 12: 265-271

Racz K, Glaz E, Kiss R, Lada G, Varga I,Vida S, Di Gleria K, Medzihradszky K, Lichtwald K, Vecsei P (1980) Adrenal cortex - a newly recognized peripheral site of action of enkephalins. Biochem Biophys Res Comm 97: 1346-1353

Ravid R, Oosterbaan HP, Hansen BL, Swaab DF (1986) Localization of oxytocin, vasopressin and parts of precursors in the human neonatal adrenal. Histochemistry 84: 401-407

Rebuffat P, Robba C, Mazzocchi G, Nussdorfer GG (1984) Inhibitory effect of somatostatin on the growth and steroidogenic capacity of rat adrenal zona glomerulosa. J Steroid Biochem 21: 387-390

Rebuffat P, Malendowicz LK, Belloni AS, Mazzocchi G, Nussdorfer GG (1988) Long-term stimulatory effect of neuropeptide-Y on the growth and steroidogenic capacity of rat adrenal zona glomerulosa. Neuropeptides 11: 133-136

Said SI, Mutt V (1970) Polypeptide with broad biological activity: isolation from small intestine. Science 169: 1217-1218

Schlatmann RJ, Jansen AP, Prenen H, van den Korst JK, Majoor, CL (1964) The natriuretic and aldosterone-suppressive action of heparin and some related polysulfated polysaccharides. J Clin Endocrinol Metab 24: 35-47

Schmitt M, Kummer W, Heym C (1988): Calcitonin gene-related peptide (CGRP)-immunoreactive neurons in the human cervico-thoracic paravertebral ganglia. J Chem Neuroanat 1: 287-292

Shima S, Komoriyama K, Hirai M, Kuoyama H (1984) Studies on cyclic nucleotides in the adrenal gland. XI. Adrenergic regulation of adenylate cyclase activity in the adrenal cortex. Endocrinology 114: 245-252

Smith EM, Blalock JE (1981) Human lymphocyte production of corticotrophin and endorphin-like substances: association with leukocyte interferon. Proc Natl Acad Sci USA 78: 7530-7534

Smith EM, Morrill AC, Mayer WJ, Blalock JE (1986) Corticotrophin releasing factor induction of leukocyte-derived immunoreactive ACTH and endorphins. Nature 321: 881-882

Spengler RN, Allen RM, Remick DG, Streiter RM, Kunkel SL (1990) Stimulation of alpha adrenergic receptor augments the production of macrophage-derived tumor necrosis factor. J Immunol 145: 1430-1434

Stern N, Ozaki L, Tuck ML (1986) Evidence for dopaminergic binding sites in the human adrenal cortex. Metabolism 35: 1154-1158

Suda T, Tomori H, Demura H, Shizume K, Mouri T, Miura Y, Sasano N (1984) Immunoreactive corticotrophin and corticotrophin releasing factor in human hypothalamus, adrenal, lung cancer and pheochromocytoma. J Clin Endocrinol Metab 58: 919-924

Szalay KS (1971) The effect of ouabain on aldosterone production in the rat. Acta Endocrinol Copenhagen 68: 477-484

Talmage EK, Pouliot WA, Schemann M, Mawe GM (1996) Structure and chemical coding of human, canine and opossum gallbladder ganglia. Cell Tissue Res 284: 289-302

Torda T, Cruciani RA, Saavedra JM (1988) Localization of neuropeptide Y binding sites in the zona glomerulosa of the bovine adrenal gland. Neuroendocrinology 48: 207-210

Unsicker K (1984) Innervation of endocrine tissues. In: Motta PM (ed) Ultrastrucure of endocrine cells and tissues. Martinus Nijjoff, The Hague, pp 321-332

Unsicker K, Heym C (1996) Innervation of the adrenal cortex. In: Unsicker K (ed) Neuro-endocrine interactions. In: Burnstock G (ed) Handbook of the autonomic nervous system. Harwood, Amsterdam, pp 257-287

Unsworth CD, Viveros OH (1987) Neuropeptides of the adrenal medulla. In: Rosencheck K, Lelkes PI (eds) Stimulus-secretion coupling in chromaffin cells. CRC Press, Boca Raton, Florida, pp 87-109

Vinson GP, Hinson JP (1992) Blood flow and hormone secretion in the adrenal gland. In: James VHT (ed) The adrenal gland. Raven Press, New York, pp 1-86

Walker SW, Lightly ER, Clyne C, Williams BC, Bird IM (1991) Adrenergic and cholinergic regulation of cortisol secretion from the zona fasciculata/reticularis of bovine adrenal cortex. Endocrine Res 17: 237-265

Whalley LJ, Borthwick N, Copolov D, Dick H, Christie JE, Fink G (1986) Glucocorticoid receptors and depression. Br Med J 292: 859

Wurtman RJ, Axelrod J (1966) Control of enzymatic synthesis of adrenaline in the adrenal medulla by adrenal cortical steroids. J Biol Chem 241: 2301

17 Neuroepithelial Bodies in Vertebrate Lungs

D. Adriaensen, J.-P. Timmermans, and D.W. Scheuermann

Laboratory of Cell Biology and Histology, University of Antwerpen (RUCA),
Groenenborgerlaan 171, B-2020 Antwerpen, Belgium;
Phone: +32-3-2180300, Fax: +32-3-2180301,
E-mail: dietsche@ruca.ua.ac.be

Abstract

The pulmonary epithelial lining of humans and of all vertebrate species studied so far has been shown to harbor a system of neuroendocrine cells, composed of solitary cells and innervated organoid cell groups called neuroepithelial bodies (NEBs). The system was first described more than 45 years ago, but data were limited for many years because NEBs are low in number and diffusely localized, and because of the lack of reliable methods to visualize them. This contribution provides an overview of the current state of knowledge of the microscopic morphology of pulmonary NEBs. Some of the most important techniques on hand to study the system, e.g., silver staining, formaldehyde-induced fluorescence, immunocytochemistry, electron microscopy and confocal laser scanning microscopy, will be touched upon briefly. Extensive comparative studies revealed the presence of a wide range of bioactive substances in pulmonary NEBs of all vertebrate classes, suggesting an essential role in the respiratory system. Structurally, NEBs are conceived as receptor-effector units and resemble known chemoreceptors in other organs. Many hypotheses have been proposed about their function, but so far, the exact role played by the NEBs in any given physiological process remains to be elucidated. Experimental data suggest that the complex innervation of NEBs, which has given rise to conflicting interpretations in the literature, may play an important role in their reaction to stimuli. It may be concluded that, even making use of the great possibilities provided by modern histological techniques, the investigation of pulmonary NEBs cells remains a challenge.

Introduction

Since certain neurons were discovered to secrete hormones (for review see Scharrer and Scharrer 1963) the boundaries between the nervous and endocrine systems that integrate body functions have gradually blurred. Several unifying concepts, such as the APUD system (Pearse 1966) and the concept of paraneurons (Fujita 1977), have been proposed. At present, the system of neuropeptide-producing and secreting cells is commonly referred to as the diffuse neuroendocrine system (DNES).

Neuroendocrine cells located in the surface epithelium of airways are included in this DNES. Members of the DNES have been assigned important roles in the peripheral control of various organs, but their function in the lung is still obscure.

There is ample evidence from comparative morphologic and cytochemical studies that neuroendocrine cells can be found in the epithelial lining of the extra- and intrapulmonary airways and in the gas-exchange area of the lungs of humans and all air-breathing vertebrates. Diffuse single cells, also known as solitary neuroepithelial endocrine (NEE) cells, and innervated organoid cell groups called neuroepithelial bodies (NEBs) form a well-organized system. It has been found that the first epithelial cell type to differentiate in the developing human lung is the NEE cell. This differentiation takes place before the 8th week of gestation. Although solitary NEE cells occur throughout the respiratory tract in mammals, the greatest numbers are found in extrapulmonary airways. NEBs, on the other hand, are primarily located at airway branching points in intrapulmonary airways and the gas-exchange area. While the secretory products of NEE cells may vary within and among species, the presence of small cytoplasmic granules (60-300 nm) that reveal characteristic electron microscopic features and store the secretion products is a constant feature of these cells.

Fröhlich (1949) and Feyrter (1953) described a system consisting of solitary or grouped cells with a clear cytoplasm in light microscopic sections of the mammalian respiratory tract after routine fixation with formaldehyde and a hematoxylin-eosin or trichrome staining (Fig. 1). A clear cytoplasm is, however, not unique to pulmonary NEE cells. After conventional light microscopic overview staining, some NEBs can be recognized by their specific localization or morphologic characteristics (Fig. 2), but not so the solitary NEE cells or small, morphologically less well-defined NEBs. The fact that NEE cells are scarcer and more diffused than other epithelial cells in normal lungs has not facilitated their study. It is only in the last 25 years that some real progress has been made with the microscopic investigation of the pulmonary endocrine system.

Methods for the Microscopic Demonstration of Pulmonary Neuroendocrine Cells

The comparative morphologic and cytochemical characteristics of the pulmonary NEE system have been extensively reviewed (Scheuermann 1987; Sorokin and Hoyt 1989; Scheuermann et al. 1992; Adriaensen and Scheuermann 1993). The present paper will focus on the most important technical approaches taken to examine the NEE system. The actual techniques can, however, only be described in brief. For more citations, we refer to the above-mentioned review articles.

Silver Staining

Argentaffin cells contain endogenous substances that can reduce silver salts from a solution, causing a local silver deposit (Masson 1914). When an extraneous reducing agent is added to the solution, the same cells stain, together with an array of other cells and tissue components. The latter phenomenon is referred to as argyrophilia. By means of the method of Grimelius (1968) or one of its many modifications, cytoplasmic argyrophilia can be demonstrated in solitary NEE cells and NEBs (Fig. 3) in formaldehyde-fixed, paraffin-embedded lung tissues of humans and nearly every vertebrate species studied. A clear argentaffinity was shown in few reports only, and was expressed as a granular, dark brown staining that was often most conspicuous in the basal cytoplasm. Silver staining methods are still regularly used in the investigation of the NEE system in normal and pathological tissues.

Formaldehyde-Induced Fluorescence and Microspectrofluorometry

By means of the sensitive and highly specific formaldehyde-induced fluorescence (FIF) technique, biogenic monoamines such as serotonin (5-hydroxytryptamine, 5-HT) or catecholamines, can be demonstrated at the cellular level. The underlying principle is that with the aid of formaldehyde, non-fluorescent monoamines can be converted into strongly fluorescent products under well-controlled circumstances (Falck et al. 1962). The initial monoamine in cells, including pulmonary NEBs, can be identified owing to the spectral characteristics of the fluorescence (Stilman and Scheuermann 1987). The FIF method has revealed that 5-HT is responsible for the fluorescence seen in the pulmonary NEE system of various species (Fig. 4), including humans. In some species, NEE cells of the lung have a low endogenous amine content, hampering a direct demonstration with the FIF technique. By administering in vivo or in vitro amino acid precursors (L-5-hydroxytryptophan or L-DOPA), the primary amine content can be increased. Consequently, the NEE cells as well as the amine uptake (APUD) characteristics of the system can be visualized with the FIF technique. Until the advent of immunocytochemical methods for the demonstration of 5-HT in NEBs (Lauweryns et al. 1982, see below), which require much less specialized equipment and expertise, FIF was the primary method to demonstrate 5-HT in the pulmonary NEE system at the light microscopic level.

Immunocytochemistry, In Situ Hybridization, Receptor Studies

Coons et al. (1941) developed a revolutionary concept called immunocytochemistry (ICC), which specifically localizes tissue elements using antigen-antibody reactions. Today's biomedical research would be unthinkable without it. For the demonstration of pulmonary NEE cells, the following approaches are commonly used. The system can be labeled with antibodies against general markers of the neural and neuroendocrine system or with antibodies against bioactive substances that are contained in pulmonary NEE cells. The list of general and specific ICC markers of pulmonary NEE cells is growing fast, most of them revealing obvious species variation and some age-related changes. Recently, receptor studies, making use of antibodies against receptors for bioactive substances, have been reported.

In the past decade, ever more refined recombinant-DNA techniques and sensitive methods for the detection of specific DNA or RNA sequences using molecular hybridization were developed. In situ hybridization (ISH), also known as hybridization histochemistry, allows a morphological demonstration of the presence of DNA or RNA sequences in microscopic preparations and is therefore an appropriate tool to demonstrate the cellular localization of a messenger, even in heterogeneous cell populations.

Immunocytochemistry with Antibodies against General Neuroendocrine Markers

Neuron-Specific Enolase (NSE). NSE, a form of the cytosolic glycolytic enzyme enolase, has been revealed to be a marker of the DNES (Schmechel et al. 1978). Immunostaining procedures employing antibodies against NSE have been widely used for the demonstration of single NEE cells and NEBs, and their innervation in the respiratory tract of humans and various animal species. Because NSE is not linked to the secretory granule compartment of the cells and is thus retained after degranulation, its presence may be a more reliable parameter than the occurrence of bioactive substances for the identification of pulmonary NEE cells.

Protein Gene-Product 9.5 (PGP 9.5). Another neuron-specific cytosolic protein which, besides in the nervous system, appears to be widely distributed in cells of the DNES is PGP 9.5 (Thompson et al. 1983). Using antibodies against PGP 9.5, the

Fig. 1. Micrograph of a paraffin section of gecko lung after a routine trichrome staining. Areas harboring neuroepithelial bodies (NEBs) conspicuously protrude into the air-space, are covered by a flattened, non-ciliated epithelium and are basally provided with a rich capillary (*c*) network. Note that the NEB cells (*arrows*) are easily recognized, having a clear cytoplasm compared with the surrounding epithelial (*e*) cells. x440. **Fig. 2.** Paraffin section of a bronchus in the cat lung after a PAS-lead hematoxylin staining. An NEB (*arrows*) is easily recognized by its distinct morphological characteristics. x260. **Fig. 3.** Argyrophilic silver staining revealing a tangentially sectioned NEB in a bronchiole of the neonatal rabbit lung. The density of dark-brown silver deposits is higher at the basal side of the NEB (*arrows*). Note also a solitary neuroendocrine cell (*arrowhead*). x590. **Fig. 4.** Fluorescence micrograph of a paraffin section of the lung of a neonatal rabbit showing formaldehyde-induced fluorescence in a bronchiolar NEB. x270. **Fig. 5.** Overview of a paraffin section of rat lung immunolabeled for PGP 9.5, revealing a bronchiolar NEB (*arrows*) and nerve fibers in the lamina propria (*arrowheads*). x50. **Fig. 6.** Serotonin (5-HT)-immunoreactive NEB in a bronchiole of the lung of a neonatal rabbit. The strongest reaction is seen at the basal pole. x260. **Fig. 7.** Branching point of a bronchiole in a paraffin section of the lung of a newborn gerbil. Note the presence of many calcitonin gene-related peptide (CGRP)-immunoreactive NEBs, solitary NEE cells (*arrows*) and varicose nerve fibers (*arrowheads*) in this area. x240. **Fig. 8.** Two CGRP-immunoreactive NEBs at the alveolar level in the lung of a newborn cat. x350

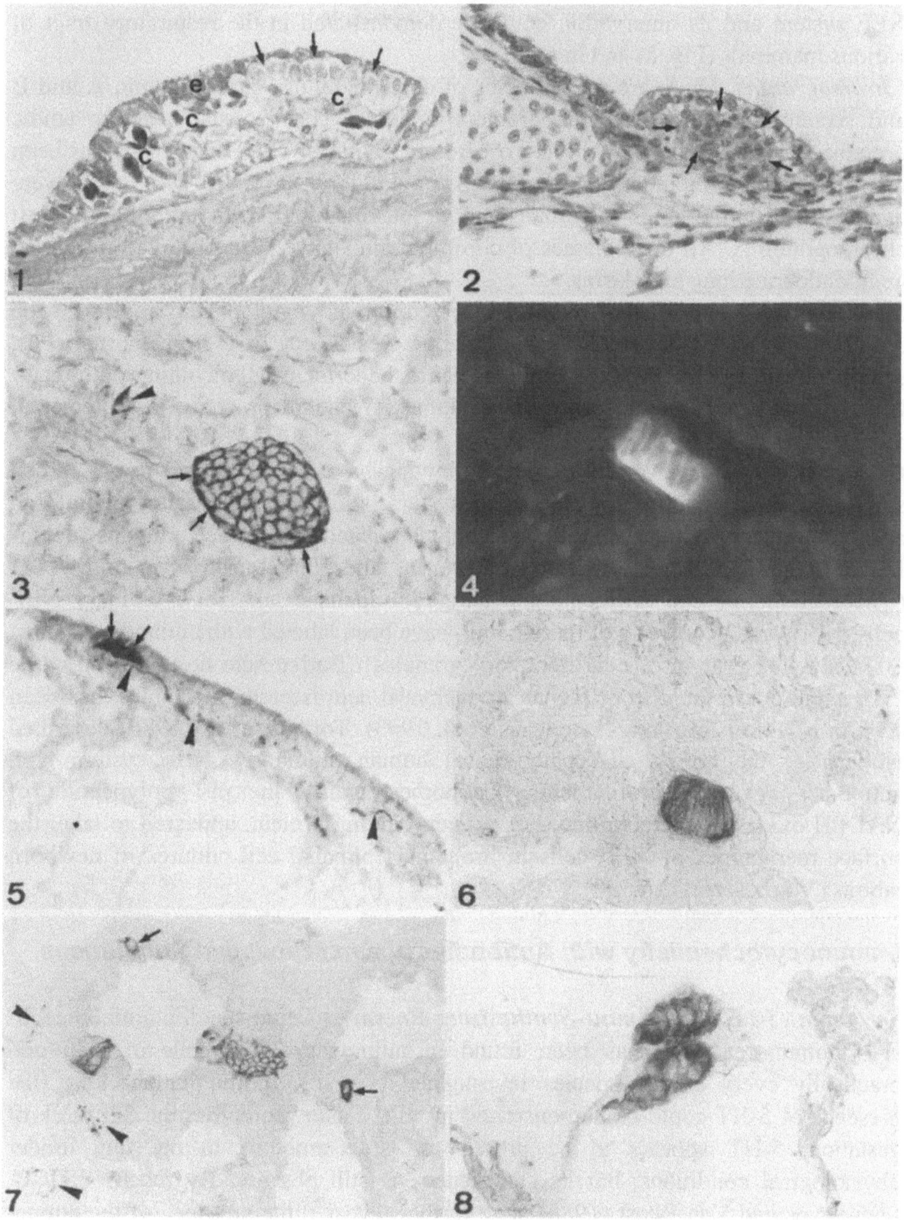

e
c
c
c
c
1
2
3
4
5
6
7
8

NEE system and its innervation could be demonstrated in the respiratory tract of various mammals (Fig. 5) and in man.

Chromogranins. There are three classes of chromogranins, chromogranin A and B and secretogranin II, with comparable properties. They consist of large acidic proteins that are associated with neurosecretory granules. Their occurrence has been reported in many members of the DNES (cf. Fujita et al. 1988). NEE cells were demonstrated in the lungs of humans and various mammals with antibodies against chromogranin A. All three classes of chromogranins have been observed in human neuroendocrine lung neoplasms.

Synaptophysin. The integral membrane glycoprotein synaptophysin appears to be specifically located in small clear synaptic vesicles. Apart from nerve cells, synaptophysin is a marker for a range of neuroendocrine cells, including pulmonary NEE cells in healthy and diseased human lungs (Wiedenmann et al. 1986; Lee et al. 1987).

Other Neuroendocrine Markers. The following have been reported to act as cell surface markers of the pulmonary NEE system: antibodies against the neural cell adhesion molecule (Ito et al. 1995) and against related cell-surface antigens such as Leu-7 (natural killer cell antigen) (Lauweryns and Van Ranst 1987) or MOC-1 (a neuroendocrine cell differentiation antigen) (Speirs et al. 1993). Solitary NEE cells, NEBs and neoplasms of human lung have been labeled with antibodies against 80 kDa Ag (a protein of neurosecretory granules) (Barbareschi et al. 1987), while 7B2, a protein isolated from human hypophysial neurosecretory granules, has been seen in NEBs of frog lungs (Bodegas et al. 1993). Torikata et al. (1986) described labeling of the normal and pathological human pulmonary NEE system with antibodies against neurofilaments. Antibodies against the p91 polypeptide of NADPH oxidase (*b*-cytochrome), an oxygen-binding protein, appeared to label the surface membranes of NEB cells in bronchial epithelial cell cultures of newborn rabbits (Youngson et al. 1993).

Immunocytochemistry with Antibodies against Bioactive Substances

Serotonin (5-HT) and Amine-Synthesizing Enzymes. Using specific antibodies, 5-HT- immunoreactivity has been found in pulmonary NEE cells of man and practically every animal species investigated (Fig. 6). In the human lung, the presence of 5-HT could be demonstrated in NEE cells even before the 8th week of gestation. 5-HT appears to be present in large amounts in the lung under physiological conditions, but its significance is still obscure. By means of ICC, Lauweryns and Van Ranst (1988) have further revealed the presence of the amine-synthesizing enzyme aromatic-*L*-amino acid decarboxylase in the neuroendocrine system of lungs.

Bombesin/Gastrin-Releasing Peptide (GRP). GRP is the mammalian homologue of the well-known amphibian peptide bombesin. It acts as a regulatory hormone and as a cell type specific growth factor, as evidenced in epithelial cells in vivo in normal and neoplastic lungs. While it is overwhelmingly present in pulmonary NEE cells in the human fetus, GRP is less marked after birth, but in disease it may return to a high level of expression. The presence of GRP in pulmonary NEE cells in man can be established during the 8th week of gestation.

Calcitonin Gene-Related Peptide (CGRP) and Calcitonin. CGRP and calcitonin are derived from a single calcitonin-CGRP gene by alternate RNA processing. Although they are widely distributed in the pulmonary NEE system (Figs. 7, 8, 11-13) and CGRP is regarded as one of the major marker substances for NEBs in many mammals including humans of all ages, their precise role is still not clear. Additionally, CGRP-immunoreactivity appears to be a marker for at least part of the NEB innervation in different species (Fig. 12a; Adriaensen et al. 1997a).

Enkephalin and Somatostatin. These neuropeptides have been reported in the pulmonary NEE system of several species. In the healthy human lung, NEE cells that stain for enkephalin and somatostatin appear solitary and are rather scarce. Their ontogenetic development and function in the lung remain to be elucidated.

Cholecystokinin (CCK). CCK is a regulatory peptide that is widely distributed in the gastrointestinal tract but also in the NEE system of the human lung at all ages and in various animal species (Wang and Cutz 1993). Its role in the lung is unclear.

Substance P (SP). There is one account of the presence of this well-known sensory neuropeptide in solitary NEE cells and NEBs of healthy human lungs at all ages (Gallego et al. 1990). In addition, SP has been observed in some solitary NEE cells and NEBs of other species (Figs. 9, 10). SP may also be used to selectively label at least part of the NEB innervation in different species (Figs. 9b, 12b) (Adriaensen et al. 1997a-c).

Other Bioactive Substances. Polypeptide YY, helodermin, pituitary adenylate cyclase-activating peptide, endothelin, sarafotoxin, human chorionic gonadotropin and pancreatic secretory trypsin inhibitor are examples of bioactive peptides that can be localized in NEE cells of healthy lungs using ICC. Pulmonary neuroendocrine neoplasms appear to express an even greater variety of neuropeptides.

Colocalization and Fine Structural Localization of Bioactive Substances. Immunoreactivity for several bioactive mediators reported in the pulmonary NEE system has been localized electron microscopically in the secretory granules (Fig. 17). Various ICC techniques have shown that two or more of the described bioactive substances are colocalized in the same NEE cells (Fig. 9a, b) and even in the same secretory granules.

Immunocytochemistry with Antibodies against Neuropeptide Receptors

The innervation of the NEE system and its content of bioactive substances are well documented but the action sites of these substances are not. A combination of ICC or ISH and molecular biological techniques for receptor cloning might yield this information. So far, it has been demonstrated that NEE cells of the healthy human lung express an intense VIP receptor immunoreactivity (Fischer et al. 1992). SP- and neurokinin 3 receptor (NK3)-like immunoreactivities are colocalized in NEB cells of

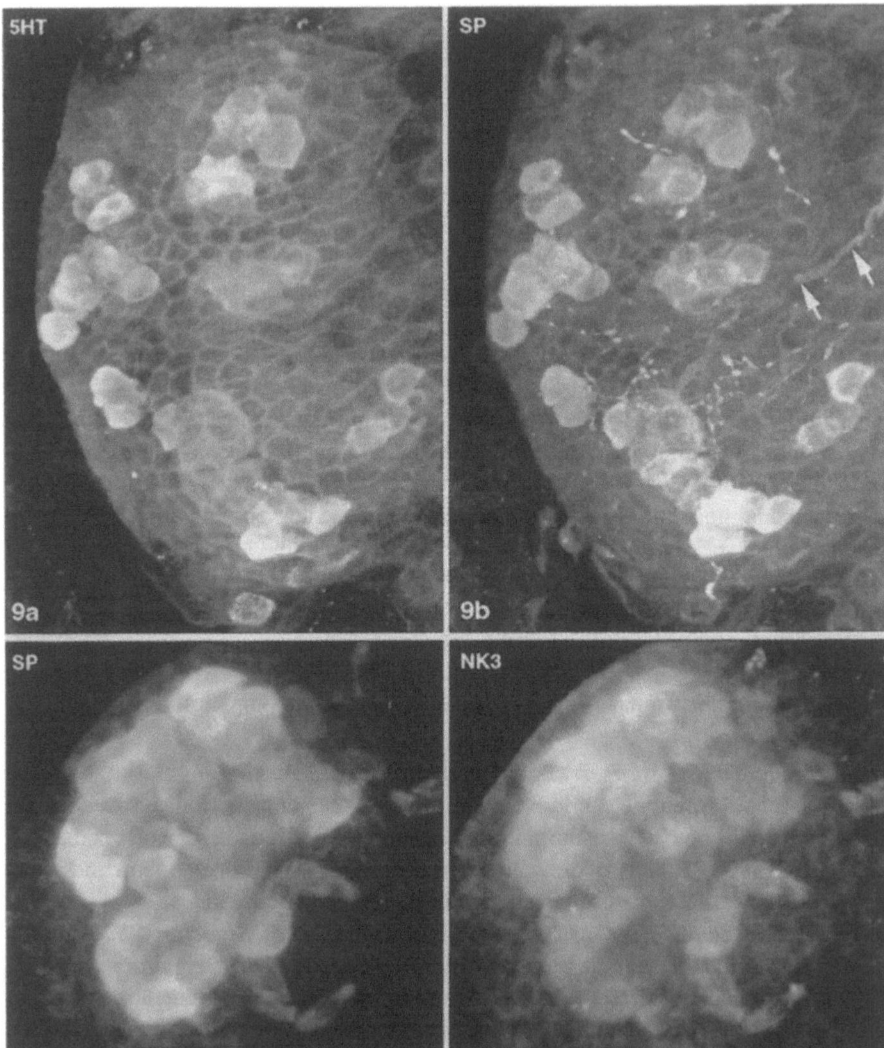

Fig. 9. Maximum intensity projection of a stack of 30 confocal optical sections (0.65 μm interval) through a whole-mount preparation of the lung of a lizard. **a** 5-HT-immunofluorescent (DTAF) NEB. **b** Substance P (SP)-immunoreactivity (Texas Red) can be seen in all serotonin-labeled NEB cells and additionally in a nerve fiber that approaches the epithelium (*arrows*) giving rise to a complex network of branching varicose nerve terminals contacting the NEB cells. **a, b**: x520. **Fig. 10.** Simultaneous immunofluorescent demonstration of SP (**a**) and neurokinin 3 receptor (NK3)-like immunoreactivity (**b**) imaged as maximum intensity projection of 20 confocal optical sections (1 μm interval) through a whole-mount preparation of the lung of a lizard. All NEB cells appear to express both SP and NK3, although not with an identical intensity distribution. x420

gecko lungs, which are innervated by SP-containing nerve terminals (Adriaensen et al. 1997b, c; Figs. 9, 10).

In Situ Hybridization

Pulmonary NEE cells contain various bioactive peptides, as shown by ICC. Since ISH-techniques visualize peptide mRNAs and hence the production of e.g. bioactive substances, they might be used to evaluate the reaction of pulmonary NEE cells to various physiological and pathological situations. As far as the pulmonary NEE system is concerned, however, data obtained with ISH are very limited.

Nerve Tracing

Experimental data suggest that the innervation of pulmonary NEBs affects their reaction to stimuli (Lauweryns and Van Lommel 1986). A clearer image of the complex innervation pattern of NEBs is beginning to emerge, which appears to consist of sensory (afferent) and motoric (efferent) components, part of which are extrinsic and part of which may be intrinsic in nature (Sorokin and Hoyt 1989). The broad majority of findings concerning NEB innervation are morphological and have given rise to conflicting interpretations in the literature. So far, the most established hypothesis favors an extensive vagal sensory innervation of pulmonary NEBs. To test this, we recently injected the fluorescent anterograde neuronal tracer DiI into the nodose ganglion of adult rats (Adriaensen et al. 1997d). In the lung, DiI-labeled nerve endings and NEBs can then be visualized in toto using ICC and confocal laser scanning microscopy on thick frozen sections. Most striking are the extensive intraepithelial terminal arborizations of DiI-labeled vagal afferents in intrapulmonary airways, apparently always co-appearing with CGRP-immunoreactive NEBs (Fig. 13a, b), thus providing support for the hypothesis of a receptor function for NEBs.

Confocal Laser Scanning Microscopy and Three-Dimensional Reconstruction of NEBs

Even though it is essential to have a good knowledge of the specific markers, this alone does not suffice for understanding the exact spatial relationships between NEE cells in an NEB or between NEB cells and their microenvironment, such as surrounding non-endocrine lung epithelial cells and contacting nerve fibers. Conventional light, fluorescence or electron microscopic sections only provide a two-dimensional view of a part of an NEB. Pearsall et al. (1985) reconstructed one hamster lung NEB from a series of 256 transmission electron micrographs. This method is, however, too time-consuming to be applied routinely, and can therefore only provide data for a very limited number of NEBs, the more so as NEBs are diffusely localized and often hard to trace with the electron microscope. On the other hand, thick light microscopic specimens, sectioned to include NEBs and their innervation in toto, do not provide sufficiently precise spatial information because the larger part of the image is out of focus and the enhanced background staining is disturbing.

Confocal laser scanning microscopy (CLSM) remedies these problems, in that it is able to provide thin optical sections from large and thick specimens in which NEBs and their innervation are readily identified using specific fluorescently labeled immunocytochemical markers or nerve tracing. So far, CLSM data regarding the pulmonary NEE system are scanty. Some examples of the versatile possibilities of CLSM for the study of pulmonary NEBs are shown in Figs. 9-13.

Electron Microscopy

Transmission electron microscopic investigations have significantly contributed to our knowledge of the pulmonary NEE system. They allowed visualization of the typical dense-cored vesicles (DCVs). These are small secretory granules ranging in size from 60-300 nm in diameter (Figs. 14-16), showing a characteristic electron-dense core that is separated from the limiting membrane by a clear halo. The DCVs testify to the endocrine-like nature of NEE cells, and their presence is used as a criterion to identify solitary pulmonary NEE cells and NEBs at the electron microscope level. NEE cells are recognizable as intraepithelial elements having a basal pole resting on the basement membrane (Fig. 14) and an apical pole that may reach the luminal surface. Electron microscopy has also demonstrated that NEBs receive an extensive and complex innervation involving morphologically afferent (Figs. 15, 16) as well as efferent nerve endings. The above-mentioned techniques of silver staining, ICC (Fig. 17) and ISH have also proved to be useful at the electron microscopic level.

Figs. 11, 12. Confocal microscopic images of 100-μm-thick frozen sections of cat lung, double labeled for CGRP (**a**) and SP (**b**). CGRP-immunoreactivity can be seen in NEBs (*asterisks*) and fully colocalized with SP in thick varicose nerve terminals in the cilio-mucous bronchial epithelium (*arrows* in **11**). Fine varicose SP-containing nerve fibers are present in a smooth muscle bundle and around submucosal glands (*arrowheads*). Note that the basal knob (*star*) of the NEB in **Fig. 12** receives an abundantly branching CGRP/SP-immunoreactive nerve terminal (*arrows* in **12**). **11a, b**: x280; **12a, b**: x470. **Fig. 13.** Simultaneous demonstration of CGRP-immunoreactivity (**a**) and the anterograde tracer DiI in nerve terminals originating from the vagal nodose ganglion (**b**) in a 100-μm-thick frozen section of rat lung. Images reconstructed from a stack of 30 confocal optical sections (1 μm interval), showing a CGRP-labeled NEB (*asterisk*) at the alveolar level. DiI-labeled nerve fibers running along a bronchiole (*B*) enter the epithelium at the base of an NEB (*arrows*). Abundant DiI-labeled nerve varicosities are seen between the neuroendocrine cells. x570

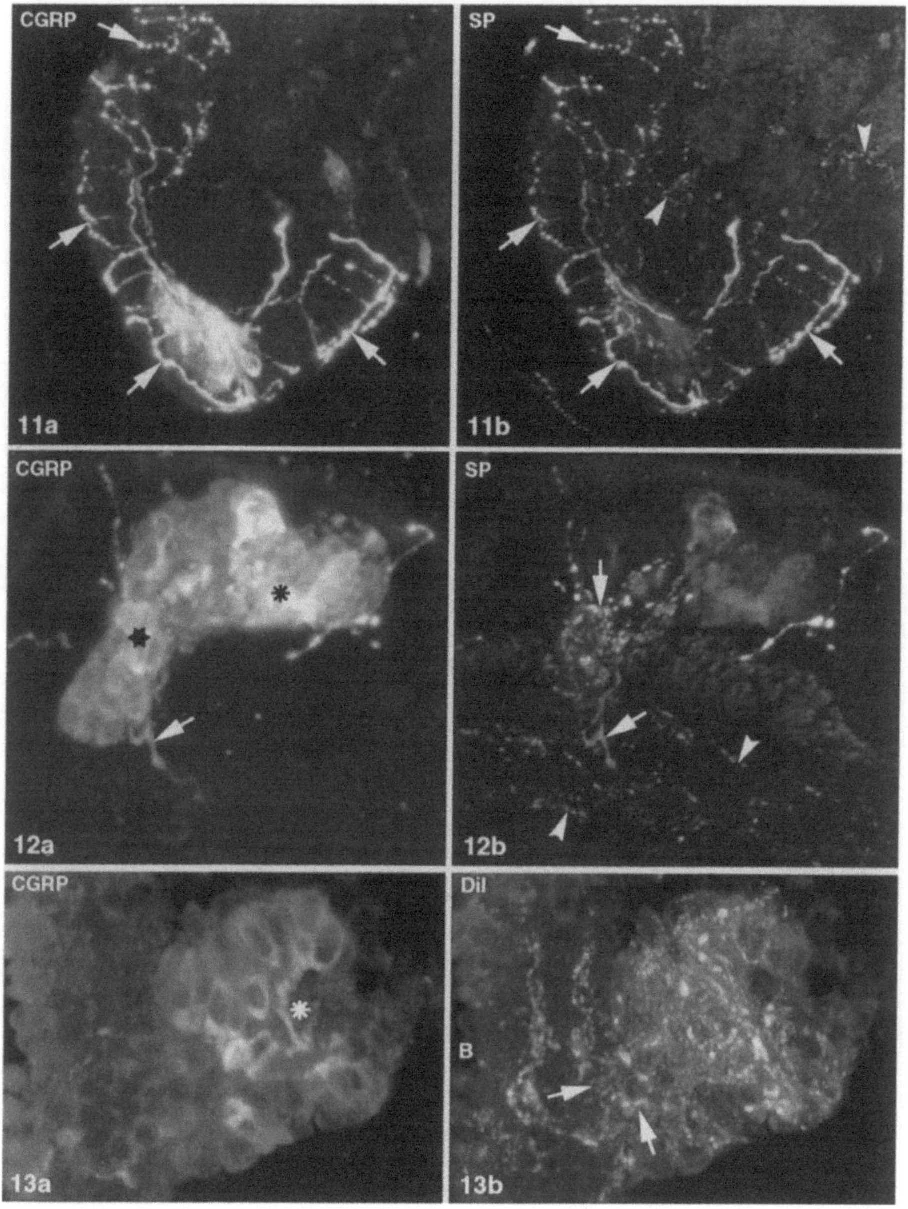
CGRP
11a
SP
11b
CGRP
12a
SP
12b
CGRP
13a
DiI
B
13b

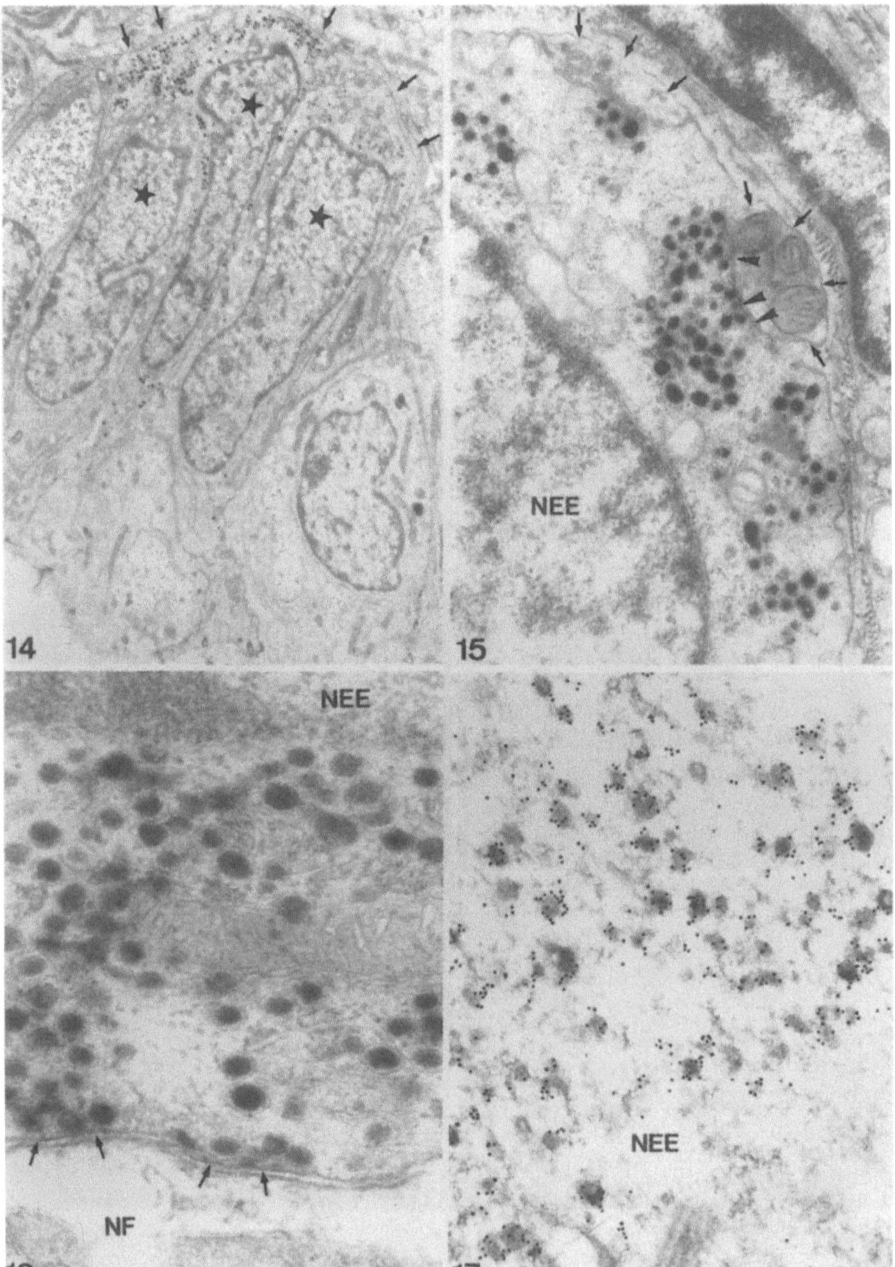
14
15
NEE
NEE
NF
16
NEE
17

Fig. 14. Transmission electron microscopic overview of an NEB in the epithelium of a small bronchus in the lung of a neonatal rabbit. The neuroendocrine cells (*stars*) are seen to rest on the basement membrane (*arrows*). The highest concentration of dense-cored vesicles (DCVs) is located basally. x4620. **Fig. 15.** Detail of the basal part of a neuroendocrine cell (*NEE*) and contacting nerve terminals (*arrows*) in the neonatal rabbit lung. One of the nerve profiles is packed with mitochondria, and DCVs are seen accumulated near cone-shaped presynaptic thickenings (*arrowheads*) in the NEE cell. x18780. **Fig. 16.** Transmission electron microscopic detail of a pulmonary neuroendocrine (NEE) cell in an amphibian lung. The NEE cell is contacted by a nerve fiber (*NF*) containing many mitochondria and some small clear vesicles. Note the presence of abundant DCVs and microfilaments in the NEE cell. In two spots, the DCVs can be seen accumulating in the NEE cell near afferent-like synaptic specializations (*arrows*). x60700. **Fig. 17.** 5-HT-immunoreactivity represented by 10-nm gold particles is specifically localized over the DCVs in a pulmonary neuroendocrine (*NEE*) cell of a quail. x43880

Organ Cultures, Cell Cultures and Cell Lines

NEE cells of cultured lung tissues were shown to display FIF, argyrophilia and the presence of DCVs. Some authors have tried to study the pulmonary NEE system in vitro, using various isolation methods that all have specific advantages and disadvantages (for review, see Speirs and Cutz 1993). Cultured NEBs are suitable for direct electrophysiological measurements and manipulation (Youngson et al. 1993), and will certainly turn out to be a valuable tool for future investigations.

Concluding Remarks

Over the past 25 years, detailed information has been provided about the morphology of pulmonary NEE cells, including their light and electron microscopic structure and immunocytochemical coding. Mainly on the basis of morphological data, several hypotheses about the function of pulmonary NEE cells have been proposed (Sorokin and Hoyt 1989). They are now believed to have different functions during different periods in the prenatal, early postnatal and adult life, and the status of NEE cells in the lung has progressed from a mere morphological curiosity to a potentially major player in pulmonary physiology and pathophysiology. It is now well established that NEE cells are capable of sensing changes in the oxygen tension of the airways. They respond to hypoxic conditions by releasing secretory substances, which in turn could influence specific targets, either via synaptic contacts, local action or a vascular route. While experimental data suggest that the innervation of NEBs plays a crucial functional role, recent morphological data suggest that our knowledge of this innervation is still far from complete. It may be concluded that the investigation of the pulmonary NEE system remains a challenge, even when modern histological techniques are used optimally. Therefore, it is important that every new technical possibility is explored to the full. Most promising for the near future are (1) CLSM, particularly for further elucidation of the complex innervation pattern and the chemical coding of the system, (2) methods for the isolation and culture of NEE cells, allowing direct measurements and manipulation of the cells, and (3) the localization of receptors for the bioactive substances present in pulmonary NEE cells and their innervation. A multidisciplinary approach, combining modern histological and microscopic techniques with cell biological, electrophysiological and molecular biological methods, will be crucial to achieve further progress.

Acknowledgments

The authors wish to thank H. De Pauw, D. De Rijck, A. Hertog, R. Spillemaeckers, G. Svensson, F. Terloo, J. Van Daele and G. Vermeiren for excellent assistance during the preparation of this work. Neurokinin 3-receptor antiserum was kindly provided by N. Mizuno and R. Shigemoto (Kyoto University, Japan). Gratitude is due to W.L. Neuhuber (University of Erlangen, Germany) and H.-R. Berthoud (Louisiana State University, Baton Rouge, USA) for their help with the nerve tracing experiments, and G. Zaccone (University of Messina, Italy) for providing specimens of *Gekko gecko*. The research was financially supported by a Concerted Action Program (1992-1997) of the Flemish Government, by a grant from the National Bank of Belgium to D.W.S. and by an NFWO grant (G.0302.95) to J.-P.T.

References

Adriaensen D, Scheuermann DW (1993) Neuroendocrine cells and nerves of the lung. Anat Rec 236: 70-85

Adriaensen D, Timmermans J-P, Brouns I, Scheuermann DW (1997a) Dual sensory innervation of neuroepithelial bodies in the cat lung? J Auton Nerv Syst 65: 153

Adriaensen D, Timmermans J-P, Mizuno N, Scheuermann DW (1997b) Confocal imaging of substance P- and neurokinin 3 receptor-like immunoreactivity in neuroepithelial bodies of gecko lung whole mounts. Faseb J 11: A207

Adriaensen D, Timmermans J-P, Mizuno N, Shigemoto R, Zaccone G, Scheuermann DW (1997c) Substance P (SP)- and neurokinin 3 receptor (NK3)- like immunoreactivity in neuroepithelial bodies of gecko lungs. J Auton Nerv Syst 65: 141

Adriaensen D, Timmermans J-P, Brouns I, Berthoud H-R, Neuhuber WL, Scheuermann DW (1997d) Vagal afferent innervation of pulmonary neuroepithelial bodies (NEBs) revealed by anterograde tracing and confocal microscopy. J Auton Nerv Syst 65:153

Barbareschi M, Frigo B, Muscarà M, Mauri FA (1987) Neuroendocrine structures in human lung: immunohistochemical evaluations. Med Biol Environ 15: 665-672

Bodegas M.E, Montuenga LM, Polak JM, Sesma P (1993) Immunohistochemical colocalization of 7B2 and 5HT in the neuroepithelial bodies of the lung of *Rana temporaria*. Cell Tissue Res 273: 137-140

Coons AH, Creech HJ, Jones RN (1941) Immunological properties of an antibody containing a fluorescent group. Proc Soc Exp Biol Med 47: 200-202

Falck B, Hillarp N-Å, Thieme G, Torp A (1962) Fluorescence of catecholamines and related compounds condensed with formaldehyde. J Histochem Cytochem 10: 348-354

Feyrter F (1953) Über die peripheren endocrinen (parakrinen) Drüsen des Menschen. Wilhelm Maudrich, Wien

Fischer A, Kummer W, Couraud J-Y, Adler D, Branscheid D, Heym C (1992) Immunohistochemical localization of receptors for vasoactive intestinal peptide and substance P in human trachea. Lab Invest 67: 387-393

Fröhlich F (1949) Die "Helle Zelle" der Bronchialschleimhaut und ihre Beziehungen zum Problem der Chemoreceptoren. Frankf Z Pathol 60: 517-559

Fujita T (1977) Concept of paraneurons. Arch Histol Jpn 40 (Suppl): 1-8

Fujita T, Kanno T, Kobayashi S (1988) The paraneuron. Springer, Tokyo

Gallego R, Garcia-Caballero T, Roson E, Beiras A (1990) Neuroendocrine cells of the human lung express substance-P-like immunoreactivity. Acta Anat 139: 278-282

Grimelius L (1968) The argyrophil reaction in islet cells of adult pancreas studied with a new silver nitrate procedure. Acta Soc Med Uppsala 73: 271-294

Ito T, Nozawa A, Usuda Y, Kitamura H, Kanisawa M (1995) Hamster pulmonary endocrine cells with neural cell adhesion molecule (NCAM) immunostaining. Histochem Cell Biol 104: 357-362

Lauweryns JM, Van Lommel A (1986) Effect of various vagotomy procedures on the reaction to hypoxia of rabbit neuroepithelial bodies: modulation by intrapulmonary axon reflexes. Exp Lung Res 11: 319-339

Lauweryns JM, Van Ranst L (1987) Leu-7 immunoreactivity in human, monkey, pig bronchopulmonary neuroepithelial bodies and neuroendocrine cells. J Histochem Cytochem 35: 687-691

Lauweryns JM, Van Ranst L (1988) Immunocytochemical localization of aromatic-*L*-amino acid decarboxylase in human, rat and mouse bronchopulmonary and gastrointestinal endocrine cells. J Histochem Cytochem 36: 1181-1186

Lauweryns JM, de Bock V, Verhofstad AAJ, Steinbusch HWM (1982) Immunohistochemical localization of serotonin in intrapulmonary neuro-epithelial bodies. Cell Tissue Res 226: 215-223

Lee I, Gould VE, Moll R, Wiedenmann B, Franke WW (1987) Synaptophysin expressed in the bronchopulmonary tract: neuroendocrine cells, neuroepithelial bodies, neuroendocrine neoplasms. Differentiation 34: 115-125

Masson P (1914) La glande endocrine de l'intestin chez l'homme. C R Acad Sci Paris 158: 59-61

Pearsall AD, Hoyt RF jr, Sorokin SP (1985) Three-dimensional reconstruction of a small-granule paracrine cell cluster in an adult hamster bronchus. Anat Rec 212: 132-142

Pearse AGE (1966) Common cytochemical properties of cells producing polypeptide hormones, with particular reference to calcitonin and the thyroid C cells. Vet Rec 79: 587-590

Scharrer E, Scharrer B (1963) Neuroendocrinology. Columbia Univ Press, New York

Scheuermann DW (1987) Morphology and cytochemistry of the endocrine epithelial system in the lung. Int Rev Cytol 106: 35-88

Scheuermann DW, Adriaensen D, Timmermans J-P, De Groodt-Lasseel MHA (1992) Comparative histological overview of the chemical coding of the pulmonary neuroepithelial endocrine system in health and disease. Eur J Morphol 30: 101-112

Schmechel D, Marangos PJ, Brightman M (1978) Neurone-specific enolase is a molecular marker for peripheral and central neuroendocrine cells. Nature 276: 834-836

Sorokin SP, Hoyt RF jr (1989) Neuroepithelial bodies and solitary small-granule cells. In: Massaro D (ed) Lung cell biology. Marcel Dekker, New York, pp 191-344

Speirs V, Cutz E (1993) An overview of culture and isolation methods suitable for in vitro studies on pulmonary neuroendocrine cells. Anat Rec 236: 35-40

Speirs V, Eich-Bender S, Youngson CR, Cutz E (1993) Localization of MOC-1 cell surface antigen in small-cell lung carcinoma cell lines: an immunohistochemical and immunoelectron microscopic study. J Histochem Cytochem 41: 1303-1310

Stilman C, Scheuermann DW (1987) Computer-operated microspectrofluorometry to identify formaldehyde-induced fluorophores of biogenic monoamines and precursor substances in models and tissue sections. Histochemistry 86: 255-267

Thompson RJ, Doran JF, Jackson P, Dhillon AP, Rode J (1983) PGP9.5 - a new marker for vertebrate neurons and neuroendocrine cells. Brain Res 278: 224-228

Torikata C, Mukai M, Kawakita H, Kageyama K (1986) Neurofilaments of Kultschitsky cells in human lung. Acta Pathol Jpn 36: 93-104

Wang Y-Y, Cutz E (1993) Localization of cholecystokinin-like peptide in neuroendocrine cells of mammalian lungs: a light and electron microscopic immunohistochemical study. Anat Rec 236: 198-205

Wiedenmann B, Franke WW, Kuhn C, Moll R, Gould VE (1986) Synaptophysin: a marker protein for neuroendocrine cells and neoplasms. Proc Natl Acad Sci USA 83: 3500-3504

Youngson C, Nurse C, Yeger H, Cutz E (1993) Oxygen sensing in airway chemoreceptors. Nature 365: 153-155

18 Control of the Release of Plasma Catecholamines

A. Epple, H. Al-Kharrat, B. Abele, C.B. Hathaway, B. Nibbio, S. Scholz, and Q. Tran

Department of Pathology, Anatomy and Cell Biology, Thomas Jefferson University, Philadelphia, PA 19107, USA; Phone: +215 503 7850, Fax: +215 923 3808, E-mail: epple1@jeflin.tju.edu

Abstract

Using the eel (*Anguilla rostrata*) as a model, recent studies revealed that the plasma catecholamines dopamine (DA), norepinephrine (NE) and epinephrine (E) originate from different sources. Plasma DA derives mainly from walls of large blood vessels, but its precise tissue origin is unknown. DA release is stimulated by both catecholamines and endogenous opioids. Plasma NE and E derive mainly from the chromaffin cells (adrenomedullary equivalent). The control mechanisms of NE and E secretion are multiple: "direct" sympathetic fibers affect NE and E release by (stimulatory) nicotinergic and (both stimulatory and inhibitory) muscarinergic mechanisms, and NE and E stimulate each other's secretion. The latter phenomenon is probably an inheritance from the early vertebrates (cyclostome stage) where NE and E coordinate their release from non-innervated chromaffin cells by mutual stimulation. The picture is further complicated by autocrine/paracrine co-secretions of the chromaffin cells: met-enkephalin inhibits NE and E release; it is counteracted by endogenous morphine and codeine. The interactions of the catecholamines, opioids and cholinergic substances with further co-secretions of both the chromaffin cells and their innervation await analysis.

Introduction

Despite a large number of in vivo and in vitro studies, there are many gaps in our knowledge of the three circulating catecholamines, i.e., dopamine (DA), norepinephrine (NE) and epinephrine (E). For instance, the physiological roles of free and conjugated plasma DA are not understood. Vasodilatory and renal effects of free plasma DA have been reported (Landsberg and Young 1992; Goldstein 1995; Hoffman and Lefkowitz 1996; Stamler et al. 1996), but its spectrum of functions is by no means clear. There is circumstantial evidence that conjugated DA, or at least a fraction of it, may not be destined for metabolic breakdown but rather have active functions (cf. Tyce et al. 1986; Epple et al. 1993b, 1995b), somewhat in analogy to conjugated androgens. Another unresolved question is the physiological role of E, especially in the human. A seemingly unshakeable dogma links this messenger substance to glucose metabolism (cf. Navarro and Epple 1993). However, more than 60 years ago Cori and Buchwald (1930) showed that the vascular system of the human is particularly sensitive to E when compared with other species. We confirmed this in phylogenetic perspective unusual sensitivity of the human heart to E, and furthermore showed that it makes a metabolic role of E, at least under physiological conditions, unlikely (Epple et al. 1989a). In the following, we will discuss recent findings concerning two further poorly understood phenomena in basic catecholamine physiology: the origin and control of circulating free dopamine and the interaction of neural and autocrine/paracrine factors in the control of catecholamine release from chromaffin cells.

Materials and Methods

Much of the work reported here was carried out in an ideal model, the American eel (*Anguilla rostrata*). This species is very hardy under experimental conditions, does not require feeding for months when kept at a temperature below 18 °C, and is available in convenient sizes (100 to about 800g body weight) for work both in vivo and in vitro (Epple and Nibbio 1994; Kraschinski et al. 1996). The chromaffin cells (adrenomedullary equivalent) of the eel are not organized as a distinct adrenal medulla, but form a subendothelial layer of the posterior cardinal veins, close to the heart (Hathaway and Epple 1990). However, like their mammalian homologue, these cells contain and release catecholamines and co-secretions (Epple et al. 1993a, 1994, 1996; Reid et al. 1995), and they are innervated (Hathaway et al. 1989; Reid et al. 1995).

Another "lower" vertebrate model we studied is the sea lamprey (*Petromyzon marinus*). The adults of its fresh water populations from the Great Lakes region of North America are of a similar size as the adult eel, reasonably hardy under experimental conditions (Dashow and Epple 1985), and do not feed for months during the last stage of their life cycle (Hardisty 1979), when they are available in large numbers. The chromaffin cells of the lampreys are located in the heart (especially the sinus venosus and atrium) and adjacent veins (Dashow and Epple 1985; Epple et al. 1985). The special advantage of the lampreys is that their chromaffin cells are **not** innervated, which allows the study of autocrine/paracrine regulations without neural interference. To some extent, the lamprey's chromaffin cell system resembles the scattered paraganglia of the prenatal human, which control the circulation before the development of the adrenal medulla (Phillippe 1983).

In situ studies e.g., collection of blood samples, were carried out using anesthetized (MS 222 or urethane) lampreys and eels (Dashow and Epple 1985; Abele et al., submitted). Chronically cannulated lampreys and eels fitted with cardiovascular cannulas were used for serial blood sampling from unanesthetized animals after the stress effect of the surgery had worn off (Dashow et al. 1982; Epple et al. 1982). In vitro perifusions of cardiovascular regions of the eel were carried out in a simple dripping system at room temperature without O_2/CO_2 gassing (Epple et al. 1993a; Kraschinski et al. 1996). The chromaffin cells of the eel were stimulated indirectly, using a Grass S88 stimulator and an external hook electrode that was applied to the hindbrain close to the spinal cord (Abele et al., submitted).

Other species studied were the human and the rat. The veins of the forearms of male volunteers (between 20 and 35 years of age) were fitted with cannulas for injections and blood withdrawal. The electrocardiogram was continuously monitored and recorded with a Physio-Control model 011-ST cardiac monitor (Epple et al. 1989a). Albino rats of different strains with one cannulated jugular vein were used for repeated blood sampling (Epple et al. 1988), or decapitated with or without preceding anesthesia, depending on the experimental protocol (Epple et al. 1982; Kraschinski et al. 1996).

The catecholamines were determined using a radioenzymatic assay, the opioid substances by HPLC that was followed by radioimmunoassays (cf. Epple et al. 1993a, 1996). For the identification of conjugated catecholamines, plasma and perifusates were treated with a sulfatase/glucuronidase mixture (Epple et al. 1995a).

Results

The circulating catecholamines of the eel originate mainly from two sources: DA is released from a component of the wall of large arteries and veins, probably the endothelium, but not from microvessels (Milakofsky et al. 1995; Kraschinski et al. 1996). Plasma NE and E are released from the chromaffin cells (Epple et al. 1993a, 1994, 1996; Al-Kharrat et al. 1997). There are considerable stores of DA in the chromaffin cells (Hathaway and Epple 1989); however, it now appears that they serve as precursors of NE and E rather than as a direct secretion. There may be minor admixtures of plasma catecholamines from other sources, e.g., epinephrinergic neurons (Nilsson 1983), under certain conditions (Abele et al., submitted); however, this matter remains to be clarified.

The macrovascular DA release of the eel seems to be under complex controls. In vivo it is stimulated by NE and E (Epple and Nibbio 1985; Epple et al. 1989b); in vitro, it is also stimulated by met-enkephalin, codeine and morphine. In the caudal vein, but not the ventral aorta, there also was a "paradoxical" stimulation of DA release (Fig. 1) by the general opioid antagonist naloxone (Scholz et al. 1997). These findings show that (1) macrovascular DA release is sensitive to both catecholamines and opioids, and (2) the antagonistic effects between alkaloid opiates and met-enkephalin, seen in chromaffin cells and other tissues (see below), are absent in macrovessels.

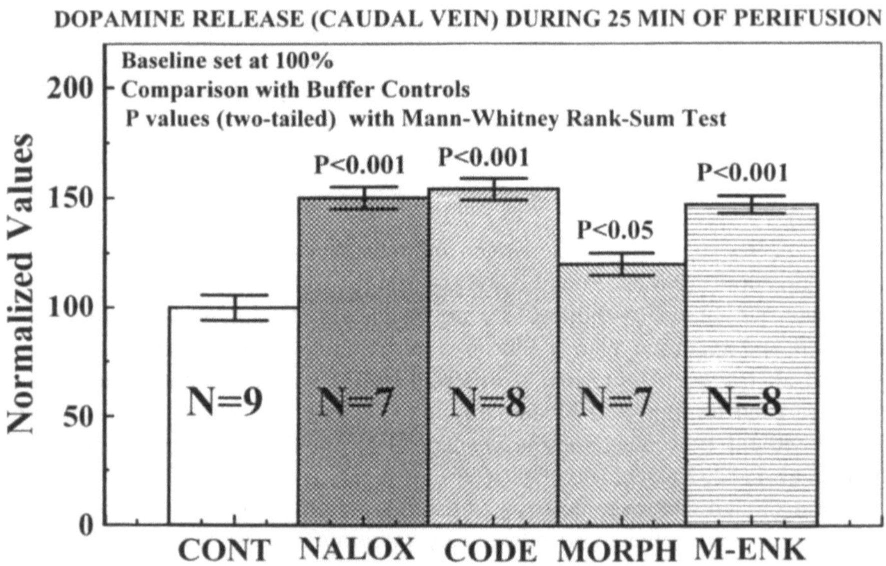

Fig. 1. The effects of the general opiate antagonist naloxone (*NALOX*), the alkaloid opiates codeine (*CODE*) and morphine (*MORPH*), and the opioid peptide met-enkephalin (*M-ENK*) on dopamine release from the perifused caudal vein of the eel (*Anguilla rostrata*). Note that all substances have a stimulatory effect even though (1) naloxone usually antagonizes the other three opioids, and (2) both codeine and morphine counteract the inhibitory effect of met-enkephalin in other systems (for details, see Results and Discussion). Data from Scholz et al. (1997)

Besides NE and E, the chromaffin cells of the eel release met-enkephalin, codeine and morphine (Epple et al. 1993a, 1994, 1996). The release of NE and E is from two different cell types (Hathaway and Epple 1990). Under most conditions, both in vivo and in vitro, it is parallel (Fig. 2). This phenomenon is due to the mutual stimulation of NE and E release (Epple and Nibbio 1985). The α-adrenergic receptor antagonist phentolamine abolished electrostimulated NE and E release (Abele et al., submitted), and the β-adrenergic receptor antagonist propranolol abolished both the acetylcholine-stimulated and the electrostimulated release of NE and E (Al-Kharrat et al. 1997; Abele et al., submitted). Thus, the "catecholaminotropic effect of catecholamines" (Epple and Nibbio 1985) is now firmly established. Originally, this effect was seen in sea lampreys with indwelling cardiac cannulas (Dashow and Epple 1983); since then, it has also been demonstrated in the rat (Porta et al. 1985; Epple et al. 1988) and in an amphibian (Hanke et al. 1989). However, we were unable to show the effect in the human because of the unusual sensitivity of this species to neural and cardiac effects of E (Epple et al. 1989a).

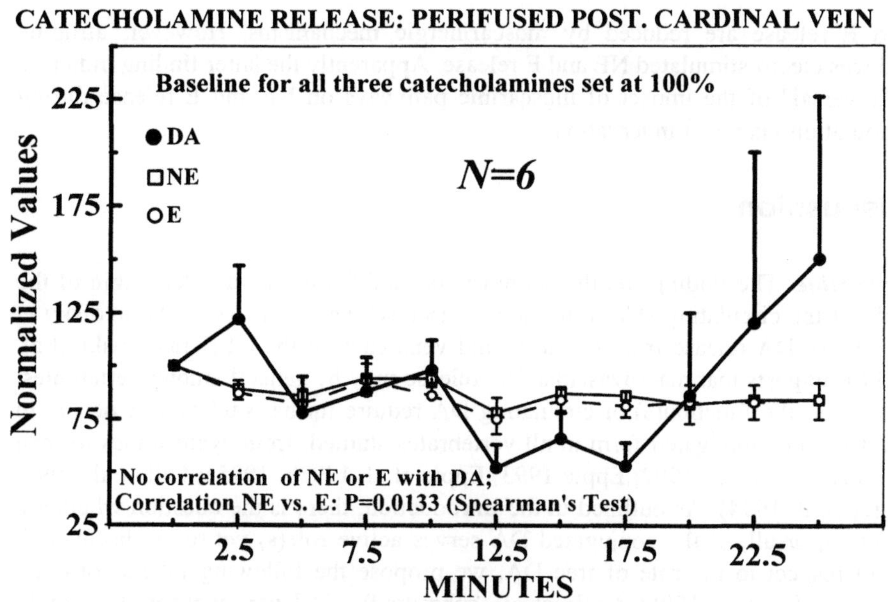

Fig. 2. Spontaneous release of catecholamines from the posterior cardinal vein of the eel in vitro. This region contains the chromaffin cells (adrenomedullary equivalent) of this species. Note the correlated release of norepinephrine (*NE*) and epinephrine (*E*), and the differing release pattern of dopamine (*DA*). Data from Al-Kharrat et al. (1997)

Under non-stimulated conditions, the release of DA in the eel is usually not correlated with that of NE and E (Fig. 2). However, the release of the three catecholamines is correlated during stress responses (Epple and Nibbio 1985; Epple et al. 1989b), and under several experimental conditions (Abele et al., submitted).

As in the rat (Jarry et al. 1989), met-enkephalin also inhibits the spontaneous release of NE and E from chromaffin cells of the eel (Epple et al. 1996; Al-Kharrat et al. 1997); furthermore, it inhibits the electrostimulated NE and E release in situ (Abele et al., submitted). In the perifusion system, codeine antagonizes this inhibitory effect partly, and morphine abolishes it completely (Epple et al. 1996). The weaker effect of codeine may be due to its dual action on both naloxone-sensitive opiate receptors (probably of the mu type) and non-opiate receptors; the former stimulate, and the latter suppress NE and E release (Epple et al. 1994, 1996).

Acetylcholine, the natural neurotransmitter, stimulates NE and E release, and so does its agonist nicotine (Epple et al. 1996; Weiss et al. 1996; Al-Kharrat et al. 1997). Muscarine, another cholinergic agonist, inhibits non-stimulated NE release in vitro, while the muscarine antagonist atropine has the opposite effect. Muscarine also reduces nicotine-stimulated release of NE and E (Al-Kharrat et al. 1997). Thus, non-stimulated (basal) NE release, as well as nicotine-stimulated NE and E release are reduced by muscarinergic mechanisms. However, atropine reduces electrostimulated NE and E release. Apparently the latter finding indicates a "reversal" of the impact of muscarinic pathways on NE and E release during stimulation via intact innervation.

Discussion

Dopamine. The findings on the eel have clarified the macrovascular origin of the bulk of the circulating DA in this species (Kraschinski et al. 1996; Milakofsky et al. 1995). DA release from the aorta and vena cava of the rat (Kraschinski et al. 1996) suggests that macrovascular DA release may be general among vertebrates. However, the function(s) of circulating DA require further studies. DA occurs in the free and conjugated form in all vertebrates studied, from cyclostomes to man (Randall and Perry 1992; Epple 1993; Epple et al. 1993b, 1995a; Gill et al. 1994; Porta et al. 1994). As outlined in the Introduction, there is the question whether a fraction, or all, of the conjugated DA serves active role(s) yet to be discovered. With respect to the role of free DA, we propose the following interactions (cf. Kraschinski et al. 1996; Abele et al., submitted): (1) Upon appropriate stimuli, "direct" sympathetic fibers (cf. Epple et al. 1995b) stimulate the chromaffin cell secretion of NE and E, (2) NE and E coordinate their releases by stimulatory adrenergic mechanisms, (3) NE and E, once released into the circulation, stimulate the DA release from the macrovascular endothelium in an abluminal direction, and (4) this DA fraction counteracts the vasoconstrictory impact of NE and E, functioning like a gentle brake (Epple et al. 1989b).

This simplified picture must certainly be fleshed out as further information becomes available. In particular, the interactions of macrovascular DA with endothelial factors acting on the vascular muscle, such as nitric oxide (Goldstein 1995; O'Rourke and Vanhoutte 1996), endothelins (Olson et al. 1991; Poder et al. 1991) and prostanoids (Olson and Villa 1991; Campbell and Halushka 1996) must be included. Also, the interactions of adrenergic and opioid factors (Scholz et al. 1997; Al-Kharrat et al. 1997) in the DA release must be elucidated, and differing responses of different vessels must be explained.

Norepinephrine and Epinephrine. As outlined above, the control of NE and E release from the chromaffin cells of the eel involves both neural and autocrine/paracrine mechanisms. With respect to neural mechanisms, there are so far only data on the cholinergic mechanisms involved (Reid and Perry 1995; Al-Kharrat et al. 1997; Abele et al., submitted). However, the presence of substances other than acetylcholine in the fibers innervating the chromaffin cells, i.e., vasoactive intestinal peptide and pituitary adenylyl cyclase activating peptide (Reid et al. 1995) suggests that the neural control is multifactorial. Possibly, the peculiar switch of the impact of muscarinergic mechanisms from inhibitory to stimulatory, as described above, is modulated by neural co-secretions.

The autocrine/paracrine control mechanisms of the chromaffin cells appear rather complex. Probably, our knowledge is just the tip of the iceberg. The interactions between NE and E, as analyzed in the eel, require adrenergic mechanisms (Al-Kharrat et al. 1997; Abele et al., submitted). However, it is by no means clear whether "classical" α- and β-adrenergic receptors are involved. There are considerable phylogenetic variations in adrenergic mechanisms (Nilsson 1983; Johansson 1984), and one should not be surprised if there are adrenoreceptors that do not fit either category - at least in "lower" vertebrates. After all, the number of known adrenoreceptors is continuously growing (Hoffman and Lefkowitz 1996).

Probably, the ideal model to pursue the question of potential "non-conventional" adrenoreceptors is the lamprey, where we first discovered the stimulatory effect of E on DA and NE release (Dashow and Epple 1983). In lampreys, the chromaffin cells occur in high concentrations in the sinus venosus (Epple et al. 1995a), and the normal absence of innervation should greatly simplify the experimental approach. We suggest that the catecholaminotropic effect of catecholamines in "higher" vertebrates with innervated chromaffin cells is an inheritance from the cyclostome stage, where it serves to correlate the release of the three catecholamines. Even at the teleost and amphibian level, only a fraction of chromaffin cells is innervated (Accordi 1991; Abelli et al. 1996; Imagawa et al. 1996).

Perhaps, the functional importance of the adrenomedullary innervation in mammals (Landsberg and Young 1992) is related to the switch in the blood supply: up to the amphibian level, the adrenomedullary equivalent is located in the venous drainage close to and/or inside the heart; this checkpoint-like location allows its chromaffin cells to directly respond to the overall state of the "spent" blood; e.g., to increased CO_2 concentrations with catecholamine release, as shown in the lamprey (Dashow and Epple 1985). Once the "adrenal medulla" cells are supplied with arterial blood, as occurs to varying degrees in the amniotes, they become dependent on neural messages (Epple et al. 1995b).

In this context, it may be interesting to see whether the chromaffin cell system of the lamprey can yield data relevant to the (mostly?) non-innervated paraganglia (other than the "sensory bodies") of the prenatal human (Böck 1982; Phillippe 1983). In particular, one wonders whether the paraganglia interact and/or control the circulation via adrenoreceptors that differ from the α- and β-adrenoreceptors of adult mammals.

In the chromaffin cells of the eel, the inhibitory impact of the opioid peptide met-enkephalin on NE and E release is opposed by alkaloid opiates, i.e., morphine, and its biosynthetic precursor codeine (Epple et al. 1996). At a first glance, an antagonism between opioid substances that are all antagonized by naloxone (Reisine and Pasternak 1996) appears paradox. However, an antagonism between met-enkephalin and alkaloid opiates has also been seen in two very different systems, i.e., in vitro pancreatic islet secretion (Kanter et al. 1980) and endocytotic trafficking of opioid receptors (Murray et al. 1994). Thus, an antagonism between an inhibitory system, linked to delta receptors (met-enkephalin), and an opposing system linked to mu receptors (alkaloid opiates) may be a basic mechanism in neuro-endocrine regulations. This issue is of particular interest since (1) Bianchi et al. (1993, 1994) have shown immunohistologically that endogenous codeine and morphine are stored in specific brain neurons, and (2) enkephalin fibers inhibit catecholamine release in the locus coeruleus (Van Bockstaele et al. 1995).

Besides opioids, the chromaffin cells of the eel contain further substances likely to be co-secretions. Reid et al. (1995) identified neuropeptide Y, peptide YY and serotonin-like immunoreactivity. One may suspect that there are more co-peptides yet to be identified (see, e.g., Wolfensberger et al. 1995). Whether all co-secretions are indeed of functional significance is a question yet to be answered. However, in the case of serotonin, its stimulatory impact on catecholamine release in the trout (Fritsche et al. 1993) suggests that it is indeed an active component of the secretory cocktail of the eel's chromaffin cells.

In summary, studies using "lower" vertebrate models, the eel and the lamprey, revealed highly complex regulations of plasma catecholamines by neural and autocrine/paracrine factors. The bulk of plasma DA derives from the wall of large arteries and veins, and the control of its release appears to involve, in addition to circulating NE and E, non-neural opioid factors. On the other hand, the bulk of plasma NE and E derives from the chromaffin cells (adrenal medulla and its equivalent). The control of the NE and E release from the chromaffin cells is via both neural and autocrine/paracrine factors. Among the latter, the mutual stimulation (coordination) of the release of NE and E by adrenergic mechanisms plays a pivotal role. Probably, this "catecholaminotropic effect of catecholamines", which is also seen in the rat, evolved in the cyclostomes, whose chromaffin cells are not innervated. Overall, the chromaffin cells of "lower" vertebrates have been ideal tools for basic work on the role of chromaffin co-secretions.

Acknowledgments

The investigations carried out in the laboratory of A. Epple were made possible by PHS grants 5 RO1 AGO1148, 1 T35 HLO7497 and 5 SO7 RR05414, and NSF grants PCM 8209263, DCB 8510585, IBN 8823085 and IBN 9408427.

References

Abele B, Hathaway CB, Nibbio B, Epple A (1997) Neural stimulation of catecholamine release in the eel: modulation by autocrine agonists and antagonists. (submitted)

Abelli L, Gallo VP, Civinini A, Mastrolia L (1996) Immunohistological and ultrastructural evidence of adrenal chromaffin cell subtypes in sea bass *Dicentrarchus labrax* (L). Gen Comp Endocrinol 102: 113-122

Accordi F (1991) The chromaffin cells of urodele amphibians. J Anat 179: 1-8

Al-Kharrat H, Weiss U, Tran Q, Nibbio B, Scholz S, Epple A (1997) Cholinergic control of catecholamine release in the eel. Gen Comp Endocrinol (in press)

Bianchi E, Alessandri C, Guarna M, Tagliamonte A (1993) Endogenous codeine and morphine are stored in specific brain neurons. Brain Res 627: 210-215

Bianchi E, Guarna M, Tagliamonte A (1994) Immunocytochemical localization of endogenous codeine and morphine. Adv Neuroimmunol 4: 83-92

Böck P (1982) The Paraganglia. In: Oksche A, Vollrath L (eds) Handbuch der mikroskopischen Anatomie des Menschen, vol VI/8. Springer, Berlin, Heidelberg, New York, pp 1-315

Campbell WB, Halushka PV (1996) Lipid-derived autacoids: eicosanoids and platelet-activating factor. In: Hardman JG, Limbird LE, Molinoff PB, Ruddon RW, Gilman AG (eds) The pharmacological basis of therapeutics, 9th edn. McGraw-Hill, New York, pp 601-616

Cori CF, Buchwald KW (1930) Effect of continuous intravenous injection of epinephrine on the carbohydrate metabolism, basal metabolism and vascular system of normal men. Am J Physiol 95: 71-78

Dashow L, Epple A (1983) Effects of exogenous catecholamines on plasma catecholamines and glucose in the sea lamprey, *Petromyzon marinus*. J Comp Physiol B 152: 35-41

Dashow L, Epple A (1985) Plasma catecholamines in the lamprey: intrinsic cardiovascular messengers? Comp Biochem Physiol 82C: 119-122

Dashow L, Epple A, Nibbio B (1982) Catecholamines in anadromous lampreys: baseline levels and stress-induced changes, with a note on cardiac cannulation. Gen Comp Endocrinol 46: 500-504

Epple A (1993) Adrenomedullary catecholamines. In: Schreibman MP, Scanes CG, Pang PKT (eds) The endocrinology of growth, development and metabolism in vertebrates. Academic Press, San Diego, pp 327-343

Epple A, Nibbio B (1985) Catecholaminotropic effects of catecholamines in a teleost fish, *Anguilla rostrata*. J Comp Physiol B 155: 285-290

Epple A, Nibbio B (1994) Surgery: selective islet removal and total pancreatectomy. In: Hochachka PW, Mommsen TP (eds) Biochemistry and molecular biology of fishes, vol 3. Analytical techniques. Elsevier, Amsterdam, pp 37-48

Epple A, Vogel WH, Nibbio B (1982) Catecholamines in head and body blood of eels and rats. Comp Biochem Physiol 71C: 115-118

Epple A, Hilliard RW, Potter IC (1985) The cardiovascular chromaffin cell system of the Southern hemisphere lamprey, *Geotria australis* Gray. J Morphol 184: 253-261

Epple A, Heim K, Vogel WH (1988) Single doses of catecholamines in the rat: catecholaminotropic, but not hyperglycemic. Comp Biochem Physiol 89C: 83-86

Epple A, Bertoni J, Hathaway CB (1989a) Physiological doses of epinephrine in the human: chronotropic, but not hyperglycemic or catecholaminotropic. J Exp Zool 250: 67-72

Epple A, Hathaway CB, Nibbio B (1989b) Circulatory catecholamines in the eel: origins and functions. Fish Physiol Biochem 7: 273-278

Epple A, Navarro I, Horak P, Spector S (1993a) Endogenous morphine and codeine: release by the chromaffin cells of the eel. Life Sci 52: PL117-121

Epple A, Porta S, Nibbio B, Leitner G (1993b) Release of conjugated catecholamines by the adrenal medulla equivalent of the American eels (*Anguilla rostrata*). Gen Comp Endocrinol 90: 58-63

Epple A, Nibbio B, Spector S, Brinn JE (1994) Endogenous codeine: autocrine regulator of catecholamine release from chromaffin cells. Life Sci 54: 695-702

Epple A, Porta, S, Nibbio B (1995a) Release of conjugated catecholamines from the Sinus venosus of a lamprey (*Petromyzon marinus*). Comp Biochem Physiol 111C: 45-48

Epple A, Brinn JE, Gill TS (1995b) The evolution of the adrenal medulla: unconventional aspects. In: Kramer B, Rawdon BB (eds) Embryos, endocrine cells and the neural crest. Witwatersrand University Press, Johannesburg, pp 126-140

Epple A, Nibbio B, Horak P, Spector S, Dores RM (1996) Codeine, morphine and met-enkephalin: endogenous regulators of catecholamine release. In: McCarty R, Aguilera G, Sabban E, Kvetnansky R (eds) Stress: molecular genetic and neurobiological advances. Gordon and Breach, New York, pp 327-331

Fritsche R, Reid SG, Thomas S, Perry SF (1993) Serotonin-mediated release of catecholamines in the rainbow trout *Oncorhynchus mykiss*. J Exp Biol 178: 191-204

Gill TS, Porta S, Nibbio B, Epple A (1994) Sulfate conjugates of catecholamines in the allantoic fluid of the chicken embryo. Gen Comp Endocrinol 96: 255-258

Goldstein DS (1995) Stress, catecholamines, and cardiovascular disease. Oxford University Press, New York

Hanke W, Albrecht S, Gerhard A, Kloas W (1989) Effects of excess amounts of catecholamines in fish and amphibia. Acta Endocrinol 129 (Suppl 1): 96

Hardisty MW (1979) Biology of the cyclostomes. Chapman and Hall, London

Hathaway CB, Epple A (1989) The sources of plasma catecholamines in the American eel, *Anguilla rostrata*. Gen Comp Endocrinol 74: 418-430

Hathaway CB, Epple A (1990) Catecholamines, opioid peptides and true opiates in the chromaffin cells of the eel: immunohistological evidence. Gen Comp Endocrinol 79: 393-405

Hathaway CB, Brinn JE, Epple A (1989) Catecholamine release by catecholamines in the eel does not require presence of brain or anterior spinal cord. J Exp Zool 249: 338-342

Hoffman BB, Lefkowitz RJ (1996) Catecholamines, sympathomimetic drugs, and adrenergic receptor antagonists. In: Hardman JG, Limbird LE, Molinoff PB, Ruddon RW, Gilman AG (eds) The pharmacological basis of therapeutics, 9th edn. McGraw-Hill, New York, pp 199-248

Imagawa T, Kitagawa H, Uehara M (1996) The innervation of the chromaffin cells in the head kidney of the carp, *Cyprinus carpio*; regional differences of the connections between nerve endings and chromaffin cells. J Anat 188: 149-156

Jarry H, Dietrich M, Barthel A, Giesler A, Wuttke W (1989) In vivo demonstration of a paracrine, inhibitory action of met-enkephalin on adrenomedullary catecholamine release in the rat. Endocrinology 125: 624-629

Johansson P (1984) Alpha-adrenoceptors: recent development and some comparative aspects. Comp Biochem Physiol 78C: 253-261

Kanter RA, Ensinck JW, Fujimoto WY (1980) Disparate effects of enkephalin and morphine upon insulin and glucagon secretion by islet cell cultures. Diabetes 29: 84-86

Kraschinski S, Epple A, Nibbio B (1996) Macrovascular dopamine release. Am J Physiol 270: R1244-1249

Landsberg L, Young JB (1992) Catecholamines and the adrenal medulla, In: Wilson JD, Foster DW (eds) Williams textbook of endocrinology, 8th edn. Saunders, Philadelphia, pp 621-705

Milakofsky L, Epple A, Nibbio B (1995) Release of amino acids and related compounds from the adrenal equivalent and caudal vein of the eel in vitro. Comp Biochem Physiol 112C: 275-283

Murray S, Keith D, Zaki P, Evans C, von Zastrow M (1994) Morphine and enkephalin exert opposing effects on rapid endocytotic trafficking of opioid receptors. Mol Biol Cell 5 (Suppl): 196a

Navarro I, Epple A (1993) Plasma catecholamines do not respond to insulin-induced hypoglycemia in a teleost, *Anguilla rostrata*. Am J Physiol 265: E20-23

Nilsson S (1983) Autonomic nerve function in vertebrates. Springer, Berlin, Heidelberg, New York

Olson KR, Villa J (1991) Evidence against non-prostanoid endothelial-derived relaxing factor(s). Am J Physiol 260: R925-933

Olson KR, Duff DW, Farrell AP, Keen J, Kellogg MD, Kullman D, Villa J (1991) Cardiovascular effects of endothelin in trout. Am J Physiol 260: H1214-H1223

O'Rourke ST, Vanhoutte PM (1996) Vascular pharmacology. In: Loscalzo J, Creager MA, Dzau VJ (eds) Vascular medicine, 2nd edn. Little, Brown and Co, Boston, pp 117-139

Phillippe M (1983) Fetal catecholamines. Am J Obstet Gynecol 146: 840-855

Poder TC, Silberberg SD, Rampe D (1991) Contraction of reptile, amphibian and fish blood vessels by endothelin-1. Can J Physiol Pharmacol 69: 215-217

Porta S, Rangetiner B, Rinner I, Ertl U, Sadjak A, Supanz S (1985) Long-term application of some catecholamines elevates levels of other catecholamines in rats. Exp Pathol 28: 181-186

Porta S, Gebru G, Slawitsch P, Liebman P, Epple A, Zirm B (1994) Magnesium pretreatment alters ratio of free to conjugated plasma dopamine after exhaustive stress. Magnesium Bull 16: 77-80

Randall DJ, Perry SF (1992) Catecholamines. In: Hoar WS, Randall DJ, Farrell AP (eds) Fish physiology XII/B: the cardiovascular system. Academic Press, San Diego, pp 255-300

Reid SG, Perry SF (1995) Cholinoceptor-mediated control of catecholamine release from chromaffin cells in the American eel, *Anguilla rostrata*. J Comp Physiol B 165: 464-470

Reid SG, Fritsche R, Jönsson A-C (1995) Immunohistological localization of bioactive peptides and amines associated with the chromaffin tissue of five species of fish. Cell Tiss Res 280: 499-512

Reisine T, Pasternak G (1996) Opioid analgesics and antagonists. In: Hardman JG, Limbird LE, Molinoff PB, Ruddon RW, Gilman AG (eds) The pharmacological basis of therapeutics, 9th edn. McGraw-Hill, New York, pp 521-555

Scholz S, Al-Kharrat H, Nibbio B, Epple A (1997) Opioid effects on macrovascular dopamine release. Gen Comp Endocrinol 105: 239-245

Stamler JS, Dzau VJ, Loscalzo J (1996) The vascular smooth muscle cell. In: Loscalzo J, Creager MA, Dzau VJ (eds) Vascular medicine, 2nd edn. Little, Brown and Co, Boston, pp 69-116

Tyce GM, Messick JM, Yaksh TL, Byer DE, Danielson DR, Rorie DK (1986) Amine sulfate formation in the central nervous system. Fed Proc 45: 2247-2253

van Bockstaele EJ, Branchereau P, Pickel VM (1995) Morphologically heterogenous met-enkephalin terminals form synapses with tyrosine hydroxylase-containing dendrites in the rat nucleus locus coeruleus. J Comp Neurol 363: 423-438

Weiss U, Al-Kharrat H, Nibbio B, Porta S, Epple A (1996) Magnesium and nicotinergic effects on catecholamine release from chromaffin cells. Magnesium-Bull 18: 63-65

Wolfensberger M, Forssmann WG, Reinecke M (1995) Localization and coexistence of atrial natriuretic peptide (ANP) and neuropeptide Y (NPY) in vertebrate adrenal chromaffin cells immunoreactive to TH, DBH and PNMT. Cell Tissue Res 280: 267-276

19 Lysenin, a Novel Bioactive Protein Isolated from Coelomic Fluid of the Earthworm *Eisenia foetida* - Structure, Secretion and Biological Activity

H. Kobayashi[1], Y. Sekizawa[1], S. Shioda[2], S. Natori[3], T. Nakajima[4], and M. Umeda[5]

[1]Research Laboratory, Zenyaku Kogyo Co., Ltd., 2-33-7 Ohizumi-machi, Nerima-ku, Tokyo 178, Japan; Phone: +81-3-3922-5131, Fax: +81-3-3922-5065
[2]Department of Anatomy, Showa University School of Medicine, 1-58 Hatanodai, Shinagawa-ku, Tokyo 142
[3]Faculty of Pharmaceutical Sciences, University of Tokyo, 7-3-1 Hongo, Bunkyo-ku, Tokyo 113
[4]Suntory Institute for Bioorganic Research, 1-1-1 Wakayamadai, Shimamoto-cho, Osaka 618
[5]Department of Inflammation Research, Tokyo Metropolitan Institute of Medical Science, Honkomagome, Bunkyo-ku, Tokyo 113

Abstract

A novel protein, which we named lysenin, was purified from the coelomic fluid of the earthworm *Eisenia foetida*. It caused contraction of strips of isolated rat aorta. The cDNA for lysenin was cloned from an *E. foetida* cDNA library, and the complete amino acid sequence of lysenin was determined. The lysenin cDNA encoded a protein of 297 amino acids with a molecular mass of 33 kDa. The expression of this cDNA in *Escherichia coli* yielded a protein with contractile activity that was quite similar to that of native lysenin. Immunoreactive lysenin was localized in chloragocytes and cells of the pharyngeal gland of the earthworm. Immuno-electron microscopy revealed that gold particles indicating the presence of lysenin, were found in the cells of the pharyngeal gland. Both chloragocytes and pharyngeal gland cells seem to secrete lysenin into the coelomic fluid. Lysenin was discharged together with chloragocytes from the worms through the dorsal pores in response to chemical or physical stimuli. Lysenin induced lysis of human, rat and sheep erythrocytes, the latter being most sensitive to lysenin. When lysenin was preincubated with vesicles that contained sphingomyelin, it did not induce hemolysis of sheep erythrocytes, whereas lysenin preincubated with vesicles that contained other phospholipids caused hemolysis. Immunoblotting analysis showed that lysenin did not cross-react with cellular proteins of the erythrocyte membrane. These findings suggest that lysenin might bind specifically to sphingomyelin on erythrocyte membranes, thereby inducing hemolysis. Lysenin did not bind to sphingomyelin analogs, ceramide, sphingosine, sphingosine-1-phosphate or sphingosylphosphorylcholine. Accordingly, it appears that lysenin strictly recognized the molecular structure of sphingomyelin.

Introduction

Cellular self-defense mechanisms in annelids, in particular the earthworm, were of great interest with regard to regeneration and transplantation experiments during the 1960s. However, the involvement of several kinds of coelomocytes other than phagocytes in such defense mechanisms has not been fully elucidated (Šíma 1994). Nevertheless, the following biological activities of the coelomic fluid (CF) of the earthworm, most notably *Eisenia foetida*, have been demonstrated as specific humoral defense activities: antibacterial (Cameron 1932; Avel 1959; Valembois et al. 1982a), hemolytic (Du Pasquier and Duprat 1968), cytotoxic (Kauschke and Mohrig 1987; Cooper and Roch 1994), and hemagglutinating (Cooper et al. 1974) activities. Since the discovery of these activities, the biological and chemical nature of the substances responsible for the activities has been investigated extensively for almost 20 years, mainly by French investigators (Roch 1979; Roch et al. 1981, 1991; Valembois et al. 1982a, 1984; Lassègues et al. 1989; Hirigoyenberry et al. 1992; Milochau et al. 1997). They proved that all the above mentioned activities are mainly mediated by two proteins with molecular masses of 40 and 45 kDa, respectively, in CF (Valembois et al. 1982a; Hirigoyenberry et al. 1992; Cooper and Roch 1994; Milochau et al. 1997). Recently, Milochau et al. (1997) have purified the two proteins and reported the amino acid sequence of the 40 kDa protein.

While we were engaged in an immunocytochemical study of biologically active peptides (Fujii et al. 1989; Hasegawa and Kobayashi 1991) in the nervous system of *E. foetida*, we found that intravenous (i.v.) injection of a small amount of CF of *E. foetida* was lethal in the mouse, rat and Japanese quail. This remarkable phenomenon prompted us to attempt to purify lethal substance(s) in the CF of *Eisenia*.

Biological Actions of the CF from *Eisenia*

Some biological actions of the CF of *Eisenia* were examined in order to develop an appropriate bioassay for purification of the active substance(s) in the CF. For the experiments, we used CF from which coelomocytes had been removed by centrifugation.

1. Lethal Effects. Minimum lethal doses of intravenously (i.v.) injected CF were approximately 10 µl/kg, 6-12 µl/kg and 20-50 µl/kg for adult rats, mice and Japanese quail, respectively. Heated CF (i.v.) or CF administered intraperitoneally, subcutaneously or per os had no effect.

*2. **Effects on Smooth Muscles and Blood Pressure.*** CF caused contraction of smooth muscles of the rat thoracic aorta and uterus, as well as the guinea pig ileum, trachea and tenia coli in vitro. CF had a positive inotropic effect on the rat right atrium. CF at 10 µl/ml was sufficient to cause these effects in vitro. In rats, blood pressure was decreased by a dose of 1.0 µl/kg body weight CF (i.v.).
*3. **Antifungal and Antibacterial Effects.*** CF had no effects on the growth of *Candida albicans*, *Trichophyton asteroides*, *Aspergillus niger* and *Penicillium chrysogenum*. Lassègues et al. (1981) and Valembois et al. (1982a) found that only soil bacteria pathogenic for *E. foetida*, such as *Bacillus megaterium* and *Aeromonas hydrophila* were sensitive to the CF.
*4. **Cytotoxic Effects.*** CF had cytotoxic effects on mouse P388 leukemia cells, human KB tumor cells and mouse normal splenic cells. It probably damaged cell membranes, as pointed out by Kauschke and Mohrig (1987) from observations of the cytotoxic actions of CF on chicken fibroblasts, guinea pig polymorphonuclear leucocytes and insect hemocytes.
*5. **Hemolytic Effects.*** Hemolysis by CF was confirmed in sheep erythrocytes (see below).

To test the activity of the purified substances we selected the rat thoracic aorta as a bioassay system because this preparation could be handled relatively easily and provided reproducible and sensitive results.

Purification and Structure of Lysenin

Purification of Lysenin from CF of *Eisenia*

The supernatant obtained after centrifugation of the coelomic fluid induced contraction of strips of rat aorta in vitro at final concentrations > 0.005% (v/v). The active agent was purified by anion-exchange HPLC on a column of DEAE-2SW and gel-filtration HPLC on a G2000SW column (Fig. 1). The active fraction G2 from the G2000SW column yielded a single band of protein after SDS-PAGE. We named the protein in this fraction lysenin (lysis + *Eisenia*; Sekizawa et al. 1996a). The protein in peak G1 from the G2000SW column had only weak contractile activity. Sequencing of the two proteins showed that the G1 protein had a primary structure similar to that of lysenin (Fig. 4).

Lysenin caused a dose-dependent contraction of the isolated rat aorta at concentrations > 0.03 µg/ml (Fig. 2). The contraction reached a maximum within 15 min and lasted for a relatively long time, up to 40 min. Lysenin also caused the lysis of erythrocytes from several animal species (see below).

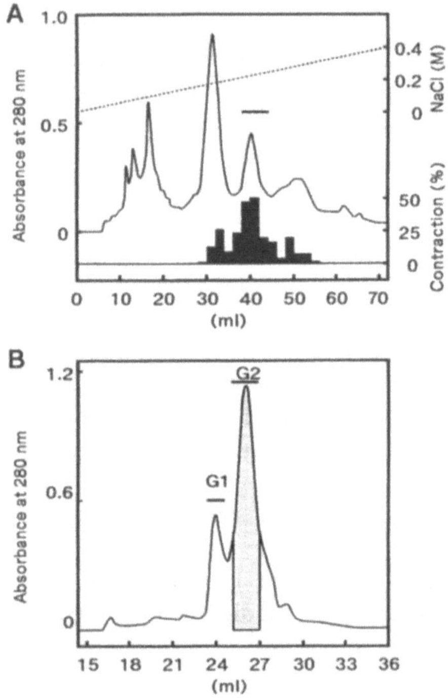

Fig. 1. Purification of an active substance from the CF of *Eisenia foetida*. **A** Fractionation of a crude extract by anion-exchange HPLC. A crude extract obtained by precipitation of proteins with ammonium sulfate was applied to a column of DEAE-2SW (Tosoh, Tokyo). The contractile activity of each fraction was determined with strips of the isolated rat aorta. The activity was expressed as a percentage of the maximum contraction induced by 60 mM KCl. The *bar* indicates the most active fraction. **B** Purification of an active substance by gel-filtration HPLC. The active fraction obtained from the column of DEAE-2SW was applied to a G2000SW column (Tosoh), and it was separated into two peaks of protein, *G1* and *G2*. The active substance was mainly concentrated in the *shaded area* (*G2*). (From Sekizawa et al. 1996a, with permission)

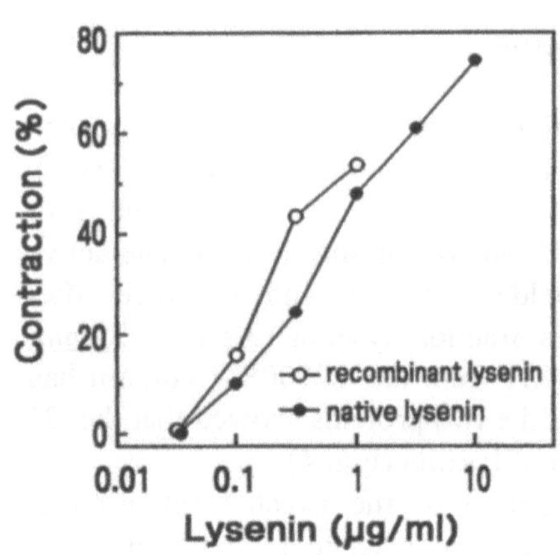

Fig. 2. Concentration-response curves for the effects of lysenin on isolated rat aorta. *Filled circle* native lysenin; *open circle* recombinant lysenin. Data are mean responses, expressed as percentages of the maximum contraction induced by 60 mM KCl (n = 4-7). (From Sekizawa et al. 1997, with permission)

Fig. 3. The nucleotide sequence of lysenin cDNA and the deduced amino acid sequence. The amino acid sequences obtained from fragments of lysenin are underlined. The polyadenylation signals are *double-underlined*. The termination codon is indicated by *asterisks*. (From Sekizawa et al. 1997, with permission)

```
CTCAGATGCAGTTATTGTATCTAATAGGCATCTCGGTCTTCCGCTGGTTTCTCGCAGATACATCAAAAGACTAGAAGAGGCCATCAAGATCCTGTTGAATTGAGTGCACTCCTGTTTGCT  120
TCAGAAACTTGTGAGCGATGTCGGCTAAAGCAGCAGAGGGATATGAACAGATAGAAGTAGATGTGGTGGCAGTATGGAAAGAAGGCTATGTATACGAAAATCGCGGAAGCACCAGTGTGG  240
                 M  S  A  K  A  A  E  G  Y  E  Q  I  E  V  D  V  V  A  V  W  K  E  G  Y  V  Y  E  N  R  G  S  T  S  V  D   35
ATCAGAAGATCACAATAACAAAAGGCATGAAAAATGTGAATTCAGAAACAAGGACAGTGACTGCTACGCATAGTATTGGTTCTACTATTAGCACTGGAGATGCATTCGAAATTGGAAGCG  360
 Q  K  I  T  I  T  K  G  M  K  N  V  N  S  E  T  R  T  V  T  A  T  H  S  I  G  S  T  I  S  T  G  D  A  F  E  I  G  S  V   75
TGGAGGTTAGCTACAGCCATTCACATGAAGAATCCCAAGTTAGCATGACGGAAACTGAAGTTTATGAATCAAAGGTGATCGAACACACTATAACGATTCCACCTACTTCAAAATTCACAA  480
 E  V  S  Y  S  H  S  H  E  E  S  Q  V  S  M  T  E  T  E  V  Y  E  S  K  V  I  E  H  T  I  T  I  P  P  T  S  K  F  T  R  115
GATGGCAACTGAATGCTGACGTTGGTGGAGCGGATATTGAATACATGTATTTGATTGATGAAGTCACACCCATAGGAGGGACTCAGAGTATTCCACAGGTCATCACAAGTCGGGCTAAAA  600
 W  Q  L  N  A  D  V  G  G  A  D  I  E  Y  M  Y  L  I  D  E  V  T  P  I  G  G  T  Q  S  I  P  Q  V  I  T  S  R  A  K  I  155
TTATAGTTGGCCGACAGATAATCCTTGGAAAAACAGAAATTCGAATTAAGCATGCAGAAAGGAAGGAGTACATGACAGTCGTTTCAAGAAAAAGTTGGCCAGCTGCAACTCTTGGACATA  720
 I  V  G  R  Q  I  I  L  G  K  T  E  I  R  I  K  H  A  E  R  K  E  Y  M  T  V  V  S  R  K  S  W  P  A  A  T  L  G  H  S  195
GCAAACTTTTCAAGTTTGTGCTCTATGAAGATTGGGGGGGATTTCGAATTAAAACGCTGAACACCATGTATTCGGGCTATGAGTATGCCTATTCCTCTGATCAAGGAGGAATCTACTTTG  840
 K  L  F  K  F  V  L  Y  E  D  W  G  G  F  R  I  K  T  L  N  T  M  Y  S  G  Y  E  Y  A  Y  S  S  D  Q  G  G  I  Y  F  D  235
ATCAGGGTACTGATAATCCGAAACAGCGCTGGGCAATCAATAAGTCATTGCCTCTTCGTCATGGTGACGTAGTCACCTTCATGAATAAGTACTTCACTCGCAGTGGGCTGTGCTACGATG  960
 Q  G  T  D  N  P  K  Q  R  W  A  I  N  K  S  L  P  L  R  H  G  D  V  V  T  F  M  N  K  Y  F  T  R  S  G  L  C  Y  D  D  275
ATGGACCGGCAACAAACGTGTACTGTCTGGACAAACGTGAAGACAAGTGGATTTTGGAAGTGGTTGGTTAACCCCTGAAATGGGGTGTCATGATTGGTCAGCAAGCATCACCATTTCGCA 1080
 G  P  A  T  N  V  Y  C  L  D  K  R  E  D  K  W  I  L  E  V  V  G  ***                                                   297
TTTGCTTCTGTCATATATTTCACCTTTAGCACAATGATGAGGTGCAACTAAAGCGGTGTAGATTGTAATAGCGACTCATTTATGTATTATTCAGCATTATTTACATAAGACATTAAGAAC 1200
ATTCGGTTTATTTATTTATATGCTTTTGTAGGCCTTATATTGATTATTAACAGGCGACCTAATTCAGCAAACATGCAGGGTCTGCAAACATTTGTAGGCCGATGTCAATGTTTATATTCA 1320
CGAGAACTTTAAACCTTATGATAGCTTCGCTTTGCTTAACTAGGCCTAAACTAATATAGGTCAACCAGCTATGTATCTAAACTAACCGAGGACTATATTTAACCATTAGCATTAGCATTC 1440
TTGTTAACCCAAGTCCTAAAATGTCTTGTACGGAGATGCAGTTTGTTGTTCCAATTGTTCTGCAAAAAAACAAGGATAGGCCATGACATGTTAAACTAGTATATCTCAAAGCTATTGAAA 1560
AATAAATGGAGATCAATAAATCAAAAAAAAAAAAAAAAAA                                                                                   1600
```

Structure of Lysenin

To determine the complete amino acid sequence of lysenin, we isolated a cDNA for lysenin. A cDNA library of *E. foetida* was constructed in lambda ZAP II (Stratagene, La Jolla, California, USA). A partial cDNA for lysenin was generated by the polymerase chain reaction (PCR) with a primer designed by reference to an internal amino acid sequence of lysenin. For cloning of a full-length cDNA for lysenin, the cDNA generated by PCR was used as a probe for screening a cDNA library of *E. foetida*. The sequencing of isolated cDNAs revealed at least three kinds of cDNA (efL1, efL2 and efL3), which were very similar to one another. Clone efL1 consisted of 1600 bp and contained an open reading frame that encoded 297 amino acid residues (Fig. 3). The molecular mass of the putative protein encoded by efL1 was estimated to be 33 440 Da. The predicted sequence contained all the sequences of the peptides that had been chemically or enzymatically derived from lysenin. The apparent molecular mass (41 kDa) of recombinant lysenin, expressed in *E. coli*, coincided with that of lysenin purified from the earthworm, as determined by SDS-PAGE. The concentration-response relationship for the action of the purified recombinant lysenin on the isolated rat aorta was quite similar to that of native lysenin (Fig. 2; Sekizawa et al. 1997).

Clones efL2 and efL3 encoded proteins of 300 amino acid residues, in which 76% (efL2) and 90% (efL3) of the amino acid residues were identical to those in the deduced sequence of lysenin (Fig. 4). These putative proteins were named lysenin-2 and lysenin-3. Lysenin-3 was the same as the protein in fraction G1 that was also purified from the coelomic fluid. The contractile activity of lysenin-3 was lower than that of lysenin. At the present time, an attempt is being made to determine the active site(s) responsible for the contractile activity of lysenin. Lysenin-2 might be present in the coelomic fluid also (Sekizawa et al. 1997).

Several peptides that cause contraction of mammalian muscles have been demonstrated immunohistochemically in the earthworm: substance P (Aros et al. 1980), vasopressin and oxytocin (Kinoshita and Kawashima 1986), FMRFamide (Fujii et al. 1989) and endothelin-1 (Hasegawa and Kobayashi 1991). Moreover, three novel peptides that cause contraction of the earthworm gut, annetocin (Oumi et al. 1994), the tetradecapeptide GFKDGAADRISHGF-NH_2 from *Eisenia* and the tetradecapeptide GFRDGSADRISHGF-NH_2 from *Pheretima* (Ukena et al. 1995) have been isolated from earthworms. However, the sequence of lysenin did not show any homology to the sequences of these peptides. Moreover, computer analysis revealed that there were no proteins or peptides with sequences homologous to that of lysenin in the standard databases. Furthermore, lysenin did not have any obvious internal functional domains. These results suggest that lysenin has a novel mode of action on the rat aorta.

```
Lysenin-1  MSAKA--AEGYEQIEVDVVAVWKEGYVYENRGSTSVDQKITITKGMKNVN  48
Lysenin-2  ··SSTVM·D·F·E······S······A·····NS··Q····M······L·  50
Lysenin-3  ··SR·GI·····························E···K·····R·L·  50

Lysenin-1  SETRTVTATHSIGSTISTGDAFEIGSVEVSYSHSHEESQVSMTETEVYES  98
Lysenin-2  ···K·L····TL·R·LKV··P···A······TF··QK······Q····S· 100
Lysenin-3  ···K·L··S···········L···AT·D····Y················· 100

Lysenin-1  KVIEHTITIPPTSKFTRWQLNADVGGADIEYMYLIDEVTPIGGTQSIPQV 148
Lysenin-2  Q·····V····NK·····K·······TG···········A··ADLT··E· 150
Lysenin-3  ·E··········································L····· 150

Lysenin-1  ITSRAKIIVGRQIILGKTEIRIKHAERKEYMTVVSRKSWPAATLGHSKLF 198
Lysenin-2  NK·····L·····H··E················I···········N·N·· 200
Lysenin-3  ·K·····L···E·Y··E········D·······················Y 200

Lysenin-1  KFVLYEDWGGFRIKTLNTMYSGYEYAYSSDQGGIYFDQGTDNPKQRWAIN 248
Lysenin-2  ····F··SS·I·········P···W············ESS········LS 250
Lysenin-3  ·······MY······························S·········· 250

Lysenin-1  KSLPLRHGDVVTFMNKYFTRSGLCYDDGPATNVYCLDKREDKWILEVVG  297
Lysenin-2  ·AM··········R·NF··N··M·············E···········NT 300
Lysenin-3  ·························Y·····D················KP 300
```

Fig. 4. Comparison of the amino acid sequences of lysenin-1 (designated lysenin in this report), lysenin-2 and lysenin-3. *Dots* indicate identical amino acids. (From Sekizawa et al. 1997, with permission

Two hemolytic and antibacterial proteins, named fetidins, with apparent molecular masses of 40 and 45 kDa, were recently purified from the CF of *E. foetida* (Milochau et al. 1997). The amino acid sequence of the 40-kDa fetidin is identical to that of lysenin-3. Lysenin was also shown to have hemolytic activity, as discussed below. It appears that lysenin is very similar to the fetidins. However, lysenin did not exhibit antibacterial activity to *Bacillus megaterium,* unlike fetidins.

Production and Secretion of Lysenin

Expression of Lysenin mRNA

When Northern blotting analysis was performed with the cDNA for lysenin as the probe, lysenin mRNA was detected in coelomocytes but not in the intestine and body wall. Thus, the gene for lysenin seems to be expressed in coelomocytes (Sekizawa et al. 1997). However, it had to be determined which type(s) of coelomocyte are responsible for the expression, since several of them have been described (see Valembois et al. 1985; Cooper and Roch 1994).

Light Microscopy

Among coelomocytes, only chloragocytes reacted with the lysenin antibody (Sekizawa et al. 1996b). Chloragocytes are packed in masses in the chloragogen tissue, which is attached to the peritoneum (Fig. 5a). They are mostly elongated in shape. A positive immunoreaction was also observed in independent

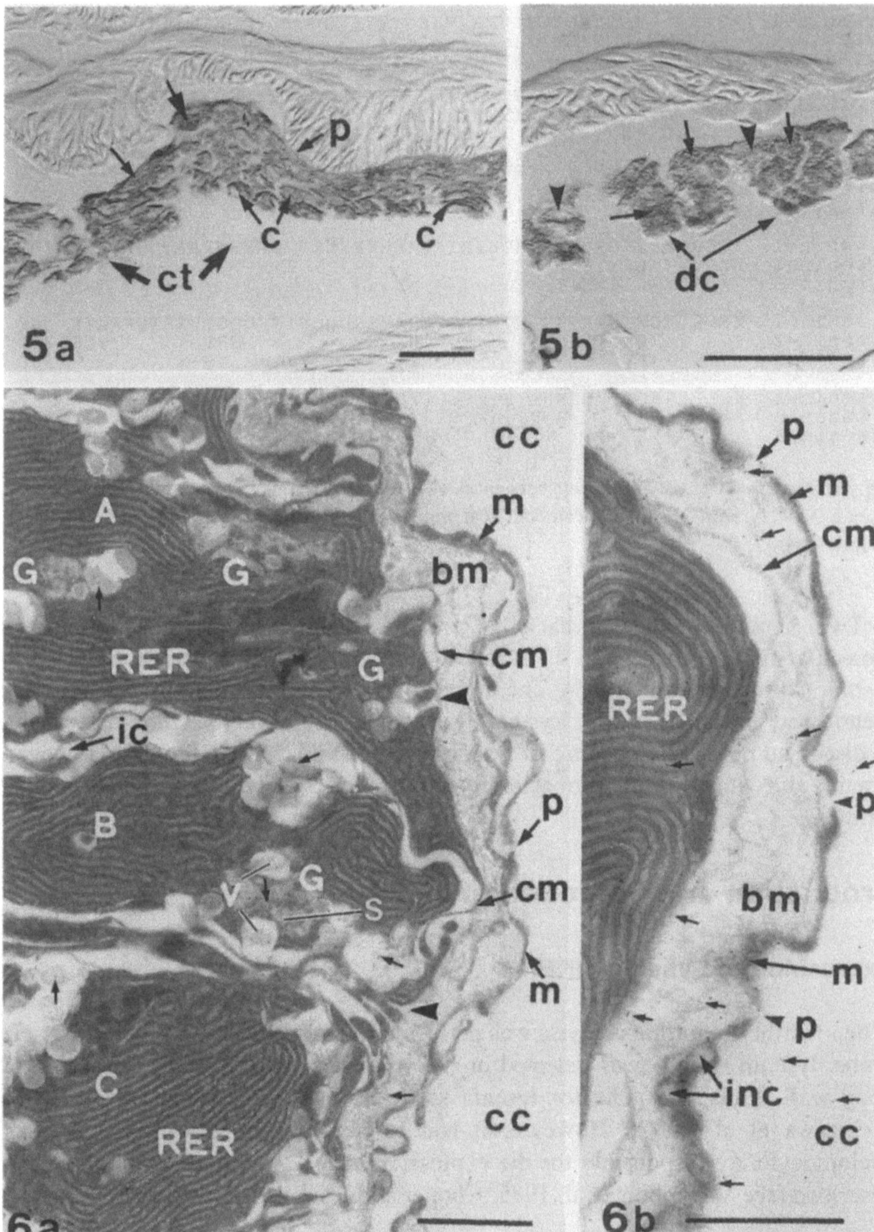

Fig. 5. Lysenin-immunoreactivity in chloragogen tissue and chloragocytes. **a** Chloragogen tissue (*ct*) attached to the peritoneum (*p*). Chloragocytes (*c*) are small and elongated. **b** Chloragocytes detached (*dc*) from the tissue. They are large and round. Chloragocytes contain secretory chloragosomes (*arrows*). Some cells (*arrowheads*) are disintegrating to release chloragosomes. *Bars* 50 μm. **Fig. 6.** Lysenin-immunoreactivity in the pharyngeal gland. Gold particles indicating

chloragocytes gathered as loose aggregates in the coelomic cavity (Fig. 5b). They are large and round or irregular in shape. Immunoreactive cells contained many immunostained vesicles, which are called chloragosomes (Fig. 5b). Larger independent chloragocytes were occasionally found to release chloragosomes in the process of disintegration (Fig. 5b). After the discharge of chloragosomes, chloragocytes showed a weak immunoreaction only in their peripheral region. Cooper and Stein (1981) also observed some chloragocytes that were partially disintegrated after staining with hematoxylin and eosin. These observations suggest the release of lysenin from chloragocytes into the coelomic cavity. In *Pheretima communissima*, no tissues exhibited any immunoreaction with the lysenin antibody.

In response to chemical or physical stimulation, the coelomic fluid with numerous coelomocytes, including chloragocytes, is expelled from the worm's body through the dorsal pores. Immediately after the discharge, many intact chloragocytes with chloragosomes were found in the exudate and showed a positive immunoreaction. Some chloragocytes were disintegrated or swollen, indicating the discharge of chloragosomes. These cells were only weakly immunostained.

Since lysenin and fetidin have highly homologous amino acid sequences, it seems probable that lysenin and fetidin are produced in the same chloragocytes. Valembois et al. (1982b) proposed that both hemolytic and bacteriostatic factors are secreted from the chloragocytes.

In addition to the chloragocytes, cells of a glandular tissue mass, which is located just above the pharynx, reacted with the lysenin antibody. Because of its location, we may call this tissue pharyngeal gland. Details of the immunostained cells of this gland will be discussed below.

Electron Microscopy

Cells producing lysenin in *E. foetida* were studied by immuno-electron microscopy with the lysenin antibody. The pharyngeal gland was surrounded by a thick basement membrane and mesothelium that faced the coelomic cavity (Figs. 6, 8). Each gland cell was almost totally occupied by well-developed rough endoplasmic reticulum (RER). The Golgi apparatus did not show typical lamellae, but it consisted of many small vesicles and larger vacuoles with electron-dense contents that formed the Golgi complex (Figs. 6 and 7). These vesicles and vacuoles showed lysenin immunoreaction (Fig. 7). Therefore, the small

the presence of lysenin are indicated by *small arrows*. **a** Three gland cells (*A*, *B*, *C*) facing the coelomic cavity (*cc*). Secretory substances seem to be moving (*arrowheads*) into the basement membrane (*bm*). **b** An enlarged electron micrograph of the peripheral region of the gland. *cm* Cell membrane; *G* Golgi complex; *ic* intercellular space; *inc* inclusion; *m* mesothelium; *p* micropore; *RER* rough endoplasmic reticulum; *s* secretory vesicle; *v* secretory vacuole. *Bars* 1.5 μm

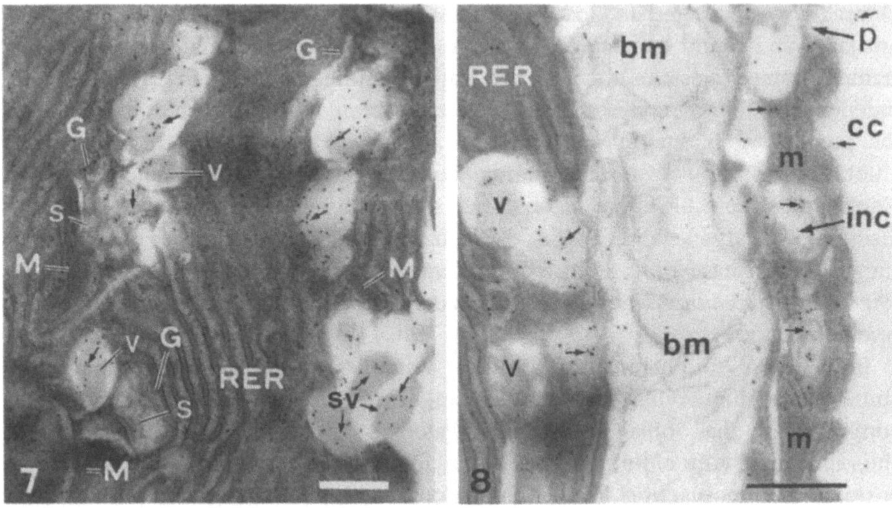

Fig. 7. Part of a pharyngeal gland cell. Golgi complex (*G*) including secretory vesicles (*s*) and vacuoles (*v*) containing gold particles (*small arrows*); *sv* secretory vacuole being released into the intercellular space; *M* mitochondrion; *RER* rough endoplasmic reticulum. *Bar* 500 nm
Fig. 8. Part of a gland cell facing the coelomic cavity (*cc*). Gold particles indicating the presence of lysenin are labeled by *small arrows*. *bm* Basement membrane; *inc* inclusion in mesothelium; *m* mesothelium; *p* micropore; *RER* rough endoplasmic reticulum; *v* secretory vacuole. *Bar* 500 nm

vesicles and the variously sized vacuoles in the Golgi complex were identified as secretory organelles that contained lysenin. Mitochondria were always associated with these complexes (Fig. 7).

In gland cells, smaller secretory vacuoles seemed to be transferred to the periphery of the cells (Figs. 6a, 8). Then the vacuoles were released into the intercellular space by exocytosis to reach the peripheral portion of the gland (Figs. 6a, 8). Here, the secretory vacuoles seemed to be released into the basement membrane and then into the coelomic cavity, passing through the mesothelium, since the immunogold labeling was frequently observed both in the basement membrane and in the mesothelium (Figs. 6a, 8). In the mesothelium, micropores, through which the secrete could be released into the coelomic cavity, were often encountered (Figs. 6, 8). Gold particles were occasionally observed in the coelomic fluid near the micropores. These observations suggest that the gland secretes lysenin into the coelomic cavity.

Lysenin is considered to be produced by chloragocytes and the so-called pharyngeal gland. At the moment, the biological significance of lysenin secreted into or from the body is unclear. Valembois et al. (1982a, 1985) demonstrated selective antibacterial activity of CF against bacteria isolated from *Eisenia* breeding medium and they proposed the coelomic fluid might protect the worms against specific environmental pathogens. If so, lysenin might be one of the agents responsible for this activity of the CF.

Biochemical Aspects of Lysenin

We have attempted to explore the cellular mechanisms that induce the biological actions of the CF of *Eisenia*. We focused our attention on the lysis of erythrocytes, which is known to be induced by the CF.

Hemolysis by Lysenin and Its Inhibition by Sphingomyelin

Lysenin induced hemolysis of human, rat and sheep erythrocytes in a concentration-dependent manner (Fig. 9). The sheep erythrocytes were most sensitive to lysenin: 5 ng/ml lysenin was required to induce 50% hemolysis, and this concentration was 20 times lower than that required for 50% hemolysis of human and rat erythrocytes. A total lipid extract from sheep erythrocyte membranes inhibited lysenin-induced hemolysis. We examined the binding of lysenin to various phospholipid vesicles by performing a hemolysis inhibition assay. We preincubated lysenin with various phospholipid vesicles and then measured the hemolytic activity of each lysenin-phospholipid vesicle mixture

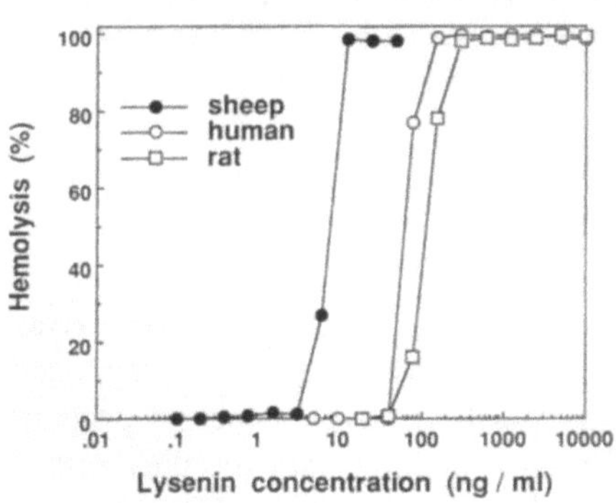

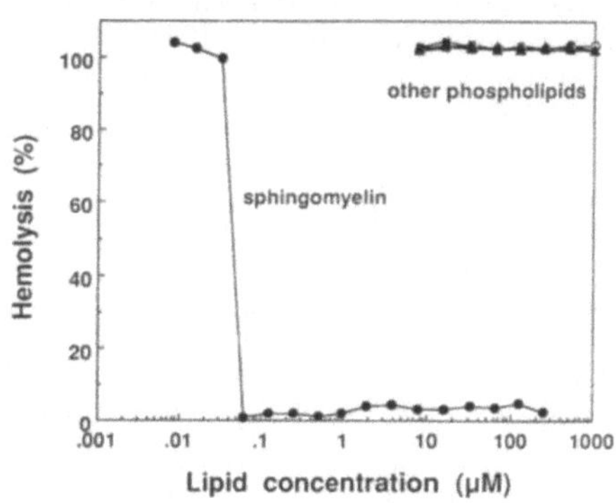

Fig. 9. (**Left**) Lysenin-induced hemolysis. Erythrocytes (10^7 cells/ml) from sheep (*filled circle*), human (*open circle*) and rat (*open square*) were incubated with lysenin at various concentrations at 37 °C for 30 min, and the percentage of erythrocytes hemolyzed was determined by measuring the amount of hemoglobin released. **Fig. 10.** (**Right**) Inhibition of lysenin-induced hemolysis by sphingomyelin. Lysenin was preincubated with vesicles at various concentrations that were composed of 50 mole% of each phospholipid and 50 mole% of cholesterol and the hemolytic activities of the resulting mixtures were measured. *Filled circle* Sphingomyelin; *open circle* dipalmitoylphosphatidylcholine; *open triangle* phosphatidylethanolamine; *filled triangle* phosphatidic acid; *open square* phosphatidylserine; *filled square* phosphatidylinositol

(Choung et al. 1988). As shown in Fig. 10, vesicles containing sphingomyelin inhibited hemolysis, whereas those composed of other phospholipids, such as phosphatidylcholine, phosphatidylethanolamine, phosphatidylserine and phosphatidylinositol, had no effect on hemolysis. Immunoblotting analysis showed that lysenin did not cross-react with cellular proteins of the erythrocyte membrane, suggesting that lysenin might bind specifically to sphingomyelin on erythrocyte membranes to induce hemolysis.

The high sensitivity of sheep erythrocytes to lysenin might be attributable to the high level of sphingomyelin in sheep erythrocyte membranes: sphingomyelin is the predominant phospholipid (51% of total phospholipids) in these erythrocyte membranes, whereas human and rat erythrocyte membranes contain much less sphingomyelin (about 15-20% of total phospholipids; Barenholz and Gatt 1982). Although some bacterial sphingomyelinases have been shown to cause hemolysis (Ikezawa et al. 1980), lysenin-induced hemolysis did not result from the degradation of sphingomyelin, since no degradation of sphingomyelin in either erythrocytes or sphingomyelin vesicles was observed after prolonged incubation with lysenin.

Specific Binding of Lysenin to Sphingomyelin

The binding of lysenin to various phospholipids was evaluated by enzyme-linked immunosorbent assay (ELISA; Umeda et al. 1989) and a thin-layer chromatography-immunostaining (TLC-immunostaining) assay (Igarashi et al. 1995). In the ELISA, the wells of microtiter plates were coated with 50 μl of a solution of phospholipid (10 μM) in ethanol with evaporation at room temperature. After incubation with lysenin at various concentrations, bound lysenin was detected by sequential incubation with the lysenin antiserum, biotinylated antibodies against rabbit IgG and peroxidase-conjugated streptavidin. The results showed that lysenin bound specifically to sphingomyelin and did not cross-react with any other phospholipids (Fig. 11A). In order to define further the specificity of lysenin, we performed TLC-immunostaining since, in the ELISA, sphingomyelin analogs such as sphingosine-1-phosphate had been washed from the surface of the microtiter plates during the assay. In the TLC-immunostaining assay, the lipids coated onto high-performance TLC plates were fixed by treatment with polyisobutylmethacrylate, and almost all the lipids remained after the assay procedure. As shown in Fig. 11B, lysenin bound specifically to sphingomyelin, but not to the sphingomyelin analogs, ceramide, sphingosine, sphingosine-1-phosphate, or sphingosylphosphorylcholine. These results indicated clearly that lysenin strictly recognized the molecular structure of sphingomyelin.

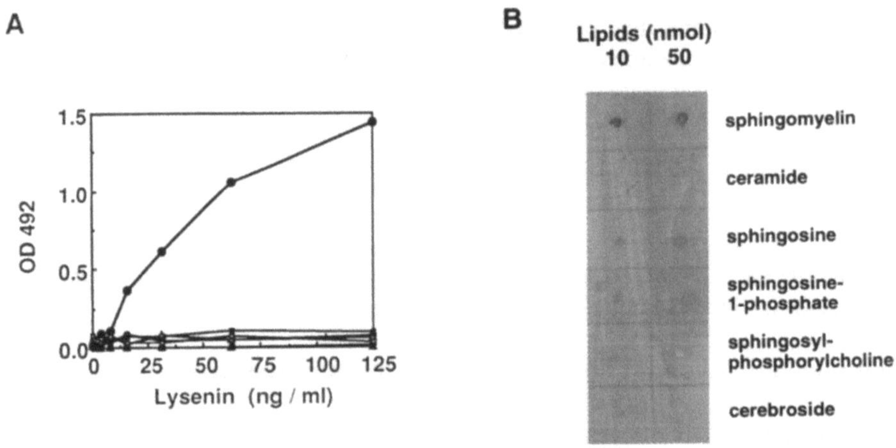

Fig. 11. Specific binding of lysenin to sphingomyelin. **A** ELISA. The wells of a microtiter plate were coated with 50 pmol/well of phospholipid and incubated with lysenin at various concentrations. Binding of lysenin to phospholipids was detected by sequential incubation with anti-lysenin antiserum, biotinylated antibodies against rabbit IgG and peroxidase-conjugated streptavidin. *Filled circle* Sphingomyelin; *open circle* dipalmitoylphosphatidylcholine; *open triangle* phosphatidylethanolamine; *filled triangle* phosphatidic acid; *open square* phosphatidylserine; *filled square* phosphatidylinositol; *cross* cardiolipin. **B** TLC-immunostaining assay. Ten and 50 nmol of *sphingomyelin* and its analogs were spotted onto high-performance TLC plates and incubated with 100 ng/ml lysenin. The bound lysenin was detected with the lysenin antiserum, biotinylated antibodies against rabbit IgG and peroxidase-conjugated streptavidin

Conclusions

The coelomic fluid (CF) of *E. foetida* has cytotoxic, antibacterial, hemolytic, agglutinating and smooth-muscle contractile activities. Milochau et al. (1997) and Sekizawa et al. (1996a, 1997) independently purified novel bioactive proteins using different bioassay methods. We named our protein lysenin and they named theirs fetidin. Lysenins and fetidins seem to belong to one family. The chemical and biological identities or differences in their structures and actions have not yet been fully clarified. However, mechanisms responsible for the biological activities of CF from *Eisenia* will be explained by studies of lysenin or fetidin. Lysenin is produced by chloragocytes and cells of the pharyngeal gland and is released into the coelomic cavity. In response to chemical and physical stimuli, lysenin is discharged from the worm's body through its dorsal pores. We found that lysenin binds specifically to sphingomyelin. Since sphingomyelin is one of the phospholipid components of the cell membrane, lysenin should be a good tool for studies of the biology of cell membranes. The biological significance of lysenin in *E. foetida* itself remains to be clarified.

Acknowledgments

The authors extend their appreciation to Drs. M. Kurokawa (Tokyo Institute of Psychiatry, Tokyo), K. Kurahashi (Peptide Institute, Inc., Osaka), K. Nagai (Mitsubishi Kasei, Institute of Life Science, Tokyo) and K. Wake (Tokyo Medical and Dental University, Tokyo) for their kind and helpful suggestions, advice and encouragement. They are also indebted to Mr. K. Hashimoto (President of Zenyaku Kogyo Co., Ltd.) for his continued support and encouragement.

References

Aros B, Wenger T, Vigh B, Vigh-Teichmann I (1980) Immunohistochemical localization of substance P and ACTH-like activity in the central nervous system of the earthworm *Lumbricus terrestris* L. Acta Histochem 66: 262-268

Avel M (1959) Classe des annélides oligochètes. In: Grassé P-P (ed) Traité de Zoologie, vol 5. Masson, Paris, pp 224-270

Barenholz Y, Gatt S (1982) Sphingomyelin: metabolism, chemical synthesis, chemical and physical properties. In: Hawthrone JN, Ansell GB (eds) Phospholipids. New comprehensive biochemistry, vol 4. Elsevier, Amsterdam, pp 129-177

Cameron GR (1932) Inflammation in earthworms. J Pathol Bacteriol 35: 933-972

Choung S-Y, Kobayashi T, Inoue J, Takemoto K, Ishitsuka H, Inoue K (1988) Hemolytic activity of a cyclic peptide Ro09-0198 isolated from *Streptoverticillium*. Biochim Biophys Acta 940: 171-179

Cooper EL, Roch P (1994) Immunological profile of annelids: transplantation immunity of annelids. In: Vetvicka V, Šíma P, Cooper EL, Bilej M, Roch P (eds) Immunology of annelids. CRC Press, Boca Raton, pp 201-243

Cooper EL, Stein EA (1981) Oligochaetes. In: Ratcliffe NA, Rowley AF (eds) Invertebrate blood cells. Academic Press, London, p 106

Cooper EL, Lemmi CAE, Moore TC (1974) Agglutinins and cellular immunity in earthworms. Ann N Y Acad Sci 234: 34-50

Du Pasquier L, Duprat P (1968) Aspects humoraux et cellulaires d'une immunité naturelle non spécifique chez l'oligochète *Eisenia fœtida* Sav. (Lumbricinae). C R Acad Sci, Paris 266: 538-541

Fujii K, Ohta N, Sasaki T, Sekizawa Y, Kobayashi H (1989) Immunoreactive FMRFamide in the nervous system of the earthworm, *Eisenia foetida*. Zool Sci 6: 951-961

Hasegawa K, Kobayashi H (1991) Immunohistochemical localization of endothelin-1 in the nervous system of the earthworm *Eisenia foetida*. Gen Comp Endocrinol 81: 433-441

Hirigoyenberry F, Lassègues M, Roch P (1992) Antibacterial activity of *Eisenia fetida andrei* coelomic fluid: immunological study of the two major antibacterial proteins. J Invertebr Pathol 59: 69-74

Igarashi K, Asai K, Kaneda M, Umeda M, Inoue K, (1995) Specific binding of a synthetic peptide derived from an antibody complementarity-determining region to phosphatidylserine. J Biochem 117: 452-457Ikezawa H, Mori M, Taguchi R (1980) Studies on sphingomyelinase of *Bacillus cereus*: hydrolytic and hemolytic actions on erythrocyte membranes. Arch Biochem Biophys 199: 572-578

Kauschke E, Mohrig W (1987) Cytotoxic activity in the coelomic fluid of the annelid *Eisenia foetida* Sav. J Comp Physiol B 157: 77-83

Kinoshita K, Kawashima S (1986) Differential localization of vasopressin- and oxytocin-immunoreactive cells and conventional neurosecretory cells in the ganglia of the earthworm *Pheretima hilgendorfi*. J Morphol 187: 343-351

Lassègues M, Roch P, Valembois P, Davant N (1981) Action de quelques souches bactériennes telluriques sur le lumbricien *Eisenia fetida andrei*. C R Acad Sci Paris (Série III) 292: 731-734

Lassègues M, Roch P, Valembois P (1989) Antibacterial activity of *Eisenia fetida andrei* coelomic fluid: specificity of the induced activity. J Invertebr Pathol 54: 28-31

Milochau A, Lassègues M, Valembois P (1997) Purification, characterization and activities of two hemolytic and antibacterial proteins from coelomic fluid of the annelid *Eisenia fetida andrei*. Biochim Biophys Acta 1337: 123-132

Oumi T, Ukena K, Matsushima O, Ikeda T, Fujita T, Minakata H, Nomoto K (1994) Annetocin: an oxytocin-related peptide isolated from the earthworm, *Eisenia foetida*. Biochem Biophys Res Commun 198: 393-399

Roch P (1979) Protein analysis of earthworm coelomic fluid: 1) Polymorphic system of the natural hemolysin of *Eisenia fetida andrei*. Dev Comp Immunol 3: 599-608

Roch P, Valembois P, Davant N, Lassègues M (1981) Protein analysis of earthworm coelomic fluid - II. Isolation and biochemical characterization of the *Eisenia fetida andrei* factor (EFAF). Comp Biochem Physiol 69B: 829-836

Roch P, Lassègues M, Valembois P (1991) Antibacterial activity of *Eisenia fetida andrei* coelomic fluid: III - Relationship within the polymorphic hemolysins. Dev Comp Immunol 15: 27-32

Sekizawa Y, Hagiwara K, Nakajima T, Kobayashi H (1996a) A novel protein, that causes contraction of the isolated rat aorta: its purification from the coelomic fluid of the earthworm, *Eisenia foetida*. Biomed Res 17: 197-203

Sekizawa Y, Ohta N, Natori S, Kobayashi H (1996b) Immunocytochemical localization of lysenin, a novel protein isolated from the coelomic fluid of the earthworm *Eisenia foetida*. Biomed Res 17: 327-330

Sekizawa Y, Kubo T, Kobayashi H, Nakajima T, Natori S (1997) Molecular cloning of cDNA for lysenin, a novel protein in the earthworm *Eisenia foetida* that causes contraction of rat vascular smooth muscle. Gene 191: 97-102

Šíma P (1994) Annelid coelomocytes and hemocytes: Roles in cellular immunity. In: Vetvicka V, Šíma P, Cooper EL, Bilej M, Roch P (eds) Immunology of annelids. CRC Press, Boca Raton, pp 115-165

Ukena K, Oumi T, Matsushima O, Ikeda T, Fujita T, Minakata H, Nomoto K (1995) A novel gut tetradecapeptide isolated from the earthworm, *Eisenia foetida*. Peptides 16: 995-999

Umeda M, Igarashi K, Nam KS, Inoue K (1989) Effective production of monoclonal antibodies against phosphatidylserine: stereo-specific recognition of phosphatidylserine by monoclonal antibody. J Immunol 143: 2273-2279

Valembois P, Roch P, Lassègues M, Cassand P (1982a) Antibacterial activity of the hemolytic system from the earthworm *Eisenia fetida andrei*. J Invertebr Pathol 40: 21-27

Valembois P, Roch P, Lassègues M, Davant N (1982b) Bacteriostatic activity of a chloragogen cell secretion. Pedobiologia 24:191-195

Valembois P, Roch P, Lassègues M (1984) Simultaneous existence of hemolysins and hemagglutinins in the coelomic fluid and in the cocoon albumen of the earthworm *Eisenia fetida andrei*. Comp Biochem Physiol 78A: 141-145

Valembois P, Lassègues M, Roch P, Vallien J (1985) Scanning electron-microscopy study of the involvement of coelomic cells in earthworm antibacterial defense. Cell Tissue Res 240: 479-484

20 Receptors and Effects of the Inhibitory Neuropeptide Somatostatin in Glial Derivatives

J. Feindt, B. Krisch, and R. Mentlein

Anatomisches Institut der Universität zu Kiel,
Olshausenstrasse 40, D-24098 Kiel, Germany;
Phone: +49 431 880 2460, Fax: +49 431 880 1557

Abstract

Somatostatin is one of the most widespread neuropeptides in the brain. Receptors are found not only on neurons, but also on astroglia and meningeal cells in vitro and in situ and on microglia cells at least in vitro. Moreover, somatostatin receptors are overexpressed in malignant derivatives of these cells, human gliomas and meningiomas. In human gliomas and glial cell lines the receptors are located mainly on the processes of the cells and only rarely on the somata. Interestingly, the pattern of somatostatin receptor subtypes (sst) differs between normal astrocytes and gliomas or glioma cell lines. Whereas astrocytes express multiple subtypes (sst_1, sst_2, sst_4) the subtype sst_2 is predominant in gliomas (often with low levels of other subtypes). Somatostatin inhibits the basal or agonist-stimulated proliferation of microglia cells, but not of astrocytes or gliomas. The reason for this lack of inhibitory action in astrocytes or gliomas might be (in part) a desensitization of the receptor, e.g. by phosphorylation. An increased phosphorylation of the receptor sst_2, after stimulation with its agonist, was observed in U343 glioma cells metabolically labeled with radioactive phosphate. These results show that not only neurons but also non-neuronal cells have to be considered as target cells for neuropeptides in the brain.

Introduction

Somatostatin (SS) is a cyclic peptide composed of 14 or 28 amino acids which is produced in neurons and endocrine cells. Both forms (SS-14 and SS-28) are derived from a common precursor. The 28 residue peptide is an N-terminal extended form of the (predominant) 14 residue peptide. The sequences of SS-14 and SS-28 are conserved in all mammals, two different variants of SS-14 have been isolated from fishes (Fig. 1). Recently, two structurally similar peptides named cortistatin-14 and cortistatin-29 have been cloned from rat brain (de Lecca et al. 1996). The existence of SS variants has already been suggested from early

The somatostatin peptide family

Mammalian somatostatin

SS-28

S A N S N P A M A P R E R K A G C K N F F W K T F T S C

SS-14

* * * * * * * * * * * * * *

Cortistatin-29

Q E R P P L Q Q P P H * D K * P * * * * * * * * * * * * K

Cortistatin-14

P * * * * * * * * * * * * K

Catfish somatostatin

D N T V * S *K P* L A * M * Y * * * S S * A *

Anglerfish somatostatins

A * S G G * L L *

S V D * T N N L P * * * * * * * * * * * * * * * * * * *

Fig. 1. Sequences of somatostatins and related peptides. Amino acids are given in the one letter code. *Asterisks* denote identical residues, the *solid line* indicates the disulphide bond between the two cystine residues. *Double-underlined* are pairs of basic residues that represent cleavage points for convertases in the maturation of the precursors

immunohisto- and -cytochemical comparative results using different polyclonal SS-antisera (Krisch 1980), but the precise distribution and functions of cortistatins are less well investigated.

Originally, SS was purified from sheep hypothalami as a neuroendocrine peptide inhibiting the secretion of growth hormone from pituitary cells (somatostatin or somatotropin release-inhibiting factor; Brazeau et al. 1972). Later, SS was detected in various regions of the central and peripheral nervous systems and in endocrine cells, e.g., in the gastrointestinal tract. In accord with its wide anatomical distribution, SS acts on multiple targets including the brain, pituitary, gut, endocrine and exocrine pancreas, adrenals, thyroid and kidneys. SS regulates a variety of physiological functions locally as a neurotransmitter, neuromodulator, or paracrine/autocrine regulator, or systemically as a true hormone. At the cellular level, the actions of SS are mostly inhibitory, antisecretory and antiproliferative.

SS induces its biological actions by interacting with a family of receptors, of which five have been cloned (human sequences: Yamada et al. 1992a, b; Rohrer et al. 1993; Yamada et al. 1993) and have been named sst_{1-5} according to their order of identification. All belong to the G-protein-coupled seven transmembrane receptors that are located in the plasma membrane. Subtypes of sst vary in size, for the human sequences, from 418 amino acids (sst_3) to 391 (sst_1), 388 (sst_4), 369 (sst_2) and 364 (sst_5). The human sst genes have been localized on separate chromosomes (sst_1: chromosome 14; sst_2: 17; sst_3: 22; sst_4: 20; sst_5: 16). However, all sst subtypes share a high similarity of almost 50% between their sequences; sst_1 and sst_4 have the most similar sequences with more than 60% identity (Yamada et al. 1993).

The degree of sequence similarity is probably also the structural basis for a pharmacological subdivision of sst into two groups according to the ability to bind certain somatostatin analogs, e.g. the metabolically stable agonist octreotide (or SMS 201-995, Sandostatin[R]; Fig. 2). While sst_2, sst_3 and sst_5 bind octreotide at low (nanomolar) concentrations, sst_1 and sst_4 have affinities only at higher (micromolar) concentrations (Patel and Srikant 1994). The native peptide SS-14 binds to all receptor subtypes (sst_{1-5}) and the N-terminal extended form SS-28 to sst_{1-4} with similar affinities in the nanomolar range. The subtype sst_5, however, has about a tenfold higher selectivity for SS-28 than SS-14 (Patel and Srikant 1994).

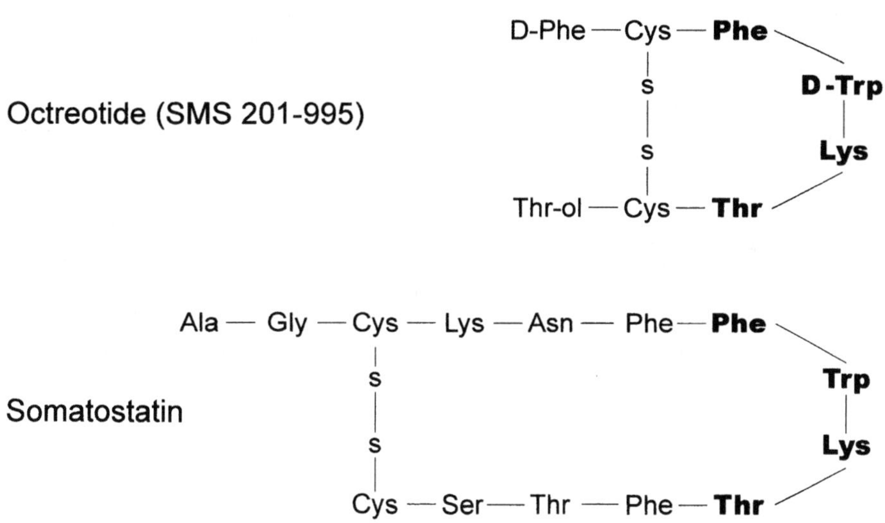

Fig. 2. Primary structures of *somatostatin-14* (SS-14) and the metabolically stable analog, *octreotide* (SMS 201-995). Amino acids printed in *bold* are identical in octreotide and somatostatin

Somatostatin Receptor Subtypes on Normal Brain Cells

Despite their widespread distribution in various organs, sst subtypes show no distinct tissue specificity. In the central nervous system, all subtypes have been detected, although with varying intensities in certain regions (Table 1). In the rat brain, sst_1 and sst_2 are most widely expressed, high levels of sst_3 are dominant in the cerebellum and of sst_4 in the hippocampus (Bruno et al. 1993). Less information is available for the distribution of sst subtypes in the human brain (Schindler et al. 1996, for review).

Table 1. Distribution of somatostatin receptor subtypes (sst) in different areas of rat brain

Brain area	sst				
	1	2	3	4	5
Cerebral cortex	++++	++++	++	++	++
Cerebellum	(+)	(+)	++++	-	-
Hypothalamus	++	++	+	(+)	++++
Hippocampus	++	++	++	++	++
Striatum	(+)	+	++	+	++
Amygdala	++++	++	+++	+	++
Thalamus	++	+	+	(-)	-
Olfactory bulb	++	++	++	++	++

Modified after Bruno et al. (1993). Data from in situ hybridization studies denoting the intensities of the signal by the number of +

Little is known about the localization of sst subtypes at the level of single brain cells. In previous studies we could demonstrate that somatostatin-binding sites are not only present on neurons, but also on astrocytes and meningeal cells in situ and in vitro (Mentlein et al. 1990; Krisch et al. 1991, 1993; Krisch 1994). Direct demonstration of somatostatin receptors on astrocytes strengthens several biochemical observations concerning effects of somatostatin on astrocytes in culture (Krisch and Mentlein 1994, for review). In cultivated astrocytes, somatostatin reduces cyclic AMP levels elevated by co-stimulation with other agonists and stimulates phospholipase A_2 which generates arachidonic acid and modulates several other cellular effector systems. Here we summarize and extend our findings on the occurrence of somatostatin receptor subtypes on astrocytes and other non-neuronal brain cells.

Since sst subtypes are low copy genes, we tested their expression by the sensitive reverse transcription-polymerase chain reaction (RT-PCR) in cultures that were immunocytochemically pure, as established by markers specific for the cell type investigated. Cultivated astroglia cells were more than 98% pure for the marker glial fibrillary acidic protein (GFAP), microglia cells for the lysosomal protein CD68, and meningeal cells for vimentin. After isolation of mRNA, a DNA-digest and a reverse transcription, several PCRs with subtype-specific sense and antisense primers were performed (cf. Feindt et al. 1995). Controls for the correct amplification of rat sst subtypes were carried out by RT-PCR of total rat brain mRNA, for the human sst with human genomic DNA, resulting in the demonstration of all subtypes. The length of the different PCR-products was determined by agarose gel electrophoresis and comparison with a base-pair standard, and their identity was tested by cycle sequencing and comparison with the published sequences. β-actin served as a reference mRNA, but it should be kept in mind that the following results are only semi-quantitative ones.

Both cultivated human and rat cortical astrocytes express mRNA of three subtypes, sst_1, sst_2 and sst_4 (Table 2). In contrast, in rat microglia cells the subtypes sst_2 as well as sst_3 and - only after stimulation with lipopolysaccharide - sst_4 can be detected. Rat meningeal cells (as well as total meninges from rat) express the subtypes sst_2 and sst_3. Thus, multiple receptor subtypes are present in astroglial, microglial and meningeal cells.

Table 2. Expression of somatostatin receptor subtypes (sst) in cultivated brain cells

Cell type	Species	sst 1	2	3	4	5
Astrocytes	Rat	+	(+)	-	+	-
	Human	+	+	-	(+)	-
Microglia cells	Rat	-	+	+	-	-
Microglia cells + LPS	Rat	-	+	+	++	-
Meningeal cells	Rat	-	+	+	-	-

The number of + roughly indicates the intensity of sst expression. *LPS* microglia cells treated for 4 h at 37 °C with 1 μg/ml lipopolysaccharide. Data for astrocytes from Feindt et al. (1995), for microglia cells from Mentlein, Feindt and Schmidt (unpubl.), and for meningeal cells from Feindt et al. (1997a)

Somatostatin Receptors on Gliomas

Several types of peripheral tumors, including neuroendocrine tumors (pituitary adenomas, carcinoids, islet cell carcinomas, paragangliomas, pheochromocytomas), lymphomas, breast tumors, or small cell lung carcinomas, express somatostatin-binding sites (Reubi et al. 1992, for review) which can be visualized in vivo by receptor scintigraphy using a radiolabeled somatostatin analog, ^{111}In-DTPA-D-Phe^1-octreotide (DTPA, diethylenetriaminepentaacetic acid). Meningiomas and part of glioblastomas can also be visualized by this technique (Hildebrandt et al. 1994; Luyken et al. 1994). Moreover, scintigraphic negative glioma patients (WHO grade I and II) express somatostatin-binding sites when analyzed after surgery in vitro (Luyken et al. 1994). Apparently, the intact blood-brain-barrier in the gliomas of these patients hinders the visualization of the somatostatin-binding sites in vivo.

Thus, astrocytes as well as their malignant derivatives in gliomas that comprise more than half of all brain tumors express somatostatin-binding sites. We therefore evaluated whether or not the sst subtypes in normal and malignant cells are different or identical. Since RT-PCR is a very sensitive method, contaminating cells (e.g. endothelial cells, fibroblasts) need to be removed from the gliomas obtained by surgery. Therefore, we first dissociated the gliomas and cultivated them under conditions that removed most of the contaminating cells (Feindt et al. 1996). Usually after 2-4 subcultures the cells were immunocytochemically more than 95% pure for the astroglial marker glial fibrillary acidic protein (GFAP), and they were then analyzed by RT-PCR for sst subtypes. Part of the tumors were graded histologically, directly after surgery, according to the WHO classification.

Generally, RT-PCR signals of gliomas were stronger than those of normal astrocytes. The most interesting difference is, however, a strong expression of the sst_2 alone (rarely observed) or in combination with sst_1 - a pattern which is quite different from that of non-malignant astroglial cells (Table 3). Figure 3 illustrates,

Table 3. Somatostatin receptor subtypes (sst) in human astrocytes and gliomas

Cell type			sst		
	1	2	3	4	5
Astrocytes	+	+	-	(+)	-
Gliomas					
WHO I, A. K.	+	++	-	-	-
WHO II, J. K	+	++	-	+	-
WHO II, U. L.	+	++	-	(+)	-
WHO III, K.-P. W.	+	++	-	-	-
WHO IV, M. E.-H.	n. d.	++	n. d.	n.d.	n. d.
WHO IV, E. C.-L.	n. d.	++	-	-	-
WHO IV, E. C. E.	n. d.	++	n. d.	n. d.	n. d.
WHO IV, E. D.	n. d.	++	n. d.	n. d.	n. d.
WHO IV, E. P.	+	++	-	-	-
WHO IV, H.-P. B.	n. d.	++	n. d.	n. d.	n. d.
WHO IV, H. C.	-	++	-	-	-
WHO IV, H. D.	+	++	-	(+)	-
WHO IV, K. H.	n. d.	++	n. d.	n. d.	n. d.
WHO IV, N. H.	+	++	-	-	-

The number of + roughly indicates the intensity of sst expression. Data for normal astrocytes represent identical results from 3 different patients. *n. d.*, not determined. Data from Feindt et al. (1995 and unpubl. results)

with representative examples, the strong expression of sst_2 in gliomas, and the minute signal in astrocytes in comparison to an equivalent signal of the β-actin standard. This pattern was detected in all of the 15 tumors investigated.

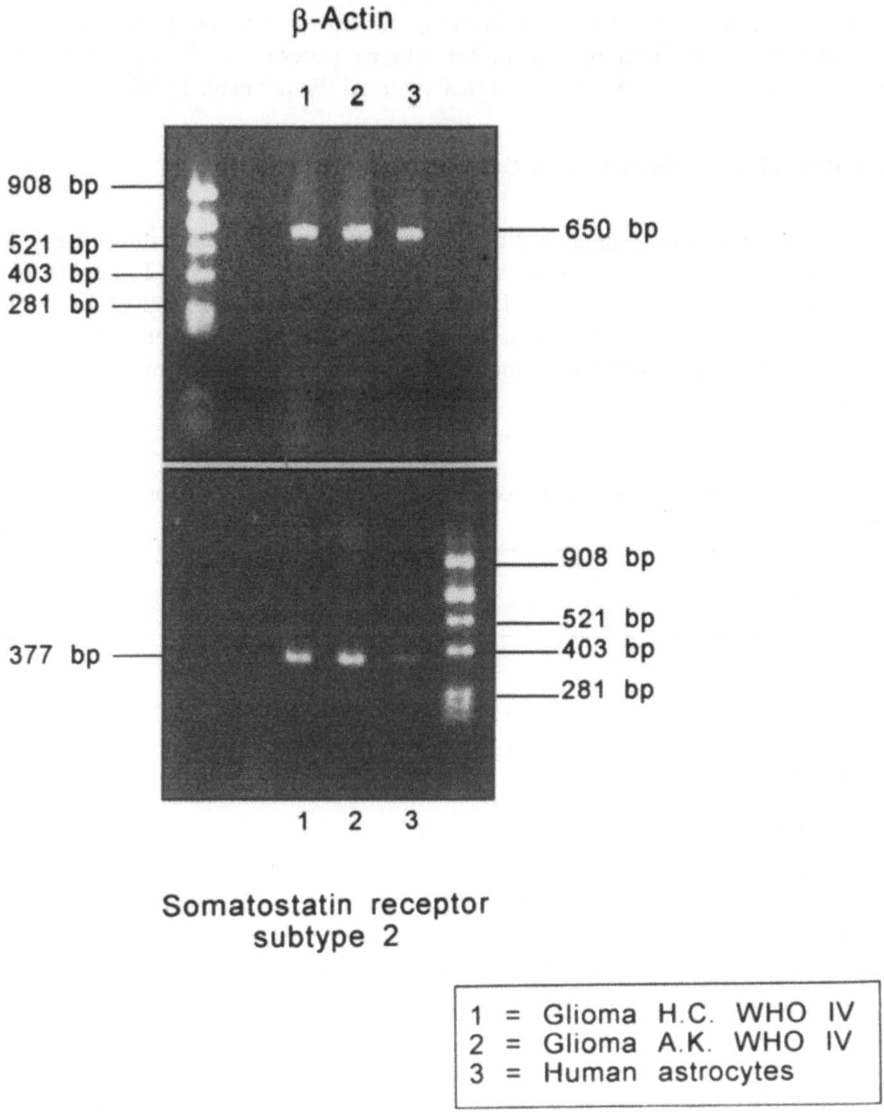

Fig. 3. RT-PCR of somatostatin receptor subtype sst_2 (*lower panel*) in normal human astrocytes as compared to human gliomas with β-actin (*upper panel*) as reference. RNA was isolated from cultivated cells, digested with DNAse, reversed transcribed, cDNA amplified by polymerase chain reaction with primers specific for human sst_2 or β-actin (35 cycles), and amplified DNA was separated on a 2% agarose gel that was stained with ethidium bromide. *bp* Base pairs

In some cases, another test was performed to find out whether the cultivation process had an influence on the expression of sst subtypes. For this, mRNA was directly isolated from part of the tumor - regardless of contamination by other cell types - as well as from subcultivated (and thus purified) glioma cells derived from the same tumor. In three tumors investigated and a non-malignant astrocyte surgical sample, no difference in the sst subtype pattern was detected between cultivated, purified cells and crude surgical material (Feindt et al. 1996).

Somatostatin Receptors on Glioma Cell Lines

Not only solid human gliomas but also nearly all glia cell lines - often derived from tumors - predominantly expressed the receptor subtype sst_2 (Table 4). An exception was the glioma cell line U118 which resembled normal astrocytes in its sst-pattern. These results show that experiments with glia cell lines should be interpreted with care, because some of their properties result from malignant transformation.

Table 4. Somatostatin receptor subtypes (sst) in rat/human glia cell lines as compared to normal astrocytes

Cell type	Species	sst 1	2	3	4	5
Astrocytes	Rat	+	+	-	+	-
	Human	+	+	-	(+)	-
Glia cell lines						
C6 glioma cells	Rat	(+)	+++	+	(+)	-
A172 glioblastoma	Human	-	+++	-	+	-
U118 glioma	Human	+	++	-	+	-
U138 glioblastoma	Human	+	+++	-	-	-
U343 glioma	Human	n. d.	+++	n. d.	n. d.	n. d.
U373 glioma	Human	-	+++	-	+	-
H4 neuroglioma	Human	+	+++	-	(+)	-

The number of + roughly indicates the intensity of sst expression. Data from Feindt et al. (1995 and unpubl. results). *n.d.*, not determined

Morphological Distribution of sst in Normal and Malignant Glial Cells

The morphological distribution of SS receptors was investigated by two methods: affinity-labeling with a somatostatin-gold ligand (cf. Mentlein et al. 1990, Krisch et al. 1991, Krisch 1994) and by immunostaining with an antibody against a synthetic peptide derived from the extracellular N-terminus of the receptor subtype sst_2 (Krisch, unpubl.). Light microscopy renders identical results with both methods. When simultaneously incubated, normal astrocytes are always less intensely stained than human glioma cell lines and cultivated human tumor material. In the latter, a selective staining of individual cells is observed, the intensity of which varies. A considerable number of cells constantly remains unstained without obvious differences in general morphology. Whereas normal astrocytes are diffusely labeled on the surface of their somata and processes, malignant glia cells have a tendency to carry the label more densely on their processes than on their somata (Fig. 4). This pattern does not change over a period of three months and five to six subcultivation procedures (Schneider 1997). Electron microscopically, the gold-labeled somatostatin ligand is localized exclusively at the cell surface provided that the incubation temperature has been low enough to impede internalization (Fig. 5).

Functional Significance of Somatostatin Receptor Subtypes

Somatostatin inhibits the proliferation in many, but not all normal or malignant target cells. Among the non-neuronal cells from rat brain, this antiproliferative effect was obvious only in microglia cells (Fig. 6). After longer application, somatostatin or its metabolically stable, but subtype selective analog octreotide did not inhibit the basal or serum-stimulated proliferation of astrocytes (Fig. 7; Feindt et al. 1995) or of meningeal cells (Feindt et al. 1997a, b). In contrast, in microglia cells the basal (Fig. 6) or stimulated proliferation was significantly reduced by somatostatin as well as by octreotide (Mentlein, Feindt and Schmidt, unpubl.).

Somatostatin and octreotide also failed to inhibit the proliferation of human astrocytes or gliomas cells after longer application (48 h) irrespective of whether they were stimulated by serum or not (Fig. 7). Thus, somatostatin or its analogs appear to be less effective in regulating the growth of gliomas, although they are useful hormonal regulators for the control of tumor growth in peripheral systems (Weckbecker et al. 1993).

The contrasting effects of somatostatin (octreotide) on microglia cells on the one hand and astroglial cells and gliomas on the other, may be explained by different sst subtype expression, different signal transduction mechanisms or downregulation of other stimuli, or by desensitization of the receptors.

Via G_i- or G_0-proteins, sst subtypes may transduce different intracellular signals in their target cells (see Florio and Schettini 1996 for a review). sst has been shown to inhibit adenylate cyclase (all subtypes) and Ca^{2+}-channels (sst_2), or to activate K^+-channels (sst_2), phospholipase A_2 (sst_4), mitogen-activated protein (MAP) kinase (sst_4) and protein phosphatases (sst_1, sst_2 and sst_4). Activation of protein phosphatases and reduction of intracellular cAMP are regarded as direct

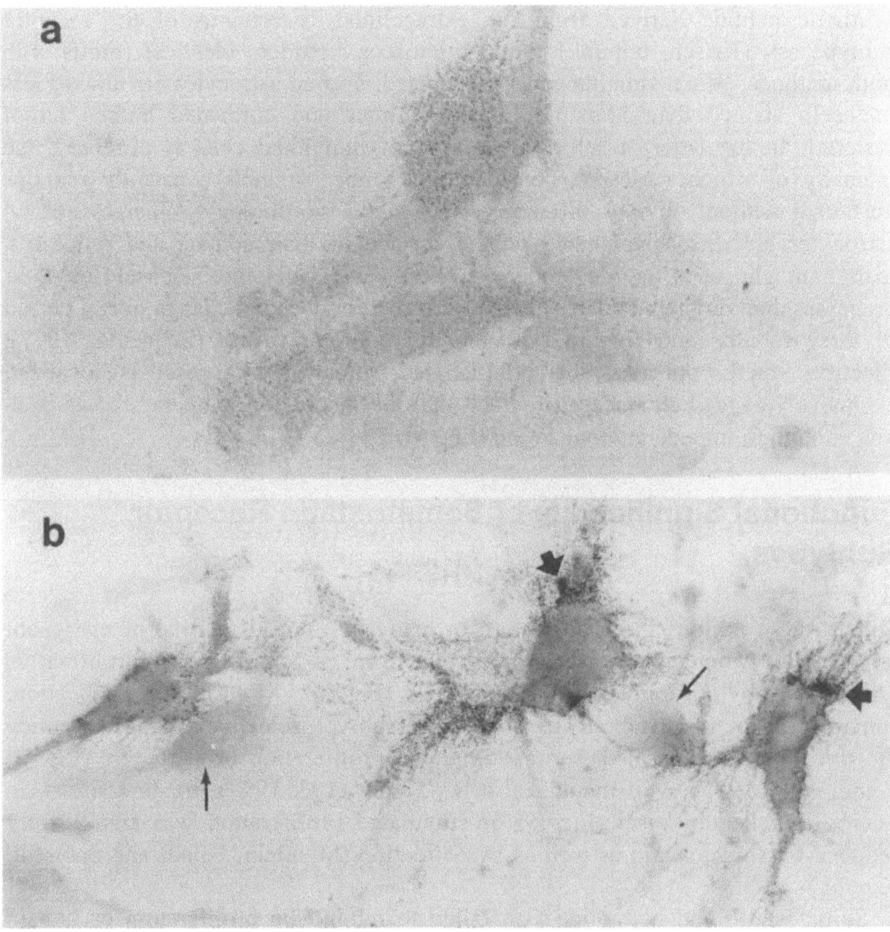

Fig. 4. Light microscopic immunogold histochemistry of somatostatin receptors (sst_2) on the surface of human astrocytes (**a**) as compared to glioma cells (**b**). Astrocytes or glioma cells (WHO IV) were stained with a polyclonal antibody (1:750) directed to the extracellular N-terminus of the cloned SS receptor subtype sst_2 followed by a gold-labeled secondary antibody and silver intensification. Incubation was done simultaneously under identical conditions. Note the even but delicate distribution of reaction product on some of the astrocytes (**a**). As in astrocyte preparations, cultivated human glioma (**b**) include unstained cells (*thin arrows*). On stained cells, however, the reaction product is densely but unevenly distributed and preferentially concentrated on the base of the processes (*thick arrows*). x440

Fig. 5. Electron microscopy of somatostatin-binding sites on the surface of tangentially sectioned cultivated human glioma cells. Human glioma cells were affinity-labeled with a somatostatin-gold ligand (non-selective for a receptor subtype) at 4 °C for 60 min, washed, fixed and silver intensified. Note that intense staining can be observed on the processes, whereas the somata bear less receptors. The results of the affinity labeling of the receptors totally correspond to those observed by immunostaining of the sst_2 subtype (Fig. 4). **a** x5880; **b** x8830

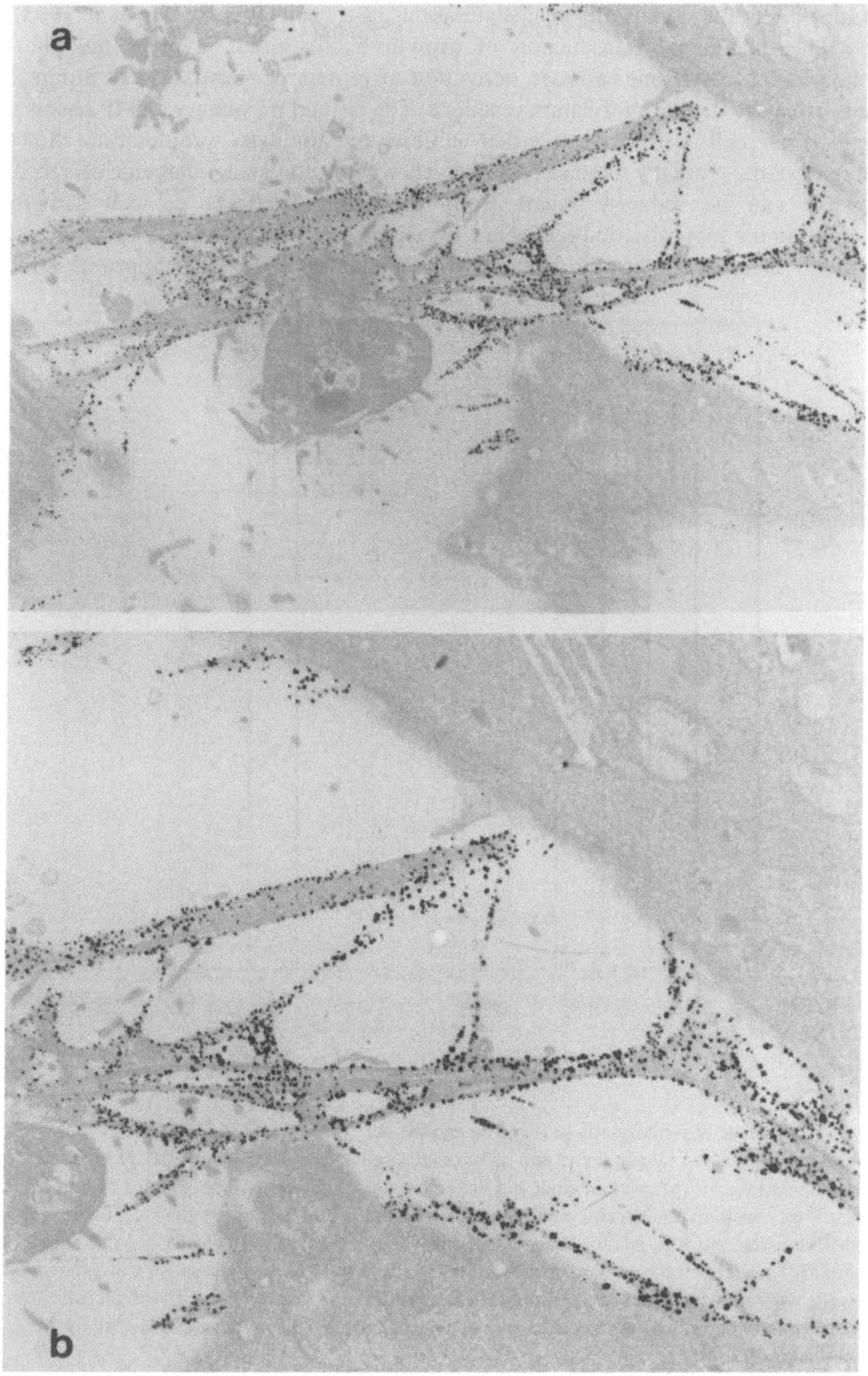
a
b

antiproliferative signals of somatostatin on its target cell (Weckbecker et al. 1993).

Since the signal transduction of growth factors often involves the phosphorylation of tyrosine residues, activation of protein phosphatases can diminish this effect by dephosphorylation reactions. The second messenger cAMP has dual effects on cell growth as it either inhibits or stimulates proliferation. Since somatostatin generally reduces cAMP levels when stimulated, mitogenic effects of cAMP can be reduced. Apart from these direct effects on cell growth, somatostatin may also indirectly act by suppressing the synthesis of autocrine growth factors, e.g. insulin-like growth factor-1 (IGF-1). The antiproliferative

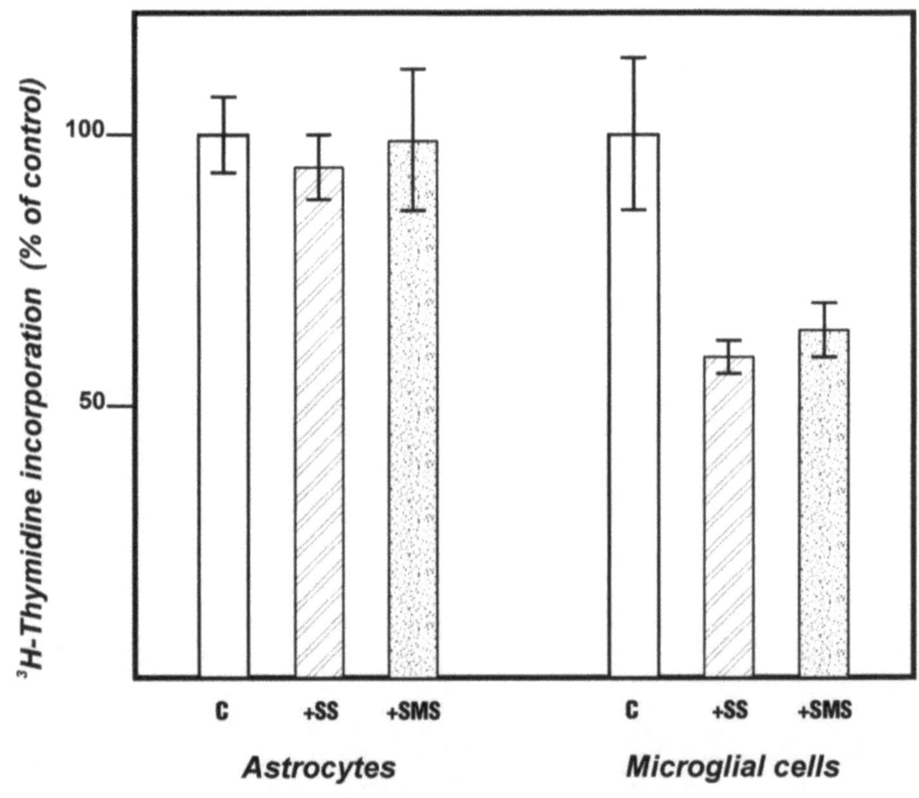

Fig. 6. Influence of somatostatin (*SS*) and its agonist octreotide (*SMS*, SMS 201-995) on the basal proliferation of rat astrocytes and rat microglia cells. Data from Feindt et al. (1995) and unpublished results (Mentlein, Feindt and Schmidt). Astrocytes were stimulated for 48 h in a serum-free medium (Dulbecco's modified Eagle's medium, DMEM), microglia cells for 24 h in DMEM containing 10% fetal calf serum without (control, *C*) or with 10 nM *SS* or 1 nM octreotide (*SMS*), and the incorporation of tritiated thymidine into DNA was measured. Corresponding results were also obtained in cells stimulated for proliferation (astrocytes by 10% fetal calf serum, microglia cells by granulocyte-macrophage colony-stimulating factor, GM-CSF, or by interleukin-3, IL-3)

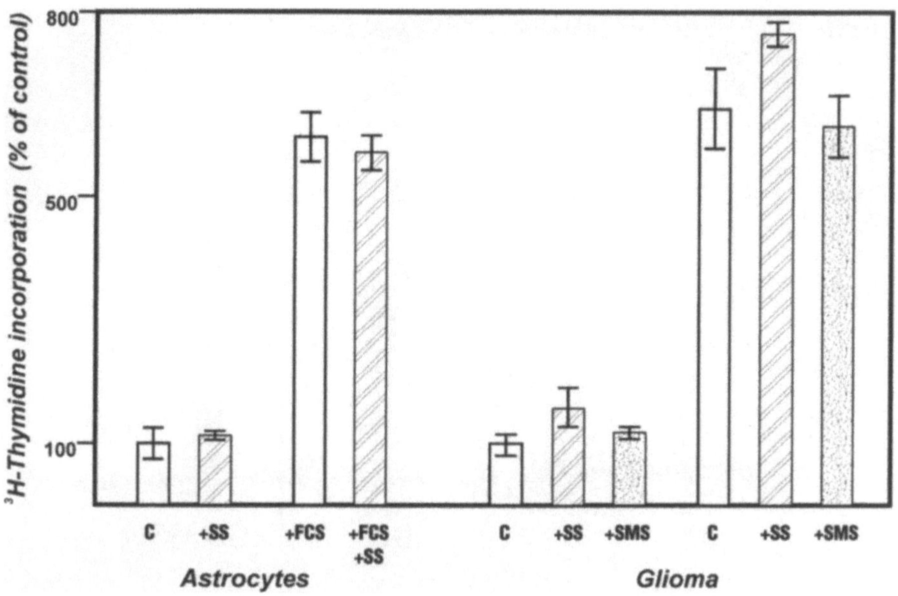

Fig. 7. Influence of somatostatin (*SS*) and its agonist octreotide (SMS 201-995, *SMS*) on the basal and serum-stimulated (10% fetal calf serum, *FCS*) proliferation of human astrocytes and glioma cells (WHO grade IV) after 48 h in culture. *C* Unstimulated control. Data from Feindt et al. (1995). For experimental details, see Fig. 6

effect of somatostatin on microglia cells may thus be explained by reduction of synthesis of autocrine hematopoietic growth factors (e.g., interleukin-3) or by dephosphorylation of activated proteins.

Since astrocytes and gliomas express sst subtypes that are normally involved in antiproliferative signal transducing effects of somatostatin (namely sst_1, sst_2 or sst_4), they may either lack the transducing mechanism, or the receptors may be desensitized after longer incubations. To test this latter hypothesis, we measured the effects of somatostatin and octreotide on incorporation of tritiated thymidine into DNA after short incubations. Indeed, after 2-5 h, but not after 12 h, an antiproliferative effect was clearly detectable (Feindt et al. 1997b). It depended on the dose of the agonist as expected for the sst affinities and, thus, must be receptor-mediated. Therefore, desensitization of sst can partly explain the lack of antiproliferative effects in astrocytes and gliomas (after longer incubations).

Possible mechanisms involved in receptor desensitization include: receptor phosphorylation by G-protein coupled receptor kinases (GRK) or protein kinase A (PKA), resulting in uncoupling from their signal transducing mechanism, increased receptor internalization, increased receptor degradation, or de-stabilization of receptor mRNA. To test one of these possibilities, we started to investigate the phosphorylation of the activated sst_2 in glioma cells (Fig. 8).

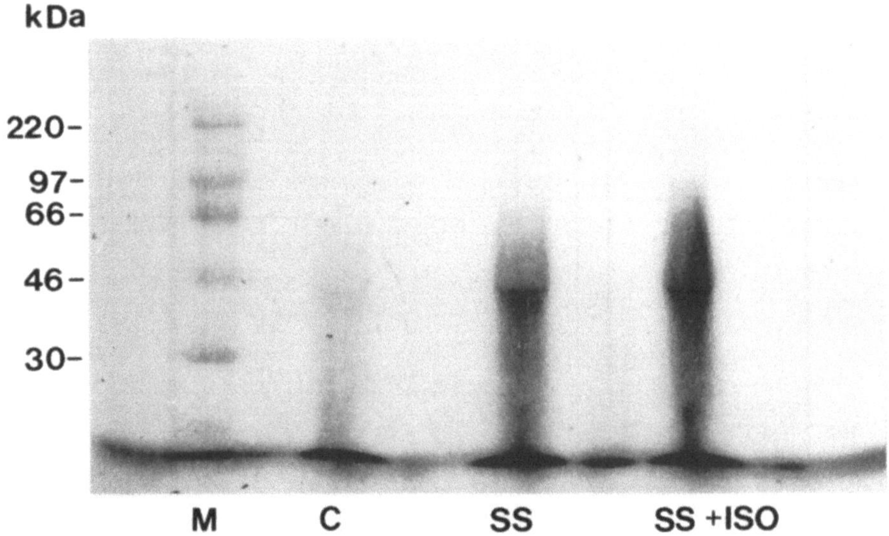

Fig. 8. Autoradiogram of an electrophoresis gel showing the phosphorylation of the somatostatin receptor sst_2 in U343 cells without (*C*) or with stimulation with somatostatin (10 nM) alone (*SS*) or in combination with the β-adrenergic agonist isoproterenol (*SS* + *ISO*). *M* Marker proteins and their molecular masses in kDa. We metabolically labeled the glioma cell line U373 with radioactive (^{33}P) phosphate for 3 h, stimulated the cells for 15 min without or with somatostatin alone, or in combination with the β-adrenergic agonist isoproterenol (1 μM), lysed the cells and immunoprecipitated the sst_2 with a specific antibody (directed to a synthetic peptide corresponding to the last 15 amino acids of the human/rat/mouse sst_{2A}, ETQRTLLNGDLQTSI). The immunoprecipitates were separated by dodecylsulfate polyacrylamide gel electrophoresis, the gel was dried and subjected to autoradiography. Note labeled bands in the lanes of the agonist-stimulated cells and only a rather faint one in that of the control. In further experiments we have observed the existence of the G-protein coupled receptor kinase 2 (GRK 2) in the glioma cell line U373 which may be responsible for this phosphorylation process (unpubl.)

Conclusions and Perspectives

Receptors for the neuropeptide somatostatin are expressed not only in neurons but also in non-neuronal cells of the brain. Multiple receptor subtypes are present in astroglia, microglia and meningeal cells in culture. Glioma cells are characterized by a strong (over-)expression of the subtype sst_2 alone, or in combination with others. However, for the diagnosis and treatment of gliomas this finding is of limited value: depending on the integrity of the blood-brain barrier, only part of the gliomas can be visualized by receptor scintigraphy in vivo, and due to receptor desensitization, SS or its analogs fail to inhibit glioma proliferation.

The detection of receptors for SS and some other neuropeptides on astrocytes (or meningeal cells) in situ (Krisch and Mentlein 1994, Wilkin et al. 1995, Porter and McCarthy 1997, for reviews) provide evidence for a peptidergic paracrine regulation of non-neuronal cells by neuronal activity. The physiological consequences of this action are, however, not yet understood and may be the subject of future research.

The reasons for and the consequences of the overexpression of the receptor subtype sst_2 in gliomas (or other SS receptor positive tumors) are even less understood. The overexpression of sst_2 might be caused by autocrine growth factors of the tumors or paracrine factors of the tumor microenvironment (e.g. cytokines from invading immune cells), endogenous influences within the tumor cells like activation or repression of transcription factors or tumor suppressor genes (e.g. by mutation), or mutations within the sst_2 gene promotor region resulting in an increased expression of the receptor. The consequences of sst_2 overexpression should result in an increased interaction of its transducing pathway with those of other receptors, e.g. dephosphorylation of activated (autophosphorylating) growth factor receptors, activated kinases or activated transcription factors.

From these results it seems that research on neuropeptides in the brain should not only focus on signal transmission between neurons, but also consider a paracrine neuron-glia network.

Acknowledgments

The experimental work described in this review was supported by grants from the Deutsche Forschungsgemeinschaft (Kr 569/7-1; B. K.) and from the Stiftung Volkswagenwerk in the project "Neuroimmunology, Mood and Behaviour" (R. M.).

References

Brazeau P, Vale W, Burgus R, Ling N, Rivier J, Guillemin R (1972) Hypothalamic polypeptide that inhibits the secretion of immunoreactive pituitary growth hormone. Science 129: 77-79

Bruno JF, Xu Y, Song J, Berelowitz M (1993) Tissue distribution of somatostatin receptor subtype messenger ribonucleic acid in the rat. Endocrinology 133: 2561-2567

Feindt J, Becker I, Blömer U, Hugo H-H, Mehdorn HM, Krisch B, Mentlein R (1995) Expression of somatostatin receptor subtypes in cultured astrocytes and gliomas. J Neurochem 65: 1997-2005

Feindt J, Hugo H-H, Mentlein R, Krisch B (1996) Somatostatin receptor subtypes in human astrocytes and gliomas: influence of cultivation process. In: Krisch B, Mentlein R (eds) The peptidergic neuron. Birkhäuser Verlag , Basel, pp 141-150

Feindt J, Krisch B, Lucius R, Mentlein R (1997a) Meningeal cells are targets and inactivation sites for the neuropeptide somatostatin. Mol Brain Res 44: 293-300

Feindt J, Krisch B, Mentlein R (1997b) Time-dependent influence of the somatostatin analog octreotide on the proliferation of rat astrocytes and glioma cells. Brain Res 746: 309-313

Florio T, Schettini G (1996) Multiple intracellular effectors modulate physiological functions of the cloned somatostatin receptors. J Mol Endocrinol 17: 89-100

Hildebrandt G, Scheidhauer K, Luyken C, Schicha H, Klug N, Dahms P, Krisch B (1994) High sensitivity of the in vivo detection of somatostatin receptors by ^{111}indium-[DTPA-ocreotide]-scintigraphy in meningioma patients. Acta Neurochir (Wien) 126: 63-71

Krisch, B. (1980) Differing immunoreactivities of somatostatin in the cortex and hypothalamus of the rat. Cell Tissue Res 212: 457-464

Krisch B (1994) Somatostatin binding sites in functional systems of the brain. Progr Histochem Cytochem 28: 1-40

Krisch B, Mentlein R (1994) Neuropeptide receptors and astrocytes. Int Rev Cytol 148: 119-169

Krisch B, Buchholz C, Mentlein R (1991) Somatostatin binding sites on rat diencephalic astrocytes. Light-microscopic study in vitro and in vivo. Cell Tissue Res 263: 253-263

Krisch B, Buchholz C, Mentlein R, Turzynski A (1993) Visualization of neuropeptide-binding sites on individual telencephalic neurons of the rat. Cell Tissue Res 272: 523-531

de Lecca L, Criado JR, Prospero-Garcia O, Gautvik KM, Schweitzer P, Danielson PE, Dunlop CL, Siggins GR, Henriksen SJ, Sutcliff JG (1996) A cortical neuropeptide with neuronal depressant and sleep-modulating properties. Nature 381: 242-245

Luyken C, Hildebrandt G, Scheidhauer K, Krisch B, Schicha H, Klug N (1994) ^{111}Indium (DTPA-octreotide) scintigraphy in patients with cerebral gliomas. Acta Neurochir (Wien) 127: 60-64

Mentlein R, Buchholz C, Krisch B (1990) Somatostatin-binding sites on rat telencephalic astrocytes. Light- and electron-microscopic studies in vitro and in vivo. Cell Tissue Res 262: 431-443

Mentlein R, Feindt J, Schmidt A (1997) Somatostatin receptor subtypes and their effects in microglia cells. J Neurochem (submitted)

Patel YC, Srikant CB (1994) Subtype selectivity of peptide analogs for all five cloned human somatostatin receptors (hsstr 1-5). Endocrinology 135: 2814-2817

Porter JT, McCarthy KD (1997) Astrocytic neurotransmitter receptors in situ and in vivo. Progr Neurobiol 51: 439-455

Reubi JC, Laissue J, Krenning E, Lamberts SWJ (1992) Somatostatin receptors in human cancer: incidence, characteristics, functional correlates and clinical implications. J Steroid Biochem Mol Biol 43: 27-35

Rohrer L, Raulf F, Bruns C, Buettner R, Hofstaedter, Schüle R (1993) Cloning and characterization of a fourth human somatostatin receptor. Proc Natl Acad Sci USA 90: 4196-4200

Schindler M, Humphrey PPA, Emson PC (1996) Somatostatin receptors in the central nervous system. Progr Neurobiol 50: 9-47

Schneider M (1997) Somatostatinrezeptoren an humanen Gliatumoren in vitro. Dissertation Medizinische Fakultät der Christian-Albrechts-Universität zu Kiel

Weckbecker G, Raulf F, Stolz B, Bruns C (1993) Somatostatin analogs for the diagnosis and treatment of cancer. Pharmacol Therapeut 60: 245-264

Wilkin GP, Marriott DR, Pearce B (1995) Peptide receptors on astrocytes. In: Kettenmann H, Ransom RB (eds) Neuroglia. Oxford University Press, Oxford, pp 367-376

Yamada Y, Post SR, Wang K, Tager HS, Bell GI, Seino S (1992a) Cloning and functional characterization of a family of human and mouse somatostatin receptors expressed in brain, gastrointestinal tract, and kidney. Proc Natl Acad Sci USA 89: 251-255

Yamada Y, Reisine T, Law SF, Ihara Y, Kubota A, Kagimoto S, Seino M, Seino Y, Bell GI, Seino S (1992b) Somatostatin receptors, an expanding gene family: cloning and functional characterization of human SSTR3, a protein coupled to adenylyl cyclase. Mol Pharmacol 6: 2136-2142

Yamada Y, Kagimoto S, Kubota A, Yasuda K, Masuda K, Someya Y, Ihara Y, Li Q, Imura H, Seino S, Seino Y (1993) Cloning, functional expression and pharmacological characterization of a fourth (hSSTR4) and fifth (hSSTR5) human somatostatin receptor subtype. Biochem Biophys Res Commun 195: 844-852

21 Specific Glycoproteins of the Subcommissural Organ/Reissner's Fiber Complex: Molecular and Functional Evidence of Their Involvement in Developmental Events

A. Meiniel

INSERM U384 et Laboratoire de Biochimie médicale, Faculté de Médecine,
28 Place Henri Dunant, 63001 Clermont-Ferrand cédex, France;
Phone: +33 4 73 60 80 53, Fax: +33 4 73 27 61 32,
E-mail: annie.meiniel@inserm.u-clermont1.fr

Abstract

The subcommissural organ (SCO) is a phylogenetically old structure of the vertebrate brain made up of ependymal cells which secrete glycoproteins basally towards the leptomeningeal spaces and apically into the ventricular cavity. The latter mode of secretion forms a particular structure called Reissner's fiber (RF). The functional significance of the SCO/RF complex is still discussed, in spite of a wide range of experimental data. Here we present new molecular and biological data favoring a putative morphogenetic activity of the SCO/RF complex. The glycoproteins secreted by the SCO exhibit sequence homologies with proteins expressed in the developing nervous system, including thrombospondins 1 and 2, semaphorins F and G and F-spondin. These glycoproteins were called SCO-spondin because of their high expression in the SCO and the presence of conserved domains (thrombospondin type 1 repeats). Molecules exhibiting such domains have been shown to interfere with cell adhesivity and neuritic outgrowth. Besides the molecular features, we analyzed the biological effect of RF and solubilized material from RF on neuronal cells in primary cell cultures. In these in vitro systems, RF markedly enhanced cell survival and interfered with cell aggregation and neuritic outgrowth. Thus, both molecular and functional data give new insights into a putative morphogenetic activity of the SCO/RF complex.

Introduction

The subcommissural organ/Reissner's fiber (SCO/RF) complex has been investigated by many neurobiologists since the discovery of an "axon-like" structure running in the central canal of the spinal cord by Reissner (1860). This intriguing structure was given the name of Reissner's fiber and was subsequently shown to be secreted by an intracerebral "gland", the subcommissural organ

(Studnicka 1900). The secretory activity of the subcommissural organ is now well documented, but in spite of a wide range of work, the function of the SCO/RF complex is still under discussion (Oksche et al. 1993). Clarification of how the SCO/RF complex exerts its function within the central nervous system (CNS) remains a great challenge.

The ependymal secretion of the SCO/RF complex was originally considered as a "neurosecretion" because of its stainability with various histochemical reagents. Aldehyde fuchsin (Wislocki and Leduc 1952), alcian blue, periodic acid schiff and Gomori (Stutinsky 1950) staining procedures labeled material within the intracellular compartment, i.e., the SCO cells and also the secreted compound, i.e., RF. These histochemical analyses also suggested a putative glycoproteic nature of the secretory material. The knowledge of these glycoproteins was broadened by studies with lectins or phyto-agglutinins, allowing the characterization of carbohydrate moieties. These tools were first used by our research group (Meiniel and Meiniel 1985) to study the SCO secretory glycoproteins and have helped to determine (1) the synthesis of complex-type glycoproteins showing N-linked (or Asn-linked) oligosaccharides, (2) the conservation of the carbohydrate moieties in the course of evolution, (3) the probable structure of the sugar chains at different steps of the glycosylation process, and (4) the various cellular compartments of the secretory ependymal cells implicated in glycoprotein biosynthesis (Rodríguez et al. 1986; Meiniel et al. 1988, 1996).

Several polyclonal and monoclonal antibodies raised either against Reissner's fiber or the SCO proteins were produced, and their application gave complementary information, revealing the great specificity of the SCO secretory products (see Meiniel et al. 1996 for details). Sterba and collaborators (Sterba et al. 1981, 1982) were the first to prepare polyclonal antibodies against a water soluble fraction of the isolated bovine Reissner's fiber and to demonstrate convincingly that the SCO ependymal cells were the source of Reissner's fiber since they showed that both the bovine SCO and RF were immunoreactive with these antibodies. Subsequently, Rodríguez and colleagues (Rodríguez et al. 1984a, b) as well as Meiniel and colleagues (see Meiniel et al. 1996 for details) prepared antibodies against bovine RF solubilized in urea. The anti-bovine RF IgG revealed the secretory material of the SCO and Reissner's fiber throughout the scale of the vertebrates (Sterba et al. 1982; Rodríguez et al. 1984b), except for anthropoid apes and humans. Furthermore, the infundibular organ and fiber of prochordates (Sterba et al. 1983; Olsson et al. 1994) as well as the flexural organ of vertebrates (Schoebitz et al. 1993) were immunolabeled with these antibodies. From these studies, it was concluded that (1) the SCO was the main source of RF synthesis, (2) other areas in the brain could participate in RF formation, particularly the flexural organ, (3) these secretions were highly conserved in the course of phylogeny, and (4) two different routes of secretory release occurred: a ventricular one leading to the formation of RF and a basal one that could correspond to a leptomeningeal and/or vascular route of secretion (Lösecke et al. 1984, 1986; Meiniel et al. 1991).

Because of the particular pattern in the expression of glial markers (Didier et al. 1986; Didier-Bazes et al. 1993) and the presence of a secretory activity, it has been suggested that the ependymal cells in the SCO could have maintained embryonic features. Drawing a parallel between the secretory ependyma of the floor plate in the spinal cord and that of the SCO in the brain, a putative morphogenetic role of the SCO/RF secretions was hypothesized (see Meiniel et al. 1996). Analyzing domains present in the SCO/RF glycoproteins and the effect of Reissner's fiber material on neurons in primary cell cultures, we provided direct molecular and functional evidence for a putative developmental role of the SCO/RF glycoproteins.

Molecular Analysis of the SCO/RF Glycoproteins

The search for homologies of the secreted glycoproteins with other already known factors appeared essential to elucidate the SCO/RF function. In order to analyze the nucleic sequence encoding the SCO/RF glycoproteins, we constructed a bovine SCO cDNA library in the vector λgt 11; the latter was screened using an anti-RF antibody directed against the SCO/RF secretions (Meiniel et al. 1996). With this approach, several cDNA clones were isolated and analyzed (see Meiniel et al. 1995; Gobron et al. 1996; Creveaux et al. 1997 for details). In situ hybridization histochemistry revealed that the mRNAs corresponding to these clones are specifically expressed in the SCO (Fig. 1).

The size of the major transcript encoding the SCO/RF glycoproteins was found to be 14 kb in Northern blots (Creveaux et al. 1997) and 6 kb have as yet been analyzed. Figure 2 shows various consensus domains deduced from the nucleic sequence, which could account for at least part of the functional properties of the SCO/RF glycoproteins.

In fact, this sequence exhibited consensus domains called "thrombospondin type 1 repeats" (TSP type 1 repeats, Lawler and Hynes 1986), previously identified in proteins of the developing vertebrate CNS. These include thrombospondins 1 and 2 (Tucker 1993; Adams et al. 1995), a component of the extracellular matrix (ECM), semaphorins F and G (Adams et al. 1996) and F-spondin (Klar et al. 1992) secreted by the floor plate. Thus, we called the protein synthetized by the SCO SCO-spondin to refer to these homologies (Gobron et al. 1996).

When compared with its relatives (see Gobron et al. 1996, for detailed comparison), SCO-spondin exhibits numerous TSP type 1 repeats (nine so far). In thrombospondin, six consensus domains of this type were identified, which are suspected to participate in cell adhesive mechanisms (O'Shea et al. 1991; see Frazier 1991 and Bornstein and Sage 1994 for reviews). These domains may also promote neuritic outgrowth because antibodies directed against a 50-kDa proteolytic fragment comprising three TSP type 1 repeats and the procollagen domain were able to inhibit neuritic outgrowth that was found to be enhanced in vitro in the presence of thrombospondin (Osterhout et al. 1992). A neuronal

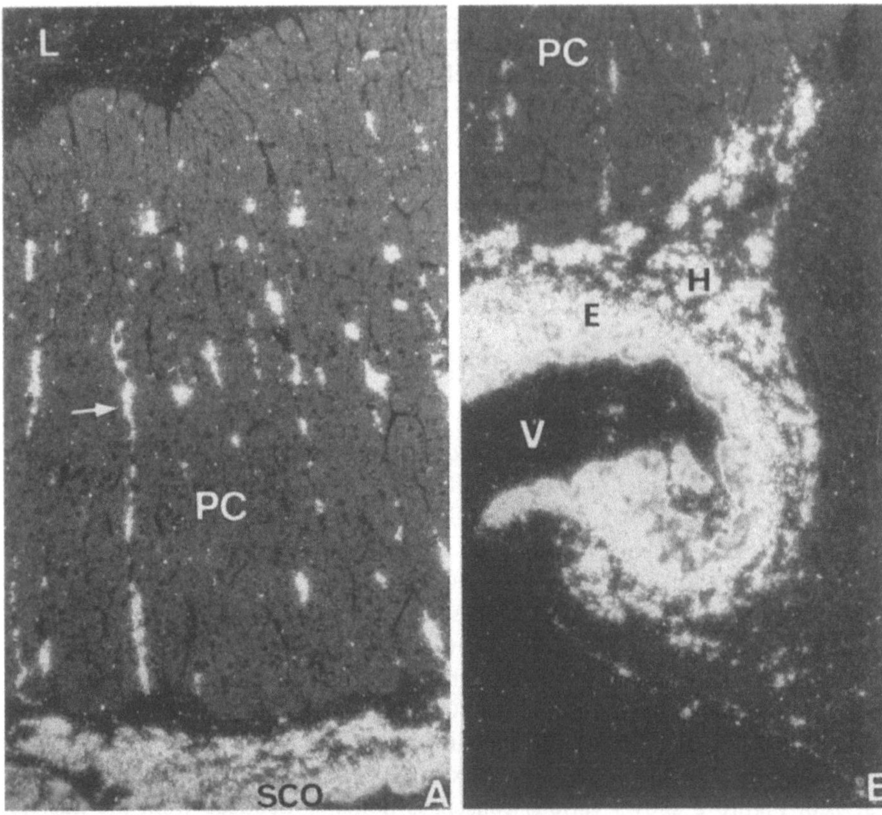

Fig. 1. Demonstration of mRNA encoding for SCO-spondin by means of in situ hybridization and the antisense probe λ RF 103 in the bovine. Dark field images. **A** Strong radioactive labeling of the subcommissural organ (*SCO*) and of cells (*arrow*) distributed in rays in the posterior commissure (*PC*). *L* Leptomeningeal spaces. x30. **B** Caudal end of the subcommissural organ. Note the strong expression in the ependymal cells (*E*) located at the border of the ventricular cavity (*V*), in hypendymal cells (*H*) arranged in rosettes and in cells located in the posterior commissure (*PC*). x70. Photographs by R. Meiniel

thrombospondin receptor, $\alpha_3\beta_1$ integrin, was shown to mediate this interaction in vitro (DeFreitas et al. 1995). Similarly, F-spondin in the floor plate of the vertebrate CNS (Klar et al. 1992) and unc-5 in the CNS of *Caenorhabditis elegans* (Hedgecock et al. 1990), exhibit TSP type 1 repeats and are held as factors influencing neuritic outgrowth.

In the presently identified SCO-spondin sequence, two of the TSP type 1 repeats were found to contain potent binding sites to cytokines and growth factors of the class 2 receptor family. Mutations in this domain abolished transduction events induced by cytokines (Miyazaki et al. 1991). Regarding SCO-spondin, these consensus sequences could participate in regulatory processes involved in the enhanced neuronal survival observed in the presence of RF (Monnerie et al.

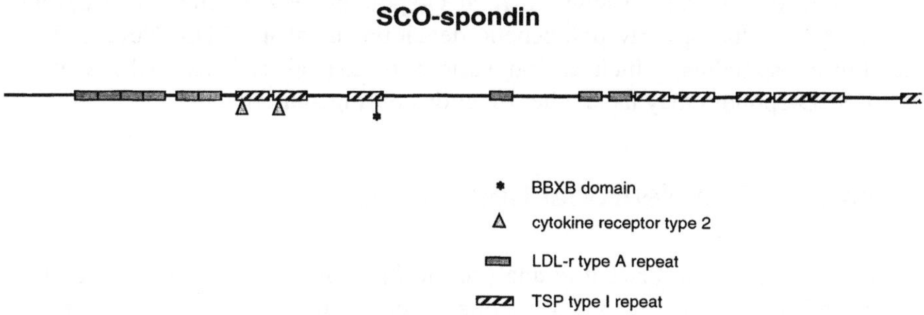

Fig. 2. Schematic drawing of the sequenced portion of SCO-spondin, expressed in the bovine subcommissural organ. Various consensus domains are indicated that could account for the function of this protein. Courtesy of S. Gobron

1995). The actual SCO-spondin sequence also contains potent binding sites to proteoglycans that may be present either on the cell surface or in ECM and comprises consensus domains called "low density lipoprotein receptor type A repeats" (LDLr type A repeats), which are found in a number of extracellular proteins acting as morphogens. The functional significance of these consensus domains is not yet completely understood but studies with the nudel protein have suggested that they participate in the regulation of extracellular protease activity (Hong and Hashimoto 1995) known to play a crucial role during morphogenesis (Hecht and Anderson 1992). Finally, the presence of several potent N-glycosylation sites in the SCO-spondin sequence is in accordance with previous studies demonstrating that the SCO/RF proteins exhibit the same lectin binding properties as complex-type glycoproteins (Meiniel et al. 1988). Since, up until now, only a part of the encoding sequence has been analyzed, the presence of other consensus domains is highly probable. Their identification will also contribute to further clarification of the functional significance of the SCO-spondin.

The present molecular analysis allows the following comments:

1. Homologies with morphoregulatory molecules suggest the involvement of SCO-spondin in developmental events of the CNS.
2. On the basis of the consensus domains analyzed, this molecule appears as a member of the ECM family.
3. Like analogous proteins expressed in the developing CNS (thrombospondins, semaphorins and F-spondin), SCO-spondin could modulate cell adhesion and neuritic outgrowth.
4. SCO-spondin exhibits a particular structure "in mosaic" because domains of the same type are not aligned in clusters.
5. SCO-spondin may bind to some as yet unidentified cell receptors, but all the consensus domains are potent sites of protein-protein interaction and represent multiple direct or indirect means of cellular modulation.

These molecular features suggest that SCO-spondin may interfere with several cellular events during early ontogenetic development of the CNS. Nevertheless, like thrombospondins, which act on various biological systems (Adams et al. 1995), SCO-spondin may have other functions during adult life.

Effect of RF on Neuronal Cells in Vitro

Biological assays were needed to analyze the functional significance of the SCO glycoproteins in the CNS and to establish a correlation between molecular and functional data. In vitro bioassays were expected to be a suitable paradigm to elucidate modulatory mechanisms that, due to regulatory processes, could remain undetected in in vivo experiments. Thus, we developed primary cell cultures to test the effect of isolated RF, or material solubilized from RF, on the development of neuronal cells.

In primary mixed cortical cell cultures, where neuron/glia interactions are partly maintained, both RF and soluble RF material markedly enhanced neuronal survival. This effect is highly significant, abolished by specific IgG directed against RF, dose-dependent and independent of the presence of serum in the culture medium (Monnerie et al. 1995). Probably, the RF activity influences neuronal survival in primary mixed cell culture indirectly because its effects depend on the presence of glial cells in the culture system. The molecular basis of this effect, however, remains to be determined. In an attempt to correlate molecular and biological data, we can hypothesize that growth factors released by glial cells in the culture medium may bind to the potent binding sites for cytokines present in SCO-spondin. SCO-spondin would thus be implicated in a regulatory process for the delivery or presentation of growth factors to neuronal cells. Another effect of SCO-spondin could be the functional activation of growth factors as described for the activation of TGFβ by thrombospondins (Feige et al. 1996).

Experiments with primary cultures of enriched cortical neurons showed that RF was also able to modify neuronal behavior in the absence of glial cells, suggesting that RF can also act on neurons directly. In such a culture system, soluble RF material displayed an opposite effect to that of RF itself on cell aggregation mechanisms. This modulatory effect of RF on neuronal aggregation was demonstrated in a system in which neurons could be kept under controlled conditions, i.e., at low plating density (neurons remained unaggregated during the whole time course of the culture) or at high density (neurons aggregated as soon as day 2 post-plating, see Monnerie et al. 1996). In high density cultures, soluble RF material was shown to prevent neuronal aggregation. This effect was specific as antibodies directed against RF were able to restore cell aggregation (Monnerie et al. 1996; Gobron et al. 1996). A portion of RF added simultaneously to soluble RF material restored aggregative mechanisms in an area centered around the RF (Monnerie et al. 1997). The dual and antagonistic activity of soluble and un-

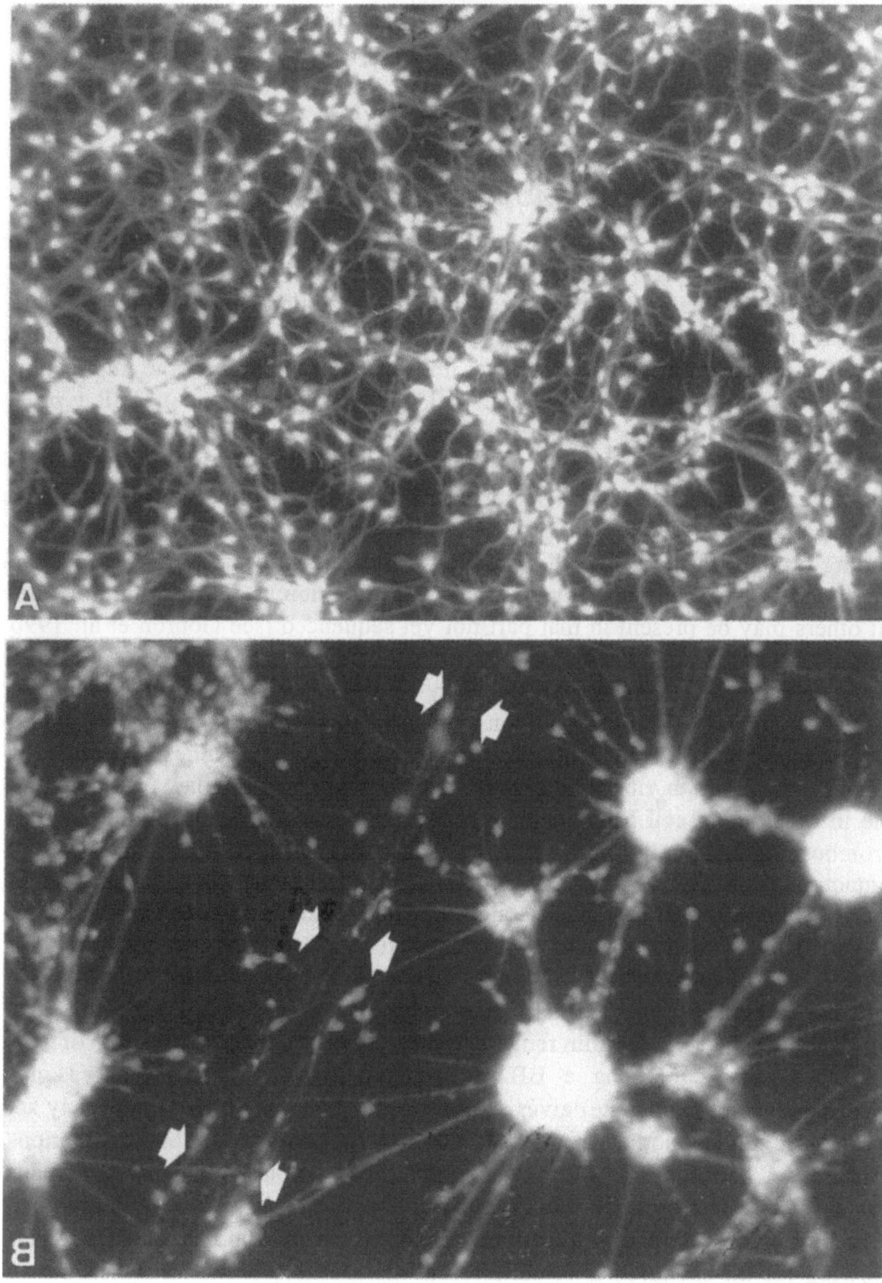

Fig. 3. Effect of Reissner's fiber on neuronal behavior in vitro. **A** Cortical neurons plated at low density on poly-L-lysine remain scatteredly distributed in controls. **B** In the presence of RF (*arrows*) neuronal cells constituted dense aggregates and developed neuritic outgrowth along RF. CFDA labeling. x120. Photographs by H. Monnerie

soluble RF material is reminiscent of that of tenascin, and could reflect two facets of a modulatory mechanism on cell adhesivity driven by proteins present in RF. In low density cultures of cortical neuronal cells, RF could induce neuronal aggregation and also promote neuritic outgrowth (Fig. 3; Monnerie et al. 1997).

SCO-Spondin and Developmental Processes

The exact mechanisms by which RF-proteins interfere with cell-cell and/or cell-matrix contacts remain to be demonstrated, but we can underline that the TSP type 1 repeats present in SCO-spondin contain several consensus domains that could influence adhesive mechanisms. Several lines of evidence support the implication of this domain in cell adhesion. The VTCG sequence, present in TSP type 1 repeats of the CS proteins from circumsporozoites, has been shown to confer the adhesive property of plasmodium to its target (Rich et al. 1990). In a similar way, this consensus sequence and its variants (VSCG, ATCG) appear to be important for the biological function of properdin (Prater et al. 1991), the complement binding protein. In the actual SCO-spondin sequence, only one VSCG sequence was found in the TSP type 1 repeats that could function as a potent binding site, but others may be present in that part not yet sequenced (see Gobron et al. 1996 for details).

Due to molecular homology, the activity of SCO-spondin may be extrapolated from the function of thrombospondins which has been most extensively studied (cf. Adams et al. 1995). Thrombospondins exhibit both adhesive and anti-adhesive properties in vitro, and it has thus been suggested that they play a unique role in modulating cell behavior in vivo. In fact, several potent binding sites of thrombospondin to cellular receptors are suspected to interfere with cell to cell contact (cf. Frazier 1991). These include (1) an RGDA sequence which could bind to $\alpha_V\beta_3$ integrin, (2) the VTCG sequences which could bind to CD36 or platelet gpIV, (3) the carboxy-terminal domain which could bind to the heterodimeric 105/80 kDa receptor in a calcium-dependent manner, and (4) the BBXB sequences (where B is a basic amino acid) which could bind to proteoglycans. The exact function of SCO-spondin remains to be determined, but the presence of TSP type 1 repeats (containing a BBXB sequence that can function as a glycosaminoglycan binding site) gives support to the idea that SCO-spondin may act as an extracellular modulator of cell-cell surface and/or cell-matrix interactions just like thrombospondins. Thrombospondins form complexes with proteoglycans (Prater et al. 1991), and this type of interaction may serve to strengthen thrombospondin binding. Alternatively, it could prevent binding of CD36, thus allowing for the modulation of adhesion at these sites (Frazier 1991). In our in vitro assay, RF-material had modulatory effects on neuronal cell aggregation. Nevertheless, production of peptides from SCO-spondin is now needed to demonstrate that SCO-spondin is indeed involved in this activity and to determine the functional domains involved in such mechanism(s).

Extracellular proteins that foster both cell attachment and spreading include thrombospondins and tenascins (Sage and Bornstein 1991; Chiquet-Ehrismann 1991). Among these, only thrombospondins exhibit TSP type 1 repeats suggesting the involvement of multiple molecular mechanisms supporting cell-cell and/or cell-matrix modulation. Nevertheless, thrombospondins as well as tenascins are known to interfere with cell attachment, to activate the intracellular signaling pathway, to influence Ca^{2+} flux, to present or sequester growth factors and cytokines, and to modulate extracellular protease activity (Sage and Bornstein 1991). Regarding the promoting activity of RF on neuritic outgrowth, one must emphasize that a number of factors constitutive of the ECM, including proteoglycans and ECM molecules, were shown to influence neuritic outgrowth. ECM molecules appear to be effective in this process, regardless of whether they exhibit TSP type 1 repeats like SCO-spondin, thrombospondin 2 and F-spondin or not, like tenascin. In fact, the strength of the interaction between neurons and substrate is an important parameter for neuritic outgrowth (Lochter et al. 1995), and we can postulate that this process is also dependent upon the forces of cell-cell interaction. Our results suggest that medium strength cell-cell interactions favor neuritic outgrowth (Monnerie et al. 1997).

Adhesion molecules, including neural cell adhesion molecule (N-CAM) and N-cadherin, have been shown to be primarily implicated in cell aggregative mechanisms in the developing CNS. The exact molecular interaction of RF material with adhesion molecules remains to be analyzed in detail. In mouse and rat strains presenting spontaneous hydrocephalus, the SCO secretory activity is reduced and RF is lacking (Takeuchi et al. 1987, 1988), suggesting a key role of the SCO/RF complex in CSF (cerebrospinal fluid) dynamics. Moreover, X-linked hydrocephalus in humans, associated with stenosis of the Sylvian aqueduct (HSAS), was recently shown to be correlated with mutations in the gene encoding L1CAM (Van Camp et al. 1993; Jouet et al. 1993), a cell surface glycoprotein, expressed by post-mitotic neurons and located on their axonal outgrowth. All data may suggest molecular interactions between L1CAM and the SCO secretions, which are in need of experimental clarification.

Secretory Ependyma and CNS Development

During development, the ependyma not only forms the germinative layer generating the various neuronal cells but also plays a crucial role in morphogenetic events via its secretory capacity (reviewed in Sarnat 1992; Del Bigio 1995). The best documented secretory ependymal differentiation is the floor plate generating polarizing signals that control cell identity in the neural tube (Korzh 1994). Numerous morphogenetic molecules have been reported to be secreted by the floor plate cells. These include F-spondin (Klar et al. 1992) which favors adhesion and neuritic outgrowth, and netrins (Kennedy et al. 1994; Serafini et al. 1994) which have a chemoattractive effect on the commissural axons. In certain prochordates, the most cephalic part of the floor plate, called the "infundibular organ", contributes to the formation of RF (Olsson 1993). This reinforces the hypothesis that the secretory ependyma in the brain, similarly to the secretory ependyma in the spinal cord, could participate in morphogenetic processes.

The most obvious ependymal secretion in the CNS is that of the subcommissural organ. This secretion is detected at early ontogenetic stages (Didier et al. 1992), but in contrast to the floor plate, ependymal cells of the SCO maintain their secretory ability in adults, leading to the formation of Reissner's fiber. The SCO/RF glycoproteins have been located by immunocytochemical techniques in the leptomeningeal spaces close to the hypendymal vessels of the posterior commissure (basal secretion) as well as in Reissner's fiber (ventricular secretion). Some SCO secretory products also may be soluble in the CSF, since a 54 kDa protein was repeatedly immunopurified from the cerebral hemispheres by use of specific antibodies against the SCO/RF glycoproteins (Karoumi 1990; Creveaux et al. 1997). This suggests that the neural tissue could be a target for SCO secretion.

It remains to be determined whether the SCO ependymal secretions have a **local** effect, acting on the guidance of the posterior commissural axons when secreted basally, and/or establishing a gradient of diffusible molecules for the spinal cord when secreted apically, forming Reissner's fiber. When transported via the CSF, the secretions may also have a more **general** effect on the differentiation of neuronal cells. Both adhesive and anti-adhesive mechanisms involving attractive and repulsive interactions between cells and their ECM may be important for cell migration and directional guidance of axon growth during neural development (Keynes and Cook 1990; Edelman et al. 1990; Faissner and Kruse 1990). Participation of SCO-spondin in these regulatory processes remains an interesting hypothesis. SCO-spondin is also expressed and synthesized during adult life. Like thrombospondins that interfere with several biological events (angiogenesis, platelet aggregation, tumor growth, wound healing, neural crest cell migration and aggregation, peripheral nerve development, growth and process formation; cf. Sage and Bornstein 1991), SCO-spondin may have a wide range of biological activities that have yet to be determined.

Spinal Cord Differentiation

The most striking information from phylogenetic studies is the simultaneous occurrence of RF and a neural tube with a central lumen in urochordates (Olsson 1993). In the last few years, developmental biologists have focussed on the specific genes that drive morphogenetic events. The molecular basis of the neural axis morphogenesis has recently been broadened by the discovery of genes (e.g., *Hox* and *Pax*, McGinnis and Krumlauf 1992; Krumlauf 1994) implicated in the identity of the antero-posterior segmental pattern. In addition, a dorso-ventral longitudinal pattern has been demonstrated, which depends on a ventral signal emanating from the notochord and the floor plate. Patterning of the neural axis is only one facet of the developmental events implicated in the differentiation of the spinal cord. Others may be constituted by neurotrophins (Snider 1994) that help neurons to grow in the developing embryo and ensure their survival in the adult or by recognition molecules (Schachner 1994) and diffusible cues that are able to direct axons. A major discovery in this field is the recent characterization of netrins which, as shown in the chick, are secreted by the floor plate cells and have a chemoattractive effect on the commissural axons of the spinal cord (Kennedy et al. 1994; Serafini et al. 1994).

A huge amount of work remains to be done to elucidate the molecular basis of spinal cord differentiation, including the characterization of axonal guidance factors, their receptors and their distribution. Improving our knowledge of directed axonal growth over long distances is an important challenge in spinal cord differentiation and regeneration (Iwashita et al. 1994; Schnell et al. 1994). The presence of RF in the spinal cord during early embryogenesis has to be considered, and the search for its targets is necessary to define a putative molecular interaction of RF with developmental genes implicated in spinal cord differentiation.

Conclusion and Perspective

A number of substances, including growth factors, cell-adhesion molecules, components of the extracellular matrix and guidance molecules are known to modulate cell-cell and/or cell-matrix interactions during ontogenesis of the CNS and neuritic outgrowth. In this study, we present molecular and functional evidence that SCO-spondin, a novel member of the thrombospondin family present in Reissner's fiber, could also be involved in neuronal development by modulating cell aggregation mechanisms and neuritic outgrowth. Knowledge of the whole SCO-spondin sequence will provide additional information regarding other conserved domains that may contribute to the function of this protein. In addition, new in vitro assays must be developed to examine the molecular interactions that take place between SCO-spondin and neurons during differentiation. Both approaches will help to understand the exact role of SCO-spondin in morphoregulatory processes.

Acknowledgments

By their skillful work and fruitful discussions, all members of the research team "SCO" contributed to this work: R. Meiniel, R. Didier, I. Creveaux, S. Gobron, H. Monnerie, A. Herbet.

References

Adams JC, Tucker RP, Lawler J (1995) The thrombospondin gene family. In: Landes RG (ed) Molecular biology intelligence unit. Springer, Berlin, Heidelberg, New York, pp 1-188

Adams RH, Betz H, Püschel AW (1996) A novel class of murine semaphorins with homology to thrombospondin is differentially expressed during early embryogenesis. Mech Dev 57: 33-45

Bornstein P, Sage H (1994) Thrombospondins. Methods Enzymol 245: 62-85

Chiquet-Ehrismann R (1991) Anti-adhesive molecules of the extracellular matrix. Curr Opin Cell Biol 3: 800-804

Creveaux I, Meiniel R, Gobron S, Dastugue B, Meiniel A (1997) Specific glycoproteins of the bovine SCO: Identification and characterization of corresponding cDNAs. (submitted)

DeFreitas MF, Yoshida CK, Frazier WA, Mendrick DL, Kypta RM, Reichardt LF (1995) Identification of integrin $\alpha3\beta1$ as a neuronal thrombospondin receptor mediating neuritic outgrowth. Neuron 15: 333-343

Del Bigio MR (1995) The ependyma: a protective barrier between brain and cerebrospinal fluid. Glia 14: 1-13

Didier M, Harandi M, Aguera M, Bancel B, Tardy M, Fages C, Calas A, Stagaard M, MØllgard K, Belin MF (1986) Differential immunocytochemical staining for glial fibrillary acidic (GFA) protein, S-100 protein and glutamine synthetase in the rat subcommissural organ, non specialized ventricular ependyma and adjacent neuropil. Cell Tissue Res 245: 343-351

Didier-Bazes M, Chouaf L, Lepetit P, Aguera M, Belin MF (1993) The subcommissural organ of the rat: an in vivo model of neuron-glia interactions. In: Oksche A, Rodríguez EM, Fernandez-Llebrez P (eds) The subcommissural organ: an ependymal brain gland. Springer, Berlin, Heidelberg, New York, pp 161-168

Didier R, Meiniel A, Meiniel R (1992) Monoclonal antibodies as probes for the analysis of the secretory ependymal differentiation in the subcommissural organ of the chick embryo. Dev Neurosci 14: 44-52

Edelman GM, Cunnigham BA, Thiery JP (1990) Morphoregulatory molecules. Wiley, New York, pp 1-648

Faissner A, Kruse J (1990) J1/Tenascin is a repulsive substrate for central nervous system neurons. Neuron 5: 627-637

Feige JJ, Quirin N, Souchelnitskiy S (1996) TGFβ, un peptide biologique sous contrôle: formes latentes et mécanismes d'activation. Médecine/Sciences 12: 929-939

Frazier WA (1991) Thrombospondins. Curr Opin Cell Biol 3: 792-799

Gobron S, Monnerie H, Meiniel R, Creveaux I, Lehmann W, Lamalle D, Dastugue B, Meiniel A (1996) SCO-spondin: a new member of the thrombospondin family secreted by the subcommissural organ is a candidate in the modulation of neuronal aggregation. J Cell Sci 109: 1053-1061

Hecht PM, Anderson KV (1992) Extracellular proteases and embryonic pattern formation. Trends Cell Biol 2: 197-202

Hedgecock EM, Culotti JG, Hall DH (1990) The unc-5, unc-6, and unc-40 genes guide circumferential migrations of pioneer axons and mesodermal cells on the epidermis in *C. elegans*. Neuron 2: 61-85

Hong Ch C, Hashimoto C (1995) An unusual mosaic protein with a protease domain, encoded by the nudel gene, is involved in defining embryonic dorsoventral polarity in *Drosophila*. Cell 82: 785-794

Iwashita Y, Kawaguchi S, Murata M (1994) Restoration of function by replacement of spinal cord segments in the rat. Nature 367: 167-170

Jouet M, Rosenthal A, Mc Farlane J, Kenwrick S A (1993) Missense mutation confirms the L1

defect in X-linked hydrocephalus (HSAS). Nature Genet 4: 331
Karoumi A (1990) Mise en évidence et caractérisation par des méthodes immunologiques de sécrétions glycoprotéiques au niveau du diencéphale dorsal. Thèse d'Université, Clermont-Ferrand II
Kennedy TE, Serafini T, de la Torre JR, Tessier-Lavigne M (1994) Netrins are diffusible chemotropic factors for commissural axons in the embryonic spinal cord. Cell 78: 425-435
Keynes R, Cook G (1990) Cell-cell repulsion: clues from the growth cone? Cell 62: 609-610
Klar A, Baldassare M, Jessell TM (1992) F-spondin: a gene expressed at high levels in the floor plate encodes a secreted protein that promotes neural cell adhesion and neurite extension. Cell 69: 95-110
Korzh VP (1994) Genetic control of early neuronal development in vertebrates. Curr Opin Neurobiol 4: 21-28
Krumlauf R (1994) Hox genes in vertebrate development. Cell 78: 191-201
Lawler J, Hynes RO (1986) The structure of human thrombospondin, an adhesive glycoprotein with multiple calcium-binding sites and homologies with several different proteins. J Cell Biol 103: 1635-1648
Lochter A, Taylor J, Braunewell K-H, Holm J, Schachner M (1995) Control of neuronal morphology in vitro: interplay between adhesive substrate forces and molecular instruction. J Neurosci Res 42: 145-158
Lösecke W, Naumann W, Sterba G (1984) Preparation and discharge of secretion in the subcommissural organ of the rat. Electron-microscopic immunochemical study. Cell Tissue Res 235: 201-206
Lösecke W, Naumann W, Sterba G (1986) Immuno-electron-microscopic analysis of the basal route of secretion in the subcommissural organ of the rabbit. Cell Tissue Res 244: 449-456
McGinnis W, Krumlauf R (1992) Homeobox genes and axial patterning. Cell 68: 283-302
Meiniel A, Molat JL, Meiniel R (1988) Complex-type glycoproteins synthesized in the subcommissural organ of mammals. Light and electron microscopic investigations by use of lectins. Cell Tissue Res 253: 383-395
Meiniel A, Meiniel R, Didier R, Creveaux I, Gobron S, Monnerie H, Dastugue B (1996) The subcommissural organ and Reissner's fiber complex. An enigma in the central nervous system? Prog Histochem Cytochem 30:1-66
Meiniel R, Meiniel A (1985) Analysis of the secretions of the subcommissural organs of several vertebrate species by use of fluorescent lectins. Cell Tissue Res 239: 359-364
Meiniel R, Duchier-Liris N, Molat JL, Meiniel A (1991) The complex-type glycoprotein secreted by the bovine subcommissural organ: an immunological study using C1B8A8 monoclonal antibody. Cell Tissue Res 266: 483-490
Meiniel R, Creveaux I, Dastugue B, Meiniel A (1995) Specific transcripts analysed by in situ hybridization in the subcommissural organ of bovine embryos. Cell Tissue Res 279: 101-107
Miyazaki T, Maruyama M, Yamada G, Hatakeyama M, Taniguchi T (1991) The integrity of the conserved 'WS motif' common to IL-2 and other cytokine receptors is essential for ligand binding and signal transduction. EMBO J 10: 3191-3197
Monnerie H, Boespflug-Tanguy O, Dastugue B, Meiniel, A (1995) Reissner's fiber supports the survival of chick cortical neurons in primary mixed cultures. Cell Tissue Res 282:81-91
Monnerie H, Boespflug-Tanguy O, Dastugue B, Meiniel A (1996) Soluble material from Reissner's fiber displays anti-aggregative activity in primary cultures of chick cortical neurons. Dev Brain Res 96: 120-129
Monnerie H, Dastugue B, Meiniel A (1997) Reissner's fibre promotes aggregation and neuritic outgrowth of chick cortical neuronal cells in culture. Cell Tissue Res 287: 285-295
Oksche A, Rodríguez EM, Fernández-Llebrez P (1993) In: Oksche A, Rodríguez EM, Fernández-Llebrez P (eds) The subcommissural organ: An ependymal brain gland. Springer, Berlin,

Heidelberg, New York, pp 1-333

Olsson R (1993) Reissner's fiber mechanisms: some common denominators. In: Oksche A, Rodríguez EM, Fernández-Llebrez P (eds) The subcommissural organ: an ependymal brain gland. Springer, Berlin, Heidelberg, New York, pp 33-39

Olsson R, Yulis R, Rodríguez EM (1994) The infundibular organ of the lancelet (*Branchiostoma lanceolatum*, Acrania): an immunocytochemical study. Cell Tissue Res 277: 107-114

O'Shea KS, Liu L-H.J, Dixit VM (1991) Thrombospondin and a 140 kd fragment promote adhesion and neurite outgrowth from embryonic central and peripheral neurons and from PC12 cells. Neuron 7: 231-237

Osterhout DJ, Frazier WA, Higgins D (1992) Thrombospondin promotes process outgrowth in neurons from the peripheral and central nervous systems. Dev Biol 150: 256-265

Prater CA, Plotkin J, Jaye D, Frazier WA (1991) The properdin-like type I repeats of human thrombospondin contain a cell attachment site. J Cell Biol 112: 1031-1040

Reissner E (1860) Beiträge zur Kenntnis vom Bau des Rückenmarks von *Petromyzon fluviatilis* L. Arch Anat Physiol 77: 545-588

Rich KA, George IV FW, Law JL and Martin WJ (1990) Cell-adhesion motif in region II of malarial circumsporozoite protein. Science 249: 1574-1577

Rodríguez EM, Oksche A, Hein S, Rodríguez S, Yulis R (1984a) Spatial and structural interrelationships between secretory cells of the subcommissural organ and blood vessels. An immunocytochemical study. Cell Tissue Res 237: 443-449

Rodríguez EM, Oksche A, Hein S, Rodríguez S, Yulis R (1984b) Comparative immunocytochemical study of the subcommissural organ. Cell Tissue Res 237: 427-441

Rodríguez EM, Herrera H, Peruzzo B, Rodríguez S, Hein S, Oksche A (1986) Light and electron-microscopic immunocytochemistry and lectin histochemistry of the subcommissural organ: evidence for processing of the secretory material. Cell Tissue Res 243: 545-559

Sage EH, Bornstein P (1991) Extracellular proteins that modulate cell-matrix interactions. J Biol Chem 266: 14831-14834.

Sarnat HB (1992) Role of human fetal ependyma. Pediatr Neurol 8: 163-178

Schachner M (1994) Neural recognition molecules in diseases and regeneration. Curr Opin Neurobiol 4:726-734

Schnell L, Schneider R, Kolbeck R, Barde YA, Schwab ME (1994) Neurotrophin-3 enhances sprouting of corticospinal tract during development and after spinal cord lesion. Nature 367: 170-173 .

Schoebitz K, Rodríguez EM, Garrido O, Del Brio-Leon MA (1993) Ontogenetic development of the subcommissural organ with reference to the flexural organ. In: Oksche A, Rodríguez EM, Fernández-Llebrez P (eds) The subcommissural organ: an ependymal brain gland. Springer, Berlin, Heidelberg, New York, pp 41-49

Serafini T, Kennedy TE, Galko MJ, Mirzayan C, Jessell TM, Tessier-Lavigne M (1994) The netrins define a family of axon outgrowth promoting proteins with homology to *C. elegans* UNC-6. Cell 78: 409-424

Snider WD (1994) Functions of the neurotrophins during nervous system development: what the knockouts are teaching us. Cell 77:627-638

Sterba G, Fredriksson G, Olsson R (1983) Immunocytochemical investigations of the infundibular organ in amphioxus (*Branchiostoma lanceolatum*; Cephalochordata). Acta Zool (Stockh) 64: 149-153

Sterba G, Kießig C, Naumann W, Petter H, Kleim I (1982) The secretion of the subcommissural organ. A comparative immunocytochemical investigation. Cell Tissue Res 226: 427-439

Sterba G, Kleim I, Naumann W, Petter H (1981) Immunocytochemical investigation of the subcommissural organ in the rat. Cell Tissue Res 218: 659-662

Studnicka FK (1900) Untersuchungen über den Bau des Ependyms der nervösen Centralorgane. Anat Hefte 15: 303-430

Stutinski F (1950) Colloïde, corps de Herring et substance Gomori positive de la neurohypophyse. C R Soc Biol 144: 1357-1360

Takeuchi IK, Kimura R, Matsuda M, Shoji R (1987) Absence of subcommissural organ in the cerebral aqueduct of congenital hydrocephalus spontaneously occurring in MT/HokIdr mice. Acta Neuropathol 73: 320-322

Takeuchi IK, Kimura R Shoji R (1988) Dysplasia of subcommissural organ in congenital hydrocephalus spontaneously occurring in Cws/Idr rats. Experientia 44: 338-340

Tucker RP (1993) The in situ localization of tenascin splice variants and thrombospondin 2 mRNA in the avian embryo. Development 117: 347-358.

Van Camp G, Vits L, Coucke P, Lyonnet S, Schrander-Stumpel C, Darby J, Holden J, Munnich A, Willems J (1993) A duplication in the L1CAM gene associated with X-linked hydrocephalus. Nat Genet 4: 421-425

Wislocki GB, Leduc EH (1952) The cytology and histochemistry of the subcommissural organ and Reissner's fiber in the rodents. J Comp Neurol 97: 515-544

22 Development and Differentiation of Blood Vessels in the Central Nervous System

J. Wilting

Anatomisches Institut II der Albert-Ludwigs-Universität,
Albertstraße 17, D-79104 Freiburg, Germany;
Phone: +49 761 203 5087, Fax: +49 761 203 5091,
E-mail: wilting@ruf.uni-freiburg.de

Abstract

The central nervous system (CNS) develops from a pseudostratified ectodermal epithelium containing neuroblasts and glioblasts. Other constituents (microglia, blood vessels) are of mesodermal origin and successively invade the neuroectoderm. Using chick-quail chimeras it is possible to study the interaction between neuroectodermal and mesodermal cells. Vascular endothelial cells start invading the CNS of birds at about day 3.5 of development. They originate from the paraxial mesoderm of the head and the trunk. Thereafter, smooth muscle cells migrate along the endothelial routes. Neuroectodermal cells secrete vascular endothelial growth factor (VEGF), which is a highly specific angiogenic and chemoattractive factor. Angioblast and endothelial cells in the paraxial mesoderm are characterized by the expression of VEGF-receptor-2. Except for the choroid plexus, VEGF and VEGF receptors are not expressed in the adult brain. The organ-typical differentiation of endothelial cells in the CNS depends on interactions with local neuroectodermal cells. Development of blood-brain barrier characteristics are obviously due to inductive signals from astrocytes. In contrast, the epithelial cells of the choroid plexus induce development of highly permeable, fenestrated capillaries. Constitutive expression of VEGF and its receptors in the choroid plexus (and the kidney glomeruli) may serve as the basis for high permeability. VEGF has been shown to increase vascular permeability in a highly potent manner.

Origin of CNS Vessels

As a result of inductive signals by the notochord, the ectoderm gives rise to the pseudostratified epithelium of the neural plate, successively being remodelled into the neural tube. During a relatively long embryonic period, the neural tube remains avascular. The first blood vessels of the neural tube have been suggested to arise from intraneural cells (Strong 1961), but earlier reports, such as that by Feeney and Watterson (1946), had already correctly described sprouting of perineural vessels into the neural tube.

Embryonic blood vessels are at first made up of only one cell type, the endothelial cell (for review, see Wilting and Christ 1996). Originating from primitive arterial tracts, endothelial cells sprout into the ventral aspect of the avian neural tube, on either side of the floor plate (Noden 1991; Kurz et al. 1996), and possess filopodia up to 100 μm in length. They are found in a random distribution, but after onset of perfusion, the intraneural vascular pattern becomes more regular (Kurz et al. 1996).

The blood vessels of the spinal cord and the brain are derived from the paraxial mesoderm of the head and the trunk (Noden 1991; Wilting et al. 1995a; Couly et al. 1995). This has been demonstrated with the chick-quail chimera technique by homotopical grafting of paraxial mesoderm (Fig. 1). The preotic paraxial head mesoderm gives rise to the vessels of the fore- and midbrain (Noden 1991), the five postotic occipital somites give rise to vessels of the hindbrain (Wilting et al. 1995b), and more caudal somites form the vessels of the spinal cord (Wilting et al. 1995a). Initially, the unsegmented and segmented (somitic) paraxial mesoderm is avascular, but precursors of endothelial cells, angioblasts, can already be observed in this tissue. Angioblasts are characterized by the expression of vascular endothelial growth factor receptor (VEGFR)-2 (Fig. 2; Eichmann et al. 1993; Wilting et al. 1997). Targeted deletion of the VEGFR-2 gene results in the development of mice which do not possess any endothelial cells and blood islands. These mice die very early during development (Shalaby et al. 1995). Mice carrying a targeted deletion of the VEGFR-1 gene possess abnormal blood vessels and also die early during development (Fong et al. 1995). VEGF is secreted by a great variety of different cell types during development (Breier et al. 1992; Flamme et al. 1995). In the developing brain, VEGF is expressed in cells of the ependymal layer, while the receptors are found in endothelial cells (Breier et al. 1992; Millauer et al. 1993). In the neural tube, strongest expression of VEGF can be observed in ependymal cells of the ventral aspect (Fig. 3), which is the region where the first blood vessels sprout into the neural tube. The expression data strongly suggest that embryonic angiogenesis is mediated by a paracrine mechanism, and the expression of VEGF and its receptors corresponds spatially and temporally with the phases of vascular development. Except for the choroid plexus, expression of both ligand and receptors is not detectable by means of in situ hybridization in the adult brain (Breier et al. 1992), but is easily detectable in human gliomas (Plate et al. 1992). Experimental studies employing differentiated avian chorioallantoic membrane have shown that VEGF induces angiogenesis but not lymphangiogenesis (Wilting et al. 1992, 1996). VEGFR-3 which binds a lymphangiogenic growth factor, VEGF-C, with high affinity (Joukov et al. 1996; Oh et al. 1997), becomes down-regulated in endothelial cells of the CNS vessels during very early stages of development (Wilting et al. 1997). This may be the reason for the fact that the CNS does not possess lymphatics.

In summary, VEGF, a highly specific vascular endothelial cell mitogen and angiogenic growth factor, induces growth of blood vessels in the CNS during normal development and under pathological conditions. During normal development, vascular endothelial cells of the CNS are derived from the paraxial mesoderm, located on either side of the neural tube.

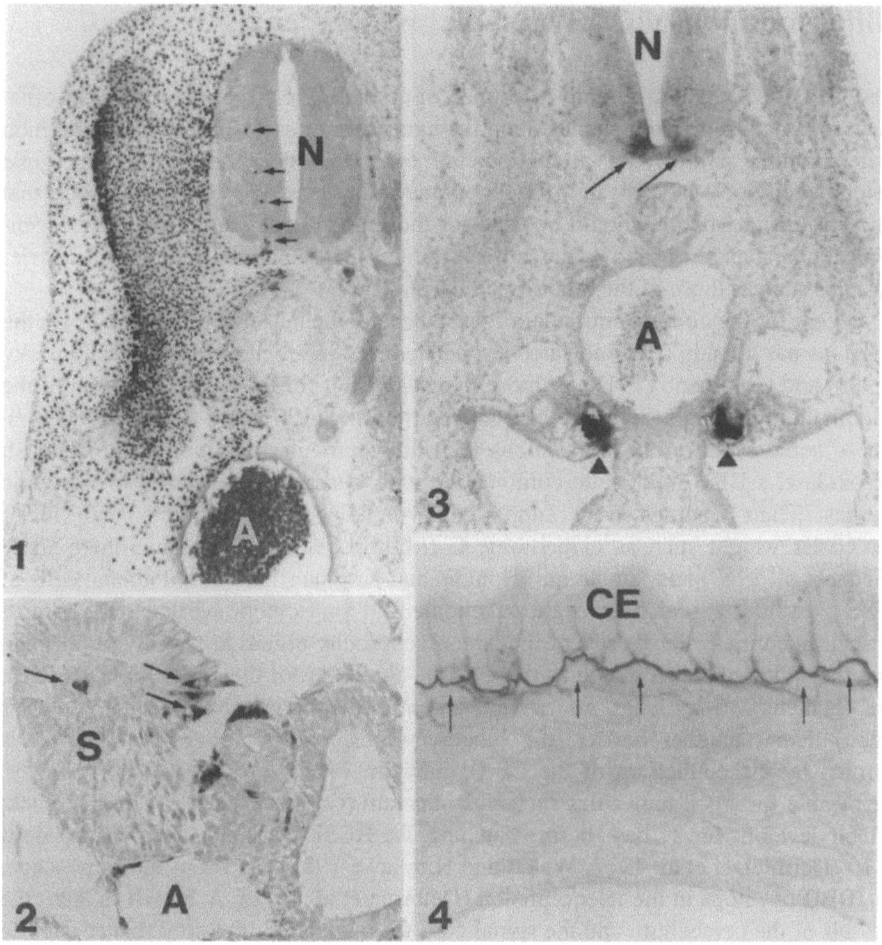

Fig. 1. Homotopical grafting of a thoracic somite of a quail embryo into a chick embryo. Reincubation 3 days and staining of quail cells with an anti-quail antibody. Note quail endothelial cells (*arrows*) in the neural tube (*N*) of the chick embryo. *A* Aorta. x44. **Fig. 2.** In situ hybridization of a 2-day-old quail embryo with a VEGFR-2 probe. Note expression in endothelial cells of the aorta (*A*) and in angioblasts (*arrows*) of the somite (*S*). x175. **Fig. 3.** In situ hybridization of a 3-day-old quail embryo with a VEGF probe. Note expression in ependymal cells (*arrows*) in the ventral aspect of the neural tube (*N*), and in the kidney tubuli (*triangles*). *A* Aorta. x44. **Fig. 4.** Induction of highly permeable capillaries. The prospective epithelium of the choroid plexus of a quail embryo was grafted into the body wall of a chick embryo. Reincubation 14 days and i.v. injection of horseradish peroxidase 45 min before fixation. The chick capillaries (*arrows*) in the vicinity of the quail choroidal epithelium (*CE*) are highly permeable for peroxidase (dark reaction product). x740

Differentiation of CNS Vessels

Endothelial cells that are lining the lumen of an embryonic blood vessel have often been regarded as differentiated. Lining the lumen of a vessel is, however, a common differentiation pathway of all endothelial cells, and does not reflect the great heterogeneity of these cells in different organs and tissues. We therefore suggest that the term differentiation should be used for those endothelial cells that show organ-typical characteristics such as pores, fenestrations, cell junctions, enzymes or carrier systems such as those of the blood-brain barrier (BBB).

Except for the circumventricular organs, the endothelial cells of the CNS form the BBB, as has already been indicated by the studies of Ehrlich (1885), Lewandowsky (1900) and Goldmann (1913). A morphological proof for the barrier function of the endothelium was published by Reese and Karnowsky (1967) and Brightman (1967). These authors performed intravenous and intraventricular injection of horseradish peroxidase, a BBB marker, in combination with transmission electron microscopic studies. Their findings were supported by studies using markers with lower molecular weight such as cytochrome C (Milhorat et al. 1973), microperoxidase (Feder et al. 1969) and lanthanum (Bouldin and Krigman 1975). Endothelial cells of CNS vessels are sealed by tight junctions, and transcytotic vesicles are almost absent. In contrast, the vessels of the circumventricular organs are highly permeable, and the endothelial cells, such as those of the capillaries of the choroid plexus (CP), are characterized by fenestrations with a central diaphragm (Meller and Wechsler 1965). Here, another barrier, the blood-cerebrospinal fluid barrier (BCSFB) is formed by the epithelium of the CP (Brightman 1967). This barrier is, however, permeable for small molecules such as lanthanum (Castel et al. 1974). In the chick, which develops for 21 days before hatching, the BCSFB becomes tight during days 9-10 (Dermietzel et al. 1977; Wakai and Hirokawa 1981). At about the same stage, the BBB develops in the telencephalon (Delorme et al. 1970). A few days later, the vessels of the cerebellum and the spinal cord form the BBB, the grey matter slightly preceding the white (Wakai and Hirokawa 1978). This shows that differentiation of vessels takes place about 6-9 days after the first vascular sprouts have migrated into the CNS of birds. In mammalian species, the BBB to endogenous albumin seems to be present from the earliest embryonic stages onward (for review see: Saunders 1992).

It has been a matter of debate whether the vessels of the CNS are derived from a specialized subpopulation of endothelial cells, or whether differentiation of these cells is dependent on local interactions with neuroectodermal cells. There is increasing evidence that the latter holds true. This has been shown most clearly in interspecific grafting experiments using quail and chick embryos. Stewart and Wiley (1981) grafted parts of avascular quail neural tubes to ectopical sites of chick embryos and found that the vessels in the grafts formed typical features of the BBB. They did not isolate the neural tissue from the surrounding mesenchyme, which contains endothelial cells that would have normally formed the BBB. However, we were able to confirm their results using isolated neural tissue (Wilting 1988).

We have furthermore studied the development of the CP and found that the choroidal epithelium induces the formation of fenestrated capillaries that are highly permeable for i.v. injected horseradish peroxidase (Wilting and Christ 1989). This has been demonstrated by grafting prospective choroidal epithelium of quail embryos into the body wall of chick embryos. The vasculature of the body wall, which does not form CNS vessels during normal development, gives rise to CP-typical vessels under these experimental conditions (Fig. 4). In the brain, except for the circumventricular organs, the astrocytes seem to be responsible for the induction of BBB characteristics in endothelial cells. The close spatial relationship of these two cells types, and also in vivo and in vitro studies, seem to support this hypothesis (De Bault and Cancilla 1980; Janzer and Raff 1987; Meyer et al. 1991). These experimental studies have been challenged by Holash et al. (1993); according to them, the in vitro studies suffer from the use of insufficient BBB markers such as alkaline phosphatase, and from the fact that mostly brain-derived endothelial cells have been used. These cells had been induced to form BBB characteristics before culturing. The in vivo grafting of astrocytes also has to be interpreted with care, since such grafts become only poorly vascularized or not at all (Holash et al. 1993). Nevertheless, the quail-chick chimera studies have clearly demonstrated that differentiation of endothelial cells in the CNS is contingent upon local interactions.

The molecular mechanisms of these interactions are only poorly understood (for review, see Risau 1994). However, constitutive expression of VEGF and its receptors may serve as a basis for high permeability of the capillaries in the CP. As shown above, expression of VEGF coincides with angiogenesis under physiological and pathological conditions. There seem to be only two organs that express VEGF and its receptors in the adult; the CP and the kidney glomeruli (Breier et al. 1992). It has been suggested that this may be important for the induction of fenestrated capillaries in the two organs; but we have to be aware of the fact that the ultrastructure of fenestrations is different in the CP and the glomeruli, as they do or do not possess a diaphragm, respectively. However, both organs are characterized by high ultrafiltration rates. VEGF has originally been isolated from tumor tissue and characterized because of its ability to induce high vascular permeability in the Miles assay. It has therefore been called vascular permeability factor (Senger et al. 1983). It appears possible that under physiological conditions VEGF induces high vascular permeability in specialized organs such as the CP and kidney glomeruli, but further investigations are needed.

In summary, the ependymo-glial elements induce organ-typical differentiation of CNS vessels during development. VEGF secreted by the epithelium of the CP may serve as a basis for high permeability of the capillaries in this organ.

Acknowledgments

We are grateful to Mrs. U. Pein, Mr. G. Frank, Mrs. S. Antoni, Mrs. L. Koschny and Mrs. M. Schüttoff for their excellent technical assistance, to Mrs. C. Micucci for photographic work and to Mrs. U. Uhl for typing the manuscript. Supported by a grant (Wi 1452/1-1) from the Deutsche Forschungsgemeinschaft.

References

Bouldin TW, Krigman MR (1975) Differential permeability of cerebral capillary and choroid plexus to lanthanum ion. Brain Res 99: 444-448

Breier G, Albrecht U, Sterrer S, Risau W (1992) Expression of vascular endothelial growth factor during embryonic angiogenesis and endothelial cell differentiation. Development 114: 521-532

Brightman MW (1967) Intracerebral movements of proteins injected into the blood and cerebrospinal fluid. Anat Rec 157: 219

Castel M, Sahar A, Erlij D (1974) The movement of lanthanum across diffusion barriers in the choroid plexus of the cat. Brain Res 67: 178-184

Couly G, Coltey P, Eichmann A, Le Douarin NM (1995) The angiogenic potentials of the cephalic mesoderm and the origin of brain and head blood vessels. Mech Dev 53: 97-112

De Bault LE, Cancilla PA (1980) γ-glutamyl transpeptidase in isolated brain endothelial cells: induction by glial cells in vitro. Science 207: 653-655

Delorme P, Gayet J, Grignon G (1970) Ultrastructural study on trans-capillary exchange in the developing telencephalon of the chicken. Brain Res 22: 269-283

Dermietzel R, Meller K, Tetzlaff W, Waelsch M (1977) In vivo and in vitro formation of the junctional complex in choroid epithelium. A freeze-etching study. Cell Tissue Res 181: 427-441

Ehrlich P (1885) Das Sauerstoff-Bedürfnis des Organismus. Eine farbenanalytische Studie. Habilitationsschrift. A. Hirschwald, Berlin

Eichmann A, Marcelle C, Bréant C, Le Douarin NM (1993) Two molecules related to the VEGF receptor are expressed in early endothelial cells during avian embryonic development. Mech Dev 42: 33-48

Feder N, Reese TS, Brightman MW (1969) Microperoxidase, a new tracer of low molecular weight. A study of the interstitial compartements of the mouse brain. J Cell Biol 43: 35a-36a

Feeney JF jr, Watterson RL (1946) The development of the vascular pattern within the walls of the central nervous system of the chick embryo. J Morphol 78: 231-304

Flamme I, Breier G, Risau W (1995) Vascular endothelial growth factor (VEGF) and VEGF receptor 2 (flk-1) are expressed during vasculogenesis and vascular differentiation in the quail embryo. Dev Biol 169: 699-712

Fong GH, Rossant J, Gertsenstein M, Breitman ML (1995) Role of the Flt-1 receptor tyrosine kinase in regulating the assembly of vascular endothelium. Nature 376: 66-70

Goldmann EE (1913) Vitalfärbung am Zentralnervensystem. Abh Preuss Akad. Wiss Phys-Math 1: 1-60

Holash JA, Noden DM, Stewart PA (1993) Re-evaluating the role of astrocytes in blood-brain barrier induction. Dev Dynam 197: 14-25

Janzer RC, Raff MC (1987) Astrocytes induce blood-brain barrier properties in endothelial cells. Nature 325: 253-257

Joukov V, Pajusola K, Kaipainen A, Chilov D, Lahtinen I, Kukk E, Saksela O, Kalkkinen N, Alitalo K (1996) A novel vascular endothelial growth factor, VEGF-C, is a ligand for the Flt4 (VEGFR-3) and KDR (VEGFR-2) receptor tyrosine kinases. EMBO J 15: 290-298

Kurz H, Gärtner T, Eggli PS, Christ B (1996) First blood vessels in the avian neural tube are formed by a combination of dorsal angioblast immigration and ventral sprouting of endothelial cells. Dev Biol 173: 133-147

Lewandowsky M (1900) Zur Lehre von der Cerebrospinalflüssigkeit. Z Klin Med 40: 480-494

Meller K, Wechsler W (1965) Elektronenmikroskopische Untersuchungen der Entwicklung der telencephalen Plexus choroides des Huhnes. Z Zellforsch 65: 420-444

Meyer J, Rauti J, Galla HJ (1991) The susceptibility of cerebral endothelial cells to astroglial induction of blood-brain barrier enzymes depends on their proliferate state. J Neurochem 57: 1971-1977

Milhorat TH, Davis DA, Lloyd BT jr (1973) Two morphologically distinct blood-brain barriers preventing entry of cytochrome C into cerebrospinal fluid. Science 180: 76-78

Millauer B, Wizigmann-Voos S, Schnürch H, Martinez R, Moller NPH, Risau W, Ullrich A (1993) High affinity VEGF binding and developmental expression suggest flk-1 as a major regulator of vasculogenesis and angiogenesis. Cell 72: 835-846

Noden DM (1991) Origins and patterning of avian outflow tract endocardium. Development 111: 867-876

Oh S-J, Jeltsch MM, Birkenhäger R, McCarthy JEG, Weich HA, Christ B, Alitalo K,Wilting J (1997) VEGF and VEGF-C: specific induction of angiogenesis and lymphangiogenesis in the differentiated avian chorioallantoic membrane. Dev Biol 188: 96-109

Plate KH, Breier G, Weich HA, Risau W (1992) Vascular endothelial growth factor is a potential tumour angiogenesis factor in human gliomas in vivo. Nature 359: 845-848

Reese TS, Karnowsky MJ (1967) Fine structural localisation of a blood brain barrier to exogenous peroxidase. J Cell Biol 34: 207-217

Risau W (1994). Molecular biology of blood-brain-barrier ontogenesis and function. (Review). Acta Neurochirurgica (Suppl) 60: 109-112

Saunders NR (1992) Ontogenic development of brain barrier mechanisms. In: Bradbury MWB (ed) Handbook of experimental pharmacology. Springer, Berlin Heidelberg New York, pp 327-369

Senger DR, Galli SJ, Dvorak AM, Perruzzi CA, Harvey VS, Dvorak HF (1983) Tumor cells secrete a vascular permeability factor that promotes accumulation of ascites fluid. Science 219: 983-985

Shalaby F, Rossant J, Yamaguchi TP, Gertsenstein M, Wu XF, Breitman ML, Schuh AC (1995) Failure of blood-island formation and vasculogenesis in flk-1 deficient mice. Nature 376: 62-66

Stewart PA, Wiley MJ (1981) Developing nervous tissue induces formation of blood-brain barrier characteristics in invading endothelial cells. Dev Biol 84: 183-192

Strong LH (1961) The first appearance of vessels within the spinal cord of the mammal: their developing patterns as far as partial formation of the dorsal septum. Acta Anat 44: 80-108

Wakai S, Hirokawa N (1978) Development of the blood-brain barrier to horseradish peroxidase in the chick embryo. Cell Tissue Res 195: 195-203

Wakai S, Hirokawa N (1981) Development of blood-cerebrospinal fluid barrier to horseradish peroxidase in the avian choroidal epithelium. Cell Tissue Res 214: 271-278

Wilting J (1988) Ultrastrukturelle und funktionelle Untersuchungen der Entwicklung des Plexus choroideus bei Vogelchimären. Dissertationsarbeit, Ruhr-Universität Bochum

Wilting J, Christ B (1989) An experimental and ultrastructural study on the development of the avian choroid plexus. Cell Tissue Res 255: 487-494

Wilting J, Christ B (1996) Embryonic angiogenesis: a review. Naturwissenschaften 83: 153-164

Wilting J, Christ B, Weich HA (1992) The effects of growth factors on the day 13 chorioallantoic membrane (CAM): a study of $VEGF_{165}$ and PDGF-BB. Anat Embryol 186: 251-257

Wilting J, Brand-Saberi B, Huang R, Zhi Q, Köntges G, Ordahl CP, Christ B (1995a) Angiogenic potential of the avian somite. Dev Dynam 202: 165-171

Wilting J, Ebensperger C, Müller TS, Koseki H, Wallin J, Christ B (1995b) Pax1 in the development of the cervico-occipital transitional zone. Anat Embryol 192: 221-227

Wilting J, Birkenhäger R, Eichmann A, Kurz H, Martiny-Baron G, Marmé D, McCarthy JEG, Christ B, Weich HA (1996) $VEGF_{121}$ induces proliferation of vascular endothelial cells and expression of flk-1 without affecting lymphatic vessels of the chorioallantoic membrane. Dev Biol 176: 76-85

Wilting J, Eichmann A, Christ B (1997) Expression of the avian VEGF receptor homologues Quek1 and Quek2 in blood-vascular and lymphatic endothelial and non-endothelial cells during quail embryonic development. Cell Tissue Res 288: 207-223

23 Melatonin Inhibits the PACAP-Induced Phosphorylation of the Transcription Factor CREB in the Rat Suprachiasmatic Nucleus

M. Kopp[1, 2], H. Meissl[2], and H.-W. Korf[1]

[1]Dr. Senckenbergische Anatomie, Anatomisches Institut II, Klinikum der Johann Wolfgang Goethe-Universität, Frankfurt am Main;
[2]Max-Planck-Institut für Physiologische und Klinische Forschung, W.G. Kerckhoff-Institut, Bad Nauheim, Germany; Phone: +49 6032 705272; Fax: +49 6032 705211;
E-mail: michael@kerckhoff.mpg.de

Abstract

The mammalian hypothalamic suprachiasmatic nucleus (SCN) is an endogenous pacemaker generating circadian rhythms. SCN activity is synchronized with environmental light/dark cycles by photic information primarily transmitted via the retinohypothalamic tract (RHT). One putative neurotransmitter/neuromodulator of the RHT is the pituitary adenylate cyclase-activating polypeptide (PACAP), which was shown to phase-shift the activity of SCN neurons during the day. In the SCN, the transcription factor cAMP response element binding protein (CREB) seems to be an important mediator between photic input and downstream transcriptional events. The pineal hormone melatonin, whose synthesis and release is under the control of the pacemaker, feeds back to the SCN, acting on high-affinity receptors. Like PACAP, melatonin is capable of affecting the biological clock by the phase-shifting of circadian rhythms. Using a brain slice technique, immunocytochemistry and immunochemistry we demonstrate that in the rat suprachiasmatic nucleus (1) CREB phosphorylation is induced during late subjective day by forskolin (10 μM) or depolarization with 60 mM KCl, resulting in an increase of intracellular concentrations of the second messengers cAMP or Ca^{2+}, (2) PACAP can induce CREB phosphorylation during this time, and (3) melatonin inhibits the PACAP-induced phosphorylation. Our data show the possibility of CREB involvement in processing of light (or darkness) information even during subjective day. The potency of PACAP, one potential neuromodulator of the RHT, to affect gene expression in the SCN via activation of a transcription factor may give an insight into the rhythm entrainment of the oscillator, whereas the antagonistic effect of melatonin may mirror a feedback loop within the circadian system.

Introduction

The hypothalamic suprachiasmatic nucleus (SCN) is the primary endogenous pacemaker of mammals which generates circadian rhythms with a period length near to, but significantly different from 24 h (Klein et al. 1991). The SCN activity is synchronized with the precise 24-h period of a day by photic inputs that regularly reset the clock and serve as a major natural phase-shifting stimulus (Zeitgeber). The SCN receives photic information directly from the retina via the retinohypothalamic tract (RHT) and also via the intergeniculate leaflet of the thalamus, the medial part of the pretectum and the raphe nuclei. Light stimuli that phase-shift circadian rhythms in the SCN (i.e. during subjective night) induce the phosphorylation of the transcription factor cAMP response element binding protein (CREB; Ginty et al. 1993; Stehle et al. 1996; Ding et al. 1997) and the transient expression of immediate early genes (IEGs; Rusak et al. 1990; Takahashi 1995; Wollnik et al. 1995). Moreover, phase-shifting of SCN activity is linked to gene transcription and requires protein synthesis (Inouye et al. 1988; Takahashi et al. 1993). CREB seems to be involved in processing light (or darkness) information and CREB phosphorylation/activation is mediated by the second messengers Ca^{2+} as well as cAMP (Gonzalez and Montminy 1989; Sheng et al. 1990, 1991; Golombek and Ralph 1995; Kako et al. 1996, 1997). The established afferent neurotransmitters, which convey photic information to the SCN have all been shown to adjust the clock by phase-shifting the circadian rhythms of the neuronal activity in vitro (Inouye and Shibata 1994). Certainly, none of them elicits a strong stimulation of adenylate cyclase (AC) activity, even though a phase-shifting effect has been shown by activation of the AC and subsequent elevation of the intracellular concentration of cyclic AMP (Prosser and Gillette 1989).

Pituitary adenylate cyclase-activating polypeptide (PACAP) is a peptide shown to stimulate adenylate cyclase and consequently enhances the intracellular cAMP concentration. The peptide is a member of the secretin/glucagon/vasoactive intestinal peptide (VIP) family and exists in two α-amidated forms, PACAP38 and PACAP27, which share the same N-terminal 27 amino acids and arise from a precursor peptide by post-translational processing (Miyata et al. 1989, 1990). Recently, PACAP has been suggested as a putative neurotransmitter/neuromodulator of the RHT, since (1) intense PACAP-immunoreactivity has been observed in retinal afferents of the rat SCN, (2) the expression of a PACAP receptor has been demonstrated in the SCN, and (3) a phase-shifting effect of PACAP in the SCN has been shown (Hashimoto et al. 1996; Hannibal et al. 1997).

Apart from photic inputs, multiple other mechanisms control the pacemaker. One is melatonin which is produced in the pineal organ with high levels during the dark-phase and low levels during the light-phase of the diurnal cycle. This rhythmicity is controlled by the endogenous pacemaker, whose efferents innervate the pineal via a multisynaptic pathway. Thus, the hormone forms a feedback loop to the SCN and is capable of affecting the biological clock by phase-shifting of circadian rhythms (Arendt 1995). These effects are mediated through high-affinity membrane receptors that are linked to an inhibitory GTP-binding regulatory protein (Reppert et al. 1994). The physiological effects of melatonin in the SCN are circadian time (CT) - dependent with maximal effects near dusk, i.e., at CT 9 - 11 (Cassone et al. 1987; Shibata et al. 1989; Stehle et al. 1989; McArthur et al. 1991; Tenn and Niles 1993; Gauer et al. 1993). Although there exist numerous data concerning melatonin's effects in the SCN, the signal transduction cascades affected by melatonin have not been clarified in detail.

In the present study we investigated in vitro, whether (1) the transcription factor CREB may be phosphorylated in the SCN during late subjective day, (2) PACAP is capable of eliciting this response, and (3) melatonin has an antagonistic effect on this phosphorylation. Due to the fact that enhanced intracellular cAMP concentrations induce phase-shifts of circadian rhythms during the day and maximal effects of melatonin are observed in the SCN near dusk, all experiments were performed during this time of day (CT 9 - 11).

Materials and Methods

Male adult Wistar rats were maintained under a photoperiod of 12:12 L/D (lights on at 7.00 h) with free access to food and water. Animals were killed by decapitation between CT 7 - 8 and coronal hypothalamic brain slices (500 µm thick) containing both SCNs, optic chiasm, third ventricle and surrounding tissue were sectioned using a tissue chopper. The slices were cultured in artificial cerebrospinal fluid (aCSF, in mM: NaCl 124.0; KCl 5.0; KH_2PO_4 1.25; $MgSO_4$ 1.3; $NaHCO_3$ 26.0; $CaCl_2$ 2.2; Glucose 10.0) adjusted to pH 7.4 and saturated with 95% O_2 and 5% CO_2 at 36.5 °C. After a period of equilibration (approximately 120 min), one group of slices was incubated with aCSF containing 60 mM KCl, 100 nM PACAP or 10 µM forskolin for 30 minutes at the end of the subjective day (CT 10 - 10.5; Table 1). A second group of slices was incubated in aCSF containing melatonin (1 nM or 1 µM) for 30 min (CT 9.5 - 10) and then treated with aCSF containing 100 nM PACAP plus melatonin for 30 min (CT 10 - 10.5). A third group of slices remained untreated and served as controls.

Following the incubation, the slices were fixed with 4% paraformaldehyde in phosphate buffer overnight, cryoprotected with 30% sucrose, frozen and cut into cryostat sections (14 - 20 µm thick). Immunoreactive p-CREB was detected with a p-CREB antibody specifically recognizing the phosphorylated form of CREB. Briefly, the sections were treated with methanol containing 0.45% peroxide to block the endogenous peroxidase and preincubated with 10% normal goat serum to reduce non-specific staining. Subsequently, sections were incubated overnight with the p-CREB antibody, diluted 1:1000 in phosphate-buffered saline containing 1% bovine serum albumin, with biotin-conjugated anti-rabbit IgG as

Table 1. Incubation protocol of the experiments

	Preincubation *(CT 7 - 9.5)*	Treatment *(CT 9.5 - 10)*	Treatment *(CT 10 - 10.5)*
Group 1	aCSF	aCSF	60 mM KCl 10 μM forskolin 100 nM PACAP
Group 2	aCSF	1 nM melatonin or 1 μM melatonin	1 nM melatonin + 100 nM PACAP or 1 μM melatonin + 100 nM PACAP
Group 3	aCSF	aCSF	aCSF

the second antibody and with peroxidase-conjugated avidin. Finally, the immunoreactions were visualized using diaminobenzidine (DAB) as a coloring reagent. For gel electrophoresis and immunoblotting, the SCN was punched out of the slices and these preparations were homogenized and treated as described (Schomerus et al. 1996). Polyclonal antibodies against total CREB and p-CREB were diluted 1:1000.

Results

After a period of 2 h in organ culture, the SCN showed no p-CREB immunoreactivity (IR) in untreated brain slices (Fig. 1a). Depolarization with 60 mM KCl, which increases the intracellular calcium ion concentration, $[Ca^{2+}]_i$, probably by increasing Ca^{2+} influx through voltage-sensitive channels, leads to phosphorylation of CREB in cells of the SCN (Fig. 1b). Treatment with forskolin (10 μM), which directly activates the adenylate cyclase and, thus, enhances the cAMP concentration also induces nuclear p-CREB IR (Fig. 1c, d).

Stimulation of hypothalamic brain slices with PACAP at a concentration of 100 nM, the concentration previously shown to elicit CREB phosphorylation in rat pinealocytes (Schomerus et al. 1996), induces the phosphorylation of CREB in the nuclei of cells of the SCN when administered at CT 10 (Fig. 1e, f). Immunoblotting shows that this IR is due to phosphorylation of a single protein band at 43 kDa obviously corresponding to CREB (Fig. 2). In contrast, PACAP application during the early night (CT 14) was ineffective (data not shown).

Application of the pineal hormone melatonin, which has been shown to inhibit adenylate cyclase in the pituitary, shows a remarkable reduction of the PACAP-induced phosphorylation of CREB in the SCN. Simultaneous incubation of PACAP and melatonin leads to a stronger suppression of p-CREB IR at a physiological concentration of 1 nM melatonin as compared to 1 μM (Fig. 1h, j).

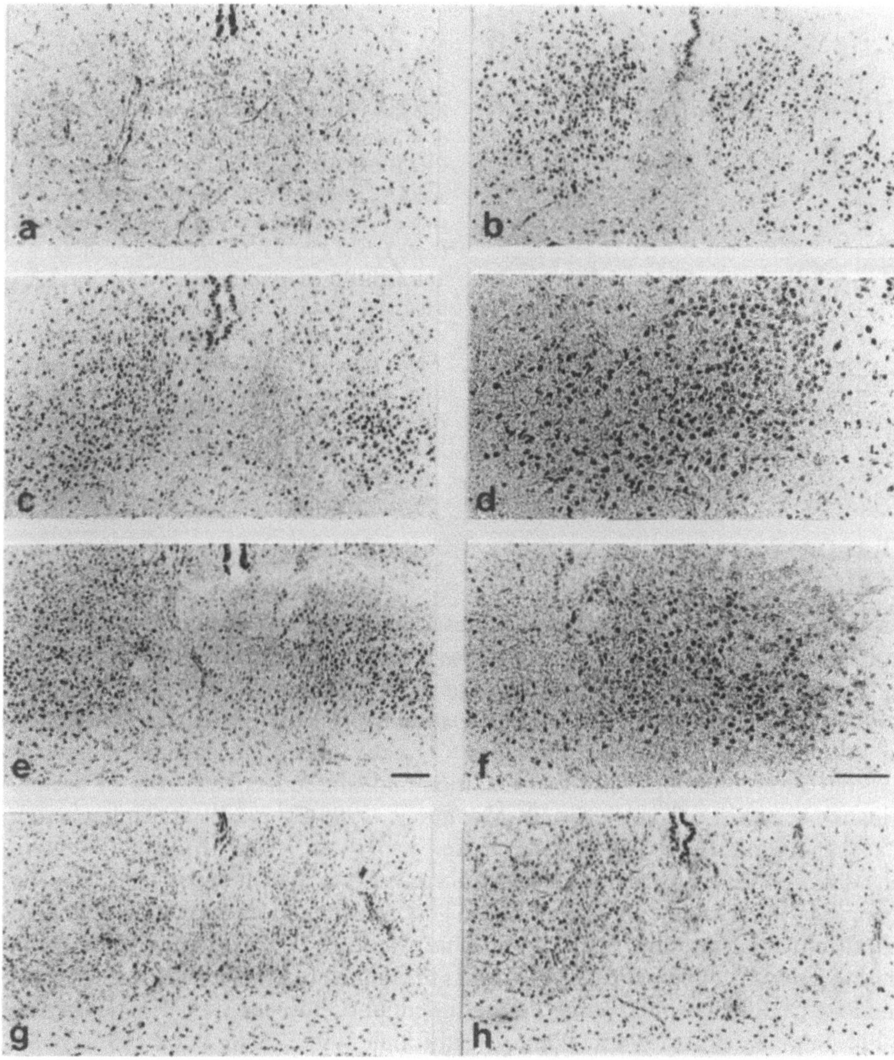

Fig. 1. p-CREB immunoreactivity in the rat suprachiasmatic nucleus. Cross sections through hypothalamic brain slices: untreated (**a**), depolarized with 60 mM KCl (**b**), treated with 10 μM forskolin (**c, d**), 100 nM PACAP (**e, f**), 100 nM PACAP plus 1 nM melatonin (**g**), 100 nM PACAP plus 1 μM melatonin (**h**). *Bars* 100 μm, the *bar* in (**e**) also applies to (**a, b, c, g, h**), the *bar* in (**f**) also to (**d**)

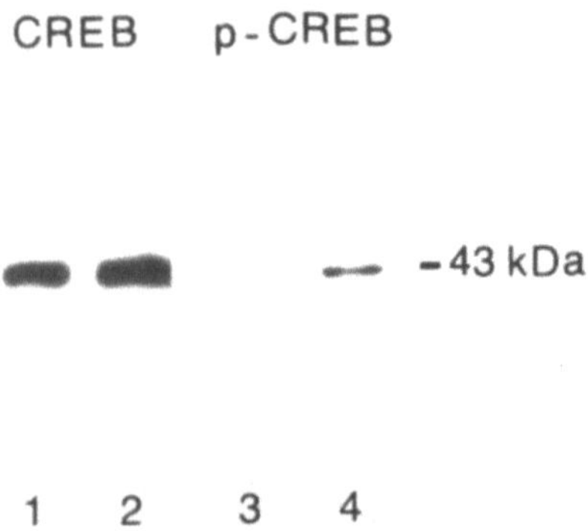

Fig. 2. Gel electrophoresis and immunoblotting of SCN punches using antibodies against total CREB (**lanes 1** and **2**) and p-CREB (**lanes 3** and **4**). The total CREB antibody detects a single protein band of 43 kDa in both control (**lane 1**) and PACAP-treated (**lane 2**) slices. The p-CREB antibody detects a single protein band of 43 kDa in PACAP-treated (**lane 4**) slices but not in controls (**lane 3**). Samples loaded on **lane 1** plus **3** and **lane 2** plus **4** were from the same preparations. (Reprinted from Kopp et al. 1997, with kind permission from Elsevier Science Ireland Ltd.)

Discussion

Previous studies have shown that the transcription factor CREB is phosphorylated in the SCN in vivo by light pulses at times when light is capable of phase-shifting circadian rhythms, i.e., during subjective night (Ginty et al. 1993; Stehle et al. 1996; Ding et al. 1997). This phosphorylation is probably mediated by activation of two different transduction cascades: (1) activation of the adenylate cyclase causing an increase in cAMP concentration and subsequent activation of protein kinase A (PKA; Gonzalez and Montminy 1989), or (2) activation of a Ca^{2+}/calmodulin-dependent protein kinase by elevated levels of $[Ca^{2+}]_i$ (Sheng et al. 1991). Here we demonstrate that in vitro direct activation of the adenylate cyclase by forskolin or elevation of $[Ca^{2+}]_i$ by membrane depolarization with high potassium concentrations leads to phosphorylation of CREB in the rat SCN even during late subjective day (CT 10). This is in contrast to previous reports, showing induced CREB phosphorylation during the night (Ginty et al. 1993; Ghosh et al. 1994; Sumova et al. 1994; Golombek and Ralph 1995; Zhang et al. 1996; Stehle et al. 1996; Ding et al. 1997). Our data are consistent with the results of Kako et al. (1997), who describe that photic-induced CREB DNA-binding activity occurs at all circadian times tested, indicating that CREB activation in the SCN is not gated by the endogenous clock. The phosphorylation of CREB is required to activate the transcription of genes containing CREs in their promotor regions like the IEGs c-fos and jun-B. These peptides are transcription factors themselves, and they are considered to be involved in the transcription and induction of late response genes (Sheng et al. 1990; Takahashi et al. 1993; Ghosh et al. 1994; Takahashi 1995). Thus, phosphorylated/activated CREB may be an important mediator between photic input and downstream transcriptional events.

As shown in this investigation, PACAP is a substance that causes phosphorylation of CREB in the rat SCN during late subjective day. PACAP immunoreactive fibers project from diffuse retinal ganglion cells to the SCN, where PACAP receptors could be demonstrated by in situ hybridization (Hashimoto et al. 1996; Hannibal et al. 1997). These PACAP type I receptors seem to be coupled to the α-subunit of the adenylate cyclase-stimulating G protein G_s and, thus, activate the cAMP signaling pathway (Hashimoto et al. 1993; Spengler et al. 1993). The PACAP type I receptor binds PACAP with a 1000-fold higher affinity than VIP, which shares 68% amino acid sequence identity with PACAP27 (Miyata et al. 1989; Hashimoto et al. 1993). In the SCN, PACAP was shown to phase-shift neuronal activity rhythms with a similar phase-dependent effectiveness like enhanced intracellular cAMP concentrations, and this effect could be blocked by a specific cAMP antagonist (Prosser and Gillette 1989; Hannibal et al. 1997). According to these data PACAP appears as a neurotransmitter or neuromodulator of the RHT, probably conveying light-information to the SCN via the cAMP signaling pathway. In contrast, the excitatory amino acid glutamate, a well described neurotransmitter of the RHT which produces light-like phase-shifts and induces CREB-phosphorylation during the night, affects the SCN via activation of ionotropic receptors (Inouye and Shibata 1994; Ding et al. 1997). The potency of PACAP to phase-shift SCN activity and to induce the phosphorylation of the transcription factor CREB may point toward the possibility that PACAP affects early gene expression in the SCN during the day, suggesting a modulatory role of the peptide in processing light information in the SCN.

Melatonin was previously shown to inhibit the phosphorylation of CREB in primary cell cultures of ovine pars tuberalis, which was induced by forskolin (McNulty et al. 1994). This effect is probably elicited by an inhibition of the PKA (Hazlerigg et al. 1991). Here we report an inhibitory effect of melatonin in the rat suprachiasmatic nucleus, i.e., a reduction of the phosphorylation of CREB induced by PACAP. Since PACAP is a potent activator of PKA (Miyata et al. 1989), melatonin's effect in the SCN may also impinge on the cAMP signaling pathway, particularly since the expression of the Mel_{1a} melatonin receptor in the SCN has been demonstrated by in situ hybridization, and this receptor has been shown to be negatively coupled to adenylate cyclase activity (Reppert et al. 1994; Godson and Reppert 1997). The effect of melatonin on CREB phosphorylation is the first report that shows an inhibitory action of the hormone on the activation of a transcription factor in the SCN, suggesting an influence on transcriptional events. This effect was found during the late subjective day (CT 10). At this time, melatonin causes a maximal inhibition of the metabolic and spontaneous electrical activity and induces a maximum phase-advance of the circadian rhythm of the neuronal firing rate (Cassone et al. 1987; Shibata et al. 1989; Stehle et al. 1989; McArthur et al. 1991). Moreover, the density of melatonin receptors in the SCN is highest during this time (Tenn and Niles 1993; Gauer et al. 1993). Thus, melatonin may exert long-term effects in the SCN like phase-shifting of circadian rhythms via the inhibition of stimulus-induced activation of transcription factors in the dusk period of the diurnal cycle.

The phosphorylation of the transcription factor CREB, subsequent early gene expression and finally the synthesis of (so far unidentified) late response genes may serve as a model to clarify how the endogenous pacemaker processes light (or darkness) information and how rhythm entrainment, phase-shifting or feedback events may take place. Our data suggest an important functional role for the cAMP signaling pathway for generating circadian rhythms. PACAP, a putative neurotransmitter/neuromodulator of the RHT, is a peptide that can affect this signal transduction cascade during late subjective day and thereby probably influences transcriptional events. The potency of the pineal hormone melatonin, the endocrine signal of darkness, to inhibit the activation of a transcription factor may mirror a feedback loop within the circadian system.

References

Arendt J (1995) Melatonin and the mammalian pineal gland, Chapman and Hall, London

Cassone VM, Roberts MH, Moore RY (1987) Melatonin inhibits metabolic activity in the rat suprachiasmatic nucleus. Neurosci Lett 81: 29-34

Ding JM, Faiman LE, Hurst WJ, Kuriashkina LR, Gillette MU (1997) Resetting the biological clock: mediation of nocturnal CREB phosphorylation via light, glutamate, and nitric oxide. J Neurosci 17: 667-675

Gauer F, Masson-Pevet M, Skene DJ, Vivien-Roels B, Pevet P (1993) Daily rhythms of melatonin binding sites in the rat pars tuberalis and suprachiasmatic nuclei; evidence for a regulation of melatonin receptors by melatonin itself. Neuroendocrinology 57: 120-126

Ghosh A, Ginty DD, Bading H, Greenberg ME (1994) Calcium regulation of gene expression in neuronal cells. J Neurobiol 25: 294-303

Ginty DD, Kornhauser JM, Thompson MA, Bading H, Mayo KE, Takahashi JS, Greenberg ME (1993) Regulation of CREB phosphorylation in the suprachiasmatic nucleus by light and a circadian clock. Science 260: 238-241

Godson C, Reppert SM (1997) The Mel_{1a} melatonin receptor is coupled to parallel signal transduction pathways. Endocrinology 138: 397-404

Golombek DA, Ralph MR (1995) Circadian responses to light: the calmodulin connection. Neurosci Lett 192: 101-104

Gonzalez GA, Montminy MR (1989) Cyclic AMP stimulates somatostatin gene transcription by phosphorylation of CREB at serine 133. Cell 59: 675-680

Hannibal J, Ding JM, Chen D, Fahrenkrug J, Larsen PJ, Gillette MU, Mikkelsen JD (1997) Pituitary adenylate cyclase-activating peptide (PACAP) in the retinohypothalamic tract: a potential daytime regulator of the biological clock. J Neurosci 17: 2637-2644

Hashimoto H, Ishihara T, Shigemoto R, Mori K, Nagata S (1993) Molecular cloning and tissue distribution of a receptor for pituitary adenylate cyclase-activating polypeptide. Neuron 11: 333-342

Hashimoto H, Nogi H, Mori K, Ohishi H, Shigemoto R, Yamamoto K, Matsuda T, Mizuno N, Nagata S, Baba A (1996) Distribution of the mRNA for a pituitary adenylate cyclase-activating polypeptide receptor in the rat brain: an in situ hybridization study. J Comp Neurol 371: 567-577

Hazlerigg DG, Morgan PJ, Lawson W, Hastings MH (1991) Melatonin inhibits the activation of cyclic AMP-dependent protein kinase in cultured pars tuberalis cells from ovine pituitary. J Neuroendocrinol 3: 597-603

Inouye SI, Shibata S (1994) Neurochemical organization of circadian rhythm in the suprachiasmatic nucleus. Neurosci Res 20: 109-130

Inouye SI, Takahashi J, Wollnik F, Turek FW (1988) Inhibitor of protein synthesis phase shifts a circadian pacemaker in mammalian SCN. Am J Physiol 255: R1055-R1058

Kako K, Wakamatsu H, Ishida N (1996) c-fos CRE-binding activity of CREB/ATF family in the SCN is regulated by light but not a circadian clock. Neurosci Lett 216: 159-162

Kako K, Banasik M, Lee K, Ishida N (1997) Regulation of cAMP response element binding protein (CREB) binding in the mammalian clock pacemaker by light but not a circadian clock. Mol Brain Res 44: 39-45

Klein DC, Moore RY, Reppert SM (1991) Suprachiasmatic nucleus. The mind's clock. Oxford University Press, New York

Kopp M, Meissl H, Korf HW (1997) The pituitary adenylate cyclase-activating polypeptide-induced phosphorylation of the transcription factor CREB (cAMP response element binding protein) in the rat suprachiasmatic nucleus is inhibited by melatonin. Neurosci Lett 227: 145-148

McArthur AJ, Gillette MU, Prosser RA (1991) Melatonin directly resets the rat suprachiasmatic circadian clock in vitro. Brain Res 565: 158-161

McNulty S, Ross AW, Barrett P, Hastings MH, Morgan PJ (1994) Melatonin regulates the phosphorylation of CREB in ovine pars tuberalis. J Neuroendocrinol 6: 523-532

Miyata A, Arimura A, Dahl RR, Minamino N, Uehara A, Jiang L, Culler MD, Coy DH (1989) Isolation of a novel 38 residue-hypothalamic polypeptide which stimulates adenylate cyclase in pituitary cells. Biochem Biophys Res Commun 164: 567-574

Miyata A, Jiang L, Dahl RR, Kitada C, Kubo K, Fujino M, Minamino N, Arimura A (1990) Isolation of a neuropeptide corresponding to the N-terminal 27 residues of the pituitary adenylate cyclase-activating peptide with 38 residues (PACAP38). Biochem Biophys Res Commun 170: 643-648

Prosser RA, Gillette MU (1989) The mammalian circadian clock in the suprachiasmatic nucleus is reset in vitro by cAMP. J Neurosci 9: 1073-1081

Reppert SM, Weaver DR, Ebisawa T (1994) Cloning and characterization of a mammalian melatonin receptor that mediates reproductive and circadian responses. Neuron 13: 1177-1185

Rusak B, Robertson HA, Wisden W, Hunt SP (1990) Light pulses that shift rhythms induce gene expression in the suprachiasmatic nucleus. Science 248: 1237-1240

Schomerus C, Maronde E, Laedtke E, Korf HW (1996) Vasoactive intestinal peptide (VIP) and pituitary adenylate cyclase-activating polypeptide (PACAP) induce phosphorylation of the transcription factor CREB in subpopulations of rat pinealocytes: immunocytochemical and immunochemical evidence. Cell Tissue Res 286: 305-313

Sheng M, McFadden G, Greenberg ME (1990) Membrane depolarization and calcium induce c-fos transcription via phosphorylation of transcription factor CREB. Neuron 4: 571-582

Sheng M, Thompson MA, Greenberg ME (1991) CREB: a Ca^{2+} regulated transcription factor phosphorylated by calmodulin-dependent kinases. Science 252: 1427-1430

Shibata S, Cassone VM, Moore RY (1989) Effects of melatonin on neuronal activity in the rat suprachiasmatic nucleus in vitro. Neurosci Lett 97: 140-144

Spengler D, Waeber C, Pantaloni C, Holsboer F, Bockaert J, Seeburg PH, Journot L (1993) Differential signal transduction by five splice variants of the PACAP receptor. Nature 365: 170-175

Stehle J, Vanecek J, Vollrath L (1989) Effects of melatonin on spontaneous electrical activity of neurons in rat suprachiasmatic nuclei: an in vitro iontophoretic study. J Neural Transm 78: 173-177

Stehle JH, Pfeffer M, Kühn R, Korf HW (1996) Light-induced expression of transcription factor ICER (inducible cAMP early repressor) in rat suprachiasmatic nucleus is phase-restricted. Neurosci Lett 217: 169-172

Sumova A, Ebling FJP, Maywood ES, Herbert J, Hastings MH (1994) Non-photic circadian entrainment in the syrian hamster is not associated with phosphorylation of the transcriptional regulator CREB within the suprachiasmatic nucleus, but is associated with adrenocortical activation. Neuroendocrinology 59: 579-589

Takahashi JS (1995) Molecular neurobiology and genetics of circadian rhythms in mammals. Annu Rev Neurosci 18: 531-553

Takahashi JS, Kornhauser JM, Koumenis C, Eskin A (1993) Molecular approaches to understanding circadian oscillations. Annu Rev Physiol 55: 729-753

Tenn C, Niles LP (1993) Physiological regulation of melatonin receptors in rat suprachiasmatic nuclei: diurnal rhythmicity and effects of stress. Mol Cell Endocrinol 98: 43-48

Wollnik F, Brysch W, Uhlmann E, Gillardon F, Bravo R, Zimmermann M, Schlingensiepen KH, Herdegen T (1995) Block of c-Fos and JunB expression by antisense oligonucleotides inhibits light-induced phase-shifts of the mammalian circadian clock. Eur J Neurosci 7: 388-393

Zhang Y, Kornhauser JM, Zee PC, Mayo KE, Takahashi JS, Turek FW (1996) Effects of aging on light-induced phase-shifting of circadian behavioral rhythms, Fos expression and CREB phosphorylation in the hamster suprachiasmatic nucleus. Neuroscience 70: 951-961

24 Transcription Factors in the Photoneuroendocrine System

J. H. Stehle

Dr. Senckenbergische Anatomie, Institute for Anatomy II,
Johann Wolfgang Goethe-University Frankfurt,
Theodor-Stern-Kai 7, D-60590 Frankfurt am Main, Germany;
Phone: +49 69 6301 6905, Fax : +49 69 6301 6017,
E-mail: stehle@em.uni-frankfurt.de

Abstract

The photoneuroendocrine system in mammals transduces environmental light stimuli into a hormonal output, the synthesis and release of melatonin into the circulation. This system serves time-keeping and time-measuring mechanisms and is comprised of three essential components that are interconnected by neural and humoral pathways: the photoreceptive retina, the endogenous oscillator in the suprachiasmatic nucleus and the neuroendocrine pineal organ. Regulation and fine tuning of rhythmic melatonin production in response to external cues requires complex interactions of transsynaptic signaling. These requirements are fulfilled by a high degree of plasticity on all levels between receptor activation and cellular response. Many receptors on the pinealocyte membrane and enzymes involved in melatonin synthesis are linked to the second messenger cAMP. Crosstalk between second and third messengers converges in the pineal gland - like in other tissues - eventually on modulated activity of transcription factors. Of fundamental importance for genes involved in the transsynaptic signaling within elements of the photoneuroendocrine system is the cAMP-inducible promoter element, the CRE (cAMP responsive element). Indeed, the CRE is shared by many physiologically important genes expressed in the pineal. The deciphering of molecular determinants regulating expression of cAMP-inducible genes in the mammalian pineal gland suggests modulation in gene transcription by a dual regulatory mechanism: posttranscriptional activation of the *early* third messenger CREB (cAMP responsive element binding protein) stimulates, *cis*-acting cAMP-induced transcriptional upregulation of the *late* third messenger ICER (inducible cAMP early repressor) inhibits genes with a CRE.

Introduction

Biological systems experience daily changes in environmental lighting conditions (photoperiod), and living organisms have to compensate or even anticipate seasonal fluctuation in the duration of light phase. Adaptation to the changing duration of light and darkness is a fundamental prerequisite for animals and plants to survive. This anticipation of the photoperiod requires some kind of a memory of the photoperiodic history the organism has experienced. In mammals, this coupling between environmental stimulus and physiological reaction is accomplished within a neuronal circuit, the photoneuroendocrine system (PNS) that is comprised of morphologically distinct structures of the central nervous system (for the first general concept, see Scharrer 1964). The main components of the PNS are the light-receptive **retina**, an endogenous **circadian oscillator** located in the suprachiasmatic nucleus of the hypothalamus and a **neuro-endocrine effector** system, the pineal gland (Korf 1994). Most interestingly, there exists a negative feedback loop, with melatonin phase-triggering the activity of SCN cells (Stehle 1995; Korf et al. 1996).

Signal transduction mechanisms required for adaptive plasticity in the PNS have to be phylogenetically old and efficient. Indeed, within the PNS, transsynaptic cAMP signaling - an ancient and one of the most important signal transduction pathways in biological systems - is of fundamental importance (Stehle 1995; Klein et al. 1996). The need for a complete recovery of cAMP signal transduction cascade in all components of the PNS within each light/dark cycle makes regulatory mechanisms in diurnal/circadian systems notably different from non-cycling systems (Armstrong et al. 1993). In particular, these systems demand a rhythmic impact of transcriptional **activators** and **inhibitors** of cAMP-inducible genes. Before reviewing our current knowledge about transcription factors active in the PNS, it is necessary to discuss how these proteins are induced in general and how they exert their effects on gene expression. The understanding of a regulated gene expression within a diurnal and/or a seasonal time frame has been greatly facilitated in recent years by the application of molecular biology.

The cAMP-Signaling Pathway

The cAMP signaling pathway is involved in almost every regulation of gene expression in the central nervous system. The effects of adenylyl cyclase stimulation and subsequent elevation of intracellular cAMP levels are brought about by the cAMP-inducible protein kinase A (PKA). This enzyme consists of catalytic and regulatory subunits, which are aggregated under resting conditions, but dissociate upon cAMP elevation. The catalytic subunits phosphorylate constitutively expressed proteins in the cytoplasm and/or - upon translocation into the nucleus - nuclear proteins, like transcription factors. Transcription factors bind upon activation to promoter and enhancer sequences located upstream of the start site of DNA transcription. As cAMP plays a pivotal role in the PNS with respect to diurnal and seasonal adaptive plasticity, a regulation of gene expression by transcription factors has been anticipated for a long time.

Signaling pathways in neurons are induced by a variety of neurotransmitters and neuroactive peptides, many of which were shown to also affect the function of mammalian pinealocytes via membrane-bound receptors (Ebadi 1993). Regardless of the transsynaptic signaling pathway activated, an important common step in gene targeting is the induction of protooncogenes, or so called immediate early genes (IEGs). IEGs are rapidly induced and their translational product functions as a transcriptional regulator. There is increasing evidence for cAMP-dependent signaling events to be mediated via the induction of IEGs through a variety of stimuli (kinases, growth factors, transforming oncogenes, transmitters). Induction of the prototypic IEG *c-fos* was shown to be rapid but transient in neurons in vivo (Sheng and Greenberg 1990; Morgan and Curran 1991). Products of the IEGs *c-fos*, *c-jun* and *fra-2* form homo- and heterodimeric DNA binding proteins of the AP-1 family, exerting a substantial transcriptional role in normal cell growth, promotion of the transforming phenotype and differentiation (ibid). The collection of homodimeric and heterodimeric AP-1 complexes, interacting with a common DNA binding site on late response genes, the serum response element (SRE; palindromic consensus sequence: TGACTCA), accounts for the versatility in the transduction of synaptic signals. Complexity is even increased by the ability of the JUN/FOS proteins to heterodimerize with transcription factors that bind to the promoter element for transcription factors of the cAMP/PKA pathway, the cAMP responsive element (CRE; palindromic consensus sequence: TGACGTCA; Masquilier and Sassone-Corsi 1992). It is not surprising that the CRE is detected on an increasing number of genes and characterized as a functional promoter element.

The wide distribution and the conserved structure of the CRE suggested that cAMP-mediated gene expression is regulated through a common and possibly widely distributed DNA binding protein, modulated in its activity by the PKA. Indeed, the cloning of **CREB** (cAMP responsive element binding protein) from PC12 cells proved this suggestion to be correct (Montminy and Bilezekjian 1987). CREB is a 43-kDa protein that is ubiquituously expressed (a typical feature of house-keeping genes), functions upon activation as a mediator of a variety of physiological events and was shown to be a prerequisite for initiation of immediate/late response gene induction in many systems (Sheng and Greenberg 1990). It is therefore not surprising that CREB mRNA is ubiquitous in the central nervous system, including the pineal gland (Stehle et al. 1993). CREB activation is accomplished in vivo by the PKA phosphorylating the amino acid serin at position 133 (Gonzales et al. 1989). Since phosphorylation induces a conformational change of the CREB protein, a specific antibody for phosphorylated CREB (pCREB) could be generated which selectively detects the activated form of this transcription factor (Ginty et al. 1993). Thus, the detection of pCREB monitors an activated cAMP signaling pathway.

It is of great functional significance that, in addition to the PKA-induced phosphorylation of CREB, crosstalks exist between the Ca^{2+}- and cAMP-pathways which can converge to activate CREB to alter transsynaptic signaling. An additional remarkable versatility is caused by heterodimerization of CREB with closely related transcription factors CREM (cAMP responsive element binding protein modulator) family (deGroot and Sassone-Corsi 1992). This versatility can even be increased by the formation of complexes between some of the transcriptionally active SRE binding AP-1 proteins and CRE binding CREB/CREM proteins, which bind to their target promoter element, although with lower affinity (Masquilier and Sassone-Corsi 1992).

The second transcription factor within the cAMP signaling pathway is **ICER** (inducible cAMP early repressor). ICER, cloned by RT-PCR and the cDNA rescued from a night-time pineal library, encodes for a very potent repressor of cAMP-inducible genes (Stehle et al. 1993). ICER is generated as a splicing product of the CREM gene by use of an internal promoter which is located downstream of the promoter which generates all other CREM isoforms known so far (Stehle et al. 1993; Molina et al. 1993). Due to the use of the ICER-specific internal promoter, this CREM transcript is a remarkably small transcription factor essentially comprising the DNA binding domain only. The ICER transcript lacks the amino terminal phosphorylation domain (P-box), although the hallmark of the CREM gene is preserved, i.e., the capacity to generate various isoforms through differential splicing (Stehle et al. 1993; Molina et al. 1993). The cAMP inducibility of ICER rests upon the presence of four CREs in the 5'-untranslated part of the transcript (Molina et al. 1993). This finding already allows the prediction of a functional role for ICER in the cAMP pathway in the rat pineal. ICER shows a superior affinity to CREs and is a far more potent inhibitor of cAMP-inducible genes than its CREM siblings (CREMα, β, γ; Stehle et al. 1993), when acting as a homodimer. On the other hand, ICER is able to heterodimerize with other CREM isoforms and CREB. By heterodimerization, ICER may impose its superior affinity to CREs on other transcription factors, but it may also scavenge potentially activating CRE-binding proteins. Thus, ICER can regulate target-gene expression in a dualistic way by either indirectly or directly inhibiting gene transcription. The omission of the phosphorylation domain from ICER indicates that posttranslational modifications do not guide its function and the activity of ICER is solely determined by the level of its expression. Most interestingly, transfection experiments in various cell systems with the CAT-reporter system revealed ICER to have a superior affinity to CRE elements compared with other CREM isoforms.

The above-mentioned general observations demonstrate the concerted action of transcriptional, transactivating proteins in serving important biological roles for the control of gene expression. The results from several recent experimental series strongly indicate that this scenario also takes place within rat PNS in a nuclear race in pinealocytes and in cells of the mammalian SCN to regulate gene expression in a diurnal and/or seasonal time frame.

Retina

In lower vertebrates, an endogenous oscillator determines a circadian rhythm in melatonin synthesis in the eye (Cahill and Besharse 1993). Although the regulatory enzymes for hormone production are still present in the mammalian retina, melatonin synthesis is low and a rhythm is so far reported to be absent (Cahill et al. 1991), indicating that evolution has silenced the endogenous oscillator and thus cAMP signaling. It could, however, recently be demonstrated, that under lowered temperature conditions (28 °C) the hamster retina displays properties of an endogenous oscillator with circadian melatonin synthesis (Tosini and Menaker 1995). It has to be determined whether melatonin can now be linked to retinal diurnal rhythms in disc shedding or to any other physiological process. Very little is known about transcription factors in the mammalian retina. A circadian expression of *c-fos* mRNA in the outer nuclear layer of the rat retina, with elevated levels during the dark period has been reported (Yoshida et al. 1993), possibly induced via phosphorylation of CREB (Hagiwara et al. 1996). As FOS encodes a transcription factor, it may possibly influence the parallel fluctuating levels of the mRNA coding for transducin (Brann and Cohen 1987) and/or opsin (Bowes et al. 1988). No fluctuation of any member of the CREM family could be detected in rat retinal extracts over a 24 h light/dark period (Stehle et al. 1995).

During phylogeny, cAMP-signaling in the mammalian retina may have been dampened due to a shift in task towards mainly photoreception. Thus, the elements within the mammalian PNS - photoreceptor, oscillating SCN cell or neuroendocrine pinealocyte - have taken over specialized functions within distinct structures.

Suprachiasmatic Nucleus (SCN)

A fundamental question in chronobiology is how circadian rhythm output is generated within the endogenous mammalian oscillator. Our understanding was greatly forwarded by experiments showing that individual neurons from rat SCN express independently phased circadian firing rhythms (Welsh et al. 1995). The molecular hardware generating circadian rhythms thus resides within a single cell. The molecular mechanisms generating circadian rhythms apparently comprise the activation of a factor that affects its own transcription directly or via a mediator in an autoregulatory feedback loop and thus possesses the basic properties of a transcription factor. Both photic input from the retina, and melatonin produce phase shifts in the activity of the mammalian SCN in a dose- and time-dependent manner. Phase-shifts in SCN activity are linked to gene transcription (Takahashi 1995; Wollnik et al. 1995), require protein synthesis (Inouye et al. 1988; Zhang et al. 1994) and thus ask for a regulated impact of transcription factors. Light-induced phase-shifts elicit the transient expression of a growing number of IEGs in the SCN (Kornhauser et al. 1990; Rusak et al. 1992; Stehle et al. 1996). In the rat SCN, *c-fos* is the most intensely studied IEG. Light-induced *c-fos* expression is restricted to the dark period, e.g. to phases where light causes phase-shifts in behavioral rhythms (Kornhauser et al. 1990). Thus, *c-fos*-expression is under clock-control and is considered to be part of a photoperiodic memory (Sumová et al. 1995). The transcription factor CREB is also light-inducible in the SCN and may be a mediator in gating the induction of IEGs (Ginty et al. 1993) as already suggested for the regulation of retinal gene expression. Moreover, there is increasing evidence that the cAMP-signaling pathway may be involved in resetting the clock and/or may be part of SCN clock function:

1. There exists a bimodal rhythm of cAMP content in rat SCN under LD entrainment (Prosser and Gillette 1991) and constant darkness (Yamazaki et al. 1994) with elevated levels during the predusk and the predawn period.
2. Treatment with cAMP analogs during the subjective day induces phase advances in the clock towards the predusk period (Prosser and Gillette 1991).
3. Stimulation of the cAMP pathway by forskolin or inhibition of phosphodiesterases leads to phase shifts, probably caused by a precocious activation of clock-intrinsic mechanisms normally induced endogenously with the rise in cAMP-content in the SCN, which is at the end of the period sensitive to cAMP-induced phase shifts.
4. Light pulses at night induce the phosphorylation of the transcriptional activator CREB (Ginty et al. 1993), which potentially regulates cAMP-inducible genes in the SCN. Moreover, recent experiments indicate that PACAP functions as an important retino-hypothalamic transmitter, conveying light information to the oscillator by phosphorylating CREB in the SCN (Kopp et al. 1997; C. von Gall, H.-W. Korf and J.H.S., unpubl. obs.). The inhibition of the PACAP-induced CREB phosphorylation by melatonin (ibid) may mirror the feedback loop within the PNS, with the pineal hormone affecting SCN cells that receive retinal efferents.
5. CREM-mRNA is present constitutively at a low but clearly detectable level in rat SCN during a 12 h light/12 h darkness period (Stehle et al. 1995) and is inducible by light stimulation during the second half of the night (Stehle et al. 1996).

These observations indicate that transcription factors of the cAMP-signaling pathway are involved in phase-shifting SCN activity and may be even part of the pacemaker mechanism itself.

Pineal Organ

It is clearly established that the circadian rhythm in melatonin production in the effector system of the PNS, the pineal gland, is pivotally linked to elevation, maintenance and downregulation of transsynaptic cAMP signaling (Klein 1985; Stehle 1995; Klein et al. 1996). The activation of this pathway is determined by neuronal signals from the circadian oscillator of the PNS located in the hypothalamic SCN. The oscillator drives the nocturnally increased release of norepinephrine (NE) from sympathetic nerve endings in the pineal gland. Subsequent activation of β_1-adrenergic receptors causes a 2-10-fold increase of cAMP levels in pinealocytes (Vanecek et al. 1985), which can be potentiated in vitro by an α_1-adrenergic/ Ca^{2+}/PKC pathway (Klein 1985). In rat pineal gland the NE/cAMP/PKA connection is tightly linked - via transcriptional and translational processes - to the regulation of the recently cloned key enzyme of the melatonin biosynthesis, the arylalkylamine-*N*-acetyltransferase (AA-NAT; Borjigin et al. 1995; Roseboom et al. 1996). Interestingly, AA-NAT promoter analysis has revealed a functional binding site for cAMP-inducible transcription factors, a CRE (Foulkes et al. 1996b; Baler et al. 1997). This suggests a regulatory impact of CREB and ICER on AA-NAT transcription (Stehle 1995; Florez and Takahashi 1996; Klein et al. 1996). Increase and maintenance of AA-NAT, depends on transcriptional, translational and posttranslational mechanisms regulated by elevated intrapineal cAMP levels (Klein 1985; Borjigin et al. 1995; Stehle 1995; Roseboom et al. 1996; Klein et al. 1996). The rise in the detectable intrapineal amount of NE (Drijfhout et al. 1996) and the subsequent increase in AA-NAT mRNA at the beginning of the night (Borjigin et al. 1995; Roseboom et al. 1996) are at least 1 h apart. To bridge this time gap between adrenergic stimulation and rise in transcription in rat pineal gland, the accumulation of a cAMP-dependent intermediate, the phosphorylation of a (nuclear) protein, was postulated 20 years ago (Winters et al. 1977). The phosphorylation of CREB induced by transsynaptic cAMP signaling falls into this time gap. This transsynaptic signal amplification is rather prolonged as compared with glycogen metabolism activated by the same pathway in muscle (Cohen 1983) - 0.5 h versus 5-8 s. The different time courses can be explained by the time required for translocation of the catalytic subunit of the PKA into the nucleus (Hagiwara et al. 1993).

Two very recent studies provide additional evidence for the presence of CREB in rat pineal gland and an NE-induced phosphorylation of CREB. Using gel-mobility shift, only a CRE oligonucleotide was found to abolish the formation of a CRE-*c-fos*-promoter-oligonucleotide/pineal-protein complex (Roseboom and Klein 1995). In addition, upon elevation of the intrapineal cAMP content, a supershifted complex could be detected with the pCREB antibody. These findings were substantiated by the immunocytochemical detection of a specific nuclear pCREB immunoreactivity in single pinealocytes or whole glands upon adrenergic stimulation in vitro (Tamotsu et al. 1995). It thus seems reasonable to consider the cAMP-dependent phosphorylation of CREB as a mediator to induce gene expression leading to the increase in pineal AA-NAT activity. Indeed, we recently demonstrated in vivo a circadian rhythm in CREB phosphorylation driven by the central oscillator located in the SCN and by the subsequent activation of β-adrenergic receptors on pinealocytes (J.H.S., E. Maronde, J. Olcese, and H.-W. Korf, unpubl. obs.). This observation is, to our knowledge, the first description of

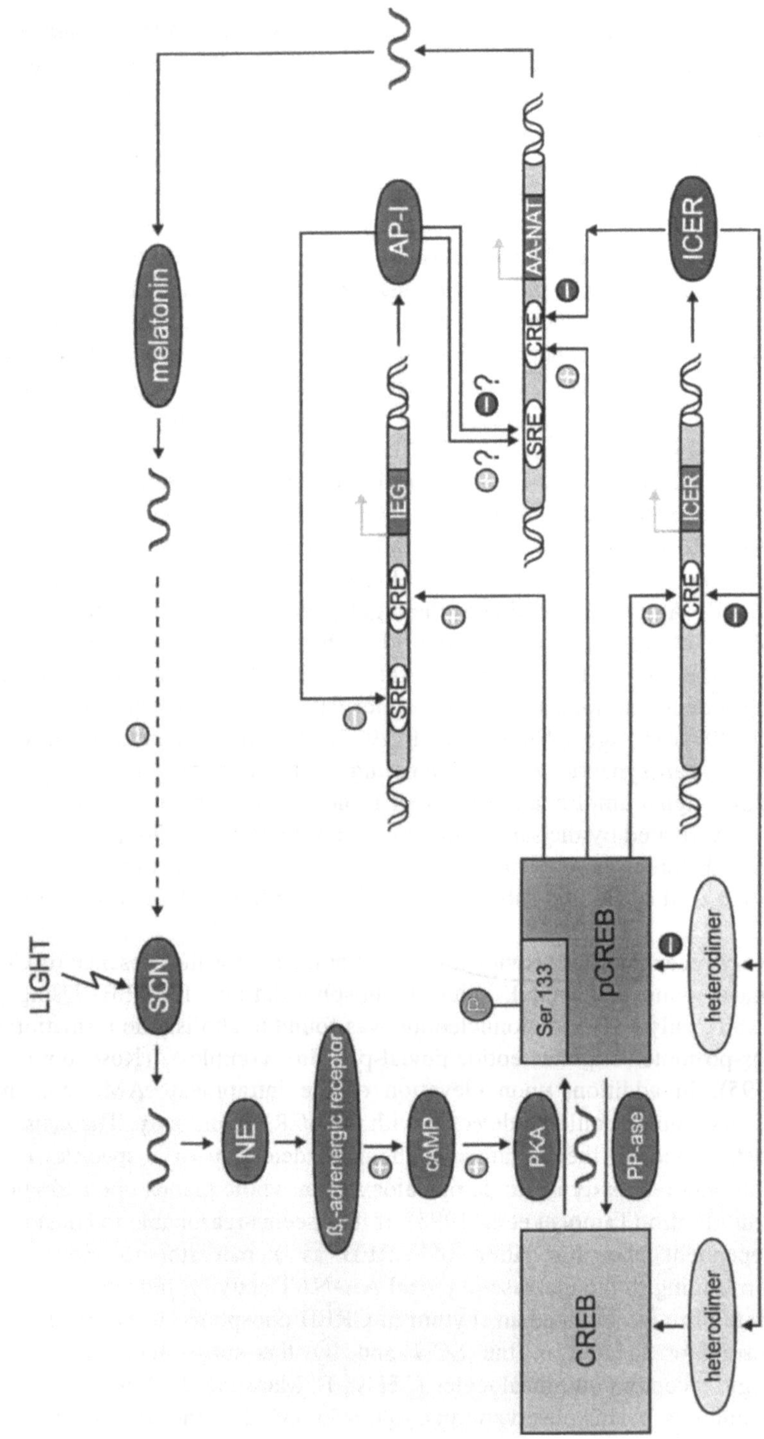
LIGHT
SCN
melatonin
NE
ß1-adrenergic receptor
cAMP
PKA
PP-ase
CREB
pCREB
Ser 133
P
heterodimer
heterodimer
SRE
CRE
IEG
AP-I
SRE
CRE
AA-NAT
?
?
CRE
ICER
ICER

an endogenous and rhythmic activation of CREB. The neuronally driven rhythmic variation in the amount of pCREB in the rat pineal gland is correlated to the cAMP-dependent nocturnal elevation of hormonal output, the synthesis of melatonin. From the molecular viewpoint these data confirm the hypothesis that pineal gene expression is controlled pivotally by the NE/cAMP/PKA connection (Stehle 1995; Maronde et al., submitted), centered around the phosphorylation of CREB. Our quantitative analysis of the pCREB induction in pinealocytes indicates that only a minor fraction of posttranslationally modified TF is needed to mediate a full transsynaptic stimulation of melatonin synthesis. This fact may explain the notorious difficulties that we and others have had in detecting the circadian pCREB rhythm in the rat pineal gland so far. The high amount of "back-up" CREB protein in the rat pineal gland may serve to compensate for the low efficiency of Ser-133 phosphorylation of CREB by the catalytic subunit of the PKA (Hagiwara et al. 1993). Additionally, the large pool of total CREB might serve to ensure a broad dynamic range in the phosphorylation of CREB, required by the environmental and physiological situations of the animals. However, a nuclear pCREB signal could always be detected at least 1 h prior to the rise in pineal melatonin synthesis and is a reliable marker for the rapid (within 0.5 h) stimulus-induced activation of the cAMP signaling pathway under in vivo and in vitro conditions (Roseboom and Klein 1995; Tamotsu et al. 1995; Schomerus et al. 1996). In vivo pCREB levels stay elevated throughout more than half of the night, indicating the need of a maintained high pool of the activated TF for nocturnally elevated pineal metabolism. As AA-NAT transcription also stays elevated throughout the night (Borjigin et al. 1995; Roseboom et al. 1996), a causal relationship between CRE-mediated transcriptional activation and the amount of pCREB - already demonstrated in PC12 cells (Hagiwara et al. 1992, 1993) - appears likely. However, maintenance of cAMP-mediated transcription during the second half of the night cannot solely be attributed to pCREB, as CREB dephosphorylation already occurs several hours before the end of the night.

Despite the intriguing evidence for CREB as the initial molecular target for NE-induced transsynaptic signaling in pinealocytes, the inducibility of other transcriptional activators like IEGs has to be integrated in this scenario. A rhythmic transcription has been demonstrated for the IEGs *c-fos*, *jun-B* (Carter 1992) and *fra-2* (Baler and Klein 1995), forming variable complexes of homo-

Fig. 1. The photoneuroendocrine system and cAMP signaling in its effector, the pineal gland. *AA-NAT* Arylalkylamine *N*-acetyltransferase; *AP-1* activating protein-1, the dimerization product of immediate early genes (*IEG*); *CRE* cAMP-responsive element, the promoter element that binds at least the transcription factors *CREB* (cAMP responsive element binding protein), and *ICER* (inducible cAMP early repressor); *NE* norepinephrine; *PKA* proteinkinase A; *Ser 133* serine residue 133 in the *CREB* amino acid sequence; *pCREB* phosphorylated CREB; *PPase* phosphatases; *SCN* suprachiasmatic nucleus, the mammalian endogenous oscillator; *SRE* 12-O-tetradecanoylphorbole-13-acetate-responsive element, the promoter element binding AP-1 proteins and heterodimeric AP-1 protein TFs (Carter 1994; Baler and Klein 1995). AP-1 and Fra-2 show low expression during the light phase and elevated levels at night, rising before or in parallel to AA-NAT mRNA increase (Borjigin et al. 1995; Roseboom et al. 1996). In the rat pineal gland, the expression of the two AP-1-forming proteins, FOS and JUN, has been shown to be differentially regulated (Carter 1992, 1993). Induction of the *c-fos* gene is primarily mediated through α_1-adrenoreceptors and the PKC pathway, while *c-jun* is regulated via a dual adrenergic mechanism with opposite effects: the α_1-adrenoceptor/PKC pathway stimulates, while the β_1-adrenoceptor/PKA pathway inhibits the expression of this IEG. It should, however, be kept in mind that the in vitro induction and regulation of pineal IEGs is not consistent with in vivo data. This differential activation and regulation of pineal IEGs may help to keep the expression level appropriate for directing downstream events in pineal late response gene expression, such as diurnal plasticity. In this respect, it is of special interest that pineal AP-1-activity is markedly enhanced during the night with JunB appearing as a major player (Carter 1994). It has to be pointed out that the transcriptional function of all IEGs so far characterized in the rat pineal gland is yet unknown and target genes remain enigmatic.

The transcriptional control of IEGs by AP-1-binding to SRE promoter elements is, however, only one half of the orchestra which plays on pineal genes and has to be considered in concert with CRE-mediated activation of gene expression. Concerted actions are also apparent in the rat pineal gland when the Ca^{2+}/PKC pathway is taken into account. However, since α_1-adrenergic receptor activation on its own has no impact on induction, maintenance, and downregulation of melatonin synthesis, analysis of the cAMP signaling pathway is essential to understand regulation of pineal gene expression. IEGs that show day/night (diurnal) or oscillator-regulated (circadian) differences in their transcription pattern, expression, structure and activity properties may, in one way or another, modulate rhythmic melatonin synthesis (Stehle 1995; Florez and Takahashi 1996). For example, it may be hypothesized that IEGs are induced upon sudden perturbations of the oscillator (SCN)-regulated melatonin synthesis, thus functioning in acute "emergency" situations. This assumption is backed by the rapid IEG induction in the SCN upon light stimulation at night-time.

The fading potency of the TF pCREB to activate gene transcription as the dark period progresses is flanked by a second transcriptional event, the active downregulation of cAMP-inducible genes: the inducible inhibitory TF ICER can function as a potent inhibitor of cAMP-inducible genes in rat pineal gland (Stehle et al. 1993, 1995). The transcriptional activity of ICER discovered and first characterized in the pineal has now been described in several other systems (developmental regulation: Stehle et al. 1995; hypothalamus: Luckman and Cox 1995; Stehle et al. 1996; cell cycle: Desdouets et al. 1995; liver: Servillo et al. 1997; thyroid gland: Lalli and Sassone-Corsi 1995; adrenal gland: Tinti et al. 1996; T-lymphocytes: Bodor et al. 1996; myometrium: Gellersen et al. 1997). Interactions between pCREB and ICER, like competition for CREs and heterodimerization (Desdouets et al. 1995; Gellersen et al. 1997), may not only shape circadian melatonin synthesis but also provide the basis for dynamic changes required for seasonal adaptation.

In vivo, a clock-driven and NE-induced ICER expression is apparent in the rat pineal gland which is dramatically elevated in the second half of the night. The following features are of specific interest with respect to ICER's role in the pineal gland and the correlation between its expression and the profile of melatonin synthesis:

1. cDNA cloning revealed the absence of the amino-terminal phosphorylation domain in the ICER transcript (P-box; Stehle et al. 1993), although the hallmark of the CREM gene, generation of modular structured isoforms by differential splicing (Foulkes and Sassone-Corsi 1992), is preserved. This indicates that, in the pineal gland, solely the level of ICER expression determines the activity of this transcriptional repressor.
2. ICER mRNA is elevated 4-6 h before the end of the dark period and shortly precedes the peak of ICER protein (Stehle et al. 1993). This pattern of expression provides at least a temporal link to the endogenous downregulation of melatonin synthesis (Takahashi 1994; Stehle 1995; Klein et al. 1996).
3. In vitro and in vivo experiments performed at different times of the day (Stehle et al. 1993) demonstrate this transcriptional repressor to be uninducible at times when ICER is expressed. Thus, ICER shows autoregulative features and inhibits its own transcription at the beginning of the light period (Stehle et al. 1993). This grants ICER the properties of a clock protein. Since the mammalian pineal gland does not harbor an endogenous oscillator, the presumed clock function of ICER may be a phylogenetic remnant. Although, in rat SCN, ICER mRNA does not fluctuate in vivo over the 24 h cycle (see above; Stehle et al. 1995, 1996), it may do so in the sauropsid pineal gland that contains an endogenous clock.
4. During early postnatal development of pineal function there is a coincidental maturation of CREM inducibility and the endogenous day-night switch in ICER transcription (Stehle et al. 1995). As soon as the sympathetic innervation gains function at the beginning of the second postnatal week, melatonin synthesis starts oscillating, as does the repressor of cAMP-inducible genes, ICER.
5. The level of the catalytic subunit of PKA in the rat pineal rises in parallel to ICER expression and melatonin rhythm (Stehle et al. 1995). This indicates an overall maturation of the cAMP-signaling pathway that is retarded in the early ontogenetic phase by an unknown transient blockade, eventually released when pups gain independence from maternal cues. It is tempting to speculate that, despite its very low expression, ICER protein level may be sufficient for this block in early postnatal life. This speculation is based on the superior affinity of ICER to CREs and its striking potency in repressing cAMP-inducible genes (Stehle et al. 1993).

In mice bearing a null mutation in the CREM locus, AA-NAT mRNA is elevated compared with control animals (Foulkes et al. 1996a). Noteworthy, however, is that this approach was performed with a mouse strain that is known to be melatonin-deficient by virtue of a genetic defect (Ebihara et al. 1986). Furthermore, this knockout not only inactivated ICER but also all activating and inhibiting transcription factors of the CREM family (deGroot and Sassone-Corsi 1992). By selectively silencing ICER in rat pinealocytes using the *antisense* technique, we have now established the final molecular link between the NE-induced activation of ICER expression and the downregulation of melatonin synthesis (Pfeffer et al., submitted).

As mentioned above, pCREB levels are not yet down to daytime levels when melatonin synthesis decreases endogenously (J.H.S., E. Maronde, J. Olcese, and H.-W. Korf, unpubl. obs.), stressing the function of ICER as a transcriptional downregulator of cAMP-inducible genes in the rat pineal gland. Our current results indicate a dual impact of ICER on the regulation of rat melatonin synthesis: (1) ICER tonically inhibits AA-NAT transcription, with attenuation modulated on the basis of the photoperiodic history (Foulkes et al. 1996b) - short photoperiods induce a prolonged and increased nocturnal upregulation in ICER expression compared with long photoperiods; (2) pCREB and ICER act as competitors for the recently characterized AA-NAT CRE (Foulkes et al. 1996a) at the beginning and the end of the night. Thus, a shift in the balance between the two transcription factors leads to an elevated impact of pCREB or ICER, respectively.

Shedding light on the regulation of pineal gene expression has revealed insights into the mechanisms prompting the circadian nature of melatonin production (Fig. 1). Since SCN efferent signals with a circadian rhythm are sufficient as regulatory input to the mammalian pineal gland, the high precision required for adaptations to environmental changes ask for timing mechanisms at all levels of the neuroendocrine transduction cascade. As pointed out, molecular biology not only unravels long hidden details of the specificity and timing of regulated gene expression, but also clarifies a growing complexity in pineal transsynaptic signaling. Still, the phylogenetically highly conserved role of the glands pivotal output, melatonin, which mediates light information to the rest of the body, may guide us to uncover an ancient strategy for this circadian activation of genes. This strategy may be maintained during evolution simply due to its efficiency. Furthermore, the pineal gland as a neuroendocrine transducer may provide a model system to understand, in general, gene expression in hormone-producing cells linked to neuronal signaling.

Acknowledgments

Supported by grants from the Deutsche Forschungsgemeinschaft (DFG) and the Paul and Cilli Weill-Stiftung.

References

Armstrong RC, Montminy MR (1993) Transsynaptic control of gene expression. Annu Rev Neurosci 16: 17-29

Baler R, Klein DC (1995) Circadian expression of transcription factor Fra-2 in the rat pineal gland. J Biol Chem 270: 27319-27325

Baler R, Covington S, Klein DC (1997) The rat arylalkylamine *N*-acetyltransferase gene promoter. J Biol Chem 272: 6979-6985

Bodor J, Spetz AL, Strominger JL, Habener JF (1996) cAMP inducibility of transcriptional repressor ICER in developing and mature human T-lymphocytes. Proc Natl Acad Sci USA 93: 3536-3541

Borjigin J, Wang MM, Snyder SH (1995) Diurnal variation in mRNA encoding serotonin *N*-acetyltransferase in pineal gland. Nature 378: 783-785

Bowes C, Veen TV, Farber DB (1988) Opsin, G-protein and 48-kDa protein in normal and rd mouse retinas: developmental expression of mRNAs and proteins and light/dark cycling of mRNAs. Exp Eye Res 47: 369-390

Brann MR, Cohen LV (1987) Diurnal expression of transducin mRNA and translocation of transducin in rods of rat retina. Science 235: 585-587

Cahill GM, Grace MS, Besharse JC (1991) Rhythmic regulation of retinal melatonin: metabolic pathways, neurochemical mechanism, and the ocular circadian clock. Cell Mol Neurobiol 11: 529-560

Cahill GM, Besharse JC (1993) Circadian clock functions localized in *Xenopus* retinal photoreceptors. Neuron 10: 573-577

Carter DA (1992) Neurotransmitter-stimulated immediate-early gene responses are organized through differential post-synaptic receptor mechanisms. Mol Brain Res 16: 111-118

Carter DA (1993) Noradrenergic regulation of *c-jun* expression in the rat pineal gland in culture: positive and negative components. Eur J Pharmacol 247: 97-100

Carter DA (1994) A daily rhythm of activator protein-1 activity in the rat pineal gland is dependent upon trans-synaptic induction of *jun-B*. Neuroscience 62:1267-1278

Cohen P (1983) Protein phosphorylation and the control of glycogen metabolism in skeletal muscle. Philos Trans R Soc Lond B 302: 13-25

deGroot RP, Sassone-Corsi P (1993) Hormonal control of gene expression: multiplicity and versatility of cyclic adenosine 3′,5′-monophosphate-responsive nuclear regulators. Mol Endocrinol 10: 145-153

Desdouets C, Matesic G, Molina C, Foulkes N, Sassone-Corsi P, Brechot C, Sobczak-Thebot J (1995) Cell cycle regulation of cyclin A gene expression by the cyclic AMP-responsive transcription factors CREB and CREM. Mol Cell Biol 15: 3301-3309

Drjifhout WJ, van der Linde AG, de Vries J, Grol CJ, Westerink BHC (1996) Microdialysis reveals dynamics of coupling between noradrenaline release and melatonin secretion in conscious rats. Neurosci Lett 202: 185-188

Ebadi M. (1993) Multiple pineal receptors in regulating melatonin synthesis. In: Yu HS, Reiter RJ (eds) Melatonin. CRC Press, London, pp 39-71

Ebihara S, Marks T, Hudson DJ, Menaker M (1986) Genetic control of melatonin synthesis in the pineal gland of the mouse. Science 231: 491-493

Florez JC, Takahashi JS (1996) Biological rhythms and the pineal gland. In: Greger R, Windhorst U (eds) Comprehensive human physiology, vol 1. Springer, Berlin, Heidelberg, New York, pp 1199-1214

Foulkes NS, Sassone-Corsi P (1992) More is better: activators and repressors from the same gene. Cell 68: 411-414

Foulkes NS, Duval G and Sassone-Corsi P (1996a) Adaptive inducibility of CREM as transcriptional memory of circadian rhythms. Nature 381: 83-85

Foulkes NS, Borjigin J, Snyder SH, Sassone-Corsi P (1996b) Transciptional control of circadian hormone synthesis via the CREM feedback loop. Proc Natl Acad Sci USA 93: 14140-14145

Gellersen B, Kempf R, Telgmann R (1997) Human endometrial stromal cells express novel isoforms of the transcriptional modulator CREM and up-regulate ICER in the course of decidualization. Mol Endocrinol 11: 97-113

Ginty DD, Kornhauser JM, Thompson MA, Bading H, Mayo KE, Takahashi JS, Greenberg ME (1993) Regulation of CREB phosphorylation in the suprachiasmatic nucleus by light and a circadian clock. Science 260: 238-241

Gonzales GA, Yamamoto KK, Fischer WH, Kerr D, Menzel P, Briggs WH III, Vale WW, Montminy M (1989) A cluster of phosphorylation sites on the cyclic AMP-regulated nuclear factor CREB predicted by its sequence. Nature 337: 749-752

Hagiwara M, Alberts A, Brindle P, Meinkoth J, Feramisco J, Deng T, Karin M, Shenolikar S, Montminy M (1992) Transcriptional attenuation following cAMP induction requires PP-1-mediated dephosphorylation of CREB. Cell 70: 105-113

Hagiwara M, Brindle P, Harootunian A, Armstrong R, Rivier J, Vale W, Tsien R, Montminy M (1993) Coupling of hormonal stimulation and transcription via the cyclic AMP-responsive factor CREB is rate limited by nuclear entry of protein kinase A. Mol Cell Biol 13: 4852-4859

Hagiwara M, Shimomura A, Yoshida K, Imaki J (1996) Gene expression and CREB phosphorylation induced by cAMP and Ca^{2+} in neuronal cells. Adv Pharmacol 36: 277-285

Inouye SIT, Takahashi J, Wollnik F, Turek FW (1988) Inhibitor of protein synthesis phase shifts a circadian pacemaker in mammalian SCN. Am J Physiol 255: R1055-R1058

Klein DC (1985) Photoneural regulation of the mammalian pineal gland. In: Evered D, Clark S (eds) Photoperiodism, melatonin and the pineal. Pitman, London, pp 38-56

Klein, DC, Roseboom PH, Coon SL (1996) New light is shining on the melatonin rhythm enzyme. The first postcloning view. Trends Endocrinol Metab 7: 106-112

Kopp M, Meissl H, Korf HW (1997) The pituitary adenylate cyclase-activating polypeptide-induced phosphorylation of the transcription factor CREB (cAMP responsive element binding protein) in the rat suprachiasmatic nucleus is inhibited by melatonin. Neurosci Lett 227: 145-148

Korf HW (1994) The pineal organ as a component of the biological clock. Ann NY Acad Sci 719: 13-42

Korf HW, Schomerus C, Maronde E, Stehle JH (1996) Signal transduction molecules in the rat pineal organ: Ca^{2+}, pCREB and ICER. Naturwissenschaften 83: 535-543

Kornhauser JM, Nelson DE, Mayo, KE, Takahashi JS (1990) Photic and circadian regulation of *c-fos* gene expression in the hamster suprachiasmatic nucleus. Neuron 248: 127-134

Lalli E, Sassone-Corsi P (1995) Thyroid-stimulating hormone (TSH)-directed induction of the CREM gene in the thyroid gland participates in the long-term desensitization of the TSH receptor. Proc Natl Acad Sci USA 92: 9633-9637

Luckman SM and Cox HJ (1995) Expression of inducible cAMP early repressor (ICER) in hypothalamic magnocellular neurons. Mol Brain Res 34: 231-238

Maronde E, Wicht H, Olcese J, Taskén K, Korf HW (1997) Melatonin biosynthesis in the rat pineal gland is regulated via cyclic AMP-dependent protein kinase type II and phosphorylation of the transcription factor CREB. J Biol Chem (submitted)

Masquilier D, Sassone-Corsi P (1992) Transcriptional cross-talk: nuclear factors CREM and CREB bind AP-1 sites and inhibit activation by *jun*. J Biol Chem 267: 22460-22466

Molina CA, Foulkes NS, Lalli, E, Sassone-Corsi P (1993) Inducibility and negative autoregulation of CREM: an alternative promotor directs the expression of ICER, an early response repressor. Cell 75: 1-20

Montminy MR, Bilezekjian LM (1987) Binding of a nuclear protein to the cyclic-AMP response element of the somatostatin gene. Nature 328: 175-178

Morgan JI, Curran T (1991) Stimulus-transcription coupling in the nervous system: Involvement of the inducible proto-oncogenes *fos* and *jun*. Annu Rev Neurosci 14: 421-451

Prosser RA, Gillette, MU (1991) Cyclic changes in cAMP concentration and phosphodiesterase activity in a mammalian circadian clock studied in vitro. Brain Res 568: 185-192

Roseboom PH, Klein DC (1995) Norepinephrine stimulation of pineal cyclic AMP response element-binding protein phosphorylation: involvement of a β-adrenergic/cyclic AMP mechanism. Mol Pharmacol 47: 439-449

Roseboom PH, Coon SL, Baler R, McCune SK, Weller JL, Klein DC (1996) Melatonin synthesis: analysis of the more than 150-fold nocturnal increase in serotonin *N*-acetyltransferase messenger ribonucleic acid in the rat pineal gland. Endocrinology 137: 3033-3044

Rusak B, McNaughton L, Robertson HA, Hunt SP (1992) Circadian variation in photic regulation of immediate early gene mRNAs in rat suprachiasmatic nucleus cells. Mol Brain Res 14: 124-130

Scharrer E (1964) Photo-neuro-endocrine systems: general concepts. Ann N Y Acad Sci 117: 13-22

Schomerus C, Maronde E, Laedtke E, Korf HW (1996) Vasoactive intestinal peptide (VIP) and pituitary adenylate cyclase-activating polypeptide (PACAP) induce phosphorylation of the transcription factor CREB in subpopulations of rat pinealocytes: immunocytochemical and immunochemical evidence. Cell Tissue Res 286: 305-313

Servillo G, Penna L, Foulkes NS, Magni MV, Dellafazia A, Sassone-Corsi P (1997) Cyclic AMP signalling pathway and cellular proliferation: induction in liver regeneration. Oncogene 14: 1601-1606

Sheng M, Greenberg ME (1990) The regulation and function of *c-fos* and other immediate early genes in the nervous system. Neuron 4: 477-485

Stehle JH (1995) Pineal gene expression: dawn in a dark matter. J Pineal Res 18:179-190

Stehle JH, Foulkes NS, Molina CA, Simonneaux V, Pevet P, Sassone-Corsi P (1993) Adrenergic signals direct rhythmic expression of transcriptional repressor CREM in the pineal gland. Nature 356: 314-320

Stehle JH, Foulkes NS, Pevet P, Sassone-Corsi P (1995) Developmental maturation of pineal gland function: Synchronized CREM inducibility and adrenergic stimulation. Mol Endocrinol 9: 706-716

Stehle JH, Pfeffer M, Kühn R, Korf HW (1996) Light-induced expression of transcription factor ICER (inducible cAMP early repressor) in rat suprachiasmatic nucleus is phase-restricted. Neurosci Lett 217: 169-172

Sumová A, Trávnícková Z, Peters R, Schwartz WJ, Illnerová H (1995) The rat suprachiasmatic nucleus is a clock for all seasons. Proc Natl Acad Sci USA 92: 7754-7758

Takahashi JS (1994) ICER is nicer at night, sir! Curr Biol 4: 165-168

Takahashi JS (1995) Molecular neurobiology and genetics of circadian rhythms in mammals. Annu Rev Neurosci 18: 531-553

Tamotsu S, Schomerus C, Stehle JH, Roseboom PH, Korf HW (1995) Norepinephrine-induced phosphorylation of the transcription factor CREB in isolated rat pinealocytes: an immunocytochemical study. Cell Tissue Res 282: 219-226

Tinti C, Conti B, Cubells JF, Kim KS, Baker H, Joh TH (1996) Inducible cAMP early repressor can modulate tyrosine hydroxylase gene expression after stimulation of cAMP synthesis. J Biol Chem 271: 25375-25381

Tosini G, Menaker M (1995) Circadian rhythms in cultured mammalian retina. Science 272: 419-421

Vanecek J, Sugden D, Weller J, Klein DC (1985) Atypical synergistic α_1- and β_1-adrenergic regulation of adenosine 3′,5′-monophosphate in cultured rat pinealocytes. Endocrinology 116: 2167-2173

Welsh DK, Logothetis DE, Meister M, Reppert SM (1995) Individual neurons dissociated from rat suprachiasmatic nucleus express independently phased circadian firing rhythms. Neuron 14: 697-706

Winters KE, Morrissey JJ, Loos PJ, Lovenberg W (1977) Pineal protein phosphorylation during serotonin N-acetyltransferase induction. Proc Natl Acad Sci USA 74: 1928-1931

Wollnik F, Brysch W, Uhlmann E, Gillardon F, Bravo R, Zimmermann, M, Schlingensiepen, KH Herdegen T (1995) Block of c-Fos and JunB expression by antisense oligonucleotides inhibits light-induced phase shifts of the mammalian circadian clock. Eur J Neurosci 7: 388-393

Yamazaki S, Maruyama M, Cagampang RA, Inouye SIT (1994) Circadian fluctuations of cAMP content in the suprachiasmatic nucleus and the anterior hypothalamus of the rat. Neurosci Lett 651: 329-331

Yoshida K, Kawamura K, Imaki J (1993) Differential expression of c-fos mRNA in rat retinal cells: regulation by light/dark cycle. Neuron 10: 1049-1054

Zhang Y, Seme M, Takahashi JS, Turek FW (1994) Effects of cycloheximide on light-induced phase-advances in the locomotor activity rhythm in hamster. Soc Res Biol Rhythms: 119

25 Expression of Nitric Oxide Synthase I Determines the Strength of Adrenergic Cyclic GMP Response in the Rat Pineal Gland

R. Spessert, E. Layes, A. Schollmayer, and L. Vollrath

Department of Anatomy, Johannes Gutenberg-University,
Saarstr. 19-21, D-55099 Mainz, Germany;
Phone: + 49 6131 393718, Fax: + 49 06131 393719,
E-mail: spessert@mzdmza.zdv.uni-mainz.de

Abstract

In the rat pineal gland the adrenergic cGMP response is not static but changes during postnatal development and depends on environmental lighting conditions. Since the adrenergic cGMP transduction is mediated by stimulation of the enzyme nitric oxide synthase I (NOS I), we investigated whether changes in the strength of adrenergic cGMP response coincide with the expression of NOS I. Using immunoblot analysis, we found that the increase in adrenergic cGMP transduction in the second postnatal week coincides with enhanced NOS I expression. Similarly, we observed that changes in the intensity of adrenergic cGMP transduction under different environmental lighting conditions correspond to NOS I expression. Our findings indicate that the current concept of NOS I being constitutively expressed is not generally valid. At least in the rat pineal, regulation of NOS I expression appears to determine the intensity of cGMP transduction.

Introduction

In the rat pineal, the nocturnal increase in serotonin N-acetyltransferase (NAT) activity and melatonin synthesis are stimulated by norepinephrine (NE) released from intrapineal sympathetic nerve fibers (Klein 1985). In adult rats NE stimulates production of both cyclic adenosine 3'5'-monophosphate (cAMP) and cyclic guanosine 3'5'-monophosphate (cGMP) (Vanecek et al. 1985). Further, while β-adrenergic agonists stimulate production of both cAMP and cGMP (Sugden 1990), α_1-adrenergic agonists alone do not influence formation of the cyclic nucleotides (Klein 1985; Vanecek et al. 1985). However, α_1-adrenergic agonists potentiate the stimulatory effect of the β-adrenergic agonists and increase the production of cAMP and cGMP several-fold (Chik and Ho 1989; Klein et al. 1983; Vanecek et al. 1985). Although adrenergic cAMP and cGMP responses have several features in common, there are marked differences concerning the postnatal appearance of adrenergic cyclic nucleotide response and the dependence on environmental lighting conditions. While formation of cAMP can be induced by adrenergic agonists in the rat pineal shortly after birth (Ellison et al. 1972; Klein et al. 1981b), cGMP formation could be stimulated only after two weeks of postnatal life (Weller and Klein 1992). Further, it has been found that chronic elongation of the light phases decreases the adrenergic cGMP response but not that of cAMP (Klein et al. 1981a). It has been reported that adrenergic cGMP but not cAMP transduction requires activation of the enzyme nitric oxide synthase I (NOS I). With this in mind, we hypothesize in the present study that changes in the intensity of adrenergic cGMP response coincide with the availability of NOS I (cf. Spessert et al. 1995; Layes et al. 1996).

Materials and Methods

Animals

Male Sprague-Dawley rats (mean body weight 200-250 g) were obtained from a commercial breeder (Interfauna, Tuttlingen, Germany) or from the breeding stock of our Institute. Rats were kept under standard laboratory conditions (LD 12:12; lights on at 6.00 h; illumination with fluorescent strip lights, 200 lux at cage level; 20 ± 1 °C; water and food ad libitum) 2 weeks prior to experimentation. For the study of postnatal NOS I appearance 8-, 10-, 12-, 15-, 17- and 21-day-old rats maintained under the above standard laboratory conditions were used. For the study of light-dependent changes in NOS I-immunoreactivity, animals were exposed to the following lighting conditions for 8 days: LD 12:12, LL, or LD 20:4 (lights on at 6.00 h). Subsequently, one group of LL rats was returned to LD 12:12 for 4 days (LL/LD rats). All animals were killed at 10.00 - 11.00 h. With regard to putative differences in NOS I-immunoreactivity throughout a 24 h cycle, four groups of rats were housed under LD 12:12 with time of lights on delayed by 6 h for each group, so that all the animals could be killed at a given timepoint (11.30 - 12.00 h). Pineal glands were removed after decapitation under light ether anesthesia. All investigations were conducted on groups of four rats each.

Materials

Chemicals and antibodies were obtained from the following sources: monoclonal antibody to NOS raised against a 22.3 kDa protein fragment (amino acids 1095-1289) of human NOS I from Affiniti (Nottingham, UK); alkaline phosphatase-conjugated goat anti-mouse antibody from Dianova (Hamburg, Germany); equipment for electrophoresis and electrotransfer was from Bio-Rad Laboratories (Munich, Germany).

Partial Purification of NOS. NOS I was partially purified as described by Bredt and Snyder (1990) with slight modifications. All procedures were carried out at 4 °C. After decapitation, pineal glands were immediately homogenized by sonication in 30 µl of 50 mM TRIS-HCl, pH 7.4, containing antipain (10 µg/ml), chymostatin (10 µg/ml), dithiothreitol (1 mM), EDTA (0.1 mM), EGTA (0.1 mM), leupeptin (3 µM), mercaptoethanol (12 mM), pepstatin A (1 µM), trypsin-inhibitor (10 µg/ml), phenylmethylsulfonyl fluoride (1 mM). The homogenates were centrifuged for 1 h (10^5 *g*), and the supernatant fractions were used for SDS gel electrophoresis.

Protein Determination. Protein concentrations of the supernatant were analyzed according to the method of Lowry et al. (1951).

Gel Electrophoresis. For immunoblot analysis of rat pineal supernatant 60 µg protein were used per lane. Supernatants (and, as indicated, pellets) were dissolved in up to 100 µl sample buffer (0.15 M Tris-HCl, pH 6.8, containing 4% SDS, 22% glycerol and 0.05% bromophenol blue) and prepared by adding 10 µl-mercaptoethanol and heating for 10 min at 90 °C. SDS gel electrophoresis was performed using 7.5% acrylamide/bisacrylamide according to the method of Laemmli (1970).

Immunoblotting. Electroblotting onto nitrocellulose (0.45 µm) was carried out in a semi-dry blot system applying a 3 mA current per cm^2 for 1 h. All subsequent incubations were followed by washing in four changes of phosphate-buffered saline (10 mM PBS, pH 7.4), the second and third wash containing 0.1% Triton-X100. The membranes were blocked in 1% BSA in PBS for 30 min and incubated in primary antibody (1:250 in PBS containing 1% BSA) for 2 h. The following immunostaining was performed using an alkaline phosphatase-conjugated secondary antibody (1:50 000) (Blake et al. 1984) for 1-h incubation. Control experiments were performed by omitting the first and/or the second antibody.

Results

First, we investigated the postnatal appearance of NOS I-immunoreactivity. Weak NOS I-immunoreactivity could be detected in the pineals of 8-, 10-, and 12-day-old rats (Fig. 1, lanes a, b, c). Prominent NOS I-immunoreactivity was found in the pineals of 15- (Fig. 1, lane d), 17- (not shown) and 21-day-old (Fig. 1, lane e), and adult rats (Fig. 1, lane f).

Second, we examined NOS I-immunoreactivity under different environmental lighting conditions. To find out whether there are rhythmic changes in the amounts of NOS I, we compared NOS I-immunoreactivity at various time-points during LD 12:12. We were unable to detect clear differences in the intensities of NOS I-immunoreactivity (Fig. 2). We then investigated the effect of LL exposure on NOS I-immunoreactivity. In comparison to LD 12:12 treatment, NOS I-immunoreactivity of the pineal supernatant fraction was markedly reduced (Fig. 3). To find out whether the decline in NOS I-immunoreactivity is reversible,

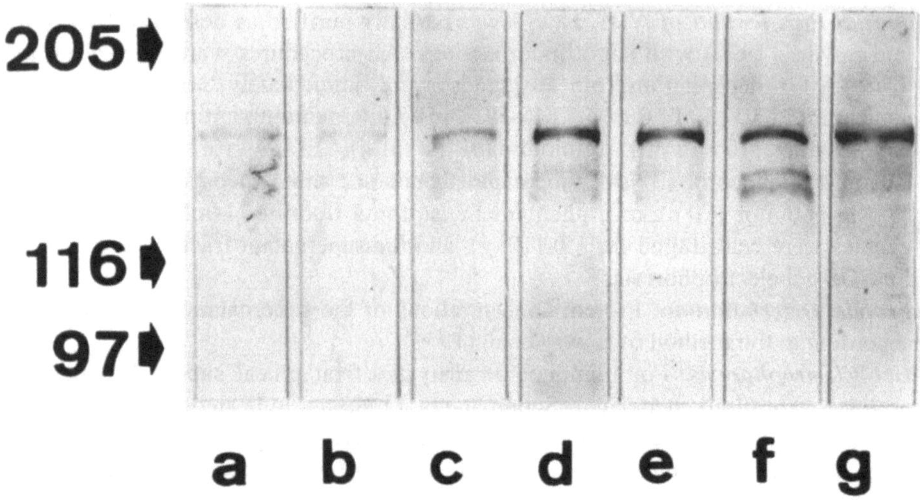

Fig. 1. Immunoblot analysis of NOS I in the pineal gland of rats at postnatal day 8 (**a**), 10 (**b**), 12 (**c**), 15 (**d**), 21 (**e**), and of 8-week-old rats (**f**). Olfactory bulb of adult rats served as a positive control (**g**). Supernatant fractions of rat pineal gland were separated on 7.5% SDS-PAGE. Immunoreaction was conducted with monoclonal antibodies against a protein fragment of human NOS I. *Arrows* molecular weight markers (kDa)

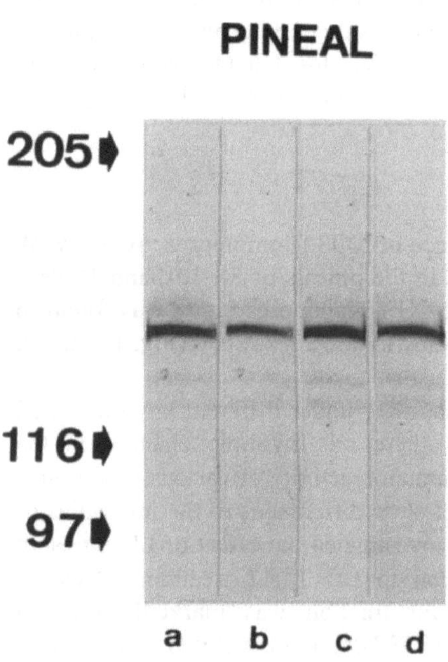

Fig. 2. Immunoblot analysis of NOS I at various time points of the light-dark cycle. Supernatant fractions were subjected to 7.5% SDS-PAGE. Immunoreaction was conducted with monoclonal antibodies raised against a protein fragment of human NOS I. Rats were kept under normal light:dark (12:12) conditions and killed at 5 h after lights on (**lane a**), at 11 h after lights on (**lane b**), 5 h after lights off (**lane c**) and 11 h after lights off (**lane d**). *Arrows* molecular weight markers (kDa)

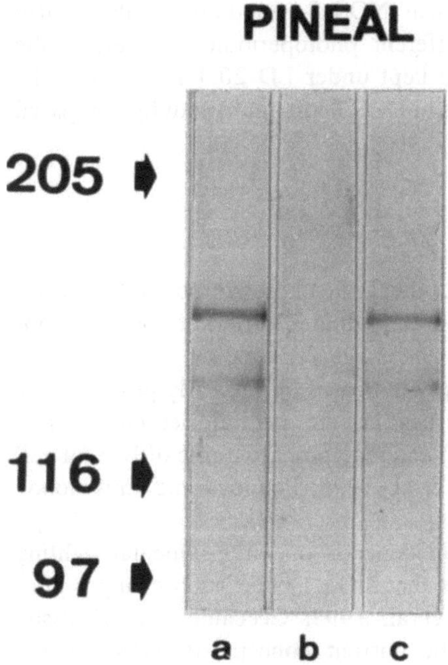

Fig. 3. Immunoblot analysis of NOS I after exposure to constant light. Supernatant fractions of rat pineal gland were separated on 7.5% SDS-PAGE. Immunoreaction was conducted with monoclonal antibodies raised against a protein fragment of human NOS I. Tissues were taken from rats killed 5 h after lights on. Animals were exposed to light:dark 12:12 (**lane a**), constant light (**lane b**) or moved back from constant light treatment to light:dark 12:12 for 4 days (**lane c**). *Arrows* molecular weight markers (kDa)

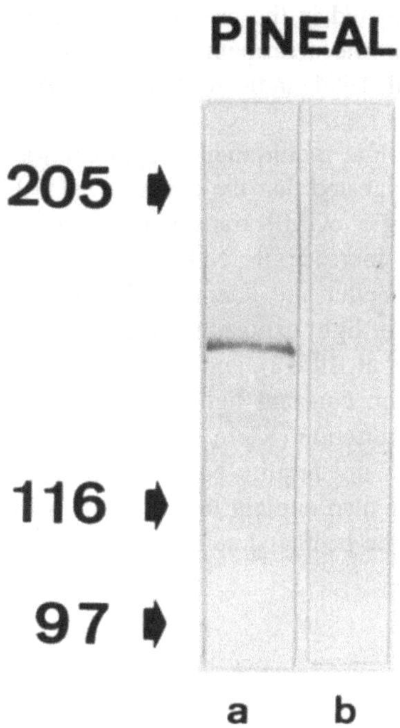

Fig. 4. Immunoblot analysis of NOS I after exposure to different photoperiods. Supernatant fractions were separated on 7.5% SDS-PAGE. Immunoreaction was carried out with monoclonal antibodies raised against a protein fragment of human NOS I. Tissues were taken from animals killed 5 h after lights on. Rats were exposed to light:dark 12:12 (**lane a**) or 20:4 (**lane b**). *Arrows* molecular weight markers (kDa)

rats were moved from LL to LD 12:12 for 4 days (Fig. 3). In comparison to animals continuously kept under LL, pineal NOS I-immunoreactivity clearly increased. Further, we examined how different photoperiodic cues affect the amount of NOS I. In pineal glands of rats kept under LD 20:4 for 8 days, the extended light period led to a clear decrease in NOS I-immunoreactivity compared with LD 12:12 animals (Fig. 4).

Discussion

We show in the present study that in the rat pineal, the expression of NOS I varies considerably during postnatal development and under different lighting conditions.

The finding that NOS I expression strongly increases during early postnatal life is in line with findings from other tissues of the central nervous system (Matsumoto et al. 1993; Ogilvie et al. 1995) and may point to a role of NO for cell differentiation or synaptogenesis (Ogilvie et al. 1995; Peunova and Enikolopov 1995).

Our observation, that NOS I expression depends on environmental lighting conditions, together with previous findings that NOS I expression changes after different experimental treatments (Calza et al. 1993; Ceccatelli and Eriksson 1993; Zhang et al. 1994), shows that the current concept of NOS I being constitutively expressed does not appear to be generally valid. Interestingly, day/night fluctuations of NOS I could not be detected in the present study. This observation, together with the finding that chronic treatment is required to demonstrate changes in NOS activity (Schaad et al. 1994) indicates that the NOS I response is rather slow.

Considering that NOS I stimulation is an absolute requirement for adrenergic cGMP transduction (Spessert et al. 1993) it is suggested that the extent of NOS I expression determines the intensity of adrenergic cGMP transduction. Three aspects support this notion: (1) the postnatal increase in NOS I expression coincides with adrenergic cGMP transduction (Weller and Klein 1992), (2) the attenuation of NOS I expression following constant light exposure corresponds to decreased adrenergic cGMP transduction (Klein et al. 1981a), and (3) the increase in NOS I availability after returning the rats from constant light to the normal photoperiod coincides with enhanced cGMP transduction (Klein et al. 1981a). In view of the finding that cAMP transduction does not require NOS I (Lin et al. 1994), the observed changes in NOS I expression also explain the differences in pineal cyclic nucleotide transduction concerning the postnatal appearance and the dependence on environmental lighting conditions.

Acknowledgements

The data contained in this study is part of theses to be presented by E. Layes and A. Schollmayer in partial fulfillment towards their degree of Dr. rer. nat. and Dr. med., respectively, at the Johannes Gutenberg-University, Mainz. The study was supported by grants from the Deutsche Forschungsgemeinschaft (Sp 403/1-3).

References

Blake MS, Johnston KH, Russel-Jones GJ, Gotschlich EC (1984) A rapid, sensitive method for detection of alkaline phosphatase-conjugated antibody on western blots. Anal Biochem 136: 175-179

Bredt DS, Snyder SH (1990) Isolation of nitric oxide synthetase, a calmodulin-requiring enzyme. Proc Natl Acad Sci USA 87: 682-685

Calza L, Giardino L, Ceccatelli S (1993) NOS mRNA in the paraventricular nucleus of young and old rats after immobilization stress. NeuroReport 4: 627-630

Ceccatelli S, Eriksson M (1993) The effect of lactation on nitric oxide synthase gene expression. Brain Res 625: 177-179

Chik CL, Ho AK (1989) Multiple receptor regulation of cyclic nucleotides in rat pinealocytes. Prog Biophys Mol Biol 53: 197-203

Ellison N, Weller JL, Klein DC (1972) Development of a circadian rhythm in the activity of pineal serotonin N-acetyltransferase. J Neurochem 19: 1335-1341

Klein DC (1985) Photoregulation of the mammalian pineal gland. In: Evered D, Clark S (eds) Photoperiodism, melatonin and the pineal. Pitman, London, pp 51-56

Klein DC, Auerbach DA, Weller JL (1981a) Seesaw signal processing in pineal cells: homologous sensitization of adrenergic stimulation of cyclic GMP response accompanies homologous desensitization of β-adrenergic stimulation of cyclic AMP. Proc Natl Acad Sci USA 78: 4625-4629

Klein DC, Namboodiri MAA, Auerbach DA (1981b) The melatonin rhythm generating system: developmental aspects. Life Sci 28: 1975-1986

Klein DC, Sugden D, Weller JL (1983) Post-synaptic α-adrenergic receptors potentiate the β-adrenergic stimulation of serotonin N-acetyltransferase. Proc Natl Acad Sci USA 80: 599-603

Layes E, Schollmayer A, Spessert R, Vollrath L (1996) Expression of nitric oxide synthase in the developing rat pineal gland coincides with the appearance of the adrenergic cyclic guanosine 3',5'-monophosphate response. Neurosci Lett 212: 71-73

Laemmli UK (1970) Cleavage of structural proteins during the assembly of the head of bacteriophage T4. Nature 227: 680-688.

Lin AMY, Schaad NC, Schulz PE, Klein DC (1994) Pineal nitric oxide synthase: characteristics, adrenergic regulation and function. Brain Res 651: 160-168

Lowry OH, Rosebrough NJ, Farr AL, Randall RJ (1951) Protein measurement with the Folin phenol reagent. J Biol Chem 193: 265-271

Matsumoto T, Pollock JS, Nakane M, Förstermann U (1993) Developmental changes in cytosolic and particulate nitric oxide synthase in rat brain. Dev Brain Res 73: 199-203

Ogilvie P, Schilling K, Billingsley ML, Schmidt HHHW (1995) Induction and variants of neuronal nitric oxide synthase type I during synaptogenesis. FASEB J 9: 799-806

Peunova N, Enikolopov G (1995) Nitric oxide triggers a switch to growth arrest during differentiation of neuronal cell. Nature 375: 68-73

Schaad NC, Vanecek J, Schulz PE (1994) Photoneural regulation of rat pineal nitric oxide synthase. J Neurochem 62: 2496-2499

Spessert R, Layes E, Vollrath L (1993) Adrenergic stimulation of cyclic GMP formation requires NO-dependent activation of cytosolic guanylate cyclase in rat pinealocytes. J Neurochem 61: 138-143.

Spessert R, Layes E, Schollmayer A, Reuss S, Vollrath L (1995) In the rat pineal gland, but not in the suprachiasmatic nucleus, the amount of constitutive nitric oxide synthase is regulated by environmental lighting conditions. Biochem Biophys Res Commun 212: 70-76

Sugden D (1990) β-Adrenergic regulation of cyclic GMP in rat pinealocytes. Biochem Biophys Res Comm 167: 835-841

Vanecek J, Sugden D, Weller J, Klein DC (1985) Atypical synergistic α_1- and β-adrenergic regulation of adenosine 3',5'-monophosphate and guanosine 3'5'-monophosphate in rat pinealocytes. Endocrinology 116: 2167-2173

Weller JL, Klein DC (1992) The pineal adrenergic → cyclic cGMP response develops two weeks after the adrenergic → cyclic AMP response. Dev Brain Res 68: 144-147

Zhang ZG, Chopp M, Gautam S, Zaloga C, Zhang RL, Schmidt HHHW, Pollock JS, Förstermann U (1994) Upregulation of neuronal nitric oxide synthase and mRNA, and selective sparing of nitric oxide synthase-containing neurons after focal cerebral ischemia in rat. Brain Res 654: 85-95

26 Noradrenergic and VIPergic Signal Transduction in Rat Pinealocytes

C. Schomerus and H.-W. Korf

Dr. Senckenbergische Anatomie, Institute for Anatomy II,
Johann Wolfgang Goethe-University Frankfurt,
Theodor-Stern-Kai 7, D-60590 Frankfurt am Main, Germany;
Phone: +49 69 6301 6905, Fax : +49 69 6301 6017

Abstract

In the present study we investigated the effects of norepinephrine and vasoactive intestinal peptide on the intracellular calcium concentration ($[Ca^{2+}]_i$) and on phosphorylation of the transcription factor CREB in isolated rat pinealocytes. Norepinephrine evoked a biphasic calcium response via activation of α_1-adrenergic receptors in virtually all pinealocytes identified by the S-antigen immunoreaction. This response involved mobilization of calcium from intracellular calcium stores and calcium influx into the cell. Moreover, norepinephrine induced phosphorylation of CREB via activation of β-adrenergic receptors. In contrast, vasoactive intestinal peptide did not cause an elevation of $[Ca^{2+}]_i$, but induced pCREB immunoreactivity in a subpopulation of the pinealocytes, suggesting their functional heterogeneity.

Introduction

The rhythmic production of melatonin in the mammalian pineal organ depends on a variety of neuronal inputs (cf. Korf 1996). Of these, the noradrenergic innervation originating from the superior cervical ganglion is most important. In the rat, norepinephrine (NE) acts upon melatonin production by stimulating both α_1- and β-adrenergic receptors (cf. Klein 1985). Activation of β-receptors causes an increase in the intracellular cAMP and cGMP concentrations and subsequent activation of serotonin-*N*-acetyltransferase (NAT), the key enzyme in melatonin synthesis, via transcriptional and translational processes which are not yet elucidated in detail. Activation of α_1-receptors potentiates the β-adrenergic effects through an increase in $[Ca^{2+}]_i$ (Sugden et al. 1986). A possible mediator of cAMP effects in pinealocytes is the transcription factor CREB, the cAMP response element binding protein. As shown for other cell systems, an increase in the intracellular cAMP concentration causes the dissociation of the regulatory and catalytic subunits of protein kinase A, translocation of the catalytic subunit into the nucleus, and phosphorylation of CREB. Phosphorylated CREB (pCREB) binds to genomic cAMP response elements and stimulates the transcription of cAMP responsive genes (De Groot and Sassone-Corsi 1993; Gonzalez and Montminy 1989; Sheng and Greenberg 1990; Yamamoto et al. 1988). This transcription factor is also expressed in the pineal organ (Stehle et al. 1993), and experiments on rat pineal extracts (Roseboom and Klein 1995) and dispersed rat pinealocytes (Tamotsu et al. 1995) have shown that NE stimulation induces pCREB immunoreactivity (IR).

Although NE is apparently the most important neuroregulator of pineal functions in mammals, melatonin biosynthesis was shown to be influenced also by neurotransmitters and neuropeptides other than NE. Among these, vasoactive intestinal peptide (VIP) was shown to stimulate both cAMP and cGMP increases, and NAT activation (Kaneko et al. 1980; Yuwiler 1983).

In the present study we analyzed, at the single-cell level, the effects of NE and VIP on $[Ca^{2+}]_i$ and on phosphorylation of the transcription factor CREB. In particular, we were interested in the question of whether the responses to NE and VIP are a characteristic feature of all pinealocytes or whether they are restricted to a subpopulation of cells.

Materials and Methods

Pinealocytes of male, adult Wistar rats were isolated by papain digestion (cf. Schaad et al. 1993; Schomerus et al. 1995) and immobilized on poly-L-lysine-coated coverslips which were provided with a tiny grid. For calcium measurements, the cells were cultured for 24 h in DMEM medium under 95% O_2/5% CO_2. The cells were then loaded with the fluorescent dye Fura-2 and superfused with saline at a constant temperature of 37 °C. The analysis of $[Ca^{2+}]_i$ upon stimulation with the indicated drugs was performed as previously described (Schomerus et al. 1995). For immunocytochemical demonstration of pCREB, the cells were incubated for 72 h and finally treated with the indicated drugs at varying concentrations. After 10, 20, 30, and 60 min, respectively, the immobilized cells were fixed with 4% paraformaldehyde, and pCREB was detected as described (Tamotsu et al. 1995). In double-labeling experiments, the S-antigen was detected by use of a specific primary antibody (Korf et al. 1986; Wicht et al. 1993) and a FITC-coupled anti-rabbit IgG. Immunocytochemical data and calcium responses of single cells were finally correlated with the help of the grid on the coverslips. For densitometric analysis of the immunostained cell preparations, the ratios between cytoplasmic and nuclear grey values of single cells were calculated by use of a VIDAS image analysis system (Tamotsu et al. 1995). Immunoblotting and densitometric analyses of the immunostained protein bands were performed as described (Schomerus et al. 1996). All experiments were conducted in accordance with guidelines on the care of experimental animals as approved by the European Communities Council Directive (86/609/EEC).

Results

Ca^{2+} Measurements

10^{-6} M NE induced a biphasic increase in $[Ca^{2+}]_i$ in isolated and immobilized rat pineal cells. After a primary phase with a transient maximum, $[Ca^{2+}]_i$ dropped to an elevated level. This secondary or plateau phase lasted until NE was removed from the superfusion buffer. Immunocytochemical analysis revealed that the vast majority of the NE-responsive cells were S-antigen immunopositive and thus represented pinealocytes. The NE-induced calcium response in rat pinealocytes was rather uniform in virtually all cells irrespective of whether they were strongly or weakly S-antigen immunoreactive. The calcium signal is apparently mediated by α_1-adrenergic receptors since essentially the same response was elicited by the α_1-adrenergic agonist phenylephrine (10^{-6} M), and both the NE- and the phenylephrine-evoked calcium responses were blocked by the α_1-adrenergic antagonist prazosine (10^{-8} M). In an attempt to further analyze the biphasic pattern of the NE-induced calcium signal, we treated pinealocytes with drugs that were shown to release calcium from intracellular calcium stores in other cell types. Depletion of intracellular calcium stores by thapsigargine prior to NE treatment totally prevented the NE-induced increase in $[Ca^{2+}]_i$. Furthermore, when cells were stimulated with NE in Ca^{2+}-free saline, the amplitude of the primary phase is reduced, whereas the secondary phase is totally abolished. Thus, both intracellular calcium stores and influx of extracellular calcium ions into the cell appear to be involved in the NE-induced calcium signal. In the following, we addressed the question of how calcium ions enter the cells during the secondary phase of the response. Voltage-gated L-type calcium channels which have been shown to be present in rat pinealocytes do not participate in the NE-evoked calcium signal, since blockade of these channels with nifedipine does not significantly affect the NE-induced calcium signal. Moreover, calcium-permeable cGMP-gated cation channels appear to be absent in rat pinealocytes because 8-Br-cGMP did not evoke an increase in $[Ca^{2+}]_i$. Thus, the identity of this calcium entry mechanism remains elusive.

In contrast to NE, treatment with VIP at concentrations ranging from 10^{-6} to 10^{-10} M did not induce an increase in $[Ca^{2+}]_i$ in any of the cells examined, irrespective of whether the papain digestion of the cells was shortened to up to 5 min or whether papain was replaced by trypsin.

pCREB Immunocytochemistry

In unstimulated control preparations, approximately 5% of isolated rat pineal cells displayed a moderate pCREB IR (Fig. 1). NE (10^{-6} M) induced a distinct nuclear pCREB IR in approximately 95% of the cells (Fig. 1). Of these cells, approximately 35% were intensely immunostained. The response to NE was dose-

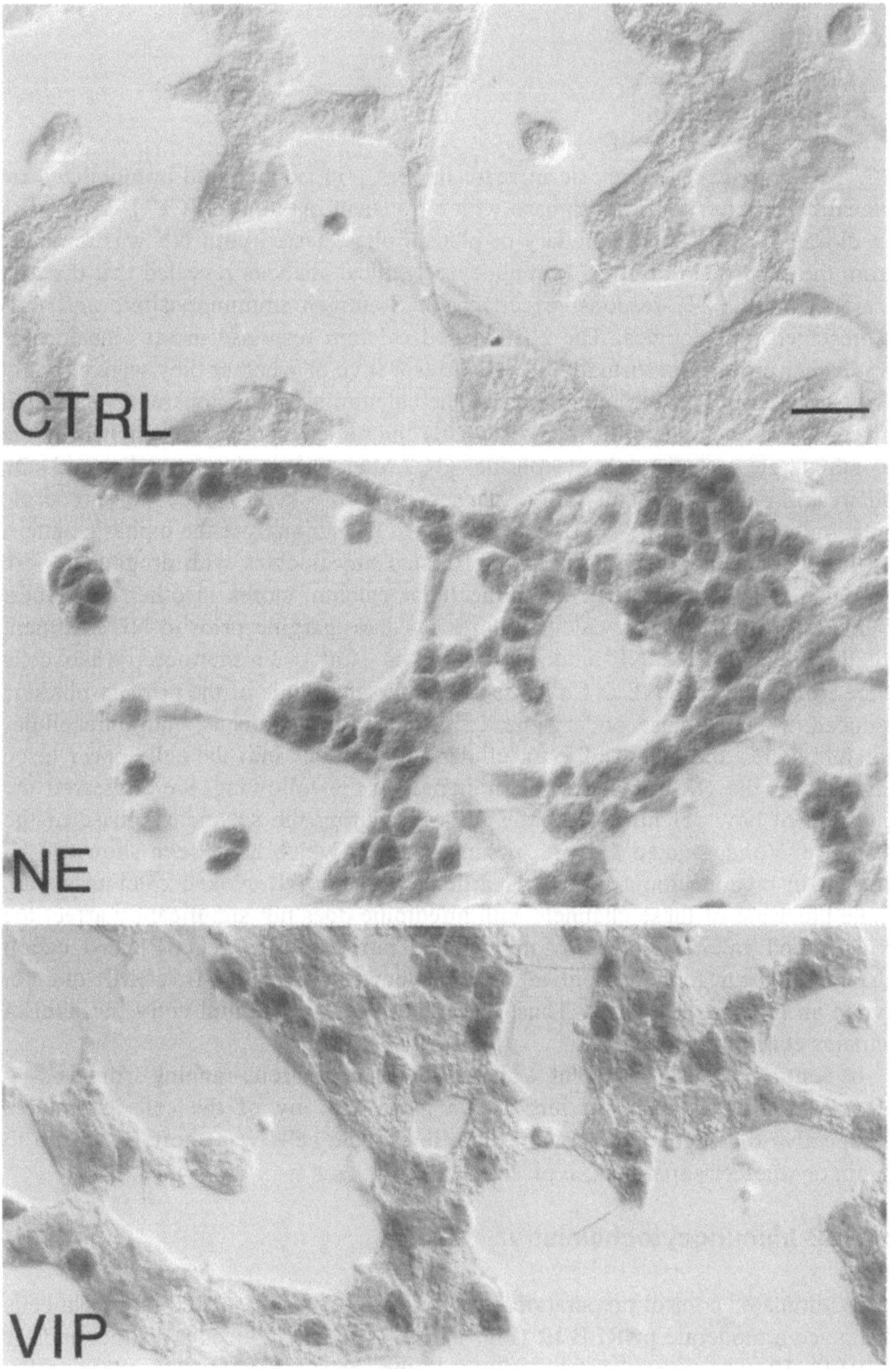
CTRL
NE
VIP

dependent, and a distinct induction of pCREB IR could be detected at NE concentrations as low as 10^{-9} M. pCREB IR was first detectable after 10 min and reached a maximum after 30 min. Immunocytochemical double-labeling experiments revealed that pCREB IR was almost exclusively induced in pinealocytes identified by the S-antigen immunoreaction. Specific activation of β-adrenergic receptors by isoproterenol (10^{-6} M) in the presence of the α_1-antagonist prazosine (10^{-7} M) virtually evoked the same response as did NE. In contrast, specific activation of α_1-adrenergic receptors by phenylephrine (10^{-6} M) in the presence of the β-adrenergic antagonist propranolol (10^{-6} M) did not induce pCREB IR.

VIP (10^{-7} M) which failed to elevate $[Ca^{2+}]_i$ in our experiments induced pCREB IR in approximately 60% - 70% of the pinealocytes (Fig. 1). The intensity of the VIP-evoked IR that was also first detectable after 10 min was far less pronounced than that caused by NE, and only approximately 20% of the cells were intensely labeled. The effects of VIP were dose-dependent and could be elicited at VIP concentrations as low as 10^{-11} M.

Immunoblotting

Immunoblotting of protein extracts from unstimulated control preparations and from cell preparations treated with NE or VIP for 30 min revealed that the amount of total CREB did not display a significant variation as shown by use of an antibody which detects both the phosphorylated and the unphosphorylated form of CREB. Using a specific pCREB antibody, however, immunoreactivities for a 42 kDa protein and for two proteins with lower molecular weight (30 and 34 kDa, respectively) were induced upon stimulation with NE or VIP, whereas in unstimulated control preparations these protein bands were undetectable with this antibody. A densitometric analysis of the immunoblot data revealed that the NE-induced immunoreaction of the 42 kDa protein is approximately three times more intense than that evoked by VIP.

Fig. 1. Nuclear pCREB immunoreactivity (IR) in isolated and immobilized rat pinealocytes. In unstimulated control preparations (*CTRL*) only very few cells display a faint pCREB IR. Stimulation with 10^{-6} M *NE* for 30 min induced a distinct pCREB IR in the vast majority of the cells. Stimulation with 10^{-7} M *VIP* for 30 min elicited pCREB IR in a subpopulation of cells only. *Bar* 20 µm

Discussion

In the present study we investigated at the single-cell level the effects of NE and VIP on both the intracellular calcium level and on the phosphorylation of the transcription factor CREB in immunocytochemically identified rat pinealocytes.

NE leads to an increase in $[Ca^{2+}]_i$ in virtually all cells. Obviously, NE acts on $[Ca^{2+}]_i$ through α_1-adrenergic receptors since NE effects on $[Ca^{2+}]_i$ are mimicked by the α_1-adrenergic agonist phenylephrine and could be blocked by the α_1-specific antagonist prazosine. The NE-induced calcium signal is marked by a distinct biphasic pattern. The primary phase is characterized by a transient maximum in $[Ca^{2+}]_i$ and is caused primarily by calcium mobilization from intracellular calcium stores because the initial peak persists when pinealocytes are kept in Ca^{2+}-free saline and then stimulated with NE. In contrast, the plateau phase is totally abolished under these conditions, indicating that it is based mainly on calcium influx into the cell (Schomerus et al. 1994; Schomerus et al. 1995). Pharmacological experiments with thapsigargine imply that α_1-adrenergic stimulation initially activates calcium release from intracellular compartments. Depletion of these stores then causes an influx of extracellular calcium ions into the cell. Thus, our data suggest the presence of a store-operated calcium influx in rat pinealocytes upon stimulation with NE. The channel which in rat pinealocytes opens upon NE stimulation and allows calcium ions to enter the cell has not yet been identified. Voltage-gated L-type calcium channels have been shown to be present in rat pinealocytes and apparently participate in acetylcholine-mediated calcium signaling in these cells (Letz et al. 1997). Acetylcholine acts upon nicotinic acetylcholine receptors and induces, via activation of an inward sodium current, a depolarization of the cell which then triggers the opening of the voltage-gated calcium channels. However, these channels are not involved in the NE-regulated signal transduction since treatment with nifedipine does not significantly affect the NE-induced calcium response. Furthermore, the influx of calcium ions does not appear to be mediated via a cGMP-gated cation channel since the cGMP analogue 8-Br-cGMP proved to be without effect on $[Ca^{2+}]_i$. Thus, the identity of this calcium entry mechanism remains elusive.

In contrast to NE, VIP did not change $[Ca^{2+}]_i$ in any of the cells examined, irrespective of whether papain digestion of the cells was shortened to up to 5 min or whether papain was replaced by trypsin. These findings are at variance with a recent report of Schaad et al. (1995), proposing that VIP elevates $[Ca^{2+}]_i$ through an increase in the intracellular cGMP level and subsequent activation of a cGMP-gated cation channel. As reported above, our results obtained with 8-Br-cGMP also did not support the findings of these authors. The reasons for these discrepancies remain to be elucidated.

In addition to elevating $[Ca^{2+}]_i$, NE and VIP induce pCREB IR in rat pinealocytes. The immunoblots show that the IR is largely caused by phosphorylation of a 42 kDa protein probably reflecting CREB (Roseboom and Klein 1995). Furthermore, two smaller proteins are phosphorylated which, with regard to the molecular weight, may be related to the transcription factor ATF-1. Thus, we conclude that treatment of pinealocytes with NE and VIP causes phosphorylation of transcription factors of the CREB/ATF-1 family with CREB being the main component. Compared with controls, the amount of total CREB remained constant after stimulation.

In experiments with preparations of isolated pinealocytes, we found that the induction of pCREB IR by NE is mediated via β-adrenergic receptors. β-Adrenergic stimulation with isoproterenol virtually evoked the same response as did NE, whereas the α_1-adrenergic agonist phenylephrine did not induce CREB phosphorylation. This approach also showed that more than 95% of immunocytochemically identified pinealocytes responded to the stimulus. In contrast, VIP elicited this response in subpopulations of pinealocytes only, and the pCREB IR in immunoreactive cells was weaker than in NE-stimulated cells. These results are compatible with the densitometric analysis of our immunoblotting results, showing that NE was approximately three times more effective than VIP in phosphorylating members of the CREB/ATF-1 family. The most probable explanation is that the VIP receptor is expressed in part of the pinealocytes only. These data suggest a functional heterogeneity among rat pinealocytes as has been implied by morphological and immunocytochemical investigations. An interesting point is the question of how VIP receptor activation is linked to CREB phosphorylation. VIP is known to elevate the intracellular cAMP concentration in rat pinealocytes via activation of adenylate cyclase (Chik et al. 1988). In a recent study it was claimed that VIP raises $[Ca^{2+}]_i$ at least in part of the cells (Schaad et al. 1995) while other authors failed to detect a change in $[Ca^{2+}]_i$ after treatment with VIP (Olcese et al. 1996). The data presented here strengthen the view that VIP acts via an elevation of the intracellular cAMP level rather than a rise in $[Ca^{2+}]_i$ because VIP never evoked an increase in $[Ca^{2+}]_i$ in cell preparations later shown to respond to NE with an induction of pCREB IR.

In summary, we have provided evidence that NE induces an increase in $[Ca^{2+}]_i$ through activation of α_1-adrenergic receptors and phosphorylation of CREB through activation of β-adrenergic receptors in virtually all immunocytochemically identified rat pinealocytes. In contrast, VIP does not affect $[Ca^{2+}]_i$ and induces pCREB IR in a subpopulation of cells only. These data support the concept of functional heterogeneity among mammalian pinealocytes. Our data indicate that the transcription factor pCREB is a possible mediator of NE and VIP effects in rat pinealocytes. Thus, they extend previous biochemical studies showing that NE and VIP increase the intracellular cAMP concentration and enhance melatonin synthesis in the mammalian pineal gland.

Acknowledgment

This investigation has been supported by the Deutsche Forschungsgemeinschaft (SFB 269, Teilprojekt B2).

References

Chik CL, Ho AK, Klein DC (1988) Dual receptor regulation of cyclic nucleotides: α_1-adrenergic potentiation of vasoactive intestinal peptide stimulation of pinealocyte cAMP. Endocrinology 122: 1646-1651

De Groot RP, Sassone-Corsi P (1993) Hormonal control of gene expression: multiplicity and versatility of cyclic adenosine 3',5'-monophosphate-responsive nuclear regulators. Mol Endocrinol 7: 145-153

Gonzalez GA, Montminy MR (1989) Cyclic AMP stimulates somatostatin gene transcription by phosphorylation of CREB at serine 133. Cell 59: 675-680

Kaneko T, Cheng PY, Oka H, Oda T, Yanaihara N, Yanaihara C (1980) Vasoactive intestinal polypeptide stimulates adenylate cyclase and serotonin N-acetyltransferase activities in rat pineal gland in vitro. Biochem Res 1: 84-87

Klein DC (1985) Photoneural regulation of the mammalian pineal gland. In: Evered D, Clark S (ed) Photoperiodism, melatonin and the pineal gland. Pitman, London, pp 38-56

Korf HW (1996) Innervation of the pineal gland. In: Burnstock G (ed) Handbook of the autonomic nervous system. Harwood, Chur, pp 129-180

Korf HW, Oksche A, Ekström P, Zigler JS, Gery I, Klein DC (1986) Pinealocyte projections into the mammalian brain revealed with S-antigen antiserum. Science 231: 735-737

Letz B, Schomerus C, Maronde E, Korf HW, Korbmacher C (1997) Stimulation of a nicotinic ACh receptor causes depolarization and activation of L-type Ca^{2+} channels in rat pinealocytes. J Physiol 499: 329-340

Olcese J, McArdle C, Mikkelsen J, Hannibal J (1996) PACAP and type I PACAP receptors in the pineal gland. Ann NY Acad Sci 805: 595-600

Roseboom PH, Klein DC (1995) Norepinephrine stimulation of pineal cyclic AMP response element-binding protein phosphorylation: involvement of a β-adrenergic/cyclic AMP mechanism. Mol Pharmacol 47: 439-449

Schaad NC, Parfitt A, Russell JT, Schaffner AE, Korf HW, Klein DC (1993) Single cell $Ca^{2+}{}_i$ analysis and biochemical characterization of pinealocytes immobilized with a novel attachment peptide preparation. Brain Res 614: 251-256

Schaad NC, Vanacek J, Rodriguez IR, Klein DC, Holtzclaw L, Russell JT (1995) Vasoactive intestinal peptide elevates pinealocyte intracellular calcium concentrations by enhancing influx: evidence for involvement of a cyclic GMP-dependent mechanism. Mol Pharmacol 47: 923-933

Schomerus C, Ruth P, Korf HW (1994) Photoreceptor-specific proteins in the mammalian pineal organ: immunocytochemical data and functional considerations. Acta Neurobiol Exp 54 (Suppl): 9-17

Schomerus C, Laedtke E, Korf HW (1995) Calcium responses of isolated, immunocytochemically identified rat pinealocytes to noradrenergic, cholinergic and vasopressinergic stimulations. Neurochem Int 27: 163-175

Schomerus C, Maronde E, Laedtke E, Korf HW (1996) Vasoactive intestinal peptide (VIP) and pituitary adenylate cyclase-activating polypeptide (PACAP) induce phosphorylation of the transcription factor CREB in subpopulations of rat pinealocytes: immunocytochemical and immunochemical evidence. Cell Tissue Res 286: 305-313

Sheng M, Greenberg, ME (1990) The regulation and function of c-fos and other immediate early genes in the nervous system. Neuroscience 4: 477-485

Stehle JH, Foulkes NS, Molina CA, Simonneaux V, Pevet P, Sassone-Corsi P (1993) Adrenergic signals direct rhythmic expression of transcriptional repressor CREM in the pineal gland. Nature 356: 314-320

Sugden AL, Sugden D, Klein DC (1986) Essential role of calcium influx in the adrenergic regulation of cAMP and cGMP in rat pinealocytes. J Biol Chem 261: 11608-11612

Tamotsu S, Schomerus C, Stehle JH, Roseboom P, Korf HW (1995) Norepinephrine-induced phosphorylation of the transcription factor CREB in isolated rat pinealocytes: an immunocytochemical study. Cell Tissue Res 282: 219-226

Wicht H, Korf HW, Schaad NC (1993) Morphological and immunocytochemical heterogeneity of cultured pinealocytes from one-week and two-month-old rats: planimetric and densitometric investigations. J Pineal Res 14: 128-137

Yamamoto KK, Gonzalez GA, Biggs WH, Montminy MR (1988) Phosphorylation-induced binding and transcriptional efficacy of nuclear factor CREB. Nature 334: 494-498

Yuwiler A (1983) Vasoactive intestinal peptide stimulation of pineal serotonin-N-acetyltransferase activity. J Neurochem 41: 146-153

27 Morphological Characterization of Rat Pineal Explant Cultures

M. Møller[1], S. Micic[2], and J.C. Reckling[3]

[1]Department of Medical Anatomy, Panum Institute, Blegdamsvej 3, DK-2200 Copenhagen, Denmark; Phone: +45 3532 7258, Fax: +45 3536 9612, E-mail: morten.m@mai.ku.dk
[2]Biostatistical Department, Statens Seruminstitut, Copenhagen Denmark
[3]Institute of Medical Physiology, University of Copenhagen, Denmark

Abstract

Pineal explants from newborn rats were cultured with the roller-tube technique for 14 days. The microscopic morphology of the pineal glands was preserved after 14 days of culture. Electron microscopic analysis showed the explants to consist of classical pinealocytes, interstitial cells and blood-vessels. Junctional complexes could be observed between the pinealocytes. Nerve fibers were not present. Only a few of the pinealocytes exhibited S-antigen immunoreactivity. The explant cultures released melatonin continuously to the culture medium. Thus, cultured pineal explants serve as a model closer to the in vivo situation than single cell cultures of pinealocytes.

Introduction

The technique of culturing single cells of the mammalian pineal gland was originally described by Klein and coworkers (1973). Such cultures have been successfully used in studies of receptors (Mess et al. 1991; Olcese 1991; Simonneaux et al. 1994), signal transduction factors, and second messengers (Baler and Klein 1995; Korf et al. 1996; Maronde et al. 1995; Spessert et al. 1995) of the pinealocytes. However, dispersed cultured pinealocytes can change their phenotypic expression (Araki et al. 1994) due to both the culture medium and possibly the lack of cellular contact complexes present in the intact organ between the pinealocytes (Møller 1976). Therefore, the development of organotypic explant cultures in mammals (Gähwiler 1981), a technique later adapted for the pineal gland by McNulty and colleagues (1995) has provided us with a new tool in pineal research, where the morphological organization of the gland is better preserved.

In this investigation we have studied the morphology of rat pineal explant cultures by immunocytochemistry and electron microscopy. In addition, we have investigated whether such explant cultures synthesize melatonin.

Materials and Methods

Culture Technique

Whole mount cultures of the pineal gland were prepared according to the roller-tube technique (Gähwiler 1984). In brief, the pineal gland was removed from 1-day-old Wistar rats, and placed on a cleaned glass coverslip, embedded in citrated chicken plasma (Sigma, lyophilized). The plasma was coagulated by addition of a drop of thrombin (Hoffmann-La Roche), 3 to 6 mg/ml in Gey's medium (GIBCO, GBSS). The pineal glands were cultured in sealed plastic tubes (flat bottom #1-56758, Nunc), placed in a rotating roller-drum (6 or 15 revolutions/h) at 36.5 °C in humidified air. The medium composed of horse serum (25%), basal medium Eagle (50%) and Hanks's balanced salt solution (25%, GIBCO). Glutamine (1 mM) and extra glucose (3.6 mM) were added. The cultures were fed once weekly (0.75 ml medium) and no anti-mycotics were used.

Immunocytochemistry

After a culture period of 14 days, the cultures were washed 3 x 5 min in phosphate buffered saline (PBS) and fixed in 4% paraformaldehyde in phosphate buffer (pH 7.4) at 4 °C for 1 h. The cultures were then reacted for S-antigen immunoreactivity by use of the ABC streptavidin-biotin-complex method in the following way: the cultures were rinsed 3 x 10 min in PBS with 0.1% Triton X-100 and 0.25% bovine serum albumin (PBS-TB) and incubated in 4% swine serum in PBS-TB for 20 min. This was followed by incubation in the specific S-antigen antibody (kindly provided by Dr. Larry A. Donoso, Wills Eye Hospital, Philadelphia) diluted 1:1000 in PBS with 0.25% bovine serum albumin and 0.3% Triton X-100 for 24 hours at 4 °C in a humid chamber. After 3 x 10 min washes in PBS-TB, the sections were then incubated in biotinylated swine anti-mouse IgG for 1 h at room temperature. This was followed by 3 x 10 min washing in PBS-TB and incubation in streptavidin-biotin-horseradish peroxidase complexes, diluted 1:200 in PBS-TB, for 1 h at room temperature. Finally, the sections were washed 2 x 10 min in PBS-TB, and 10 min in TRIS/HCl buffer (pH 7.6) followed by 5-15 min incubation in 0.05% 3,3'-diaminobenzidine/HCl with 0.005% H_2O_2 in TRIS/HCl, air dried and coverslipped with Depex.

Electron Microscopy

Fourteen-day-old cultures were washed 3 x 5 min in PBS at 4 °C and fixed for 2 h in 2% glutaraldehyde in 0.1 M phosphate buffer (pH 7.4) followed by fixation in 2% OsO_4 in distilled water for 2 h. After washing 3 x 5 min in PBS at 4 °C the specimens were dehydrated in an alcohol series, and embedded, via propylene oxide, in Epon. Thin sections with a silver to grey interference color were cut on a Reichert ultratome and poststained with uranyl acetate and lead citrate. The thin sections were examined and photographed in a Philips 400 electron microscope operated at 60 kV.

Determination of Melatonin Secretion

Fourteen-day-old cultures were used. In one control series, the explants were washed and fixed in 4% paraformaldehyde in phosphate buffer for 1 h. The fixative was then decanted and, after washing of 3 x 10 min in PBS, the culture medium was added. After 1 h of incubation (at 37 °C), 50 μl of the culture medium was assayed for melatonin by radioimmunoassay by using a commercially available kit (Nichols Institute Diagnostics, USA). In a second series, fresh medium was added to the living cultures and 50 μl of the medium was assayed for melatonin after 20 and 60 min.

Results

Light Microscopy

Two μm thick survey sections of the pineal explant cultures (Fig. 1) showed the culture to consist of a center containing pinealocytes, interstitial cells and blood vessels surrounded by a capsule of larger, less densely stained fibroblast- and macrophage-like cells. Nerve fibers were not observed in the pineal explant.

Immunocytochemistry for S-Antigen

Only a few of the pinealocytes in the explant showed S-antigen immunoreactivity. Some of the immunoreactive cells were neuron-like with a cell body from which long cellular processes with varicosities emerged (Fig. 2).

Electron Microscopy

Thin sections from the periphery of the explant (Fig. 3) showed the pinealocytes to be covered with a capsule of macrophage- and fibroblast-like cells as well as many collagen fibers. The pinealocytes exhibited an electron-lucent cytoplasm (Figs. 3-6). Between the pinealocytes, darker stained interstitial cells were observed (Fig. 6). The pinealocytes looked immature, with a high number of free ribosomes. However, rough endoplasmic reticulum was also present and many mitochondria, microtubules, filaments and some lysosomes were present. The number of synaptic ribbons was low. Several processes extended from the pinealocytes, especially in the peripheral part of the explant (Fig. 4). Often, the processes terminated in boutons with secretory granules. Several intermediate junctions were observed. The perivascular spaces of the pineal explant contained capillaries and collagen fibers. Nerve fibers were not observed.

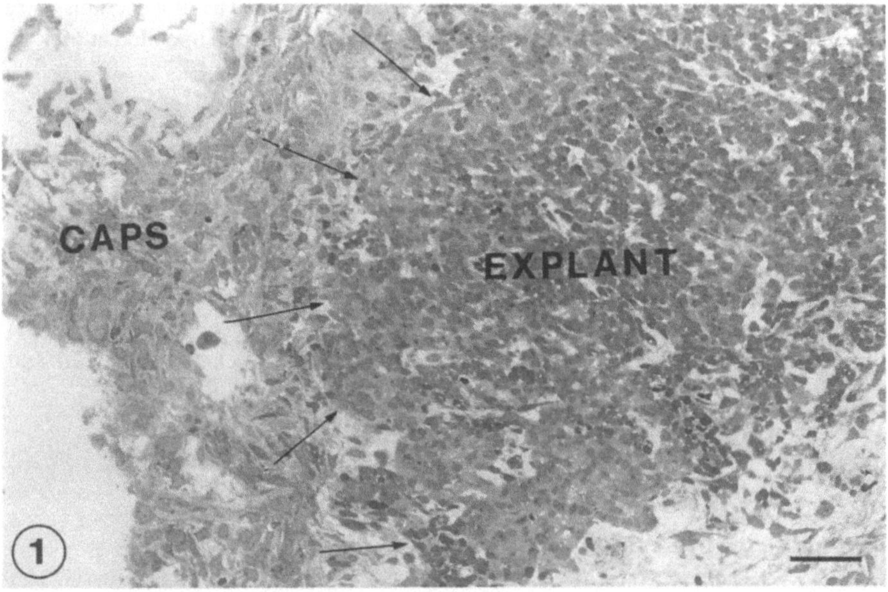

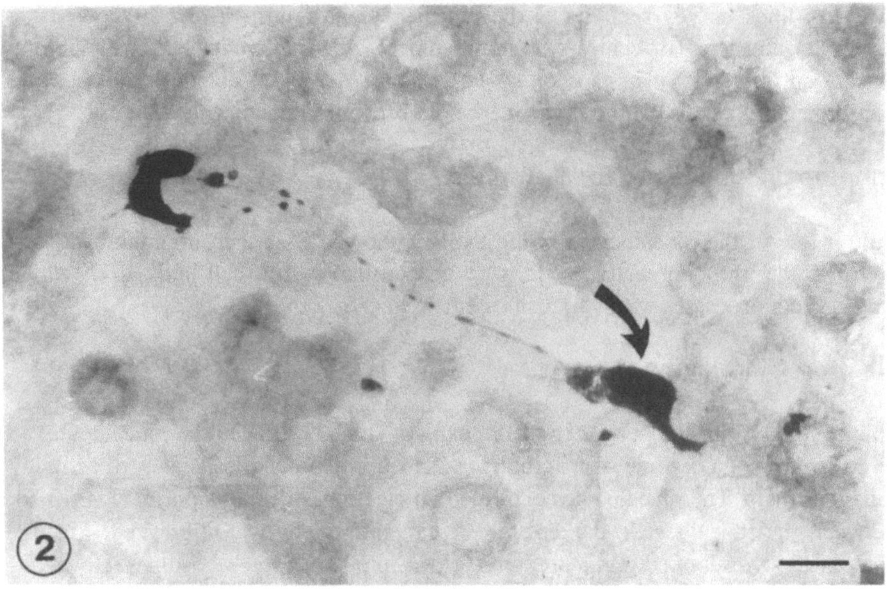

Fig. 1. Semi-thin section of a pineal explant culture. The pineal explant (*EXPLANT*) is surrounded by a capsule (*CAPS*) of fibroblast- and macrophage-like cells. The border of the explant is delineated by *thin arrows*. Toluidine blue staining. *Bar* 100 μm. **Fig. 2.** S-antigen immunoreactive cell body (*bent arrow*) with a long process endowed with varicosities in a 14-day-old pineal explant culture. *Bar* 10 μm

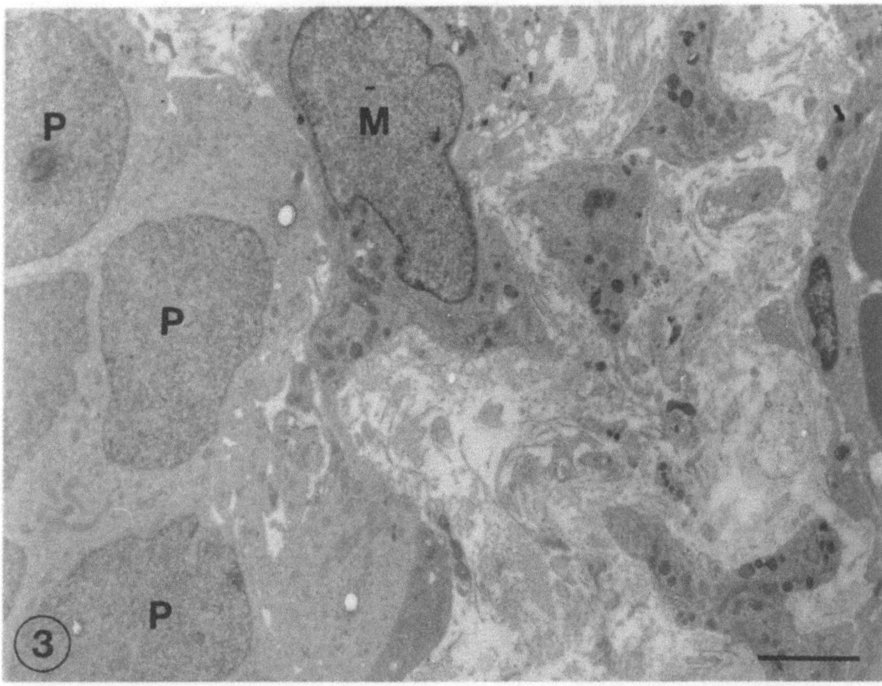

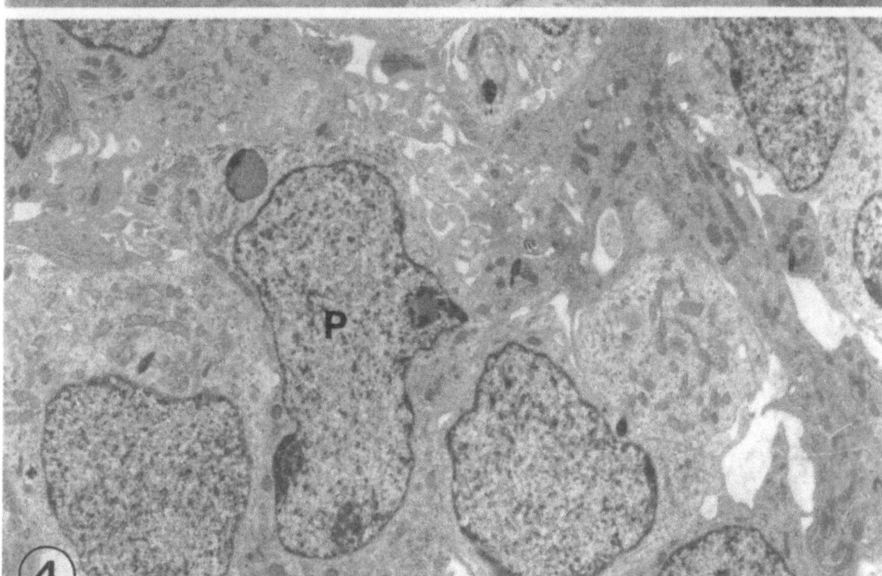

Fig. 3. Electron micrograph from the lateral part of the explant where the pinealocytes (*P*) are encapsulated by a macrophage-like cell (*M*) and a rich network of collagen fibers. *Bar* 5 µm.

Fig. 4. Electron micrograph of the lateral part of the explant containing pinealocytes (*P*) from which many small and large processes emerge. *Bar* 5 µm

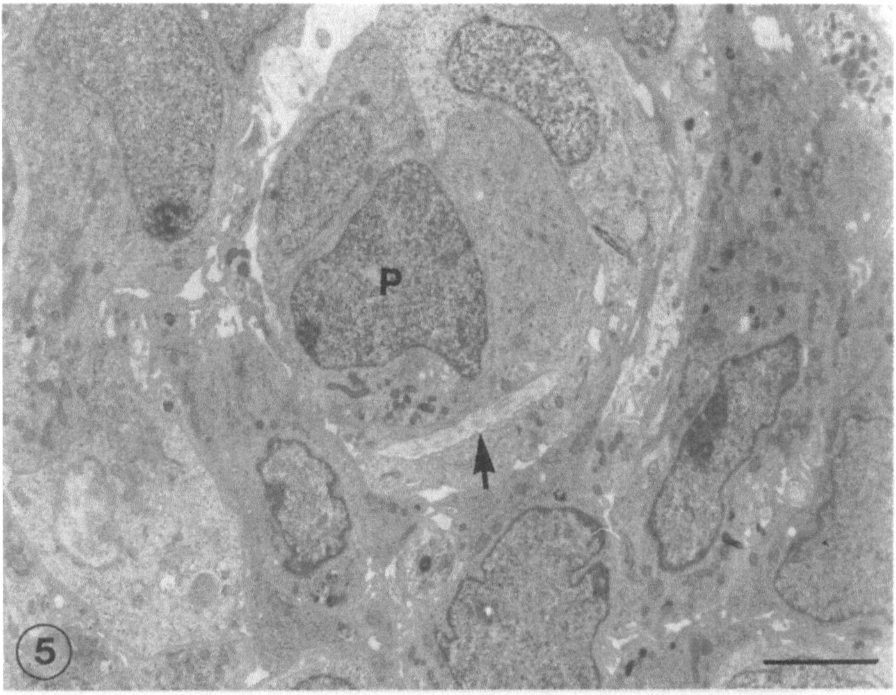

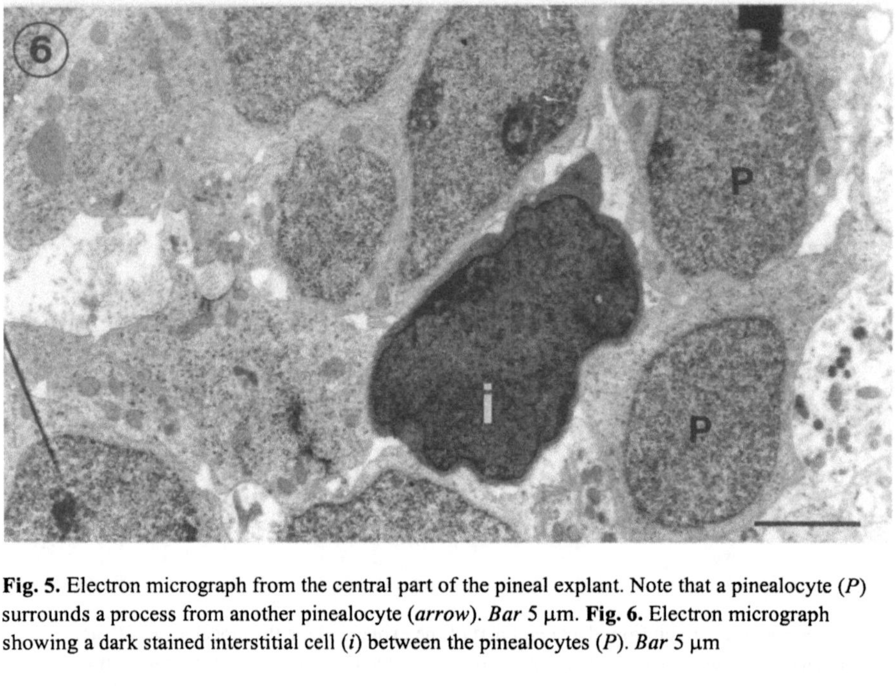

Fig. 5. Electron micrograph from the central part of the pineal explant. Note that a pinealocyte (*P*) surrounds a process from another pinealocyte (*arrow*). *Bar* 5 μm. **Fig. 6.** Electron micrograph showing a dark stained interstitial cell (*i*) between the pinealocytes (*P*). *Bar* 5 μm

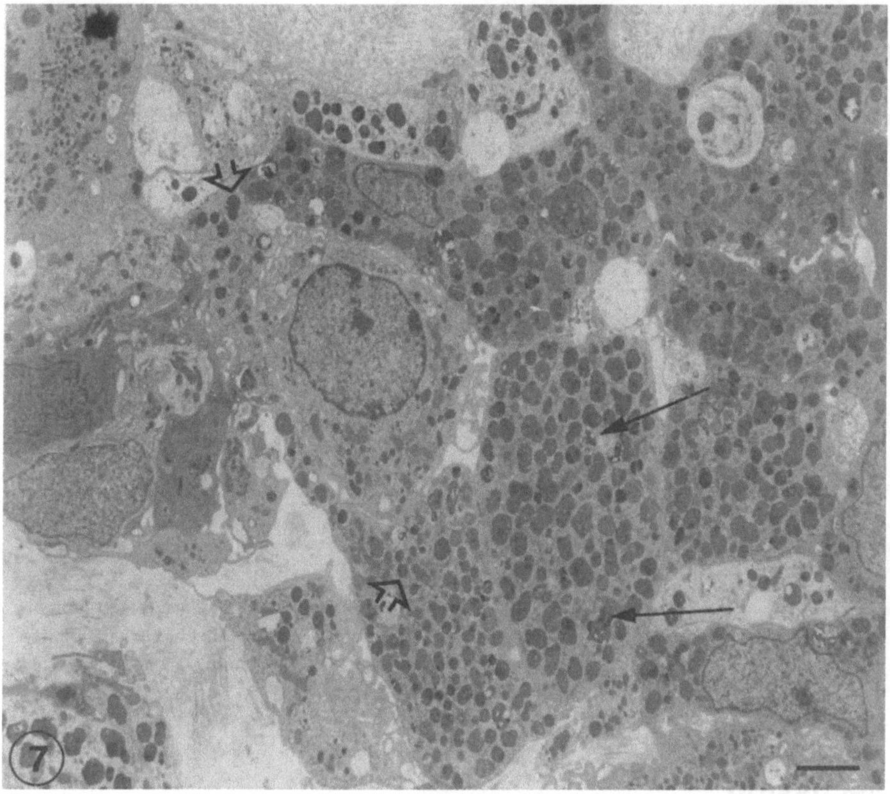

Fig. 7. Electron micrograph from the capsular area surrounding the explant. Many macrophage-like cells with primary (*open arrow*) and secondary (*thin dark arrows*) lysosomes are present. *Bar* 5 µm

The cells located in the "capsule" were endowed with a high number of mitochondria, rough endoplasmic reticulum, and many lysosomes (Fig. 7). The intercellular spaces between the capsular cells were large and contained a high number of collagen fibers. Some capillaries were also present in the peripheral part of the "capsule".

Melatonin Determination

As seen from Fig. 8, the pineal explant continuously secreted melatonin into the culture medium. After fixation of the explant with 4% paraformaldehyde no melatonin could be detected in the culture medium.

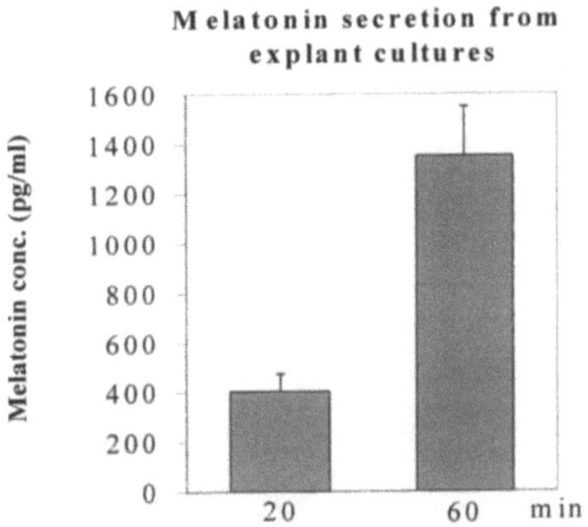

Fig. 8. Bar graph showing the concentration of melatonin in the culture medium 20 and 60 min after adding fresh medium. 14-day-old cultures; n = 5

Discussion

The present ultrastructural analysis showed the morphology of the pineal gland to be well preserved in the explant cultures after 14 days of culturing. Thus the pineal remained as a solid structure surrounded by fibroblast- and macrophage-like cells. A rich capillary network was also present in the gland.

Interstitial cells were also found in the pineal gland with location and morphology as in the adult pineal gland. The interstitial cells of the rat exhibit a morphology like fibrillar astrocytes. Whether the interstitial cell of the pineal has any functional role in pineal biochemistry, as the astrocyte has in the brain (Magistretti et al. 1993), has to be elucidated. Recent studies of the rat indicate the pineal interstitial cell to be an antigen presenting cell taking part in immunological reactions in the gland (Kelly et al. 1993). Nerve fibers were not observed in the explant. This is in accord with the observation that no innervation is present in the newborn pineal gland (Machado et al. 1968).

The pinealocytes exhibited an immature morphology with many ribosomes, very few synaptic ribbons and a low amount of rough endoplasmic reticulum. However, junctional complexes were observed. These might be important in cellular communication between the pinealocytes. In single cell cultures we have never observed junctions between the pinealocytes.

The S-antigen is a retinal antigen often used as a marker for pinealocytes (Korf et al. 1985; Redecker et al. 1996; Schomerus et al. 1996; Kroeber et al. 1997). However, this antigen was found only in a population of pinealocytes in the rat pineal explant. Therefore, the S-antigen does not appear as a good marker for the pinealocytes in the explant culture.

Compared to single cell cultures, diffusion distances to the individual pinealocytes in the explant culture are longer. However, due to the relatively small size of the explant (about 100-300 mm) chemicals applied to the culture medium will reach the center of the explant in less than 1 min. Therefore, the pineal explant culture is an excellent model for studying receptors, signal transduction and transcription factors, and second messengers involved in melatonin secretion.

Acknowledgments

This study was supported by grants from Novo's Fonds Komite, the Danish Research Foundation (#12-0236, #12-1486-1 and #12-1642-1), and by a grant from the Danish Research Foundation to the Biotechnology Centre for Cellular Communication. The authors are grateful to Mrs. Ursula Rentzmann for technical and Mrs. Grazyna Hahn for photographic assistance.

References

Araki M, Nonaka T, Akagawa K, Kimura H, Mashiko T (1994) Developing rat pineal cells manifest potential of neuronal differentiation in vitro. Neurosci Res 20: 57-69

Baler R, Klein DC (1995) Circadian expression of transcription factor Fra-2 in the rat pineal gland. J Biol Chem 270: 27319-27325

Gähwiler BH (1981) Organotypic monolayer cultures of nervous tissue. J Neurosci Methods 4: 329-342.

Gähwiler BH (1984) Slice cultures of cerebellar, hippocampal and hypothalamic tissue. Experientia 40: 235-308

Kelly JD, Fox LM, Lange CF, Bouchard CS, McNulty JA (1993) Experimental autoimmune pinealitis in the rat: ultrastructure and quantitative immunocytochemical characterization of mononuclear infiltrate and MHC class II expression. Autoimmunity 16: 1-11

Klein DC, Yuwiler A, Weller JL, Plotkin S (1973) Postsynaptic adrenergic-cyclic AMP control of the serotonin content of cultured rat pineal glands. J Neurochem 21: 1261-1271

Korf HW, Møller M, Gery I, Zigler JS, Klein DD (1985) Immunocytochemical demonstration of retinal S-antigen in the pineal organ for four mammalian species. Cell Tissue Res 239: 81-85

Korf HW, Schomerus C, Maronde E, Stehle JH (1996) Signal transduction molecules in the rat pineal organ: Ca^{2+}, pCREB, and ICER. Naturwissenschaften 83: 535-543

Kroeber S, Schomerus C, Korf HW (1997) Calcium oscillations in a subpopulation of S-antigen-immunoreactive pinealocytes of the rainbow trout (*Oncorhynchus mykiss*). Brain Res 744: 68-76

Machado CR, Wragg LE, Machado AB (1968) A histochemical study of sympathetic innervation and 5-hydroxytryptamine in the developing pineal body of the rat. Brain Res 8: 310-8

Magistretti PJ, Sorg O, Yu N, Martin JL, Pellerin L (1993) Neurotransmitters regulate energy metabolism in astrocytes: implications for the metabolic trafficking between neural cells. Dev Neurosci 15: 306-312

Maronde E, Middendorff R, Mayer B, Olcese J (1995) The effect of NO-donors in bovine and rat pineal cells: stimulation of cGMP and cGMP-independent inhibition of melatonin synthesis. J Neuroendocrinol 7: 207-214

McNulty JA, Tsai S-Y, Fox LM, Madsen TM, Silberman S, Tønder, N (1995) Neurotrophic effects of the pineal gland: role of non-neuronal cells in co-cultures of the rat pineal gland and superior cervical ganglia. J Pineal Res 19: 40-50

Mess B, Csernus V, Rékasi Z (1991) The role of different neurotransmitters and neuropeptides in the regulation of pineal hormone secretion. A dynamic in vitro study. Adv Pineal Res 6: 57-66

Møller M (1976) The ultrastructure of the human fetal pineal gland. 2. Innervation and cell junctions. Cell Tissue Res 169: 7-21.

Olcese J (1991) Neuropeptide Y: an endogenous inhibitor of norepinephrine stimulated melatonin secretion in the rat pineal gland. J Neurochem 57: 943-947.

Redecker P, Cetin Y, Korf HW (1996) Differential immunocytochemical localization of calretinin in the pineal gland of three mammalian species. J Neurocytol 25: 9-18

Schomerus C, Maronde E, Laedtke E, Korf HW (1996) Vasoactive intestinal peptide (VIP) and pituitary adenylate cyclase-activating polypeptide (PACAP) induce phosphorylation of the transcription factor CREB in subpopulations of rat pinealocytes: immunocytochemical and immunochemical evidence. Cell Tissue Res 286: 305-313

Simonneaux V, Ouichou A, Craft C, Pévet P (1994) Presynaptic and postsynaptic effects of neuropeptide Y in the rat pineal gland. J Neurochem 62: 2464-2471

Spessert R, Hill G, Vollrath L (1995) In rat pinealocytes the cyclic GMP response to NO is regulated by Ca^{2+} and protein kinase C. Brain Res 694: 207-212

28 Changes in the Three Dimensional Structure of Synaptic Ribbons in the Pineal Gland of the Guinea-Pig Caused by Constant Light

H. Jastrow[1], D. Schmanke[2], J. Weinert[3], U. Jendrysiak[3], and L. Vollrath[1]

[1]Anatomisches Institut, Johannes Gutenberg-Universität Mainz, Becherweg 13, D-55128 Mainz, Germany; E-mail: jastrow@goofy.zdv.uni-mainz.de, http://www.uni-mainz.de/~jastrow/

[2]Zentrum für Datenverarbeitung, J. Gutenberg-Universität, Bentzelweg 12, D-55128 Mainz, Germany

[3]Institut für Medizinische Statistik und Dokumentation (IMSD), J. Gutenberg-Universität, Obere Zahlbacher Str. 69, D-55101 Mainz, Germany

Abstract

Synaptic bodies (SBs) are dynamic synaptic organelles of afferent synapses of the retina, inner ear, lateral line organ and pineal gland in vertebrates. When investigated in a transmission electron microscope, their electron-dense rod-like, round or irregular profiles are surrounded by electron-lucent vesicles. The three-dimensional structure of pineal SBs is not precisely known. Pineal glands of two guinea-pigs (one kept under an LD cycle of 12:12 h; one kept in constant light for 8 weeks) were investigated. SBs were reconstructed in three dimensions to visualise morphological changes in constant light. Transmission electron microscope micrographs from up to 18 serial sections with a known distance (50 nm) of groups of SBs were scanned and processed (controlled superimposition of corresponding organelles from adjacent sections, involving rotation and shifting of the scanned image) by a self-written programme in the Interactive Data Language (IDL). The resulting 3-D array was further processed by NeurOPS (neurosurgical operation-planning and simulation software developed at the Institut für Medizinische Statistik und Dokumentation), allowing 3-D visualisation of SBs. The three-dimensional images were printed from different angles to elucidate the appearance of SBs in space. It was found that under constant light, SBs lie opposite one another in adjacent pinealocytes, closely related to the cell membrane. Further, bent plates appearing under the form of V- and U-shaped shields were seen. Smaller lumps or spherical masses of SB material were more often seen and plates orientated in parallel were less frequent.

Introduction

In vertebrates, synaptic bodies (SBs) are dynamic synaptic organelles of afferent synapses present in important sensory organs such as the retina and the inner ear (Sjöstrand 1974, 1976; Sobkowicz et al. 1982, 1986; Guth et al. 1993; Wagner 1997). They are also found in the pineal gland (Vollrath 1981; McNulty and Fox 1992; Bhatnagar 1994; Spiwoks-Becker 1995). When investigated in a transmission electron microscope (TEM), their electron-dense rod-like, round or irregular profiles are surrounded by electron-lucent vesicles. SBs of the guinea-pig pineal gland reconstructed in three dimensions were plate-like, spherical or bizarre lump-like structures in animals sacrificed at night and day (Jastrow et al. 1996) while in other animals, e.g. rhesus monkey (McNulty et al. 1986) or rats (Jastrow et al. 1997), they were mostly plate-like. Previous TEM studies have shown changes of SB profiles after animals' exposure to constant light, amongst which, most remarkably, was the occurrence of "paired synaptic ribbon fields" (Vollrath and Huss 1973; Vollrath 1986; Larisch 1989). In the present preliminary study we examined how the 3-D structure of SBs is affected by permanent illumination of the guinea-pig.

Materials and Methods

Pineal glands of two guinea-pigs (one kept under an LD cycle of 12:12 h; one kept in constant light for 8 weeks) were investigated. Pineal glands were removed in deep anaesthesia, fixed in a modified Karnovsky's (1965) solution of 2% paraformaldehyde and 2% glutaraldehyde in 0.1 M PBS (phosphate buffered saline) at 4 °C for 15 h. After washing in 0.1 M PBS, pH 7.4, with 6.8% sucrose, the glands were postosmicated for 90 min (2% OsO_4 in 0.1 M PBS, pH 7.4), block-stained in 0.5% uranyl acetate plus 1% phosphotungstic acid in 70% acetone overnight, dehydrated in acetone and embedded, randomly orientated, in Epon. Serial sections were cut with diamond knives on a Reichert Ultratome III set to a section-thickness of 50 nm, mounted on one-hole grids and stained according to Reynolds (1963). Transmission electron microscope micrographs of up to 18 serial sections of groups of SBs were scanned with an HP ScanJet 4c at a resolution of 600 dots per inch and processed. Superimposition of corresponding profiles of organelles from adjacent sections involving rotation and shifting of the scanned image under visual control was realized by a self-written programme in the Interactive Data Language (IDL). The resulting 3-D data array was further processed by the Neurosurgical Operation-Planning and Simulation software (NeurOPS) developed at the Institute for Medical Statistics and Documentation (IMSD) allowing 3-D visualization of SBs. Images were printed from different angles to elucidate the appearance of SBs in space. In total, over 50 SBs were reconstructed (~ 25 from the LD animal, ~ 25 from the guinea-pig kept under permanent illumination).

Results

Three-dimensional reconstructions in both animals showed SBs always arranged in fields. The latter consisted of at least 5 to over 20 individual SBs. Most fields were composed of a mixture of all three forms of SBs (plates, irregular lumps and spheres). In a field, up to five plate-like or irregular SBs had contact with the cell membrane via ill-defined filamentous material. In the control animal, true plate-like SBs, partly in parallel orientation, clearly predominated (Figs. 1, 2). Under a normal lighting regime, plates orientated in parallel were less frequent than under constant light. Small lumps or spherical masses of SB were more often seen in the guinea-pig kept in constant light than in the LD-animal. Furthermore, under permanent illumination, some SBs resembled gutters with protrusions at their edges (Figs. 4a, b) and very bizarre, bent plates appearing under the form of V- and U-shaped shields were present (Figs. 3a, b) often located on opposite sides of the cell membrane to which they were attached via electron-dense material. About one third of all SB-fields followed up in serial sections were paired, e.g. two SB-fields lay face to face on opposite sites of the cell membranes of two pinealocytes. Their small electron-dense connections to the membrane sometimes seemed to continue straight through the intercellular space to the opposite SB-field.

Discussion

The present results show that constant light increases the complexity of SB structure in the guinea-pig pineal. The observation that the rod-like profiles are indeed plates, and the fact that under constant light the SBs formed paired fields, indicates that the area for intercellular communication increases. If the electron-dense material of the synaptic bodies serves as a storage site for synaptic vesicles (Osborne and Thornhill 1972; Vollrath and Spiwoks-Becker 1997) or for their quick release (Bunt 1971; Gray and Pease 1971; Wagner 1997), a straight surface of SBs may be useful for quickest release of synaptic vesicles, since here a direct transport of the vesicles to their release sites is possible. In analogy to the retina where the firing rate is reduced under the influence of light (Spiwoks-Becker 1995; Juusola et al. 1996; Wagner 1997), one can expect a higher vesicle release under a normal day/night cycle than under permanent illumination. The fact that in the control animal straight SB plates predominated fits into this concept. In contrast, V- or U-like SB profiles deriving from folded (shield- or gutter-like) SBs will be more appropriate for SV storage. In this case the vesicles in the inner part of the folded ridges cannot move away as easily as those attached to a more or less flat plate.

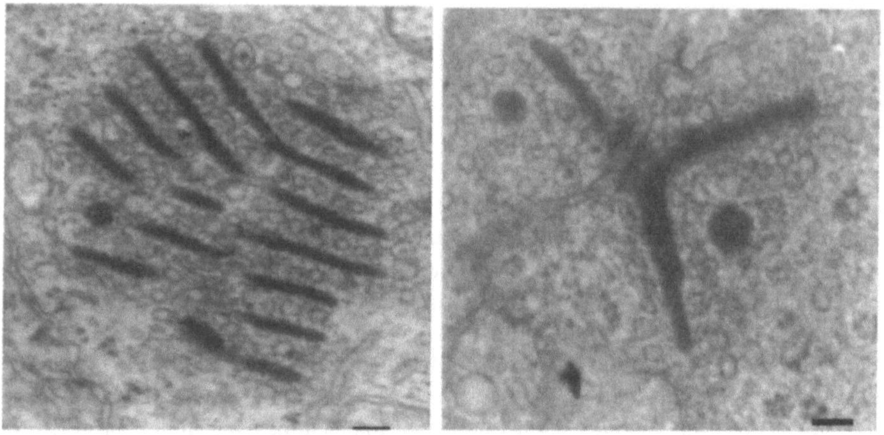

Fig. 1. Field of parallel oriented SB profiles (**left**). Paired SB field with round and bent profiles of the animal kept under LD 12:12 h (**right**). Note that SBs are attached to the cell membrane. *Bars* 100 nm

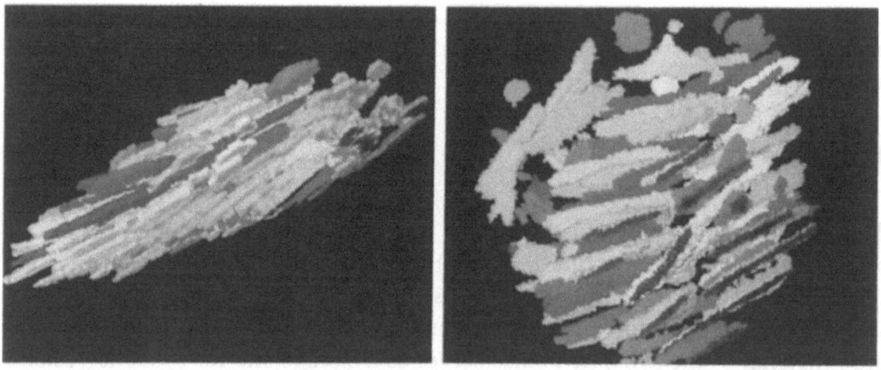

Fig. 2. Field of parallel oriented SB profiles of the animal kept under LD 12:12 h, seen from the side (**left**) and from lateral below (**right**). Note that the SBs are quite close to each other

The remarkable increase in lump-like (unorganized) SB formations under constant light also points to a disturbed arrangement of SB proteins. Synaptic bodies consist of smaller globular proteins (Usukura and Yamada 1987). Perhaps the assembly of these proteins is altered under constant light, thus preventing them reaching the highly organized state of parallel orientated plates.

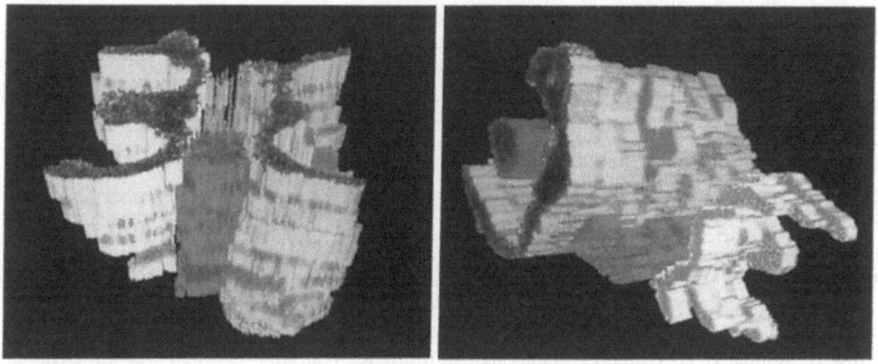

Fig. 3. Paired field of SBs on opposite sites of a cell membrane (not shown) in different views (**left, right**). Animal was kept under constant light for 8 weeks. The connections between the two SBs were also reconstructed

Fig. 4. Paired field in different views (**left, right**). Note many irregular, lump-like SBs adjacent to the gutter-like SBs. Animal was kept under constant light for 8 weeks

Acknowledgments

The authors thank Mrs. I. v. Graevenitz for her excellent technical assistance.

References

Bhatnagar KP (1994) Synaptic ribbons of the mammalian pineal gland: enigmatic organelles of poorly understood function. Adv Struct Biol 3: 47-94

Bunt AH (1971) Enzymatic digestion of synaptic ribbons in amphibian retinal photoreceptors. Brain Res 25: 571-577

Gray EG, Pease HL (1971) On understanding the origins of the retinal receptor synapses. Brain Res 35: 1-15

Guth P, Norris C, Fermin CD, Pantoja M (1993) The correlated blanching of synaptic bodies and reduction in afferent firing rates caused by transmitter-depleting agents in the frog semicircular canal. Hearing Res 66: 143-149

Jastrow H, Schmanke D, Weinert J, Jendrysiak U, Merle K, Vollrath L (1996) Three-dimensional reconstruction of pineal synaptic ribbons and spheres. Göttingen neurobiology report, vol 2. Thieme, Stuttgart, p 141

Jastrow H, von Mach MA, Vollrath L (1997) The shape of synaptic ribbons in the rat pineal gland. Cell Tissue Res 287: 255-261

Juusola M, French AS, Uusitalo RO, Weckström M (1996) Information processing by graded-potential transmission through tonically active synapses. Trends Neurosci 19: 292-297

Karnovsky MJ (1965) A formaldehyde-glutaraldehyde fixative of high osmolality for use in electron microscopy. J Cell Biol 27: 137-138

Larisch R (1989) Über den Einfluß von Dauerbelichtung auf die synaptischen Strukturen in der Meerschweinchenzirbeldrüse. Inauguraldissertation, J Gutenberg University, Mainz, Germany

McNulty JA, Fox L (1992) Pinealocyte synaptic ribbons and neuroendocrine function. Microsc Res Tech 21: 175-187

McNulty JA, Fox L, Taylor D, Miller M, Takaoka Y (1986) Synaptic ribbon populations in the pineal gland of the rhesus monkey (*Macaca mulatta*). Cell Tissue Res 243: 353-357

Osborne MP, Thornhill RA (1972) The effect of monoamine depleting drugs upon the synaptic bars in the inner ear of the bullfrog. Z Zellforsch Mikrosk Anat 127: 347-355

Reynolds ES (1963) The use of lead citrate at high pH as an electron-opaque stain in electron microscopy. J Cell Biol 17: 208-212

Sjöstrand, FS (1974) A search for the circuitry of directional selectivity and neural adaptation through three-dimensional analysis of the outer plexiform layer of the rabbit retina. J Ultrastruc Res 49: 60-156

Sjöstrand, FS (1976) The outer plexiform layer of the rabbit retina, an important data processing center. Vision Res 16: 1-14

Sobkowicz HM, Rose JE, Scott GE, Slapnick SN (1982) Ribbon synapses in the developing intact and cultured organ of Corti in the mouse. J Neurosci 7:942-957

Sobkowicz HM, Rose JE, Scott GE, Levenick CV (1986) Distribution of synaptic ribbons in the developing organ of Corti. J Neurocytol 15:693-714

Spiwoks-Becker I (1995) Untersuchungen zur Plastizität retinaler und pinealer Synaptic Ribbons. Inauguraldissertation, Fachbereich Biologie, J Gutenberg-University, Mainz, Germany

Usukura J, Yamada E (1987) Ultrastructure of the synaptic ribbons in photoreceptor cells of *Rana catesbeiana* revealed by freeze-etching and freeze-substitution. Cell Tissue Res 247:483-488

Vollrath, L (1981) The pineal organ. In: Oksche A, Vollrath L (eds) Handbuch der mikroskopischen Anatomie des Menschen. Springer, Berlin, Heidelberg, New York, pp 1-665

Vollrath L (1986) Inverse behavior of "synaptic" ribbon and spherule numbers in the pineal gland of male guinea-pigs exposed to continuous illumination. Anat Embryol 173: 349-354

Vollrath L, Huss H (1973) The synaptic ribbons of the guinea-pig pineal gland under normal and experimental conditions. Z Zellforsch 139: 417-429

Vollrath L, Spiwoks-Becker I (1997) Plasticity of retinal ribbon synapses. Microsc Res Tech 35: 472-487

Wagner HJ (1997) Presynaptic bodies ("ribbons"): from ultrastructural observations to molecular perspectives. Cell Tissue Res 287: 235-446

Annotations

IDL is a registered trademark of Research Systems, Inc. 2995 Wilderness Place, Suite 203, Boulder Colorado 80301, USA

NeurOPS is copyrighted for the Institute for Medical Statistics and Documentation of the University of Mainz, Germany

29 Immunocompetent Cells in the Rat Pineal Gland with Special Reference to Macrophages and Dendritic Cells

T. Sato

Department of Anatomy (Division I), School of Medicine, Tokyo Medical and Dental University, Yushima 1-5-45, Bunkyo-ku, Tokyo, 113 Japan;
Phone: +81 3 5803 5141, Fax: +81 3 3818 7170,
E-mail: t.sato.ana 1 @med.tmd.ac.jp

Abstract

Specific cell populations in the rat pineal gland have been stained with monoclonal antibodies (mAbs) recognizing macrophages/microglia (MRC OX42), monocytes/macrophages (ED1), tissue macrophages (ED2), and MHC class II (Ia) antigen (MRC OX6). OX42, ED2, and OX6 elicit membrane staining and ED1 a patchy granular label within the cytoplasm. At the electron microscopic level, reaction product for OX42 and OX6 is confined to the plasma membrane and the intracellular vacuoles, suggesting internalization of the antigen, while that for ED1 is confined to the lysosomal membrane within the cytoplasm. The cytoplasm of OX6-positive cells contains few organelles, variously sized empty vacuoles, a few electron-dense lysosome-like structures, and well-developed hyaloplasm, whereas OX42-labeled cells and ED1-labeled cells have more organelles, including many lysosome-like structures in the cytoplasm. Pinealocytes with synaptic ribbons have been often found adjacent to OX6-immunoreactive cells. OX42-positive cells and ED1-positive cells are much more densely distributed in the pineal than ED2-positive cells. Double-immunoperoxidase staining for OX6, OX42, and ED1 reveals that all the pineal macrophages/microglia are stained by OX42 and can be divided into two groups of $OX42^{+}ED1^{+}OX6^{-}$ and $OX42^{+}ED1^{-}OX6^{-}$. The majority of OX6-positive cells in the pineal are negative for both OX42 and ED1. The number of cells expressing MHC class II (Ia) antigen significantly increases in the pineal gland of rats after treatment with carbon tetrachloride ($p < 0.005$). The results indicate that at least two different types of immunocompetent cells, dendritic cells and macrophages, are located in the pineal gland and that they are favorable candidates for interactions of this neuroendocrine organ with the immune system.

Introduction

There is now considerable evidence for interactions between the immune and neuroendocrine systems which include the behavioral conditioning of immune responses, the effects of stimulation or lesion of neuroendocrine organs on immune system functions, the effects of stressors on immune responses and tumor growth, and physiological and neurochemical changes in the brain during immune responses. These links probably involve glucocorticoids secreted from the adrenal gland, catecholamines and neuropeptides secreted by sympathetic terminals and the adrenal medulla, certain pituitary and gonadal hormones, and polypeptides produced by cells of the immune system (Dunn 1988, 1989). It is now becoming clear that immune and neuroendocrine functions cooperate in a closely interwoven network to protect the organism against various environmental attacks. For example, antigens or pathogenic microorganisms that may be considered as a special type of environmental challenge can induce the synthesis and secretion of stress hormones, such as proopiomelanocortin gene products, both centrally and peripherally (Maestroni and Pierpaoli 1981; Smith and Blalock 1986; Besedovsky et al. 1986). Basic environmental information, such as the light cycle or temperature, is transduced by the pineal gland into signals capable of modulating most neuroendocrine mechanisms. The pharmacologic inhibition of melatonin (MEL) synthesis can be obtained by injection of the β-adrenergic blocker propranolol in the evening and by pretreatment with para-chlorophenylalanine, a blocker of serotonin synthesis. The immunologic impairment induced by these interventions is completely reversed by exogenous MEL administered in the late afternoon (Maestroni et al. 1986a, b). These findings have prompted an investigation into the role of the pineal gland and its neurohormone, MEL, in immunity. Furthermore, there is also evidence for links in the reverse direction. Immunocompetent cells can produce a variety of polypeptides that can act as communicators, not only with other immunocompetent cells, but also with other organs of the body including the brain. Such peptides, which include the interferons (IFN) and interleukins (IL), are hormones produced by cells of the immune system. They have been named immunotransmitters. Recent studies have revealed that the lymphokine, IFN-γ, modulates pineal MEL production (Withyachumnarnkul et al. 1990, 1991). Autoimmune inflammation of the pineal gland has been demonstrated in conjunction with the well-established model of autoimmune retinal uveitis in response to proteins shared by both organs (Shinohara et al. 1990; Dua et al. 1991).

Functional studies have shown that the macrophage lineage can be divided into two principal groups. The phagocytotic macrophage compartment comprising tissue macrophages such as Kupffer cells and lung alveolar macrophages has a microbicidal/phagocytotic role (Cline et al. 1978). This type is immunoreactive for mAbs to the macrophage-related antigens ED1 and ED2, and expresses other macrophage antigens such as the complement receptor type (CR3). In contrast, cells of the immune macrophage compartment, comprising follicular dendritic cells, interdigitating cells and Langerhans cells, have an accessory role in the immune response (Tew et al. 1982; Schuler and Steinman 1985), particularly in terms of presenting antigen to CD4 helper T cells. CD4 helper T cells recognize foreign antigens only in connection with major histocompatibility complex (MHC) class II molecules (Benacerraf 1981). An immunohistochemical demonstration of MHC class II antigen or Ia determinant on the cell surface characterizes cells belonging to the antigen-presenting cell family; under normal conditions, this family comprises macrophages, dendritic cells, and B lymphocytes.

The brain was commonly held to be isolated from the immune system and thus as being immunologically privileged. However, it is now recognized that this concept is no longer tenable and that any 'privilege' is only partial (Widner and Brundin 1988). The presence of a population of resident macrophages within the central nervous system (CNS) raises the question of the possible role that these cells play in immune-mediated reactions within the CNS. Of major interest is whether microglia or other cells within the normal CNS express MHC antigens, in particular class II antigens, and whether these cells present antigens to effector cells of the immune system. Mononuclear phagocytes are three times more abundant than astrocytes in the rat pineal gland (Luo et al. 1984). Sato et al. (1996) have reported that cells expressing MHC class II antigen have a morphology similar to peripheral dendritic cells in the pineal gland of the rat. The present study has sought to examine immunocytochemically the distribution and morphology of macrophages and dendritic cells in the rat pineal gland, at both the light and electron microscope level, using the monoclonal antibodies, MRC OX42, ED1, ED2, and MRC OX6. In addition, the cell type of the pineal interstitial cells that express the MHC class II antigen has been confirmed using double-immunoperoxidase staining. Furthermore, the cell population kinetics of dendritic cells in the rat pineal gland have been investigated at the inflammatory stage, induced by administering carbon tetrachloride (CCl_4).

Materials and Methods

Animals

All animals were kept in a specific pathogen-free environment, with 12 h :12 h light:dark cycles in an animal room and with access to food and water ad libitum. All procedures followed the guidelines of the Tokyo Medical and Dental University for Animal Research. Animals were anesthetized with ether and sacrificed during the middle of the light period by perfusion through the heart with periodate-lysine-paraformaldehyde (PLP) fixative. Brain tissue, including pineal glands, was obtained from six male, 7-week-old Wistar rats (200-220 g). The tissue blocks were cryoprotected with 15% sucrose (12-16 h) and embedded in Tissue-Tek O.C.T. compounds (Miles, Elkhart, Indiana, USA); 6 μm-thick sections were cut at -20 °C with a cryostat. Cryostat sections were collected in six parallel serial sets on slides coated with gelatin-chrome-alum. Male, 6-week-old Wistar rats (180-200 g) were given a single subcutaneous injection of 0.3 ml/kg body weight of 10% CCl_4 in liquid paraffin, 8 and 20 h before death. Controls were untreated animals or animals injected only with liquid paraffin. Four rats were used per stage.

Immunohistochemistry

Various immunocompetent cells were immunohistochemically identified by use of the streptavidin-biotin-peroxidase complex (SAB) method and the peroxidase anti-peroxidase (PAP) method. The latter was used only for double staining as described below. Specific cellular populations in the pineal gland were stained with monoclonal antibodies (mAbs) that recognized a cytoplasmic antigen in monocytes and in monocyte-derived macrophages (ED1; Dijkstra et al. 1985; Damoiseaux et al. 1994), membrane antigens of tissue macrophages (ED2; Dijkstra et al. 1985; Barbe et al. 1990), the complement receptor type 3 (CR3) on macrophages (MRC OX42; Robinson et al. 1986; Ling et al. 1990), MHC class II (Ia) antigen (MRC OX6; McMaster and Williams 1979), and cells in S-phase (proliferating cell nuclear antigen: PCNA; Sanders et al. 1993). ED1 and ED2 were purchased from Serotec (Blacktorn, Bicester, UK), MRC OX42 from Pharmingen (San Diego, Calif., USA), MRC OX6 from Cedarlane (Hornby, Ontario, Canada), and PCNA from Dako (Copenhagen, Denmark). Sections to be stained with mAbs were blocked with 20% normal goat serum in phosphate-buffered saline (PBS) for 30 min. Primary antibodies were used at an appropriate dilution of each mAb (OX6, 1:1000 to 1:2000; ED1 and ED2, 1:400 to 1:1000; OX42, 1:50 to 1:100; PCNA, 1:100) in 1% bovine serum albumin (BSA; Sigma, St. Louis, Missouri, USA) in PBS with 0.03% Triton X-100 (Wako, Tokyo, Japan) overnight at room temperature. Sections were washed in PBS without Triton X-100 (3 x 20 min). Endogenous peroxidase was blocked with hydrogen peroxide (0.3%) in 100% methanol. The sections were incubated in biotinylated goat anti-mouse Ig (1:600; Dako), diluted in 1% BSA in PBS, and then incubated (30 min, room temperature) with peroxidase-conjugated streptavidin (Dako), diluted 1:300. The immunoreaction was visualized using 0.025% 3,3'-diaminobenzidine tetrahydrochloride (DAB; Sigma) and 0.01% hydrogen peroxide in 0.05 M TRIS-HCl buffer (pH 7.3) for 10 min. After being counterstained with hematoxylin, slides were dehydrated and mounted in Permount. The specificity of the immunoreactive staining was controlled by incubation of adjacent sections with normal mouse ascites matched for protein concentration, in place of the primary antibodies.

Quantitative analysis was performed in two representative sections of each specimen. The number of cells immunoreactive for OX42, ED1, ED2, and OX6 per unit area was calculated in order to show the distributional density of macrophages and dendritic cells in the pineal organ. For assessment of labeled cells, only cells revealing a distinct nucleus were counted.

Double Immunohistochemical Staining

When double staining was performed, the SAB and PAP methods were sequentially used according to the method of Hsu and Soban (1982), with some modifications. In the first staining sequence, DAB-H_2O_2 was used as the chromogenic substrate and yielded a brown color. After the first sequence was completed, the sections were washed with TRIS-HCl buffer and then PBS, before proceeding to the second staining sequence. The sections were again incubated with mAb at 37 °C for 2 h. Sections were then incubated with goat anti-mouse IgG (diluted 1:50; Dako) for 30 min at room temperature, followed by incubation with a complex of horseradish peroxidase and mouse anti-horseradish peroxidase (1:100, Dako) for 30 min at room temperature. The peroxidase activity was visualized by incubation in TrueBlue peroxidase substrate (TB; KPL, Md., USA) for 10 min. The reaction products were colored blue. The sections were dehydrated in a graded ethanol series and mounted in Permount.

For double staining, the following five slides were simultaneously stained so that the specificity and reliability of the staining could be checked: (1) SAB (mAb A) - PAP (mAb B), (2) SAB (B) - PAP (A), (3) SAB (A or B) - PAP (omission of the primary antibody), (4) SAB (omission of the primary antibody) - PAP (A or B), and (5) SAB (omission of the primary antibody) - PAP (omission of the primary antibody).

Quantification of OX6-Immunoreactive Cells in the Pineal Gland of Rats Treated with CCl_4

From cleared frozen preparations at 0 (controls), 8, and 20 h after a single subcutaneous injection of CCl_4, the area of the pineal organ was measured by use of an image processor and analyzer (Nikon, Cosmozone-1 SA) and the number of MRC OX6-immunoreactive cells per unit area (mm^2) was calculated to show the distributional density of immunoreactive elements in the analyzed portion of the pineal gland of animals at each stage. For the assessment of positively stained cells, only cells revealing a distinct nucleus were counted. Three sections of four rat pineals obtained at each stage after a single injection of CCl_4 were observed. The results were statistically compared among each stage using Student's *t*-test.

Electron Microscopic Immunocytochemistry

Pineals were processed according to the protocols used for light microscopy. Vibratome sections (30-50 μm in thickness) were collected directly into cold PBS for further processing as free-floating sections for immunocytochemistry. The same protocols were used for immunolabeling, except that none of the antibody solutions or washing buffers contained Triton X-100. Following immunocytochemistry, sections were treated with 1% osmium tetroxide in 0.1 M phosphate buffer (pH 7.4) for 1 h at room temperature, dehydrated in a graded series of ethanol, and embedded in Polybed 812 epoxy resin. Ultrathin sections were collected on 150-mesh fine-bar copper grids, counterstained with 0.4% lead citrate, and examined in a JEOL 100CX electron microscope operated at 100 kV.

Results

MRC OX42, ED1, and ED2 Immunohistochemistry in the Normal Pineal Gland

Numerous cells stained with MRC OX42 or ED1 mAbs were distributed throughout the gland, whereas only a few cells were immunoreactive for ED2 (Table 1). Some cells had an irregular amoeboid appearance, and others were fusiform in shape (Fig. 1a-c). OX42 and ED2 elicited membrane staining and ED1 a patchy granular label within the cytoplasm. ED1-positive cells and ED2-positive cells were frequently located in the connective tissues close to blood vessels. In other parts of the brain, ED1-positive cells and ED2-positive cells were situated exclusively near blood vessels, whereas OX42-labeled cells showed a microglia-like morphology and were distributed in a regular pattern, with each cell occupying a distinct territory. At the electron microscopic level, the peroxidase reaction product for ED1 delineated the phagolysosome-compartment (Fig. 3a). No staining of the surface membrane for ED1 was observed, whereas OX42 showed a reaction of the membranes (Fig. 2). Some reaction product was seen on the surface of tissue immediately adjoining the intensely labeled cells (Fig. 2a), apparently remaining there when the cells lost the attachment during fixation. The shape and the surface of OX42 cells (microglia) were irregular; the cells were spindle-shaped (Fig. 2a) or had elongated cytoplasmic processes (Fig. 2b). Both types were endowed with a well-developed lysosomal system.

Table 1. Number of immunocompetent cells per unit area in the rat pineal

Cell type	Number of cells per unit area
$OX42^+$	39.9 ± 5.1
$ED1^+$	29.0 ± 8.4
$ED2^+$	3.8 ± 4.3
$OX6^+$	21.6 ± 8.4

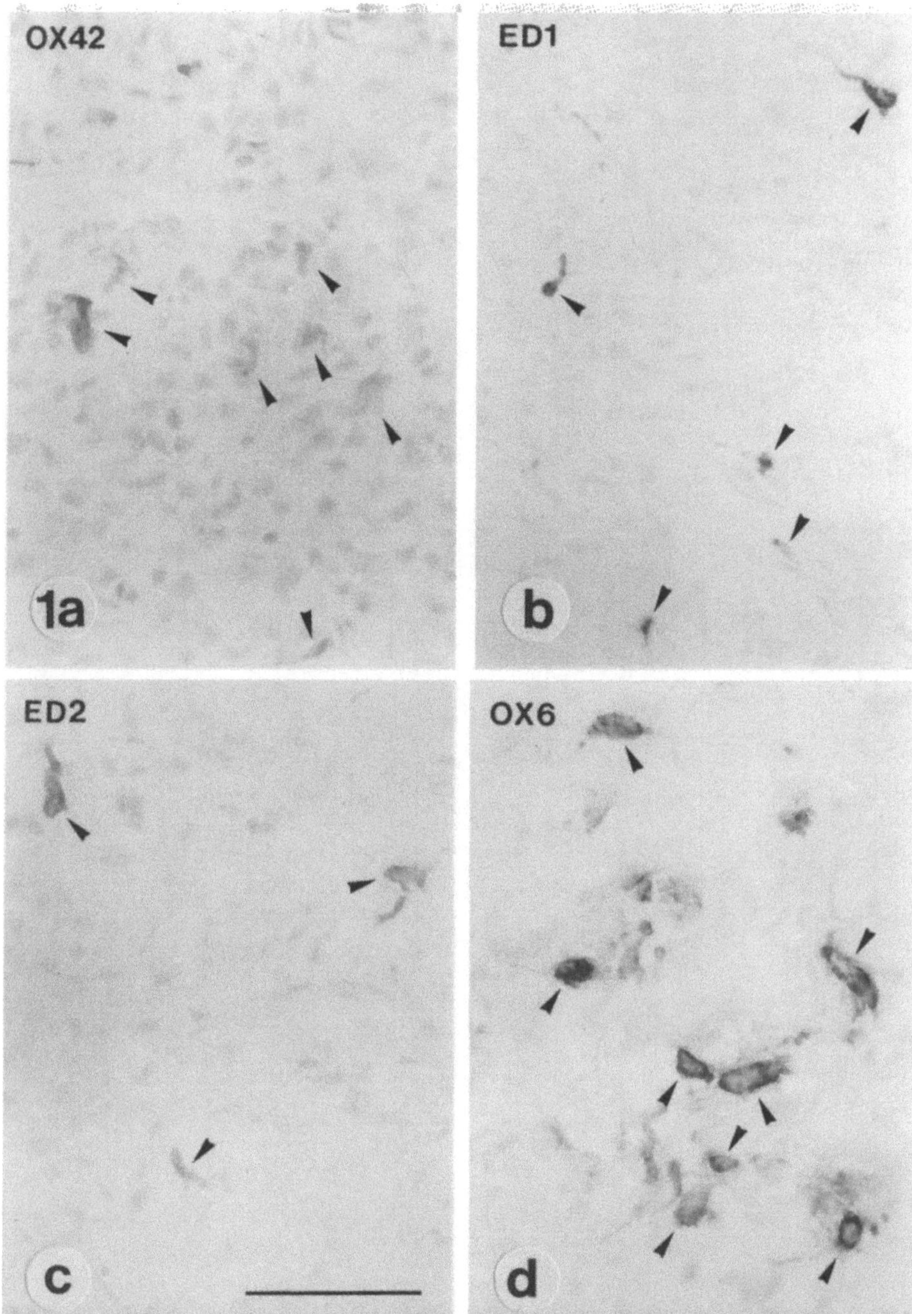

Fig. 1. Sagittal section of the pineal gland immunoreacted for *OX42* (**a**), *ED1* (**b**), *ED2* (**c**), and *OX6* (**d**). Several positive cells (*arrowheads*) are found in the pineal. OX42- and ED2-positive cells are rather weakly labeled, while ED1- and OX6-positive cells are intensely labeled. Counterstained with hematoxylin. *Bar* 50 μm for **a**, **b**, **c**, and **d**

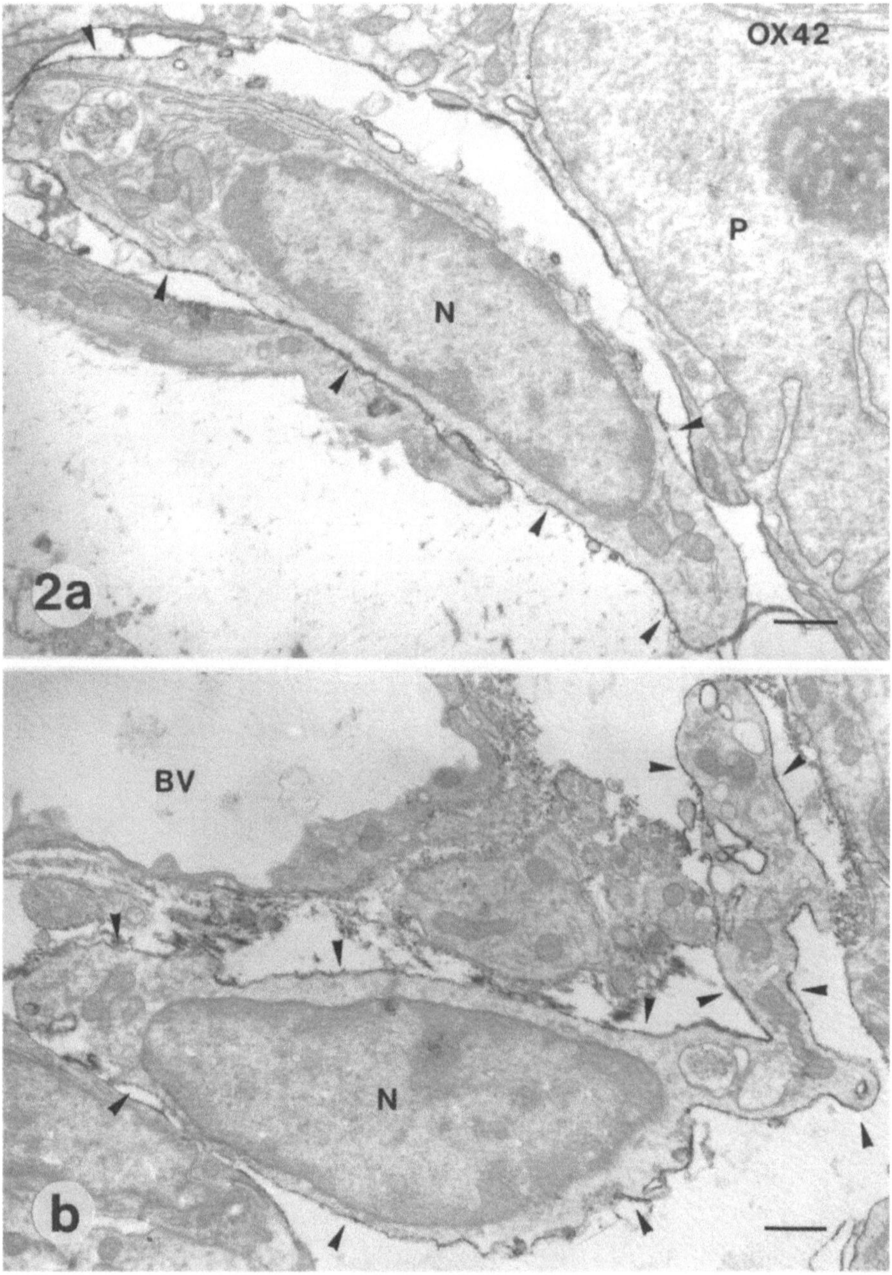

Fig. 2. Electron micrographs of cells immunolabeled for OX42. Immunolabel is in the membrane (*arrowheads*). The cells are endowed with a well-developed lysosomal system.
a A spindle-shaped *OX42*-positive cell exhibiting a smooth nuclear (*N*) envelope. Some reaction product is seen on the surface of tissue immediately adjoining the intensely labeled cell. *P* Pinealocyte. **b** The OX42-positive cell has elongated cytoplasmic processes. *BV* Blood vessel. *Bars* 1 μm

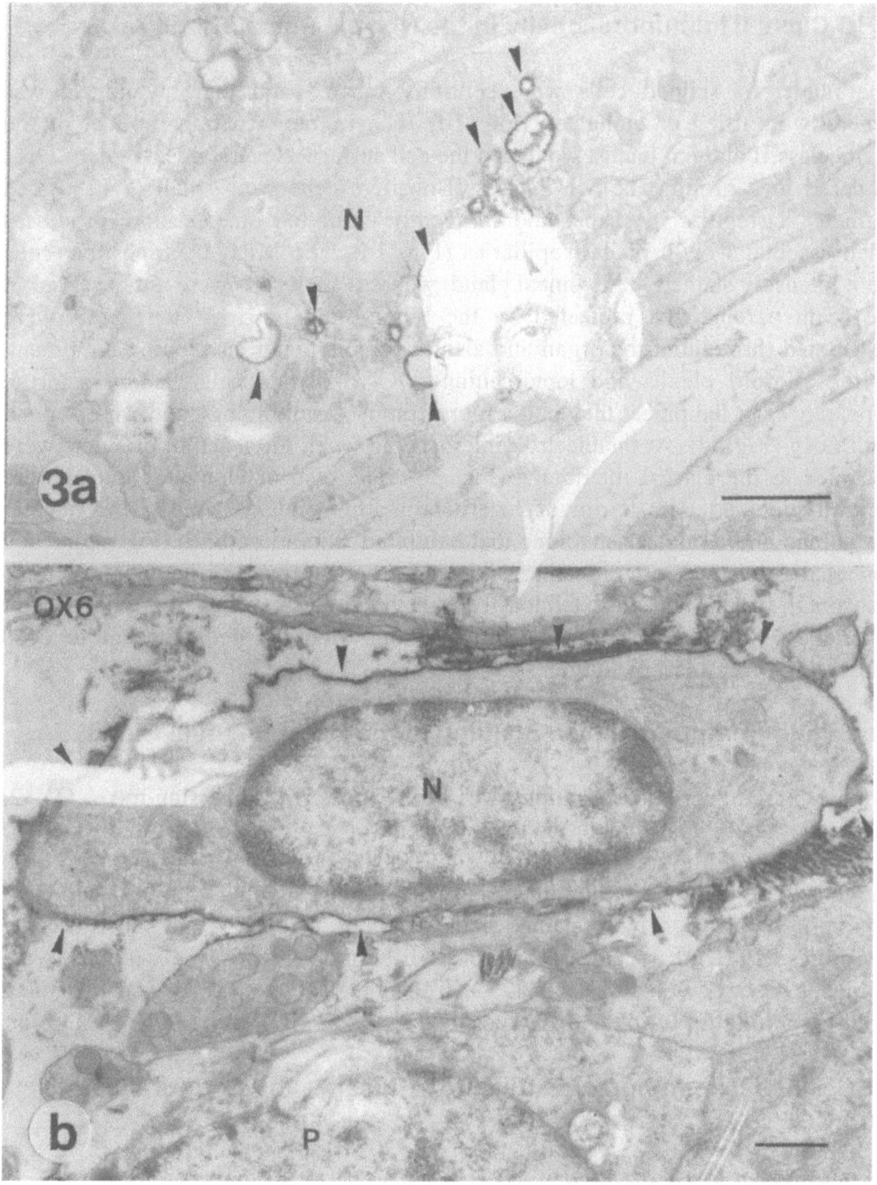

Fig. 3. Electron micrographs of an *ED1*- immunoreactive cell (**a**) and an *OX6*-immunoreactive cell (**b**). The former exhibits a segmented nucleus (*N*) and an immunoreaction of lysosomal membranes (*arrowheads*), whereas the latter shows a smooth nuclear (*N*) envelope with a comparatively thick rim of heterochromatin and an intense immunoreaction along the plasma membrane (*arrowheads*). *P* Pinealocyte. *Bars* 2 μm (**a**) and 1 μm (**b**)

MHC Class II Immunoreactivity in the Normal Pineal Gland

The nuclei of stained cells were centrally located and their cytoplasm was generally ramified or elongated (Fig. 1d). The immunoreaction product of the MHC class II antigen intensely labeled the cell surface. Several positive cells were gathered in a group. The MHC class II-positive cells were small (7-12 μm in diameter), irregular in shape, and frequently exhibited one or more processes often associated with pineal capillaries (Fig. 1d). The MRC OX6-positive cells were intensely stained in the pineal gland, whereas such cells were not observed in the brain parenchyma protected by the blood-brain barrier. These cells were distributed throughout the organ and also in the pia mater enveloping the pineal. In the choroid plexus and leptomeninges, OX6-labeled cells had a dendritic appearance. In the pineal they had a more compact morphology and were stained with OX6 (Fig. ld). At the electron microscopic level, the reaction products were confined to the plasma membrane (Fig. 3b). The oval or elongated nuclei were centrally located with a rim of heterochromatin condensed along the nuclear membrane and a nuclear envelope that exhibited smooth outlines (Fig. 3b). They had characteristic short processes and were characterized by an inconspicuous lysosomal system and an extended frequent hyaloplasm which was poor in cell organelles. Pinealocytes with synaptic ribbons were located adjacent to cells expressing MHC class II antigen.

Double-Immunoperoxidase Staining in the Normal Pineal Gland

Double-immunoperoxidase staining was further performed by using mAbs OX42, ED1, and OX6. The double staining procedure resulted in cells stained either brown, blue, or both. However, in some instances, it was difficult for both antigens to be detected in the same cell. When double staining with ED1 and OX6 was carried out, positively stained cells could be classified into the following categories: $ED1^{+}OX6^{-}$ and $ED1^{-}OX6^{+}$. Furthermore, when the sections were stained by OX42 and OX6, $OX42^{+}OX6^{-}$ and $OX42^{-}OX6^{+}$ cells were detected. Double staining for OX42 and ED1 resulted in $OX42^{+}ED1^{+}$ and $OX42^{+}ED1^{-}$ cells.

MHC Class II Immunoreactivity in the Pineal Gland of Rats Treated with CCl_4

Immunoreactivity for the MHC class II antigen was increased following treatment with CCl_4. The number of positive cells was not significantly different between controls and rats sacrificed 8 h after the injection, but MRC OX6-positive cells had significantly increased in number in the pineal gland of rats killed 20 h after injection of CCl_4: 0 h, 22.8 ± 4.0; 8 h, 25.7 ± 7.1; 20 h, 64.2 ± 7.6. At this stage, more than ten positive cells with a strongly labeled cell surface lay in a group. In addition, smaller, B lymphocyte-like, and round cells were often immunostained. Double immunostaining for PCNA and MHC class II antigen revealed double stained cells, in addition to MHC class II single-stained cells, within the pineal gland, 8 and 20 h after injection of CCl_4. Such double stained cells could not be identified in controls.

Discussion

The present investigations provide immunohistochemical evidence for immunocompetent cells in the pineal gland of the rat which represent a significant component of the interstitial cells, and are positively stained with the MHC class II (Ia)-specific or the macrophage mAbs. The MRC OX42 has been shown to label resting and activated microglia in the CNS (Robinson et al. 1986; Graeber et al. 1989; Streit et al. 1988; Ling et al. 1990; Chugani et al. 1991). The ED1 and ED2 mAbs recognize peripheral monocytes/macrophages and tissue macrophages, respectively (Dijkstra et al. 1985). The main population of macrophages in the normal rat pineal consists of two different phenotypes of macrophage/microglia-derived cells, $OX42^{+}ED1^{+}$ and $OX42^{+}ED1^{-}$. Brain macrophages and microglia play important roles in CNS development, especially during regressive events in which particular neuronal and glial constituents are eliminated (Perry and Gordon 1988; Jordan and Thomas 1988). Brain macrophages are large, round cells, 10-15 μm in diameter, with few or no cytoplasmic processes; these cells are immunopositive for OX42 and ED1. Microglia have small cell bodies with numerous, ramified cytoplasmic processes. These cells are $OX42^{+}ED1^{-}$ (Milligan et al. 1991). Amoeboid and ramified microglia in the CNS lose immunoreactivity to ED1 during development (Streit et al. 1988; Graeber et al. 1989; Chugani et al. 1991). Thus, $OX42^{+}ED1^{+}$ and $OX42^{+}ED1^{-}$ cells in the rat pineal may exhibit differences in the degree of differentiation.

The extracellular fluid of the CNS has a different composition from plasma because of the blood-brain barrier, which prevents unrestricted transfer of plasma constituents across the CNS endothelium to the extracellular space and lies at the tight junctions between the CNS endothelium (Reese and Karnovsky 1967). A major difference between the cerebrospinal fluid (CSF), which is in equilibrium with the fluids of the extracellular space, and plasma is that the CSF concentration of protein is about 200-fold less than that in plasma (Davson et al. 1987). Microglia, the resident macrophages of the CNS (Perry and Gordon, 1988, 1991; Streit et al. 1988), are not normally exposed to plasma proteins and have barely detectable or undetectable levels of CD4, while the microglia of the circumventricular organs (CVOs) exposed to plasma proteins have readily detectable levels as do the macrophages in the choroid plexus and the leptomeninges (Perry and Gordon 1987). Highly ramified microglia cells are ubiquitous, although they vary in density and morphology in different brain regions (Lawson et al. 1990). The MHC class-II-immunoreactive cells can, if sectioned through the perikaryon, be distinguished from macrophages by the rare presence of lysosomes and electron-dense inclusions, which are readily observed in macrophages. Furthermore, more than two positive cells with their cell bodies adjacent to one another have often been found in this study. The MHC class-II-labeled cells apparently have a different distributional pattern from microglia (Hume et al. 1983; Ashwell 1989; Ashwell et al. 1989). Our results are supported by several reports that MHC class I and class II antigens are largely absent from the normal CNS, irrespective of the presence of numerous microglia cells (see Perry and Gordon 1991 for details). Under normal conditions, the cell types expressing MHC class II antigen are restricted to B lymphocytes, macrophages, and dendritic cells in the lymphoid tissue. Recent papers have, however, reported that the expression of the MHC class II antigen occurs on the cellular surface of certain cells, for example Langerhans cells in the skin (Hisianoo et al. 1989), and in non-lymphoid organs such as peripheral blood (Knight et al. 1987), lung (Shah et al. 1985), gut (Mikkelsen and Rumessen 1992), liver (Hart and Fabre 1981; Yamamoto et al. 1995), and pineal gland (Pedersen et al. 1993; Sato et al. 1996). The MHC class-II-labeled cells in the rat pineal gland have several processes and only a few cytoplasmic organelles, including few multivesicular bodies and/or mitochondria. Such morphological characteristics of the immunolabeled cells in the pineal gland are similar to those of the Ia-positive peripheral dendritic cells (Sato et al. 1996).

Microglia vary in morphology depending on their location (Lawson et al. 1990). They are broadly classified into three categories. Compact cells are rounded cells, sometimes with one or two short thick limbs, bearing short processes ('bristles'). They resemble Kupffer cells of the liver and are found exclusively in sites lacking a blood brain barrier, the CVOs. Longitudinally branched cells are found in fiber tracts and possess several long processes which are usually aligned parallel to, or more occasionally perpendicular to, the longitudinal axis of nerve fibers. Radially branched cells are found throughout the neuropil of the hypothalamic magnocellular nuclei. They can be extremely elaborate and there is wide variation in the length and complexity of branching of the processes. Ultrastructurally, the nucleus of microglia is oval or elongate, sometimes indented, with clumps of chromatin beneath the nuclear membrane. A striking difference between microglia and macrophages found in the developing brain is the thin perinuclear rim of cytoplasm and the paucity of cellular organelles in the former (Perry and Gordon 1991). The Golgi apparatus is present usually at one pole of the nucleus and a prominent feature is the long narrow cisternae of the rough endoplasmic reticulum (Peters et al. 1976). Lysosomes are found in the cytoplasm, as are lipid droplets and lipofuscin granules. The ultrastructure of OX42-labeled pineal macrophage/microglia therefore resembles the amoeboid microglia described in the brain of neonatal rats (Ling et al. 1990; Pedersen et al. 1993; Sato et al. 1996). Within each structure, the microglia are distributed in a regular pattern, with each cell occupying a distinct territory (Perry and Gordon 1991). It is rare to find two cells with their cell bodies adjacent to one another or their processes intertwined. Microglia are characterized by the loss of a spectrum of antigens, and at the present time there are no antigens specifically expressed by microglia which allows one to distinguish them from other macrophage populations. However, complement type 3 receptor antigens are strongly expressed by microglia compared with macrophages in other tissues where they can be downregulated, for example, in T lymphoid regions and the liver, respectively.

Apart from their possible role in the defense of the CNS from injury and infection, the functions of microglia in normal CNS are largely unknown. Microglia in the posterior pituitary lie in a pericapillary position with their processes wrapped around the endings of magnocellular neurons. These microglia endocytose the endings of the magnocellular neurons (Pow et al. 1989). Also, it has been suggested that microglia may play a role in neurotransmitter homeostasis (Murabe and Sano 1982). These authors have shown that microglia have high levels of the enzymes thiamine pyrophosphatase and nucleoside diphosphatase on their plasma membranes and that the intensity of staining is influenced by, for example, the presence of catecholamines. This observation, coupled with the fact that microglial processes are found in close association with axodendritic synapses, led them to suggest that the microglia may play a part in neurotransmitter metabolism. It has been demonstrated that monocytes and macrophages have receptors for a number of neurotransmitters and also possess enzymes that would rapidly degrade a wide range of putative neurotransmitters (see Perry and Gordon 1988, for references), but it is not known if microglia do so.

A large body of evidence supports an important role for MHC class II (Ia) antigen in accessory or stimulatory activities, and it has suggested that the amount of Ia expressed might be critical for these processes (Nussenzweig et al. 1980). The majority of macrophages are weakly Ia-positive compared with dendritic cells and the density of Ia is not sufficiently high for these cells to function as accessories or stimulators (Klinkert et al. 1982). Klinkert et al. (1982) concluded that dendritic cells differ from macrophages in morphology, surface properties, and immunoregulatory functions on the basis of the findings that (1) dendritic cells with accessory functions are present in many rat tissues and organs, including those exposed to a variety of antigens, such as spleen, liver, and skin; (2) important attributes of macrophages such as the Fc receptor and the ability to adhere and phagocytose are not features of dendritic cells; (3) macrophage preparations highly depleted of dendritic cells do not serve as accessory cells or are inhibitory; and (4) Ia antigens playing an important role in the induction of immune reactions are present on dendritic cells, and the biological function of these cells can be inhibited with a monoclonal antibody directed against Ia antigens (Klinkert et al. 1981). Dendritic cells have been reported to be distributed in lymphoid and several non-lymphoid organs (Hart and Fabre 1981). Their function is believed to involve the collection of antigen in various tissue sites, transport of the processed antigen to the T cell areas of lymphoid tissue, presentation of the antigenic peptides and finally stimulation of the immune response of T lymphocytes (Steinman 1991). Morphologic studies have failed to demonstrate lymphocytes or lymphatics in the mammalian pineal gland (Quay 1965; Kappers 1976). However, immunoperoxidase and immunofluorescence techniques have partly succeeded in detecting cells with anti-T and anti-immunoglobulin sera in the rat pineal gland (Ueda et al. 1981). Further studies of lymphocytes or lymphatics in the pineal are needed in order to understand the possible role that dendritic cells play in immune-mediated reactions within this organ. Because of the lack of specific markers for dendritic cells, it is difficult to classify them histochemically. In order to identify dendritic cells and to clarify the difference in immunoreactivity for various surface markers between macrophages/microglia and dendritic cells in the normal pineal, we have further performed double immunoperoxidase staining for MRC OX42, ED1, and MRC OX6. All the pineal macrophages/microglia are labeled by OX42, and can be divided into groups that are $ED1^{+}OX6^{-}$ or $ED1^{-}OX6^{-}$. Even in the absence of a blood-brain barrier, at least some microglial cells remain ED1- and OX6-negative under normal conditions. The majority of MRC OX6-positive cells in the pineal are negative for both OX42 and ED1. Dendritic cells sometimes have low levels of OX42 and ED1 (Steinman 1991), indeed so low as to be immunohistochemically undetectable. Double staining experiments with a reversed sequence of antibody incubations appears to exclude the possibility that the negative results are due to blockade of OX42- and ED1-staining by the DAB color reagent used for demonstration of OX6. Based on these findings, the normal rat pineal therefore contains at least the following three different cell populations: (1) those expressing the phenotype of microglial cells ($OX42^{+}ED1^{-}OX6^{-}$) found in areas with a blood-brain barrier; (2) those with the phenotype of microglial cells ($OX42^{+}ED1^{+}OX6^{-}$); (3) those with the phenotype of peripheral dendritic cells ($OX42^{-}ED1^{-}OX6^{+}$).

If the MHC class-II-positive cells increase in number in the pineal gland after an injection of CCl_4, this must be a consequence of local cell division by the resident cells or of a significant influx from the blood. In the present study, MHC class-II-immunostained cells in S phase have been identified by PCNA labeling. This indicates that local proliferation of the MHC class-II-positive cells contributes to their expansion in the pineal gland of the rat treated with CCl_4. However, only some of the cells expressing the MHC class II antigen are immunolabeled by the monoclonal antibody for PCNA. Therefore, a numerical increase of the MHC class-II-labeled cells following CCl_4 treatment depends on both local proliferation and their influx from the blood stream.

In the normal pineal, only dendritic cells appear to express the Ia antigen but microglial cells do not. Thus, the macrophage lineage in the pineal gland is heterogeneous with regard to the expression of the MHC class II (Ia) antigen. These findings suggest that dendritic cells and microglial cells have different origins. Indeed, Langerhans cells or dendritic cells have been detected in the non-lymphoid tissues of op/op mice that lack the activity of functional macrophage-colony-stimulating factor (Takahashi et al. 1993). As part of the circumventricular system, the pineal gland is readily accessible to immune factors, because it is not isolated from the circulation by the blood-brain barrier. Our results indicate that at least two different types of immunocompetent cells, dendritic cells and macrophages, are located in the pineal gland and that they are favorable candidates for interactions of this neuroendocrine organ with the immune system.

Acknowledgment

The author is grateful to Professor K. Wake for his great encouragement.

References

Ashwell KWS (1989) Development of microglia in the albino rabbit retina. J Comp Neurol 287: 286-301

Ashwell KWS, Hollander H, Streit W, Stone J (1989) The appearance and distribution of microglia in the developing retina of the rat. Vis Neurosci 2: 437-448

Barbe E, Damoiseaux JGMC, Dopp EA, Dijkstra CD (1990) Characterization and expression of the antigen present on resident rat macrophages recognized by monoclonal antibody ED2. Immunobiology 182: 88-99

Benacerraf B (1981) Role of MHC gene products in immune regulation. Science 212: 1229-1238

Besedovsky HO, del Rey A, Sorkin E, Dinarello A (1986) Immunoregulatory feedback between interleukin-1 and glucocorticoid hormones. Science 233: 652-654

Chugani DC, Kedersha NL, Rome LH (1991) Vault immunofluorescence in the brain: new insights regarding the origin of microglia. J Neurosci 11: 256-268

Cline MHJ, Lehrer RI, Territo MC, Golde DW (1978) Monocytes and macrophages: functions and diseases. Ann Intern Med 88: 78-88

Damoiseaux JGMC, Dopp EA, Calame W, Chao D, MacPherson GO, Dijkstra CD (1994) Rat macrophage lysosomal membrane antigen recognized by monoclonal antibody ED1. Immunology 83: 140-147

Davson H, Welch K, Segal MB (1987) The physiology and pathophysiology of the cerebrospinal fluid. Churchill Livingstone, Edinburgh

Dijkstra CD, Dopp BA, Joling P, Kraal G (1985) The heterogeneity of mononuclear phagocytes in lymphoid organs: distinct macrophage subpopulations in the rat recognized by monoclonal antibodies ED1, ED2 and ED3. Immunology 54: 589-599

Dua HS, McKinnon A, McMenamin PG, Forrester JV (1991) Ultrastructural pathology of the barrier sites in experimental autoimmune uveitis and experimental autoimmune pinealitis. Br J Ophthalmol 75: 391-397

Dunn AJ (1988) Nervous system-immune system interactions: an overview. J Receptor Res 8: 589-607

Dunn AJ (1989) Psychoneuroimmunology for the psychoneuroendocrinologist: a review of animal studies of nervous system-immune system interactions. Psychoneuroendocrinology 14: 251-274

Graeber MB, Streit WJ, Kreutzberg GW (1989) Identity of ED2-positive perivascular cells in the brain. J Neurosci Res 22: 103-106

Hart DNJ, Fabre JW (1981) Demonstration and characterization of Ia-positive dendritic cells in the interstitial connective tissues of rat heart and other tissues, but not brain. J Exp Med 153: 347-361

Hisianoo L, Takaya M, Arao T, Takahasshi K (1989) An immunohistochemical and immunoelectron microscopic study of the ontogeny of rat Langerhans cell lineage with anti-macrophage and anti-Ia monoclonal antibodies. J Invest Dermatol 93: 780-786

Hsu SM, Soban B (1982) Color modification of diaminobenzidine (DAB) precipitation by metallic ions and its application for double immunohistochemistry. J Histochem Cytochem 30: 1079-1082

Hume DA, Perry VH, Gordon S (1983) Immunohistochemical localization of a macrophage-specific antigen in developing mouse retina: phagocytosis of dying neurons and differentiation of microglial cells to form a regular array in the plexiform layers. J Cell Biol 97: 253-257

Jordan FL, Thomas WE (1988) Brain macrophages: questions of origin and interrelationship. Brain Res Rev 13: 165-178

Kappers JA (1976) The mammalian pineal gland, a survey. Acta Neurochir 34: 109-149

Klinkert WEF, Wong TW, Bowers WE (1981) Rat dendritic cells from lymphoid and non-lymphoid tissues regulate T lymphocyte transformation. Immunobiology 159: 156 (Abstr)

Klinkert WEF, Labadie JH, Bowers WE (1982) Accessory and stimulating properties of dendritic cells and macrophages isolated from various rat tissues. J Exp Med 156: 1-19

Knight SC, Fryer P, Griffiths S, Harding B (1987) Class II histocompatibility antigens on the human dendritic cells. Immunology 61: 21-27

Lawson LJ, Perry VH, Dri P, Gordon S (1990) Heterogeneity in the distribution and morphology of microglia in the normal adult mouse brain. Neuroscience 39: 151-170

Ling BA, Kaur C, Yick TY, Wong WC (1990) Immunocytochemical localization of CR3 complement receptors with OX-42 in amoeboid microglia in postnatal rats. Anat Embryol 182: 481-486

Luo ZR, Schultz RL, Witter EF, Vollrath L (1984) Ultrastructural characterization of glial cells in the rat pineal gland with special reference to the pineal stalk. Anat Rec 210: 663-674

McMaster WR, Williams AF (1979) Identification of Ia glycoproteins in rat thymus and purification from rat spleen. Eur J Immunol 9: 426-433

Maestroni GJM, Pierpaoli W (1981) Pharmacologic control of the hormonally mediated immune response. In: Ader R (ed) Psychoneuroimmunology. Academic Press, New York, pp 405-425

Maestroni GJM, Conti A, Pierpaoli W (1986a) Role of the pineal gland in immunity. Circadian synthesis and release of melatonin modulates the antibody response and antagonizes the immunosuppressive effect of corticosterone. J Neuroimmunol 13: 19-30

Maestroni GJM, Conti A, Pierpaoli W (1986b) Melatonin regulates immunity via an opiatergic mechanism. Clin Neuropharmacol 9 (Suppl 4): 479-481

Mikkelsen HB, Rumessen JJ (1992) Characterization of macrophage-like cells in the external layers of human small and large intestine. Cell Tissue Res 270: 273-279

Milligan CB, Cunningham TJ, Levitt P (1991) Differential immunochemical markers reveal the normal distribution of brain macrophages and microglia in the developing rat brain. J Comp Neurol 314: 125-135

Murabe Y, Sano Y (1982) Morphological studies on neuroglia. IV. Postnatal development of microglial cells. Cell Tissue Res 225: 469-485

Nussenzweig MC, Steinman RM, Gutchinov B, Cohn ZA (1980) Dendritic cells are accessory cells for the development of anti-trinitrophenyl cytotoxic T lymphocytes. J Exp Med 152: 1070-1084

Pedersen EB, Fox LM, Castro AJ, McNulty JA (1993) Immunocytochemical and electron-microscopic characterization of macrophage/microglia cells and expression of class II major histocompatibility complex in the pineal gland of the rat. Cell Tissue Res 272: 257-265

Perry VH, Gordon S (1987) Modulation of the CD4 antigen on macrophages and microglia in rat brain. J Exp Med 165: 1218-1223

Perry VH, Gordon S (1988) Macrophages and microglia in the nervous system. Trends Neurosci 11: 273-277

Perry VH, Gordon S (1991) Macrophages and the nervous system. Int Rev Cytol 125: 203-244

Peters A, Palay SL, Webster HdeF (1976) The fine structure of the nervous system: the neurons and supporting cells. WB Saunders, Philadelphia

Pow DV, Perry VH, Morris JF, Gordon S (1989) Microglia in the neurohypophysis associate with and selectively endocytose terminal portions of neurosecretory neurons. Neuroscience 33: 567-578

Quay WB (1965) Histological structure and cytology of the pineal organ in birds and mammals. Progr Brain Res 10: 49-86

Reese TA, Karnovsky MJ (1967) Fine structural localization of a blood-brain barrier to exogenous peroxidase. J Cell Biol 34: 207-217

Robinson AP, White TM, Mason DW (1986) Macrophage heterogeneity in the rat as delineated by two monoclonal antibodies MRC OX-41 and MRC OX-42, the latter recognizing complement receptor type 3. Immunology 57: 239-247

Sanders EJ, Varedi M, French AS (1993) Cell proliferation in the gastrulating chick embryo: a study using BrdU incorporation and PCNA localization. Development 118: 389-399

Sato T, Kaneko M, Hama A, Kusakari T, Fujieda H (1996) Expression of class II MHC molecules in the rat pineal gland during development and effects of treatment with carbon tetrachloride. Cell Tissue Res 284: 65-76

Shah P, Gilbertson SM, Rowley DA (1985) Dendritic cells that have interacted with antigen are targets for natural killer cells. J Exp Med 162: 625-636

Shinohara T, Singh VK, Tsuda M, Yamaki K, Abe T, Suzuki S (1990) S-antigen: from gene to autoimmune uveitis. Exp Eye Res 50: 751-757

Schuler G, Steinman RM (1985) Murine epidermal Langerhans' cells mature into potent immunostimulatory dendritic cells in vitro. J Exp Med 161: 526-546

Smith EM, Blalock JE (1986) A complete regulatory loop between the immune and neuroendocrine systems operates through common signal molecules (hormones) and receptors. In: Plotnikoff NP et al. (eds) Enkephalins and endorphins: stress and the immune system. Plenum, New York, pp 119-127

Steinman RM (1991) The dendritic cell system and its role in immunogenicity. Annu Rev Immunol 9: 271-296

Streit WJ, Graeber MB, Kreutzberg GW (1988) Functional plasticity of microglia: a review. Glia 1: 301-307

Takahashi K, Naito M, Shultz LD, Hayashi S, Nishikawa S (1993) Differentiation of dendritic cell populations in macrophage colony-stimulating factor-deficient mice homozygous for the osteopetrosis (op) mutation. J Leukoc Biol 53: 19-28

Tew JG, Thorbecke J, Steinman RM (1982) Dendritic cells in the immune response: characteristic and recommended nomenclature (a report from the Reticuloendothelial Society Committee on Nomenclature). J Reticuloendothelial Soc 31: 371-380

Ueda T, Ishi Y, Matsuura A, Simogawara I, Kikuchi K (1981) Immunohistochemical study of lymphocytes in rat pineal gland: selective accumulation of T lymphocytes. Anat Rec 199: 239-247

Widner H, Brundin P (1988) Immunological aspects of grafting in the mammalian central nervous system. A review and speculative synthesis. Brain Res Rev 13: 287-324

Withyachumnarnkul B, Nonaka KO, Attia AM, Reiter RJ (1990) Changes in indole metabolism in organ cultured rat pineal glands induced by interferon-gamma. J Pineal Res 8: 313-322

Withyachumnarnkul B, Reiter RJ, Lerchl A, Nonaka KO, Stokkan KA (1991) Evidence that interferon-gamma alters pineal metabolism both indirectly via sympathetic nerves and directly on the pinealocytes. Int J Biochem 23: 1397-1401

Yamamoto H, Kaneko M, Sato T, Wake K (1995) Ultrastructure and distribution of the I-a positive cells in the normal liver - a new member of sinusoidal cells. In: Knook DL, Wisse E, Wake K (eds) Cells of the hepatic sinusoid, vol 5. Kupffer Cell Foundation, AK Leiden, pp 292-294

30 Melatonin Synthesis and Calcium Responses in the Pineal Gland of the Trout

S. Kroeber[1], H.-W. Korf[1], and H. Meissl[2]

[1]Dr. Senckenbergische Anatomie, Institute for Anatomy II, Johann Wolfgang Goethe-University Frankfurt, Theodor-Stern-Kai 7, D-60590 Frankfurt am Main, Germany; Phone: +49 69 6301 5661, Fax : +49 69 6301 6017

[2]Max-Planck-Institute for Physiological and Clinical Research, W.G. Kerckhoff-Institute, Parkstr. 1, D-61231 Bad Nauheim, Germany

Abstract

The present in vitro study correlates melatonin production of explanted trout pineal organs with intracellular calcium concentrations of isolated and immunocytochemically identified pinealocytes. Melatonin production of pineal organs maintained in flow-through organ culture is closely related to the incident light, showing high levels in the scotopic and mesopic range of illumination and low levels in the photopic range. Diminishing extracellular calcium (Ca^{2+}) and enhancing magnesium (Mg^{2+}) as well as application of Co^{2+} (3 mM) reduces melatonin release in the light- and dark-adapted state and causes a reversible breakdown of spontaneous oscillations in the intracellular concentration of free calcium ions $[Ca^{2+}]_i$ that are observed in approximately 20% of the isolated S-antigen-immunoreactive pinealocytes. Intracellular calcium oscillations are also reversibly blocked by application of 10 μM nifedipine. Depolarization of pinealocytes by treatment with 60 mM KCl elevates $[Ca^{2+}]_i$ in 90% of the oscillating and 50% of the non-oscillating cells. The effect of KCl is blocked by 50 μM nifedipine and 3 mM $CoCl_2$. Furthermore, application of $CoCl_2$ reduces the basal $[Ca^{2+}]_i$ in non-oscillating pinealocytes. These results suggest that voltage-gated L-type calcium channels play a major role in the regulation of $[Ca^{2+}]_i$ in trout pinealocytes. It appears that Co^{2+} and low calcium/high Mg^{2+} buffer reduces melatonin release through an action on the intracellular calcium concentration of trout pinealocytes. All these data show that the trout pineal organ synthesizes and releases melatonin in relation to the irradiance of the incident light and depends on the presence of extracellular calcium which enters the cell via voltage-gated calcium channels. Treatment of cultured pineal organs with norepinephrine (1 μM) or the β-adrenergic agonist isoproterenol has no significant effect on the light-dependent melatonin release. Furthermore, application of norepinephrine in varying concentrations (100 pM - 50 μM) that were previously shown to induce calcium release from intracellular calcium stores in rat pinealocytes has no effect on $[Ca^{2+}]_i$ of oscillating or non-oscillating trout pinealocytes. These data suggest that in the trout pineal organ a norepinephrine-sensitive pinealocyte of the mammalian type is not involved in the transduction of light signals into a melatonin response.

Introduction

The pineal organ of vertebrates represents an important component of the biological clock, despite the fact that it has undergone remarkable transformations during phylogenetic development. These phylogenetic changes are summarized by the concept of the sensory cell line (Collin 1971; Oksche 1971; Collin and Oksche 1981), which states that the secretory pinealocytes of the mammalian pineal gland are derivatives of primary sensory (photoreceptive) cells in lower vertebrates. This hypothesis is supported by the observation that mammalian pinealocytes display immunoreactions for proteins that are involved in the phototransduction cascade, like S-antigen, recoverin and opsin (Korf 1994). The direct light sensitivity of pineal photoreceptor cells of poikilothermic vertebrates has gradually been replaced by an indirect sensitivity, transmitting light-dark information from the retina to the pineal via a sympathetic neuronal innervation of the pineal gland and norepinephrine as the chemical neurotransmitter. Whereas retinal light perception and transmission of information via a sympathetic innervation of the pineal gland is mandatory to maintain rhythmic melatonin production with elevated levels during the night and low levels during the day in mammals (Arendt 1995), in poikilothermic vertebrates, light acts directly on the pineal organ without the involvement of the retina (cf. Dodt 1973; Meissl and Dodt 1981). This pineal photoreception is provided by true photoreceptors that have numerous anatomical, physiological and biochemical similarities with retinal photoreceptors (cf. Meissl 1986; Meissl and Yáñez 1994). The pineal organ of poikilothermic vertebrates is capable of translating light stimuli either into pinealofugal neuronal signals or into a neuroendocrine response, i.e., the rhythmic synthesis and release of melatonin (Collin et al. 1986; Meissl and Dodt 1981).

The signal transmission pathway involved in regulating melatonin biosynthesis in poikilothermic vertebrates has been identified only partially. In the rainbow trout, elevation of intracellular concentrations of cAMP ($[cAMP]_i$) stimulate the activity of N-acetyltransferase (NAT) and the synthesis of melatonin in cultured pineal organs (Thibault et al. 1993). In dispersed pineal cells, increased levels of extracellular calcium ions (Ca^{2+}) enhance melatonin production, whereas blocking Ca^{2+} channels by nifedipine or verapamil inhibits it, probably by their action on voltage-dependent L-type calcium channels (Begay et al. 1994). To elucidate signal transduction mechanisms and melatonin biosynthesis in pinealocytes of poikilothermic vertebrates, we determined $[Ca^{2+}]_i$ in immunocytochemically identified pinealocytes of the rainbow trout and melatonin secretion of trout pineal organs. Special emphasis has been placed on a possible role of norepinephrine in melatonin production and regulation in order to prove whether a norepinephrine-sensitive mammalian type of pinealocyte exists in the trout pineal organ.

Materials and Methods

Animals

Rainbow trout (*Onchorhynchus mykiss*) were obtained from a local hatchery and kept in large aquaria with oxygenated freshwater, under an artificial light-dark cycle of 12:12 h for at least 1 week before the experiments were started. Trout of 450-550g body weight were used for the perfusion experiments. Juvenile (approximately 1-year-old) rainbow trout were used for the dissociation of pineal cells and the analysis of $[Ca^{2+}]_i$.

Organ Culture and Melatonin Release

Pineal organs were rapidly removed from decapitated animals, washed in ice-cold Hanks' solution, and then transferred to flow-through organ culture chambers. Each chamber, which held an individual pineal organ, was continuously superfused with modified Hanks′ solution or modified Medium 199 with Hanks′ basal salt solution (Yáñez and Meissl 1996). Both superfusion solutions were buffered with 7.5 mM HEPES (pH 7.4) and gassed with a mixture of O_2/CO_2 (95%/5%). The flow rate of the superfusion solution was 0.5 ml/h , and the temperature was held constant at 17 °C, the apparent optimal range for melatonin production (Max and Menaker 1992) and for neural function of pineal photoreception in the trout in vitro (Tabata and Meissl 1993). The superfusion solution was collected at the outlet of the experimental chambers at time intervals of 1h. Illumination of the chambers was provided by a 150 W xenon arc from a photostimulator. Intensity was adjusted by neural density filters and the wavelength (520 nm) by an interference narrowband filter. An estimate of the number of photons was obtained as previously described (Meissl and Ekström 1988). Most of the superfusion experiments started with illumination of the pineal organs by the brightest light available from the photostimulator (520 nm, 0 log equal to 14.6 log photons/cm^2 per s). The intensity of the light was kept constant during the 1 h sampling intervals, the light was then attenuated by 1 log unit and the outflow of the chambers sampled. The irradiance was stepwise reduced up to 9.6 log photons/cm^2 per s, before the organs were completely dark adapted. Melatonin was separated by high-performance liquid chromatography as previously described (Yáñez and Meissl 1996).Values for melatonin were obtained as described previously (Meissl et al. 1996) and expressed in nanograms per organ per hour or normalized with respect to the maximal value (100%), usually measured in the completely dark-adapted organ. All values are presented as mean values ± SEM. For statistical analysis, mean values have been compared with the unpaired t-test or with the Mann-Whitney rank sum test. Drugs were purchased from Sigma (Deisenhofen, Germany). They were added directly to the superfusion medium for the duration of the experiments.

Cell Culture and Calcium Response

Animals were killed between 15.00 and 19.00 by decapitation. The pineal organs were collected in ice-cold saline (Hanks' balanced salt solution; HBSS) without Ca^{2+} and Mg^{2+}. Pineal organs were dissociated by papain digestion and immobilized on coated glass coverslips with an internal grid as described previously (Kroeber et al. 1997). Cells were kept in growth medium (Dulbecco's modified Eagle's medium; DMEM) supplemented with 10% fetal calf serum (FCS), 2 mM glutamine, 100 U/ml penicillin, 100 µg/ml streptomycin, 0.01% ascorbic acid, 10 mM HEPES and incubated at 18 °C in an atmosphere of 5% CO_2 and 95% air for 1 or 2 days in darkness. The viability of the cells was found to be approximately 90% as assessed by trypan-blue exclusion. For fluorescence measurements, cells were loaded with 2.5 µM fura-2 pentaacetoxymethyl ester for 15 min at 18 °C. Subsequently coverslips were placed in a perifusion chamber and superfused with saline containing 10 mM HEPES and 0.01% ascorbic acid at room temperature (19-21°C). In some experiments, calcium free saline (HBSS without Ca^{2+} and Mg^{2+}, Gibco) was used as the perifusion buffer. For stimulation, perifusion was stopped and the drug was added into the perifusion chamber. The drug was washed out by restarting the perifusion with a flow rate of 500 µl/min. All drugs were kept as stock solutions dissolved in saline (norepinephrine, KCl, $CoCl_2$) or ethanol (nifedipine) and were diluted with saline before use. The $[Ca^{2+}]_i$ of isolated trout pineal cells was analyzed with an Attofluor ratio imaging system (Zeiss, Germany; for details see Schomerus et al. 1995). Changes in $[Ca^{2+}]_i$ of single cells are expressed in a semiquantitative manner as 334/380 nm emission ratios. Following the fluorometric measurements, cells were fixed for 10 min in 4% paraformaldehyde, and pinealocytes were identified by immunocytochemical demonstration of S-antigen (a pinealocyte-specific marker) as described previously (Wicht et al. 1993). Immunocytochemical and functional data were correlated at the level of single pinealocytes whose location was determined by means of the internal grid in the coverslips. HBSS, DMEM, and FCS were obtained from Gibco (Eggenstein, Germany); papain was obtained from Boehringer (Mannheim, Germany). All other drugs and chemicals were obtained from Sigma (Deisenhofen, Germany). Coverslips provided with an internal grid were obtained from Eppendorf (Hamburg, Germany).

Results

Organ Culture and Melatonin Release

Isolated pineal organs of the trout were maintained in superfusion culture for the whole duration of the experiment using a fixed dark-adaptation protocol. For the initiation of an experiment, pineal organs were exposed for 1 h to light of the highest intensity available from the photostimulator. The irradiance was then reduced in steps of 1 log unit h^{-1}, before the light was finally turned off and the pineal organ exposed to complete darkness. This procedure was considered to mimic the slow changes in light intensities occurring under natural conditions from noon to the day-night transition in the evening. Melatonin production and release showed a highly characteristic response pattern to light under these conditions (Figs. 1, 2). Melatonin release increased with decreasing irradiances and showed maximal values in complete darkness. Exchanging the superfusion solution (Hanks' buffer) with a buffer containing low Ca^{2+} (0.13 mM) and high Mg^{2+} (6.1 mM) content reduced melatonin release considerably in the photopic range (520 nm, 12.6 log photons/cm^2 per s), but only slightly under scotopic conditions (Fig. 1A). Addition of 3 mM $CoCl_2$ significantly reduced melatonin release in the mesopic and scotopic range of illumination, but not in the photopic range when melatonin production was low (Fig. 1B). Addition of 1 μM norepinephrine (Fig. 2A) or 1 μM of the β-adrenergic agonist isoproterenol (Fig. 2B) to the superfusion medium was without effect on light-dependent melatonin release from cultured trout pineal organs. Neither the maximal release rate in darkness nor the response in the photopic or mesopic range of illumination was altered by the drugs (Fig. 2A, B).

Cell Culture and Calcium Response

Immunocytochemical demonstration of S-antigen, a specific marker for pinealocytes (Korf 1994), revealed that approximately 20% of the dispersed pineal cells were S-antigen-immunoreactive (ir), and thus represented pinealocytes (Fig. 3). With the fura-2 technique, spontaneous $[Ca^{2+}]_i$ oscillations with irregular frequencies were recorded in a distinct population (20%) of the S-antigen-ir pinealocytes (Fig. 4A-D). The remaining S-antigen-positive pinealocytes (80%) showed no spontaneous $[Ca^{2+}]_i$ oscillations but had a stable basal $[Ca^{2+}]_i$ (Fig. 5A-D). After removal of extracellular Ca^{2+} from the superfusion buffer, the spontaneous $[Ca^{2+}]_i$ oscillations were immediately blocked in virtually all pinealocytes recorded (Fig. 4A). When the preparations were again superfused with calcium-containing saline, the $[Ca^{2+}]_i$ oscillations were restored in more than 50% of the cells. Thus, external calcium plays an important role for the generation of $[Ca^{2+}]_i$ oscillations in rainbow trout pinealocytes. Treatment with 10 μM nifedipine blocked the spontaneous $[Ca^{2+}]_i$ calcium oscillations in most cells

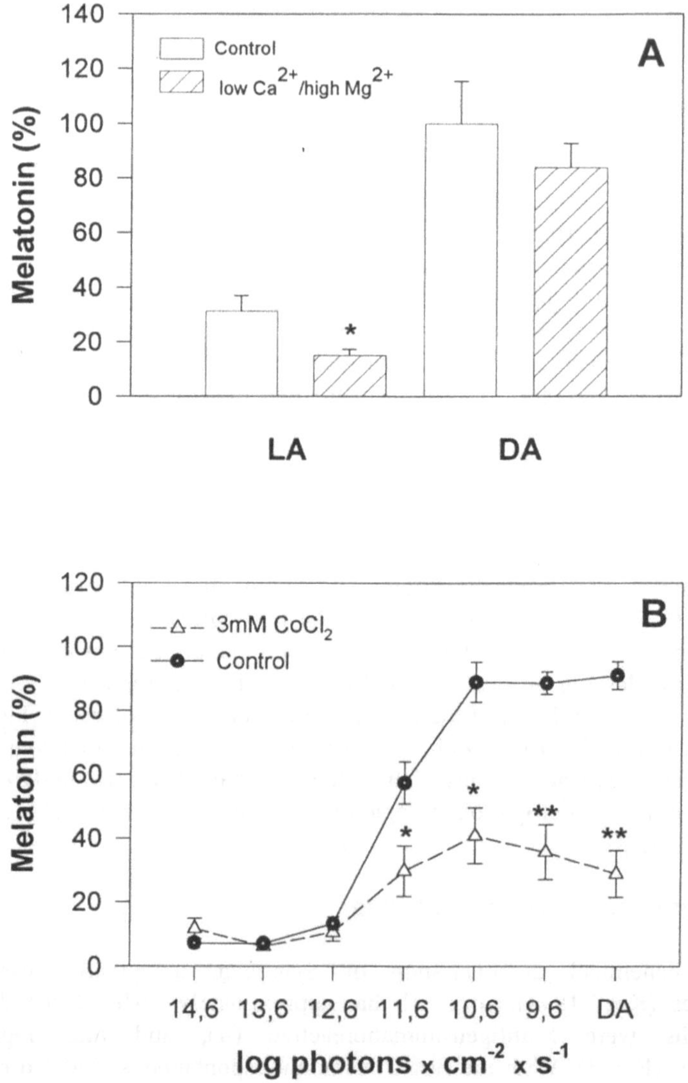

Fig. 1. Melatonin release from superfused pineal organs in response to monochromatic light (520 nm) of various intensities. Values are normalized in respect to the maximal value in darkness. **A** Treatment with Hanks' buffer with a low Ca^{2+}/high Mg^{2+}content (*hatched bars*) significantly reduced melatonin release in the light-adapted (*LA*: 12.6 log photons/cm^2 s^{-1}) and slightly in the dark-adapted (*DA*) state. **B** Exposure of the pineal gland to monochromatic light of decreasing irradiances increases melatonin production, reaching maximal values in darkness. Addition of 3 mM $CoCl_2$ to the superfusion medium significantly reduced melatonin production in the mesopic and scotopic range of illuminations, but elicited only slight changes in the light-adapted state. Mean ± SEM. * = p <0.05 vs. control; ** = p <0.01 vs. control

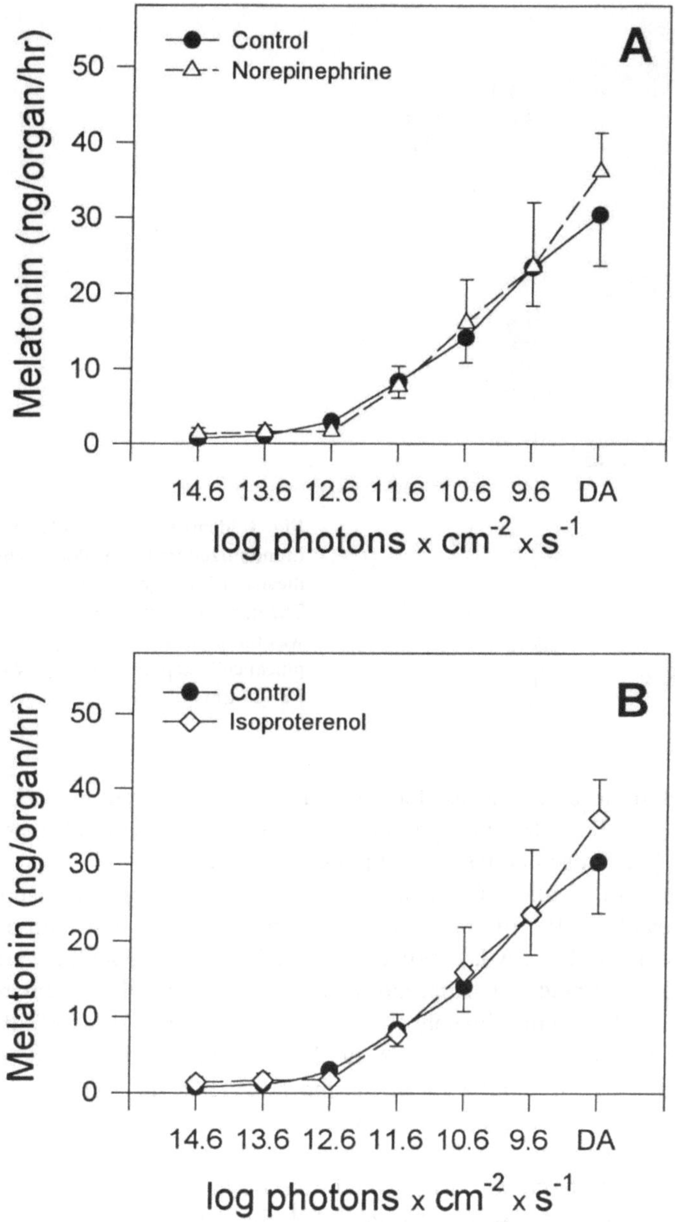

Fig. 2. Effects of (**A**) norepinephrine (1 μM) or (**B**) isoproterenol (1 μM) on the light-dependent melatonin production of superfused pineal organs. All drugs were added to the superfusion medium for the whole duration of the experiment. Controls were superfused with modified medium 199. Duration of the exposure to each irradiance was 60 min; mean ± SEM (n = 8). *DA* Dark adapted state

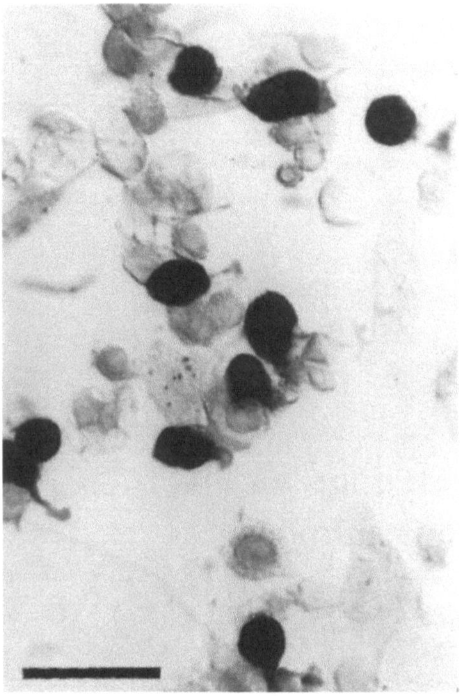

Fig. 3. Identification of isolated and immobilized trout pinealocytes by means of S-antigen immunocytochemistry. The immunoreaction (dark) was found in approximately 20% of the pineal cells kept in culture for 24 h. *Bar* 20 μm

(Fig. 4C). In some cases the oscillations reappeared after removal of the drug. These findings indicate that voltage-gated calcium channels of the L-type regulate spontaneous $[Ca^{2+}]_i$ oscillations in trout pinealocytes.

Treatment with 60 mM KCl caused an immediate and large increase in $[Ca^{2+}]_i$ in approximately 90% of the oscillating and 50% of the non-oscillating pinealocytes (Figs. 4A, 5A-D). When non-oscillating pinealocytes were treated with 10 μM nifedipine, the amplitudes of the KCl-induced calcium responses were reduced by approximately 40% in more than 70% of the cells. After application of 50 μM nifedipine, the KCl-induced calcium responses were completely blocked in about 80% of the pinealocytes (Fig. 5A). Similar results were obtained by the use of $CoCl_2$. Treatment with 3 mM $CoCl_2$ reversibly blocked the spontaneous $[Ca^{2+}]_i$ oscillations in the pinealocytes (Fig. 4B) and slightly reduced the basal $[Ca^{2+}]_i$ in several non-oscillating pinealocytes (Fig. 5C). Application of 3 mM $CoCl_2$ also inhibited the KCl-induced rise in $[Ca^{2+}]_i$ (Fig. 5B). Application of norepinephrine in concentrations between 100 pM and 50 μM had no effect on spontaneously oscillating cells (Fig. 4D) and evoked no changes in $[Ca^{2+}]_i$ in non-oscillating pinealocytes (Fig. 5D). The excitability of these cells was established by their response to KCl, which induced an increase in $[Ca^{2+}]_i$.

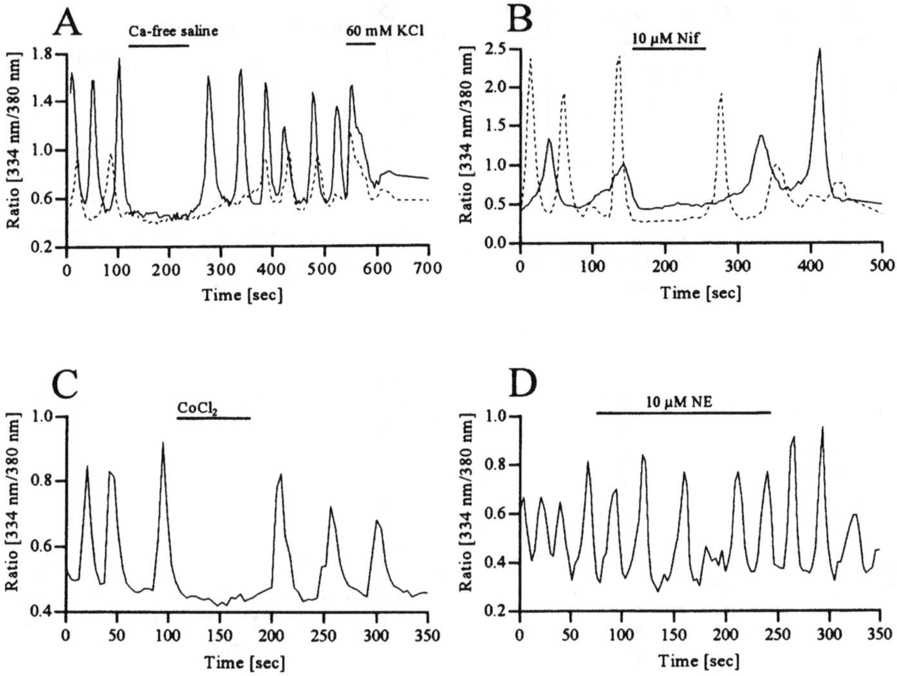

Fig. 4. Effects of different treatments on $[Ca^{2+}]_i$ in identified spontaneously oscillating S-antigen-ir trout pinealocytes. **A** The spontaneous $[Ca^{2+}]_i$ oscillations in two simultaneously recorded pinealocytes were reversibly blocked upon removal of extracellular Ca^{2+}. Application of 60 mM *KCl* induced a simultaneous increase in $[Ca^{2+}]_i$ in both recorded cells and blocked the spontaneous $[Ca^{2+}]_i$ oscillations. **B** Treatment with 10 μM nifedipine (*Nif*) blocked the spontaneous $[Ca^{2+}]_i$ oscillations in two simultaneously recorded pinealocytes. After removal of the drug the $[Ca^{2+}]_i$ oscillations started again. **C** A similar effect was obtained after application of 3 mM *$CoCl_2$* which caused a reversible breakdown of the spontaneous $[Ca^{2+}]_i$ oscillations. **D** Application of 10 μM norepinephrine (*NE*) did not influence the spontaneous $[Ca^{2+}]_i$ oscillations. (Reprinted from Kroeber et al. 1997, with kind permission from Elsevier Science, NL). Note that the two simultaneously recorded pinealocytes represented by a *dotted* and a *solid line* in **A** and **B** do not display synchronized $[Ca^{2+}]_i$ oscillations. The duration of drug application is indicated by a *bar*

Discussion

The present data demonstrate that cultured pineal organs of the trout maintain a robust irradiance-dependent secretion profile of melatonin with high levels in the scotopic range of illuminations, the strongest rise of melatonin production in the mesopic range and low secretion levels in the photopic range. These findings confirm and extend previous results obtained in vitro with the trout pineal organ (Gern and Greenhouse 1988; Gern et al. 1992; Max and Menaker 1992; Yáñez and Meissl 1996). The characteristic secretion pattern of melatonin persists even

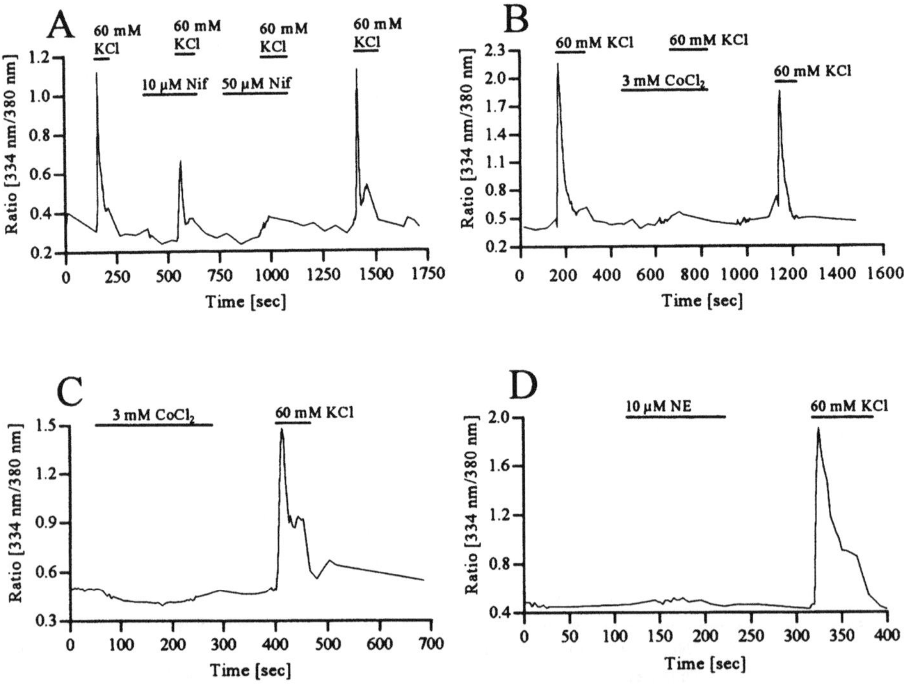

Fig. 5. Effects of different treatments on $[Ca^{2+}]_i$ in identified non-oscillating S-antigen immunoreactive trout pinealocytes. **A** Application of 60 mM *KCl* induced an increase in $[Ca^{2+}]_i$. Subsequently, the cells were preincubated with 10 and 50 μM nifedipine (*Nif*), respectively, and treated with 60 mM *KCl*. In the presence of 10 μM nifedipine, the amplitude of the depolarization-induced calcium response was reduced, and after application of 50 μM nifedipine the KCl-induced calcium response was completely blocked. After nifedipine was washed out, treatment with KCl again induced an increase in $[Ca^{2+}]_i$. (Reprinted from Kroeber et al. 1997, with kind permission from Elsevier Science, NL). **B** Treatment with 3 mM $CoCl_2$ also inhibited the KCl-induced rise in $[Ca^{2+}]_i$ and **C** slightly reduced the basal $[Ca^{2+}]_i$ in several non-oscillating pinealocytes. **D** Application of 10 μM norepinephrine (*NE*) did not evoke an increase in $[Ca^{2+}]_i$ in non-oscillating cells. The excitability of these cells was shown by depolarizing them with 60 mM KCl. The duration of drug application is indicated by a *bar*

after treatment with drugs that are known to possess physiological activity and putative transmitter function in the pineal organ as has been shown in cultured pineal organs treated with the putative photoreceptor transmitter glutamate (Meissl and George 1984). Application of glutamate, which was assumed to depolarize second-order neurons in the pineal organ and to make them completely unresponsive to light, was without effect on melatonin release (Meissl et al. 1996). Also, the clear effects of divalent ions like Mg^{2+} and Co^{2+} on melatonin production are apparently not caused by a blockade of synaptic transmission. This is shown by the analysis of the intracellular concentration of free calcium ions. Addition of 3 mM Co^{2+} reversibly blocked spontaneous $[Ca^{2+}]_i$ oscillations of dispersed trout pinealocytes and also diminished the KCl-induced rise in $[Ca^{2+}]_i$. Removal of extracellular calcium ions from the superfusion medium immediately blocked the spontaneous $[Ca^{2+}]_i$ oscillations in virtually all pinealocytes. This action was reversed in many cells when the preparations were again superfused with calcium-containing saline.

The spontaneous $[Ca^{2+}]_i$ oscillations in trout pinealocytes resembled those previously observed in isolated pinealocytes of chicken (D'Souza and Dryer 1994). $[Ca^{2+}]_i$ oscillations have never been observed in S-antigen immunonegative trout pineal cells or in mammalian pinealocytes (Saez et al. 1994; Schaad et al. 1993; Schomerus et al. 1994; Schomerus et al. 1995). Thus, it appears that they represent a characteristic feature of photoreceptive pinealocytes. However, calcium oscillations seem to be independent from light or dark adaptation conditions, because neither illumination with bright light nor dark adaptation of the cells suppressed the spontaneous $[Ca^{2+}]_i$ oscillations (Kroeber et al. 1997). Obviously, $[Ca^{2+}]_i$ oscillations depend on the influx of Ca^{2+} through dihydropyridine-sensitive voltage-gated calcium channels (Kroeber et al. 1997), but their trigger mechanism remains unknown. Interestingly, the percentage of pinealocytes exhibiting $[Ca^{2+}]_i$ oscillations increased with time after cell dissociation, showing that the dissociation procedure may affect activation of voltage-gated calcium channels. This is corroborated by findings of Begay et al. (1994) demonstrating that calcium currents could be recorded 36 h after cell isolation, but never in freshly dissociated cells.

Whether the majority of the S-antigen-ir pinealocytes which did not show spontaneous oscillations represent a special type of pinealocyte remains to be elucidated. We could not establish any correlation between morphological features and the presence or absence of $[Ca^{2+}]_i$ oscillations. Previous light and electron microscopic investigations have shown that S-antigen immunoreactivity can be demonstrated in two different types of pinealocytes in the trout pineal organ: one type gave rise to a short basal process, the other type was characterized by a long, extensively branched basal process (Ekström and Meissl 1990).

The lack of effect of catecholaminergic drugs like the β-adrenergic receptor agonist isoproterenol or norepinephrine on melatonin production in the trout pineal organ is of considerable interest with respect to the phylogenetic development of the noradrenergic (sympathetic) innervation of the pineal. As shown here, norepinephrine has no effect on $[Ca^{2+}]_i$, does not affect spontaneous $[Ca^{2+}]_i$ oscillations in any of the trout pinealocytes tested, and does not influence melatonin biosynthesis. There is also no effect of norepinephrine on cAMP levels or the activity of NAT in the trout pineal organ (Martin et al. 1991; Falcón et al. 1991). These functional data correlate well with fluorescence microscopic studies indicating that, in teleosts, only a small number of catecholaminergic fibers occur in the connective tissue surrounding the pineal organ, which apparently do not innervate the pineal parenchyma (Owman and Rüdeberg 1970; Falcón et al. 1980). In the pineal stalk of the stickleback, only a single norepinephrine-immunoreactive axon has been detected (Ekström et al. 1986). All these data suggest that in the pineal organ of teleosts a norepinephrine-sensitive pinealocyte of the mammalian type is absent.

In mammalian pinealocytes norepinephrine stimulates melatonin biosynthesis via activation of β_1-adrenergic receptors which leads to an increase of the intracellular cAMP concentration, and via activation of α_1-adrenergic receptors which causes an elevation of $[Ca^{2+}]_i$, potentiating the β_1-adrenergic effects (Klein 1985; Sugden and Klein 1988). The rise in $[Ca^{2+}]_i$ involves calcium mobilization from intracellular calcium stores as well as an influx of extracellular Ca^{2+} into the cells (Schomerus et al. 1995).

Finally, it has to be considered that some species differences exist among teleosts, which may be related to the occurrence of a special cell type, i.e., a modified photoreceptor cell type resembling those of birds, because in the pike catecholamines appear to be involved in the control of melatonin production (Falcón et al. 1991). Catecholamines also possess a weak modulatory action on electrophysiological responses in the trout pineal organ and are able to excite second-order neurons, but only in relatively high concentrations (Martin and Meissl 1992). These effects are partly reversed by specific antagonists, showing that they are receptor-mediated. The electrophysiological effects of catecholamines appear functionally separated from the melatonin synthesizing pathway because the latter remains unaffected by norepinephrine treatment.

In conclusion, this study provides evidence that norepinephrine does not cause any changes in $[Ca^{2+}]_i$ or melatonin synthesis in the trout pineal organ and supports the concept that noradrenergic mechanisms are not involved in signal transduction mechanisms in directly light sensitive pineal organs of lower vertebrates. It further demonstrates that the regulation of $[Ca^{2+}]_i$ shows conspicuous variation among pineal organs of different vertebrate classes.

References

Arendt J (1995) Melatonin and the mammalian pineal gland. Chapman & Hall, London, pp 1-331

Begay V, Bois P, Collin JP, Lenfant J, Falcón J (1994) Calcium and melatonin production in dissociated trout pineal photoreceptor cells in culture. Cell Calcium 16: 37-46

Collin JP (1971) Differentiation and regression of the cells of the sensory line in the epiphysis cerebri: In: Wolstenholme GEW, Knight J (eds) The pineal gland. Churchill-Livingstone, Edinburgh, pp 79-125

Collin JP, Meissl H, Voisin P, Brisson P, Falcón J (1986) Rhythmic signals of pineal transducers: physiological, biochemical, and cytochemical evidence. In: Reiter RJ, Karasek M (eds) Advances in Pineal Research. John Libbey, London, pp 41-50

Collin JP, Oksche A (1981) Structural and functional relationships in the mammalian pineal gland. In: Reiter RJ (ed) The pineal gland. Vol I, Anatomy and biochemistry. CRC, Boca Raton, pp 27-67

Dodt E (1973) The parietal eye (pineal and parietal organs) of lower vertebrates. In: Jung R (ed) Handbook of sensory physiology, vol VII/3B. Springer, Berlin, Heidelberg, New York, pp 113-140

D'Souza T, Dryer SE (1994) Intracellular free Ca^{2+} in dissociated cells of the chick pineal gland: regulation by membrane depolarization, second messengers and neuromodulators, and evidence for release of intracellular Ca^{2+} stores. Brain Res 656: 85-94

Ekström P, Reschke M, Steinbusch H, VanVeen T (1986) Distribution of noradrenaline in the brain of the teleost *Gasterosteus aculeatus* L.: an immunohistochemical analysis. J Comp Neurol 254: 297-313

Ekström P, Meissl H (1990) Electron microscopic analysis of S-antigen- and serotonin-immunoreactive neural and sensory elements in the photosensory pineal organ of the salmon. J Comp Neurol 292: 73-82

Falcón J, Juillard MT, Collin JP (1980) L'organe pineal du Brochet (*Esox lucius* L.) IV. Serotonine endogene et activite monoamine oxydasique: etude histochimique, ultra-cytochimique et pharmacologique. Reprod Nutr Dev 20: 139-154

Falcón J, Thibault C, Martin C, Brun-Marmillon J, Claustrat B, Collin J P (1991) Regulation of melatonin production by catecholamines and adenosine in a photoreceptive pineal organ. An in vitro study in the pike and in the trout. J Pineal Res 11: 123-134

Gern WA, Greenhouse SS (1988) Examination of in vitro melatonin secretion from superfused trout (*Salmo gairdneri*) pineal organ maintained under diel illumination or continuous darkness. Gen Comp Endocrinol 71: 163-174

Gern WA, Greenhouse SS, Nervina JM, Gasser PJ (1992) The rainbow trout pineal organ: an endocrine photometer. In: Ali MA (ed) Rhythms in fishes. Plenum, New York, pp 199-218

Klein DC (1985) Photoneural regulation of the mammalian pineal gland. In: Evered D, Clark S (eds) Photoperiodism, melatonin and the pineal gland. Pitman, London, pp 38-56

Korf H-W (1994) The pineal organ as a component of the biological clock. Phylogenetic and ontogenetic considerations. Ann NY Acad Sci 719: 13-42

Kroeber S, Schomerus C, Korf H-W (1997) Calcium oscillations in a subpopulation of S-antigen-immunoreactive pinealocytes of the rainbow trout (*Oncorhynchus mykiss*). Brain Res 744: 68-76

Martin C, Falcón J, Collin JP (1991) Catecholamines regulate cAMP levels in the photosensitive trout pineal organ. Adv Pineal Res 5: 137-139

Martin C, Meissl H (1992) Effects of dopaminergic and noradrenergic mechanisms on the neuronal activity of the isolated pineal organ of the trout, *Oncorhynchus mykiss*. J Neural Transm 88: 37-51

Max M, Menaker M (1992) Regulation of melatonin production by light, darkness, and temperature in the trout pineal. J Comp Physiol A 170: 479-489

Meissl H (1986) Photoneurophysiology of pinealocytes. In: O'Brien PJ, Klein DC (eds) Pineal and retinal relationships. Academic Press, Orlando, pp 33-46

Meissl H, Dodt E (1981) Comparative physiology of pineal photoreceptor organs. In: Oksche A, Pevét P (eds) The pineal organ: photobiology - biochronometry - endocrinology. Elsevier/North-Holland Biomedical Press, Amsterdam, pp 61-80

Meissl H, Ekström P (1988) Dark and light adaptation of pineal photoreceptors. Vision Res 28: 49-56

Meissl H, George SR (1984) Electrophysiological studies on neuronal transmission in the frog's photosensory pineal organ. The effect of amino acids and biogenic amines. Vision Res 24: 1727-1734

Meissl H, Kroeber S, Yáñez J, Korf H-W (1996) Regulation of melatonin production and intracellular calcium concentrations in the trout pineal organ. Cell Tissue Res 186: 315-323

Meissl H, Yáñez J (1994) Pineal photosensitivity. A comparison with retinal photoreception. Acta Neurobiol Exp 54 (Suppl): 19-29

Oksche A (1971) Sensory and glandular elements of the pineal organ. In: Wolstonholme GEW, Knight J (eds) The pineal gland. Churchill-Livingstone, Edinburgh, pp 127-146

Owman C, Ruedeberg C (1970) Light, fluorescence, and electron microscopic studies on the pineal organ of the pike, *Esox lucius* L., with special regard to 5-hydroxytryptamine. Z Zellforsch 107: 522-550

Saez JC, Moreno AP, Spray DC (1994) Norepinephrine induces Ca^{2+} release from intracellular stores in the rat pinealocytes. J Pineal Res 16: 57-64

Schaad NC, Parfitt A, Russel JT, Schaffner AE, Korf HW, Klein DC (1993) Single-cell $[Ca^{2+}]_i$ analysis and biochemical characterization of pinealocytes immobilized with novel attachment peptide preparation. Brain Res 614: 251-256

Schomerus C, Ruth P, and Korf HW (1994) Photoreceptor-specific proteins in the mammalian pineal organ: immunocytochemical data and functional considerations. Acta Neurobiol Exp 54 (Suppl.): 9-17

Schomerus C, Laedtke E, Korf HW (1995) Calcium responses of isolated, immunocytochemically identified rat pinealocytes to noradrenergic, cholinergic and vasopressinergic stimulations. Neurochem Int 27: 163-175

Sugden D, Klein DC (1988) Activators of protein kinase C act as a postreceptor site to amplify cyclic AMP production in rat pinealocytes. J Neurochem 50: 149-155

Tabata M, Meissl H (1993) Thermal responses of achromatic ganglion cells in the photosensory pineal organ of rainbow trout *Oncorhynchus mykiss*. Comp Biochem Physiol 105A: 453-457

Thibault C, Falcón J, Greenhouse SS, Lowery A, Gern WA, Collin JP (1993) Regulation of melatonin production by pineal photoreceptor cells: role of cyclic nucleotides in the trout (*Oncorhynchus mykiss*). J Neurochem 61: 332-339

Wicht H, Korf H-W, Schaad NC (1993) Morphological and immunocytochemical heterogeneity of cultured pinealocytes from one-week- and two-month-old rats: planimetric and densitometric investigations. J Pineal Res 14: 128-137

Yáñez J, Meissl H (1996) Secretion of the methoxyindoles melatonin, 5-methoxytryptophol, 5-methoxyindoleacetic acid, and 5-methoxytryptamine from trout pineal organs in superfusion culture: effects of light intensity. Gen Comp Endocrinol 101: 165-172

31 The Work of Ludwig Edinger and His Neurology Institute

G. Kreft

Neurologisches Institut (Edinger-Institut), Johann Wolfgang Goethe-Universität, Frankfurt am Main, Germany

Abstract

This survey of Ludwig Edinger's life, work and legacy stresses his scientific career, his foundation of modern comparative neuroanatomy and the development of his Neurology Institute. Additionally, the wide range of Ludwig and Anna Edinger's cultural activities is mentioned, especially the Ludwig Edinger Foundation. After Ludwig Edinger′s death in 1918, three further neuroscientific disciplines were founded at the Neurology Institute before 1937: neuropsychology (Kurt Goldstein), paleoneurology (Tilly Edinger) and neuroendocrinology (Ernst and Berta Scharrer). The fate of the Ludwig Edinger Foundation and the Edinger family under Nazi rule is recalled. After World War II, the reconstruction of the Neurology Institute into a center of neuropathology in Germany was related to the Max-Planck Institute for Brain Research and promoted by Wilhelm Krücke, Ekkehard Thomas and Wolfgang Schlote successively.

Ludwig Edinger (1855-1918)

Ludwig Edinger (Fig. 1) was born on 13 April, 1855 in Worms, a town on the Rhine River nearly 50 km to the southwest of Frankfurt am Main. His mother, Julie Edinger (1829-1893), was the highly enlightened daughter of Dr. Isaak Hochstätter (1798-1858), medical director (*Hauptarzt*) of the town hospital (*Bürgerhospital*) at Karlsruhe, and one of the leaders of the Jewish community in Baden (member of the *Oberrat*), who was much engaged in Jewish emancipation. Ludwig Edinger's father, Marcus Edinger (1820-1879), was the first German ready-made clothing manufacturer and a democratic representative of Worms in the parliament of Hessen-Darmstadt for more than 15 years.

Ludwig Edinger's Jewishness turned out to be one of the major determinants of his biography and the historical fate of his legacy. After having qualified as university lecturer in 1881, Ludwig Edinger faced a revived anti-semitism that made an academic career impossible. Finally, in April 1883, he decided to settle down in Frankfurt am Main as a physician and was one of the first German specialists in nerve disorders. Although Frankfurt was without a university at that

Fig. 1. Ludwig Edinger, presumably around 1910. (Archive of the Edinger Institute)

time, the city became Ludwig Edinger's second home, especially because of the high scientific standard within the famous Dr. Senckenberg Medical Society, which has been characterized as a "Frankfurt Republic of Physicians" (Mann 1972).

In his autobiographical memoirs, which were partly published in 1959 (Edinger 1959), Ludwig Edinger described impressively the atmosphere of his early Frankfurt days: after a day of medical practice, he spent the night working on microscopic investigations of fetal brains in his bedroom! The material was provided by a maternity hospital in Marburg. Whenever a parcel with fetal brains arrived, Edinger's housekeeper had to go away, because she should not know what was going on. Unintentionally, this account recalls Gothic novels like *Frankenstein* or *Dr. Jekyll and Mr. Hyde*. The latter, Stevenson's novel, appeared in 1886. As a contemporary to Edinger's secretive research, it portrays a common perception of such independent scholars in medicine at that time. Ludwig Edinger was very successful in investigating unknown neuroanatomical connections of the human brain, using the new myelin staining invented by his friend Carl Weigert (1845-1908; Fig. 2). By 1885 Edinger had presented the results of his approach to his medical colleagues in ten lectures. For the first time, he described the human brain as a whole as well as its internal anatomical structure with steady remarks towards medical efforts. The success of his lectures was overwhelming, and they were printed immediately (Edinger 1885). They were soon translated into the major languages, and Ludwig Edinger had become a scientist of international renown.

Edinger's success was related in one more respect to the pathologist Carl Weigert. Weigert was Jewish, and his career, too, had been ruined by anti-semitism. But Edinger arranged for Weigert to become head of the Dr. Senckenberg Anatomy Institute (*Dr. Senckenbergische Anatomie*) at Frankfurt in 1885. Carl Weigert then offered Ludwig Edinger a place to work in his institute, and Edinger moved the microscope out of his bedroom.

Edinger and Weigert worked very closely together in the same laboratory, avoiding separation for 17 years. During that time, a Neurology Institute grew up within the Dr. Senckenberg Anatomy Institute, directed by Weigert. An increasing number of visitors from

Fig. 2. Carl Weigert, painted in 1897. (Archive of the Edinger Institute)

Fig. 3. Ludwig Edinger surrounded by his pupils and coworkers at the old Dr. Senckenberg Anatomy Institute in March 1905. (Archive of the Edinger Institute)

all parts of the world came to study the brain with Ludwig Edinger (Fig. 3). When, in 1902, the spatial situation became totally disastrous, Weigert offered Edinger a room of his own, and Edinger equipped his first laboratory, using private funds. In 1907, at the age of 52, Edinger extended this laboratory to comprise the whole second floor of the newly built Dr. Senckenberg Pathology Institute (*Dr. Senckenbergische Pathologie*). Now the Neurology Institute (*Neurologisches Institut*) consisted of a neuroanatomy and a neuropathology department.

Ludwig Edinger can be considered as one of the founding fathers of comparative neuroanatomy, by transforming the older idealistic morphology of the brain into the paradigm of Darwin's theory of evolution (Kreft and Schlote 1995). Edinger studied the relation between form, structure and function of the vertebrate brain in fish, reptiles, birds and mammals. On the basis of this research, he formed the idea that the brain consists of one part for elementary functions (paleoencephalon) and another that evolves with animal phylogeny, attaining its highest development in man (neoencephalon; Fig. 4). In the long run, Edinger expected this approach to build bridges to an "objective psychology" (Emisch 1991). In this respect, Edinger organized a Psychology Club (*Psychologischer Verein*) in 1910/1911 and cooperated, among others, with the founders of *Gestalt*psychology Max Wertheimer, 1880-1943; Kurt Koffka, 1886-1941; and Wolfgang Köhler 1887-1867 (Kreft 1996a).

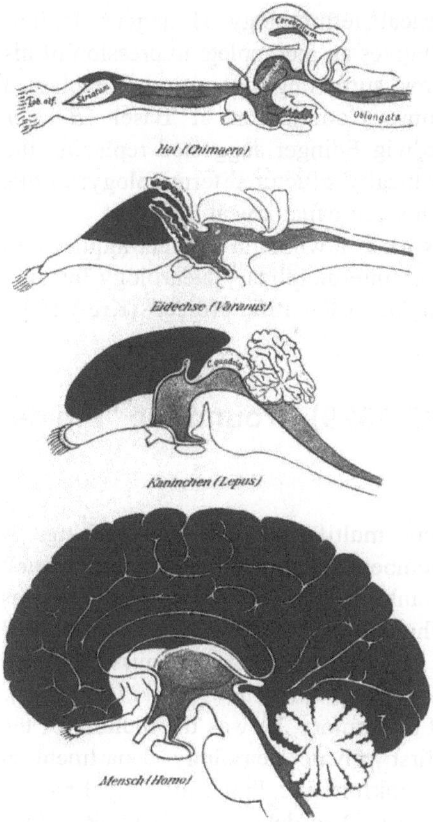

Fig. 4. The development of the neoencephalon (black) above the paleoencephalon (gray). (Edinger 1908)

Remarkable parallels exist in the development of Ludwig Edinger and his contemporary colleague Sigmund Freud (1856-1939) and reveal structures of a common *discourse* (Kreft 1989, 1996a).

According to his ambitious concepts, Ludwig Edinger pointed out the difference of his institute compared with a neuroanatomy institute, by naming it Neurology Institute. It was expected to include departments of comparative neuroanatomy and neuropathology (with possibilities for animal experiments), clinical neurology and psychology. In 1909, the International Brain Commission accepted the Neurology Institute as an Interacademic Institute for Brain Research (*Interakademisches Hirnforschungsinstitut*) (Emisch 1991). In 1912, Ludwig Edinger invited the Brain Commission to Frankfurt am Main, and an edition of a detailed atlas of the human brain was prepared (Schlote and Kreft 1997). Unfortunately, the outbreak of World War I set an end to this very productive period of cooperation. But even this catastrophe did not diminish Ludwig Edinger's international reputation. For example, the Journal of Comparative Neurology dedicated an issue to Ludwig Edinger's 60th birthday in 1915.

Edinger's presence in today's anatomical terminology (Edinger's [spino-thalamic] tract; Nucleus Edinger-Westphal) gives an incomplete impression of his importance for neuroanatomy. This seems surprising but can be explained historically. In 1892, at the Nomenclature Commission at Basel (*Baseler Nomenklaturkommission*), Switzerland, Ludwig Edinger suggested replacing the traditional use of eponyms with a topographically structured terminology; in this way, Ludwig Edinger himself anonymized his neuroanatomical lifework!

Surveying Ludwig Edinger's scientific work as a whole in a recent inquiry into the history of brain research, his vision of an interdiciplinary Neurology Institute shows remarkable outlines of a still topical "neuroscientific project" (Kreft 1995, 1996b).

Ludwig and Anna Edinger (1863-1929) Around the "Fin de Siècle"

Ludwig Edinger's cultural activities were multifarious and far-reaching. A beautifully colored *Diplôm de Mèrite* remembers Ludwig Edinger's participation at the World Fair at Brussels in 1897. At Frankfurt am Main, Ludwig Edinger was an organizer of the great Electrotechnical Exhibition (*Elektrotechnische Austellung*) in 1892. From 1908 to 1918 he belonged to the commission which built up the Frankfurt Museum for Ethnology (*Museum für Völkerkunde*). Furthermore, he was the founder of the first private neurology department at Frankfurt, the Frankfurt Polyclinic for Nerve Disorders (*Frankfurter Poliklinik für Nervenkranke*) where poor patients were treated free of charge between 1892 and 1923 (Kreft and Schlote 1995; Schlote and Kreft 1997).

Edinger's donation of his private Neurology Institute to the University of Frankfurt, which opened in 1914, is his most outstanding contribution. This is the only German university founded by private sponsors, especially by Jews.

Fig. 5. Marble bust of Anna Edinger (1894) by the artist Fritz Klimsch (1870-1960). (By courtesy of Historisches Museum der Stadt Frankfurt am Main)

Here, Ludwig Edinger became the first full professor for neurology in Germany. Unfortunately, Edinger was not able to experience satisfaction with his last donation. The Ludwig Edinger Foundation (*Ludwig Edinger-Stiftung*), dedicated to finance the work at the Neurology Institute, was only accepted by law in 1919, one year after Edinger's death.

In 1886, Ludwig Edinger married Anna Goldschmidt, the daughter of a Jewish banker and multimillionaire. In 1906, she inherited half of this fortune. Like many other women of her generation, it had not been possible for Anna Goldschmidt to realize her scientific interests in a university education. In place of this, Anna Edinger became an outstanding figure in the local coordination of social welfare and medical care (*Sozialhygiene*) at Frankfurt am Main from the late 1890s to the 1920s. Furthermore, she was one of the leaders in the German feminist movement (*Bund Deutscher Frauenvereine*). She was the first woman to receive the Honorary Medal (*Ehrenplakette*) of Frankfurt am Main in 1928 (Kreft 1996a).

Even though they were Jewish, the Edinger couple were part of Frankfurt's upper middle class. Their villa became a center for several cultural activities. For example, Ludwig Edinger organized regular meetings (*Kränzchen*) nearly every month between 1893 and 1918! Here, local representatives of science, arts, administration, politics and even protestantism gave lectures to an audience of their family members. Furthermore, Ludwig and Anna Edinger were patrons to the arts and artists (Figs. 5, 6). In 1909, Lovis Corinth (1858-1925) painted an oil portrait of Ludwig Edinger (Mann 1974). The first colored reproduction of this masterpiece was published recently (Schlote and Kreft 1997). Even his gravestone, placed centrally near the entrance of the Frankfurt main cemetry (*Hauptfriedhof*), was created by the famous Jewish artist, Benno Elkan (1877-1960). Again, a reproduction is available (Kreft and Kohring 1997). Showing a

Fig. 6. Sketches of Ludwig Edinger (Frankfurter Zeitung, April 13, 1915) and Anna Edinger (painted 1926). (Emrich 1956)

naked young man, struggling under adverse conditions, the gravestone symbolizes the cultural difference between the Edinger couple and traditional forms of religious expression - especially of Jewish orthodoxy - as well as a lasting identification with Jewish history. Without being baptized, secularized Jewish welfare and the highest ideals of German idealism (*Deutscher Idealismus*) were melting indistinguishably into each other in the lifes of these very well-acculturated German Jews.

After the death of her husband on 26 January, 1918, Anna Edinger increased the funds of the Ludwig Edinger Foundation by half a million German Reichsmark (at least 5 million German Marks in modern terms). At that time, the Ludwig Edinger Foundation financed the entire Neurology Institute, including equipment and the wages for all employees. Unfortunately, the capital of the investment became nearly worthless during the inflation years. This financial disaster crucially determined the fate of Ludwig Edinger's Neurology Institute.

Kurt Goldstein (1878-1965) and the Weimar Years

Since 1914, Kurt Goldstein had been Ludwig Edinger's last coworker and head of the neuropathology department of the Neurology Institute. In 1917, he added a third department to the Neurology Institute, the famous Institute for the Research on Consequences of Brain Injuries (*Institut zur Erforschung der Folgeerscheinungen von Hirnverletzungen*). After Edinger's death, Goldstein became the head of the Neurology Institute. From my perspective, Goldstein was a true follower of Edinger's broad interdisciplinary spirit (Kreft and Schlote 1995; Kreft 1996a, b, 1997): Goldstein, like Edinger of Jewish ancestry, was also deeply involved in German idealistic culture which was particularly influenced by the philosopher Kuno Fischer (1824-1907; Wiehl 1985). Both Edinger and Goldstein had attended Fischer's lectures during their studies at Heidelberg (Archive of the Ruprecht-Karls-University Heidelberg, letters to the author). Furthermore, the interdisciplinary interest in brain research, *Gestalt*psychology, and psychoanalysis, which lasted from Edinger to Goldstein, is striking. Together with the Jewish *Gestalt*psychologist Adhémar Gelb (1887-1936; Fig. 7), Goldstein investigated and treated brain-injured patients by experimental psychological methods for the first time. In this respect, Kurt Goldstein emerged as the "true founder of neuropsychology" (Luria 1975) and a pioneer of "psychosomatic neurology" (Kütemeyer and Schultz 1984). On the other hand, Kurt Goldstein had been a prominent figure in the institutionalization of psychoanalysis at Frankfurt am Main (Laier 1994, 1996). In the long run, Goldstein deeply influenced many - mostly Jewish - pupils who became famous scholars themselves. One of them, Walther Riese (1890-1976; Benzenhöfer and Kreft 1997), published a treatise on the structure and function of Ludwig Edinger's brain in cooperation with Goldstein (Riese and Goldstein 1950).

Fig. 7. Kurt Goldstein (left) and Adhémar Gelb (right), presumably in the late 1920s. (Kütemeyer and Schultz 1984)

Like Edinger, Goldstein had been a protagonist of German neurology as an autonomous discipline. Partly due to the lack of funds of the Edinger Foundation and the city of Frankfurt am Main, and partly due to the institutional rivalry with Karl Kleist (1869-1965), Goldstein was not successful in realizing one of Edinger's most longed for projects, the establishment of a department for clinical neurology within the Neurology Institute (Kreft 1997). In 1930, he finally decided to go to Berlin. With Goldstein's departure, Ludwig Edinger's "neuroscientific project" at the Neurology Institute came to an end. In 1933, Goldstein, a member of the Association of Socialist Physicians (*Verein sozialistischer Ärzte*) was arrested and forced to emigrate (Pross 1993). But even in the United States of America, Goldstein personified the identity and difference between Ludwig Edinger's "neuroscienitific project", the liberal Jewish bourgeoisie of the *fin de siècle* and the left wing intellectuals of the Frankfurt 1920s. For instance, he cooperated with his former pupil and lifelong friend Max Horkheimer (1895-1973), *spiritus rector* of the famous Frankfurt Institute for Social Research (*Institut für Sozialforschung*; Kreft 1996a). Goldstein's outstanding position in the cultural changes of the Weimar Republic and his later career as a leader of the "holistic approach" in the USA have been discussed recently in a tremendous investigation (Harrington 1996).

Tilly Edinger (1897-1967) and the Early 1930s

Even with Kurt Goldstein's departure, Ludwig Edinger's creative spirit was not at all dead. His youngest daughter, Tilly Edinger (Fig. 11) especially supported the Neurology Institute. During her university studies in geology, zoology and paleontology she even attended her father's lecture on form and function of the nervous system (*Bau und Verrichtungen des Nervensystems*) in 1916/1917 (Emisch 1991). During work on her doctoral thesis on nothosaurus (*Über Nothosaurus*) in 1921, Tilly Edinger saw her first natural endocast of a fossil skull. Since then, she used her father's unique collection of animal brains as an indispensable tool in establishing bridges between vertebrate paleontology and comparative neurology. Synthesizing these two fields of approach, Tilly Edinger created a new discipline, paleoneurology. In 1929, she published her first monograph on fossil brains (*Die fossilen Gehirne*), in which she investigated the relation between form and function of the brain in extinct animals.

In face of ruin of the Neurology Institute in 1931, Tilly Edinger took care of her father's collection of animal brains as a research associate. However, after Hitler's coming into power in 1933, her position became insecure. The assistant general manager of the Neurology Institute, the pathologist Bernhard Fischer-Wasels (1877-1941), whose antidemocratic speeches had already been startling in 1931 (Stuchlik 1984), urged Tilly Edinger to leave. Nevertheless, she he was able to work at the independent Senckenberg-Museum until 1938 - as a "quasi-underground curator" (Kohring 1997). Furthermore, she was able to determine her successor at the Neurology Institute personally: Ernst Scharrer (Tilly Edinger to Hans Bluntschli, August 25, 1933; Kreft and Kohring 1997).

Ernst and Berta Scharrer at the Neurology Institute (1933-1937)

I still do not know how Tilly Edinger became acquainted with Ernst Scharrer (1905-1965). She may have taken note of his classic discovery of gland-like nerve cells in the hypothalamus of the teleost fish, *Phoxinus laevis* (Scharrer 1928), during her bibliographic investigations in paleoendocrinology (Edinger 1933). Nevertheless, the decision to engage Ernst Scharrer as a research associate at the Neurology Institute turned out to be very fortunate for all concerned. This fortune was not primarily due to Ernst Scharrer's scientific reputation, still disputed at that time, but to his "aryan" genealogical tree (Albrecht Bethe to Bernhard Fischer-Wasels, June 19, 1933), which was most important in being accepted under the dictatorship. Additionally, Scharrer had to join the Nazi organization "SA" (*Sturmabteilung*) on Fischer-Wasels' suggestion (Tilly Edinger to Hans Bluntschli, March 17,1934). It is hard to imagine what this opportunism must have meant to a man like Ernst Scharrer who judged "military intelligence" to be an oxymoron

(Dr. Fruhmann, Albert Einstein College of Medicine, Department of Anatomy and Structural Biology, pers. comm., April 24, 1997). In an interview, Berta Scharrer (1906-1995) mentioned the atmosphere of "full 100 % (we called it 150 %) agreement with the new system" (Scharrer 1989) among the aspirants to university positions left by Jewish colleagues. On the other hand, this arrangement provided Ernst Scharrer and Berta Vogel, who married in 1934, with unique working conditions at the Neurology Institute as a nearly forgotten research facility. This atmosphere of the time is described impressively by Ernst Scharrer in his posthumously published autobiographical *History of the Neurological Institute 1933-1937* (Scharrer 1995). Last but not least, with Berta Scharrer the Neurology Institute got a second - nonsalaried - research associate. The congenial work of the Scharrers (Fig. 8) led to the foundation of the new discipline of neuroendocrinology: after comparative neuroanatomy, neuropsychology and paleoneurology the fourth foundation of a neuroscientific discipline at the Neurology Institute!

Outside the "deceptively sheltered milieu" (Scharrer 1975) of the Edinger Institute, "the academic environment had been reduced to a parody". Especially after being forced to join a paramilitary meeting with its indoctrinations in "this senseless and immoral philosophy of the Nazis, the idea of racial superiority, anti-Semitism, genocide, and all the rest of it" (Scharrer 1989), the conflict became intolerable for Ernst Scharrer who had presumably been a pacifist (A. Oksche, letter to the author, February 19, 1997). "We decided that is was impossible for us to be part of this system any longer (....). We are told by our colleagues, "you

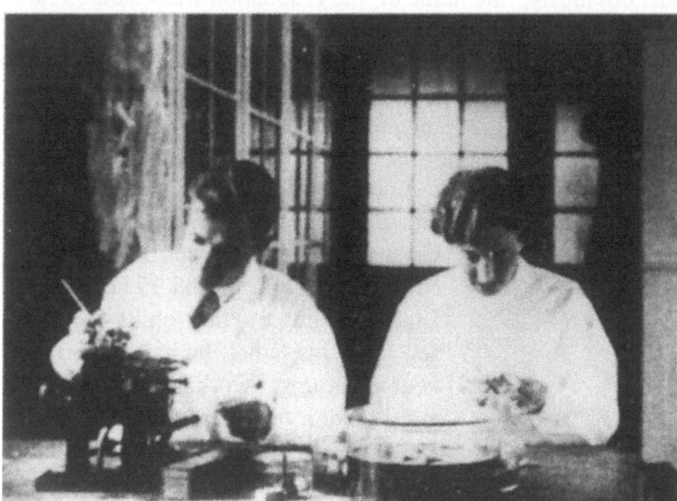

Fig. 8. Ernst and Berta Scharrer preparing serial (celloidin) sections of hypothalamus at the Neurology Institute in 1935. (Scharrer 1975)

shouldn't leave now, not even for a year, with all the nice opportunities open", because all the Jewish candidates had to leave. We were asked several times if we were out of our minds." Losing everything, the Scharrer couple left Germany with a 1-year Rockefeller fellowship in 1937.

With regard to the Scharrer Symposium, the following marginalia may be interesting: at the end of his autobiographical memoirs, Ernst Scharrer recalled "Dr. *Gerd* Peters (now director at the Department for Anatomy in the German Research Institute for Psychiatry in Munich)" as becoming his successor at the Neurology Institute. However, no connection between Gerd Peters (1906-1987) and the Neurology Institute can be confirmed (Max-Planck Institute for Psychiatry, Munich, information to the author). In fact, a man called "*Georg* Peters" (Fischer-Wasels to the Post Office, Frankfurt am Main, March 22, 1937) had been the scientific assistant at the Neurology Institute between 1940 and 1943 (University Calendar, Johann Wolfgang Goethe-University, Frankfurt am Main).

The Ludwig Edinger Foundation and the Edinger Family Under Nazi Rule (1937-1945)

According to traditional historical literature, scientific work at the Neurology Institute was completely extinguished with the emigration of the Scharrer couple (Lauche 1959). This is not equally true for the Ludwig Edinger Foundation. In 1939, it was "aryanized" (*arisiert*), i.e. the Edinger family lost all of their rights to the foundation. Most obviously, the name of the foundation was changed into "Foundation for Neurology" (*Stiftung für Neurologie*) in order to eliminate even the recollection of its Jewish founder.

In 1939, Tilly Edinger emigrated to London. Many other relatives were either unable or unwilling to leave Germany. In this respect, I want to recall Fritz Edinger (1888-1944; Fig. 9), Ludwig Edinger's first child and only son. Like his father, he had started as a neuroanatomist, working on structure and function of the frog brain (Edinger 1913) and settled down as a specialist for nerve disorders in Munich (Kallmorgen 1936). During World War I,

Fig. 9. Fritz Edinger in 1939. (By courtesy of Dr. Dieter Binswanger, Zürich)

Fritz Edinger received a decoration. After the death of his father, he became the secretary of the Ludwig Edinger Foundation. In the 1920s he obtained an additional degree in sociology (Edinger 1924), joined the German Social Democratic Party and published articles on political issues (Schairer 1929; Bosch 1989). Under Nazi pressure, Fritz Edinger abused barbiturate, which led to mental problems. Captured and hospitalized in December 1941, the tracks of his deportation end on the way to the concentration camp Izbica. In 1947, Fritz Edinger's date of death was fixed retrospectively at June 10, 1944.

Ups and Downs in Reconstructing the Neurology Institute Since 1945

By a favourable fate, the residence of the Neurology Institute stayed intact during World War II. Even its richest sources, Ludwig Edinger's library and his comparative collection of animal brains, had been saved by loyal servants (Lauche 1959). From its beginnings in 1947, the reconstruction of the Neurology Institute was promoted by Wilhelm Krücke (1911-1981), who had been one of the first German pathologists to have qualified as a university lecturer in neuropathology (Fig. 10).

Krücke reorganized Ludwig Edinger's institute as a domain of modern neuropathology. Nominally, a second department for comparative neuroanatomy was directed provisionally by Hugo Spatz (1888-1965), the head of the Max-Planck Institute for Brain Research at Giessen (Thomas 1959). A commemoration, organized on the occasion of Ludwig Edinger's 100th birthday by Krücke and Spatz, was even honored with Tilly Edinger's presence (Fig. 11).

Finally, Krücke inspired the Max-Planck Society to build up their new Institute for Brain Research in Frankfurt am Main. The location was sponsored by the city of Frankfurt am Main and the state of Hessen because of the university affiliation of the Edinger Institute. When the building was opened in 1962, the Neurology Institute and the Neuropathology Department of the Max-Planck Institute for Brain Research were merged under the direction of Wilhelm Krücke. A glori-

Fig. 10. Wilhelm Krücke, presumably in 1962. (Archive of the Edinger Institute)

Fig. 11. Tilly Edinger in the old office of her father at the Neurology Institute in April 1955. The bronze bust of Ludwig Edinger was made by the Swiss artist Fritz Huf (born 1888), presumably between 1911 and 1914. (Archive of the Edinger Institute)

ous time began. Neighbouring the Frankfurt University Hospital (*Klinikum der Johann Wolfgang Goethe-Universität*), an informal network of neuroscience departments emerged in the 1960s, the so-called *Frankfurter Nervenzentrum* (Zutt 1964). Tragically, the progressive specialization in basic research caused the Max-Planck Society to close their neuropathology department on the occasion of the retirement of Wilhelm Krücke in 1977 (Schlote 1988). Still located inside the building of the Max-Planck Institute, the Neurology Institute (Edinger Institute) became separated again.

The situation became extremely hard (Kahle 1984), and it was the very ungrateful task of Ekkehard Thomas to rebuild the Neurology Institute and even to save the Ludwig Edinger Foundation from ruin! Sparse funds reduced its activities to neuropathology services towards clinical needs. But fortunately, just like a glimpse of historical justice, an inheritance, made by a thankful patient in 1927, became available to the Ludwig Edinger Foundation in 1984. Nearly simultaneously, Wolfgang Schlote took over the directorship of the Neurology Institute, and since then, new scientific projects have been started in the field of clinical neurcpathology. Over the years, he has worked with linguists and educationalists on problems of speech and language disturbances. In this way, even aspects of Ludwig Edinger's interdisciplinary spirit have been revived. Furthermore, the Neurology Institute has been expanded, and five new rooms were occupied in 1994. Last but not least, the revival of the Neurology Institute and the Ludwig Edinger Foundation enabled the sponsoring of the Ernst and Berta Scharrer Medal awarded at this Scharrer Symposium for the first time.

References

Benzenhöfer U, Kreft G (1997) Bemerkungen zur Frankfurter Zeit (1917-1933) des jüdischen Neurologen und Psychiaters Walther Riese. In: Schriftenreihe der Deutschen Gesellschaft für Geschichte der Nervenheilkunde. K & N Königshausen & Neumann, Würzburg (in press)

Bosch, M (1989) Mit der Setzmaschine in Opposition. Auswahl aus Erich Schairers Sonntagszeitung 1920-1933. Elster, Moos, Baden-Baden

Edinger F (1913) Die Leistungen des Zentralnervensystems beim Frosch mit Rücksicht auf die Lebensweise dieses Tieres. Z Allg Physiol 15: 15-64

Edinger F (1924) Der Begriff des Bedürfnisses und des Aufwands in der ökonomischen Theorie und in der Volkswirtschaftspolitik. In: Wirtschaft und Gesellschaft. Festschrift für Franz Oppenheimer. Frankfurter Societäts-Druckerei, Frankfurt am Main 1924, pp 301-312

Edinger L (1885) Zehn Vorlesungen über den Bau der nervösen Centralorgane. Vogel, Leipzig

Edinger L (1908) Vorlesungen über den Bau der Nervösen Zentralorgane des Menschen und der Tiere. 1. Bd. 8. Aufl. FCW Vogel, Leipzig

Edinger L (1959) Aus den "Erinnerungen" von Ludwig Edinger. In: Ludwig Edinger 1855-1918. Gedenkschrift zu seinem 100. Geburtstag und zum 50jährigen Bestehen des Neurologischen Institutes (Edinger-Institut) der Universität Frankfurt am Main. Franz Steiner, Wiesbaden, pp 1-30

Edinger T (1929) Die fossilen Gehirne. Julius Springer, Berlin

Edinger T (1933) Paläo-Endokrinologie. Naturwissenschaften 21: 486-488

Emisch H (1991) Ludwig Edinger (1855-1918). Hirnanatomie und Psychologie. Gustav Fischer, Stuttgart

Emrich W (1959) Bildnisse Frankfurter Demokraten. Kramer, Frankfurt am Main.

Harrington A (1996) Reenchanted Science. Holism in German Culture from Wilhelm II to Hitler. Princeton University Press, Princeton ,New Jersey

Kahle W (1984) 80 Jahre Edinger-Institut. Zbl Allg Path Pathol Anat 129: 259-279, p 259

Kallmorgen W (1936) 700 Jahre Heilkunde in Frankfurt am Main. Moritz Diesterweg, Frankfurt am Main

Kohring R (1997) Senckenbergische Forscher: Tilly Edinger (1897-1967). Nat Mus, 1997 (in press)

Kreft (1989) Ludwig Edinger und Sigmund Freud. Ein marginaler Annäherungsversuch. Luzifer-Amor 4: 26-31

Kreft (1995) Neuroscience(s): a philological footnote to the history of brain reserach. Haussymposium 1995. Max-Planck-Institut für Hirnforschung, Frankfurt am Main

Kreft G (1996a) Zur Archäologie der Psychoanalyse in Frankfurt: Fundstücke und Perspektiven um Ludwig Edinger. In: Plänkers T, Laier M, Otto HH, Rothe HJ, Siefert H (eds) Psychoanalyse in Frankfurt am Main. Zerstörte Anfänge. Wiederannäherung. Entwicklungen. Edition Diskord, Tübingen, pp 195-234

Kreft G (1996b) Ludwig Edingers "neurowissenschaftliches" Projekt. Ein Beitrag zur Geschichte der Hirnforschung. Neuroforum 2(1): 37-38, 2(2): 39

Kreft G (1997) Zwischen Goldstein und Kleist: Zum Verhältnis von Neurologie und Psychiatrie im Frankfurt am Main der 1920er Jahre. In: Schriftenreihe der Deutschen Gesellschaft für Geschichte der Nervenheilkunde. K & N Königshausen & Neumann, Würzburg (in press)

Kreft G, Kohring R (1997) "Ich bin also sozusagen ein auserwähltes Wesen ..." Tilly Edinger (1897-1967) - Begründerin der Paläoneurologie in Frankfurt am Main. Forschung Frankfurt 4/1997 (in press)

Kreft G, Schlote W (1995) Ludwig Edinger (1855-1918) - Hirnforscher und Neurologe in Frankfurt am Main. In: Doerr HW, Korf HW (eds) Berühmte Ärzte und Forscher in Frankfurt am Main. Festschrift zur 500. Versammlung der Frankfurter Medizinischen Gesellschaft am 21. Juni 1995. Alpha Gmbh, Lampertheim, pp 38-61

Kütemeyer M, Schultz U (1984) Kurt Goldstein (1878-1965): Begründer einer psychosomatischen Neurologie? In: Pross C, Winau R (eds) Nicht mißhandeln. Das Krankenhaus Moabit, Fröhlich & Kaufmann, Berlin, pp 133-139

Laier M (1994) Kurt Goldstein - Zur Erinnerung an einen vergessenen Neurologen. In: Wilmanns JC (ed) Medizin in Frankfurt. Frankfurter Beiträge zur Geschichte, Theorie und Ethik der Medizin. Band 15. Olms Weidmann, Hildesheim, pp 176-186

Laier M (1996) Der Neurologe Kurt Goldstein (1878-1965) und seine Beziehung zu Gestaltpsychologie und Psychoanalyse. In: Plänkers T, Laier M, Otto HH, Rothe HJ, Siefert H (eds) Psychoanalyse in Frankfurt am Main. Zerstörte Anfänge. Wiederannäherung. Entwicklungen. Edition Diskord, Tübingen, pp 235-253

Lauche A (1959) Der Wiederaufbau des Edinger-Institutes. In: Ludwig Edinger 1855-1918. Gedenkschrift zu seinem 100. Geburtstag und zum 50jährigen Bestehen des Neurologischen Institutes (Edinger-Institut) der Universität Frankfurt am Main. Franz Steiner, Wiesbaden, pp 35-39

Luria A (1975) Neuropsychology: its Sources, principles and prospects. In: Worden FG, Swazey JP, Adelmann G (eds) The neurosciences: paths of discovery, vol 1. MIT Press, Cambridge, Massachussets, pp 335-362

Mann G (1972) Senckenbergs Stiftung und die Frankfurter Republik der Ärzte im 19. Jahrhundert. Medizinhist J 7: 244-263

Mann G (1974) Das Porträt des Neuroanatomen Ludwig Edinger von Lovis Corinth (1909). Medizinhist J 9:234-238

Pross C (1993) Die "Machtergreifung" am Krankenhaus. In: Bleker J, Jachertz N (eds) Medizin im "Dritten Reich". Deutscher Ärzte-Verlag, Köln, pp 97-108

Riese W, Goldstein K (1950) The brain of Ludwig Edinger. An inquiry into the cerebral morphology of mental ability and lefthandedness. J Comp Neurol 92: 133-168

Schairer E (1929) Mit anderen Augen. Jahrbuch der Sonntagszeitung 1920-1929. Verlag der Sonntagszeitung, Stuttgart

Scharrer B (1975) The concept of neurosecretion and its place in neurobiology. In: Worden FG, Swazey JP, Adelmann G (eds) The neurosciences: paths of discovery, vol 1. MIT Press, Cambridge, Massachussets, pp 231-243

Scharrer B (1989) On journeys well traveled. An occasional series. Einstein [Newsletter of the Albert Einstein College of Medicine of Yeshiva University] Spring/Summer 1989: 4-6

Scharrer E (1928) Die Lichtempfindlichkeit blinder Elritzen. Untersuchungen über das Zwischenhirn der Fische I. Z Vergl Physiol 7:1-38

Scharrer E (1995) Geschichte des Neurologischen Instituts 1933-1937. In: Doerr HW, Korf HW (eds) Berühmte Ärzte und Forscher in Frankfurt am Main. Festschrift zur 500. Versammlung der Frankfurter Medizinischen Gesellschaft am 21. Juni 1995. Alpha Gmbh, Lampertheim, pp 84-90

Schlote W (1988) Wilhelm Krücke (26.12.1911 - 7.2.1988). Verh Dtsch Ges Path 74: 647-651

Schlote W, Kreft G (1997) Der zweckentfremdete Küchentisch. Ludwig Edinger und die Anfänge der Hirnforschung in Frankfurt. Forschung Frankfurt 1/97: 47-59

Stuchlik G (1984) Goethe im Braunhemd. Universität Frankfurt 1933-1945. Röderberg, Frankfurt am Main

Thomas E (1959) Das Max-Planck-Institut für Hirnforschung. Die morphologischen Abteilungen. Hochschuldienst (Bonn) Ausg. A 12, 19: 2-3

Wiehl R (1985) Die Heidelberger Tradition der Philosophie zwischen Kantianismus und Hegelianismus. Kuno Fischer, Wilhelm Windelband, Heinrich Rickert. In: Doerr W (ed) Semper apertus. 600 Jahre Ruprecht-Karls-Universität Heidelberg 1386-1986. Band II. Springer, Berlin Heidelberg New York, pp 413-433

Zutt J (1964) Das Frankfurter "Nervenzentrum". Tagung mit Unterstützung der Dr. Thomas GmbH, Frankfurt am Main

Index

MIX
Papier aus verantwortungsvollen Quellen
Paper from responsible sources
FSC® C105338

If you have any concerns about our products,
you can contact us on
ProductSafety@springernature.com

In case Publisher is established outside the EU,
the EU authorized representative is:
Springer Nature Customer Service Center GmbH
Europaplatz 3, 69115 Heidelberg, Germany

Printed by Libri Plureos GmbH
in Hamburg, Germany